METHODS IN MOLECULAR BIOLOGY

Series Editor
John M. Walker
School of Life and Medical Sciences,
University of Hertfordshire,
Hatfield, Hertfordshire AL10 9AB, UK

For further volumes:
http://www.springer.com/series/7651

Computational Methods in Protein Evolution

Edited by

Tobias Sikosek

GlaxoSmithKline, Cellzome - a GSK company, Meyerhofstrasse 1, Heidelberg, Baden-Württemberg, Germany

Humana Press

Editor
Tobias Sikosek
GlaxoSmithKline
Cellzome - a GSK company
Meyerhofstrasse 1
Heidelberg, Baden-Württemberg, Germany

ISSN 1064-3745 ISSN 1940-6029 (electronic)
Methods in Molecular Biology
ISBN 978-1-4939-9378-9 ISBN 978-1-4939-8736-8 (eBook)
https://doi.org/10.1007/978-1-4939-8736-8

This Humana Press imprint is published by the registered company Springer Science+Business Media, LLC, part of Springer Nature.
The registered company address is: 233 Spring Street, New York, NY 10013, U.S.A.

Preface

Proteins are the most versatile kind of molecule that we know and the result of a long evolutionary process. During this process, countless rearranging, mutating, and replicating strands of DNA have managed to both encode and conserve proteins that would allow them to replicate and stay intact and on the other hand have allowed their proteins to change and ultimately help them replicate more than other strands of DNA. All cells make proteins in their protein factories called ribosomes, where the DNA of a gene is translated according to the ancient genetic code into strings of amino acids which follow the laws of thermodynamics and molecular forces to fold up into specific wobbly three-dimensional shapes. Protein evolution happens whenever an accidental "typo"—or mutation—in the gene is translated into a modified protein, and that protein is released into the busy commotion within the cell, packed within a dense soup of other molecules in water. Whatever this new protein does differently than its predecessor can determine the fate of that mutation, making it either an essential innovation, a terrible mistake that gets erased, or something that just stays around for a while without being noticed, maybe to play a role in the distant future.

This book is a compilation of methods that can be applied to various problems related to protein sequence and structure. It is a diverse collection of approaches ranging from broad conceptual ("protein space") to very specific applications ("antibody modeling"). The term "evolution" is used slightly differently in various fields of science. While evolutionary biologists think about the natural process of Darwinian evolution (and other post-Darwinian forms of evolution of organisms living in populations and environments), biochemists take a more design-oriented approach to evolution, using the evolutionary process in vitro or in silico to make proteins with certain desired properties. Physicists on the other hand use the term evolution to describe a continuous process in time that changes a system from one to another state. While physics plays a significant role in this book, it is the first two notions of evolution that will be described in the following chapters.

Evolutionary research has made extensive use of computers. While the result of evolution can be readily studied at the macroscopic, phenotypic level, evolutionary biology has always had a strong theoretical component, since the actual process had been rare to directly observe for a long time. The underlying patterns of inheritance and the interplay between geography and population dynamics have been described in mathematical terms and have always accompanied the progress made in the Molecular Biology of cells that eventually elucidated the core mechanisms of inheritance: the information stored in DNA and how it is replicated and passed on—imperfectly—to future generations. The field of Bioinformatics was born as soon as the first sequences of genes and proteins had been published at a large enough quantity to be amenable to direct sequence-to-sequence comparisons. The fields of Molecular Evolution and Phylogenetics were close companions of this development where mathematical models and computational algorithms were combined to reconstruct the most likely evolutionary history given the observed DNA sequences. Protein sequences have been a free giveaway due to the ready translatability of the amino acid sequence from DNA based on the almost universal genetic code. DNA sequences became the main source material of molecular evolution research for quite a while, further spurred by the Human Genome Project and later the advent of the next-generation sequencing data explosion. Evolutionary relationships within populations and among species were revealed in ever greater detail.

Still, no matter how much genetic sequence data has become available, there still have been many aspects of how genetics translates to observable (phenotypic) changes that cannot be understood at that level of description. Network science is another toolkit rooted in math and computation that is used to study evolution at the genotypic to phenotypic interface. There are networks representing physical and chemical molecular interactions within a cell, the flow of information and cell-level "computation" and communication, as well as more abstract networks describing the relationships and similarities between gene and protein sequences, including the entire "universe" of known proteins. While biological network science—often called systems biology—comes close to providing a working model of the cellular phenotype, the real "gap" in understanding where a mutation in the DNA sequence makes a difference to the survival and fitness of an entire organism is how physical interactions, the "edges" or connections in systems biology networks, are a result of biophysical properties of proteins, which can be altered by mutations. It is this point—where changes of DNA translate into altered protein structure and function—that most of the methods in this book are focused on.

While Molecular Evolution has been a backward-facing, almost historical, discipline in its early days, it has increasingly matured into an "applicable" science due to its intersections with Biochemistry and Biophysics. Protein evolution is therefore much more than just the description of evolutionary relationships based on sequence differences. It has become a powerful tool for interfering with the evolution of pathogens, for devising therapies against mutation-based diseases such as cancers, and for designing novel enzymes with properties that can go beyond naturally evolved functions. Methods from evolution can be easily applied whenever genetic variation is at play, and this variation is what makes all humans unique and sometimes even determines why diseases and infections affect each of us differently.

While each chapter in this book is the unique work of its authors and there is no predefined "narrative" to this book, some common themes become apparent.

The first theme is that of mutations of single amino acids, i.e. point mutations. Predicting their effect on the physical structure of a protein is an important capability that links the abundance of sequence information with the comparatively few known structures (Chapters 1 and 2). Other mutational mechanisms lead to gene duplication (Chapter 3) and even de novo emergence of new genes (Chapter 4).

Likewise, the understanding of pairwise correlated mutations can be used to reveal structure information where none is available because the fates of spatially close (and physically interacting) amino acids are evolutionarily linked and coevolve (Chapters 5, 6 and 7).

Going back into evolutionary history, the structure and function of proteins can be reconstructed and used productively, since these may bear similar functions to their extant descendants yet also may have some new functional properties (Chapters 8 and 9). Many formerly sequence-based methods such as sequence alignments and phylogenies can be improved by applying a more structural and biophysical viewpoint (Chapters 10 and 11).

Instead of exploring similar proteins along evolutionary time, one can of course also compare existing proteins based on their similarity in sequence and structure. A number of classification schemes for organizing all known proteins exist, and it is possible to explore an entire "protein universe," often by breaking full proteins into even smaller building units called domains (Chapters 12, 13, 14, 15 and 16). Homology modeling makes use of these similarities by fitting the sequences of proteins without known structure to those known structures of proteins with similar sequence (Chapter 17). This structure prediction can also be extended to protein-protein interactions (Chapter 18) and even some structural properties of proteins lacking a fixed structure, i.e., disordered/unstructured proteins can be predicted (Chapter 19). Another important aspect related to disorder is the intrinsic dynamic nature of folded proteins that always exist as an ensemble of conformations, some of which become favored or disfavored with evolutionary changes (Chapter 20).

Finally, evolutionary principles are at work in shaping such versatile proteins as antibodies or enzymes, which can also be designed to have certain properties in silico by applying directed evolution, i.e., where the evolutionary endpoint, but not its path, is determined by the researcher (Chapters 21 and 22).

The book covers a wide range of computational approaches, including the dynamic programming techniques of sequence alignments, the clustering methods of phylogenies, physics-based approaches such as molecular dynamics simulations, and a range of statistical, graph-based, and machine learning methods. While the authors take the time to give some background and references in the introductory sections, this book is not a textbook, and more detailed descriptions of underlying theory and algorithms may have to be found elsewhere. Nevertheless, I think that there is a lot to be learned from this book for an interdisciplinary readership.

I sincerely hope that this book offers many useful workflows and techniques that help many researchers and students working with proteins computationally. I also strongly encourage the reader to go beyond the individual protocol and mix and match the different methods to come up with new innovative solutions. That's what evolution would do.

Heidelberg, Germany — ***Tobias Sikosek***

Contents

Contributors

KELSEY AADLAND • *Department of Microbiology & Cell Science, Institute for Food and Agricultural Sciences, University of Florida, Gainesville, FL, USA*
MATTEO ALDEGHI • *Max Planck Institute for Biophysical Chemistry, Computational Biomolecular Dynamics Group, Göttingen, Germany*
BEAT ANTON AMREIN • *Associate Scientist, Tecan Schweiz AG, Männedorf, Switzerland*
MIGUEL ARENAS • *Department of Biochemistry, Genetics and Immunology, University of Vigo, Vigo, Spain*
MARIANO AVINO • *Department of Pathology and Laboratory Medicine, Western University, London, Canada*
UGO BASTOLLA • *Centre for Molecular Biology, Severo Ochoa (CSIC-UAM), Madrid, Spain*
NIR BEN-TAL • *Department of Biochemistry and Molecular Biology, George S. Wise Faculty of Life Sciences, Tel Aviv University, Tel Aviv, Israel*
MARTINO BERTONI • *Biozentrum, University of Basel and SIB Swiss Institute of Bioinformatics, Basel, Switzerland*
STEFAN BIENERT • *Biozentrum, University of Basel and SIB Swiss Institute of Bioinformatics, Basel, Switzerland*
LORENZA BORDOLI • *Biozentrum, University of Basel and SIB Swiss Institute of Bioinformatics, Basel, Switzerland*
ERICH BORNBERG-BAUER • *Institute for Evolution and Biodiversity, University of Münster, Münster, Germany*
CARLES CORBI-VERGE • *Terrence Donnelly Centre for Cellular and Biomolecular Research, University of Toronto, Toronto, ON, Canada*
LILIANA M. DÁVALOS • *Department of Ecology and Evolution, Stony Brook University, Stony Brook, NY, USA*
CHARLOTTE M. DEANE • *Department of Statistics, University of Oxford, Oxford, UK*
MARIA SILVINA FORNASARI • *Departamento de Ciencia y Tecnología, Universidad Nacional de Quilmes, CONICET, Bernal, Argentina*
NICHOLAS FURNHAM • *London School of Hygiene and Tropical Medicine, London, UK*
VYTAUTAS GAPSYS • *Max Planck Institute for Biophysical Chemistry, Computational Biomolecular Dynamics Group, Göttingen, Germany*
NICK V. GRISHIN • *Department of Biophysics, University of Texas Southwestern Medical Center, Dallas, TX, USA; Howard Hughes Medical Institute, University of Texas Southwestern Medical Center, Dallas, TX, USA*
BERT L. DE GROOT • *Max Planck Institute for Biophysical Chemistry, Computational Biomolecular Dynamics Group, Göttingen, Germany*
EMINE GUVEN-MAIOROV • *Cancer and Inflammation Program, Leidos Biomedical Research, Inc., Frederick National Laboratory for Cancer Research, National Cancer Institute, Frederick, MD, USA*
JOSEPH L. HERMAN • *Department of Biomedical Informatics, Harvard Medical School, Boston, MA, USA*
KRISTINA STRAUB • *Institute of Biophysics and Physical Biochemistry, University of Regensburg, Regensburg, Germany*

Xianli Jiang • *Department of Biological Sciences, University of Texas at Dallas, Richardson, TX, USA*

Shina Caroline Lynn Kamerlin • *Department of Chemistry, BMC, Uppsala University, Uppsala, Sweden*

Carsten Kemena • *Institute for Evolution and Biodiversity, University of Münster, Münster, Germany*

Philip M. Kim • *Department of Computer Science, University of Toronto, Toronto, ON, Canada; Terrence Donnelly Centre for Cellular and Biomolecular Research, University of Toronto, Toronto, ON, Canada; Department of Molecular Genetics, University of Toronto, Toronto, ON, Canada*

Bryan Kolaczkowski • *Department of Microbiology & Cell Science, Institute for Food and Agricultural Sciences, University of Florida, Gainesville, FL, USA; Genetics Institute, University of Florida, Gainesville, FL, USA*

Rachel Kolodny • *Department of Computer Science, University of Haifa, Haifa, Israel*

Lukasz Kurgan • *Department of Computer Science, Virginia Commonwealth University, Richmond, VA, USA*

Jinwoo Leem • *Department of Statistics, University of Oxford, Oxford, UK*

Rosalba Lepore • *Biozentrum, University of Basel and SIB Swiss Institute of Bioinformatics, Basel, Switzerland*

David A. Liberles • *Department of Biology and Center for Computational Genetics and Genomics, Temple University, Philadelphia, PA, USA*

Liang Liu • *Department of Statistics and Institute of Bioinformatics, University of Georgia, Athens, GA, USA*

Buyong Ma • *Cancer and Inflammation Program, Leidos Biomedical Research, Inc., Frederick National Laboratory for Cancer Research, National Cancer Institute, Frederick, MD, USA*

Joseph A. Marsh • *MRC Human Genetics Unit, MRC Institute of Genetics and Molecular Medicine, University of Edinburgh, Edinburgh, UK*

Leandro Martínez • *Institute of Chemistry, University of Campinas (UNICAMP), Campinas, SP, Brazil*

Aoife McLysaght • *Department of Genetics, Trinity College Dublin, Smurfit Institute of Genetics, University of Dublin, Dublin, Ireland*

Rainer Merkl • *Institute of Biophysics and Physical Biochemistry, University of Regensburg, Regensburg, Germany*

Alexander Miguel Monzon • *Departamento de Ciencia y Tecnología, Universidad Nacional de Quilmes, CONICET, Bernal, Argentina*

Faruck Morcos • *Department of Biological Sciences, University of Texas at Dallas, Richardson, TX, USA; Center for Systems Biology, University of Texas at Dallas, Richardson, TX, USA*

Aya Narunsky • *Department of Biochemistry and Molecular Biology, George S. Wise Faculty of Life Sciences, Tel Aviv University, Tel Aviv, Israel*

Ruth Nussinov • *Cancer and Inflammation Program, Leidos Biomedical Research, Inc., Frederick National Laboratory for Cancer Research, National Cancer Institute, Frederick, MD, USA; Department of Human Genetics and Molecular Medicine, Sackler Institute of Molecular Medicine, Sackler School of Medicine, Tel Aviv University, Tel Aviv, Israel*

Christopher J. Oldfield • *Department of Computer Science, Virginia Commonwealth University, Richmond, VA, USA*

CHRISTINE M. ORENGO • *Institute of Structural and Molecular Biology, University College London, London, UK*

GUSTAVO PARISI • *Departamento de Ciencia y Tecnología, Universidad Nacional de Quilmes, CONICET, Bernal, Argentina*

FRANK J. POELWIJK • *cBio Center, Department of Biostatistics and Computational Biology, Boston, MA, USA*

ART F. Y. POON • *Department of Pathology and Laboratory Medicine, Western University, London, Canada*

CHARLES PUGH • *Department of Microbiology & Cell Science, Institute for Food and Agricultural Sciences, University of Florida, Gainesville, FL, USA*

ASHISH RUNTHALA • *Indian Institute of Science, Bangalore, India*

RICARDO NASCIMENTO DOS SANTOS • *Institute of Chemistry, University of Campinas (UNICAMP), Campinas, SP, Brazil*

DUSTIN SCHAEFFER • *Department of Biophysics, University of Texas Southwestern Medical Center, Dallas, TX, USA*

TORSTEN SCHWEDE • *Biozentrum, University of Basel and SIB Swiss Institute of Bioinformatics, Basel, Switzerland*

IAN SILLITOE • *Institute of Structural and Molecular Biology, University College London, London, UK*

ALEXEY STROKACH • *Department of Computer Science, University of Toronto, Toronto, ON, Canada; Terrence Donnelly Centre for Cellular and Biomolecular Research, University of Toronto, Toronto, ON, Canada*

GABRIEL STUDER • *Biozentrum, University of Basel and SIB Swiss Institute of Bioinformatics, Basel, Switzerland*

GERARDO TAURIELLO • *Biozentrum, University of Basel and SIB Swiss Institute of Bioinformatics, Basel, Switzerland*

JOAN TEYRA • *Terrence Donnelly Centre for Cellular and Biomolecular Research, University of Toronto, Toronto, ON, Canada*

JANET M. THORNTON • *EMBL-EBI, Wellcome Genome Campus, Hinxton, Cambridge, UK*

CHUNG-JUNG TSAI • *Cancer and Inflammation Program, Leidos Biomedical Research, Inc., Frederick National Laboratory for Cancer Research, National Cancer Institute, Frederick, MD, USA*

JONATHAN D. TYZACK • *EMBL-EBI, Wellcome Genome Campus, Hinxton, Cambridge, UK*

VLADIMIR N. UVERSKY • *Department of Molecular Medicine and USF Health Byrd Alzheimer's Research Institute, Morsani College of Medicine, University of South Florida, Tampa, FL, USA; Institute for Biological Instrumentation, Russian Academy of Sciences, Moscow Region, Russia*

NIKOLAOS VAKIRLIS • *Department of Genetics, Trinity College Dublin, Smurfit Institute of Genetics, University of Dublin, Dublin, Ireland*

ANDREW MARK WATERHOUSE • *Biozentrum, University of Basel and SIB Swiss Institute of Bioinformatics, Basel, Switzerland*

JONATHAN N. WELLS • *MRC Human Genetics Unit, MRC Institute of Genetics and Molecular Medicine, University of Edinburgh, Edinburgh, UK*

LAUREL R. YOHE • *Department of Geology & Geophysics, Yale University, New Haven, CT, USA*

DIEGO JAVIER ZEA • *Structural Bioinformatics Unit, Fundación Instituto Leloir, CONICET, Buenos Aires, Argentina*

Chapter 1

Predicting the Effect of Mutations on Protein Folding and Protein-Protein Interactions

Alexey Strokach, Carles Corbi-Verge, Joan Teyra, and Philip M. Kim

Abstract

The function of a protein is largely determined by its three-dimensional structure and its interactions with other proteins. Changes to a protein's amino acid sequence can alter its function by perturbing the energy landscapes of protein folding and binding. Many tools have been developed to predict the energetic effect of amino acid changes, utilizing features describing the sequence of a protein, the structure of a protein, or both. Those tools can have many applications, such as distinguishing between deleterious and benign mutations and designing proteins and peptides with attractive properties. In this chapter, we describe how to use one of such tools, ELASPIC, to predict the effect of mutations on the stability of proteins and the affinity between proteins, in the context of a human protein-protein interaction network. ELASPIC uses a wide range of sequential and structural features to predict the change in the Gibbs free energy for protein folding and protein-protein interactions. It can be used both through a web server and as a stand-alone application. Since ELASPIC was trained using homology models and not crystal structures, it can be applied to a much broader range of proteins than traditional methods. It can leverage precalculated sequence alignments, homology models, and other features, in order to drastically lower the amount of time required to evaluate individual mutations and make tractable the analysis of millions of mutations affecting the majority of proteins in a genome.

Key words Computational biology, Structural biology, Bioinformatics, Protein stability, Mutations, Protein engineering

1 Introduction

Proteins usually fold into specific, stable, three-dimensional structures, which allow them to interact with other molecules and carry out their biological function. Protein-protein interactions play a critical role in the regulation of numerous biological processes. The functionality of the protein interaction network is grounded on the specificity that proteins have for each other, which arises from the complementarity in shape and the physicochemical composition of the protein interaction interface. Mutations can perturb the interaction profile of a protein by changing the shape or the composition of the protein interaction interface, leading to a loss,

Tobias Sikosek (ed.), *Computational Methods in Protein Evolution*, Methods in Molecular Biology, vol. 1851, https://doi.org/10.1007/978-1-4939-8736-8_1,

or a gain, of new interaction partners. Mutations can also change the stability of a protein, causing changes in protein conformation, solubility, or other attributes that dictate protein function.

Although most mutations are destabilizing and detrimentally affect the activity of a protein, this effect is usually modest, and only a small fraction of all possible mutations entirely abolish protein function [1]. This allows for a neutral drift in the amino acid sequence of a protein, which can eventually, under selective pressure, evolve into new functionality [2]. An immediate consequence of the accumulated variation in the human genome is the diversity in responses that people can have to drugs or other environmental stimuli. For instance, different patients with the same disease often do not react in a homogeneous manner to the same drug. A treatment usually is more effective in a subpopulation of patients, and it may be completely ineffective or even detrimental to others [3]. A better understanding of the effect that the variation in the human genome has on protein folding and protein-protein interaction would help us to predict how a patient might respond to a specific treatment. It would also improve our ability to detect and treat genetic diseases and to develop targeted therapies against cancer and drug-resistant pathogens [4].

Experimental approaches for evaluating the effect of mutations on protein folding and protein-protein interaction, such as isothermal titration calorimetry (ITC) [5], luminescence-based mammalian interactome mapping (LUMIER) [6], phage display [7], and deep mutational scanning [8], are laborious, time-consuming, and expensive. Accordingly, many computational methods have been developed to predict the thermodynamic and phenotypic consequences of mutations. Those computational tools can broadly be categorized as sequence-based tools, structure-based tools, and tools which use both sequential and structural information in order to make their predictions.

1.1 Sequence-Based Tools

Sequence-based tools usually rely on some form of a conservation score, describing the frequency with which a particular nucleotide or amino acid is found at the given position in domain-, protein-, or genome-level alignments, in order to make their prediction [9–14]. Due to their speed and scalability, sequence-based tools are the de facto standard for annotating newly discovered variants. However, they remain limited in their accuracy and the type of information that they can provide [15]. In particular, they only predict whether or not a particular mutation is likely to be deleterious and provide no information as to *why* that mutation may be deleterious. This makes it difficult to act upon those predictions, for example, by designing drugs that would curtail the effect of disease-causing mutations or would take advantage of mutations found in cancer.

1.2 Structure-Based Tools

Structure-based tools predict the effect of mutations on protein structure and/or function using features describing the three-dimensional structure of the protein. They range from accurate but computationally expensive alchemical free energy calculations, which involve modeling the structural transition from the wild type to the mutant protein and using different integration techniques to calculate the energy of the transition [16], to quicker but more approximate techniques, which use semiempirical or statistical potentials and assume that the backbone of the protein remains fixed [17–21]. In theory, structure-based tools should be able to offer more insight into the effect of missense mutations than sequence-based tools, since the effect is directly caused by changes in protein structure and function and not by changes in the DNA sequence. However, since existing structure-based tools require manual setup and a crystal structure of the protein being mutated, they are not being used systematically to evaluate the effect of newly discovered mutations.

1.3 Combination of Sequence and Structure

Several tools have been developed that attempt to combine sequence- and structure-based information in order to make more accurate predictions about the deleteriousness [22] and the structural impact [21, 23, 24] of mutations. Those tools generally are "meta-predictors" which integrate the results of several sequence- and structure-based tools using machine learning algorithms trained on an appropriate dataset [25, 26]. Most of those tools remain limited in their coverage because only a small fraction of all proteins and protein-protein interactions have an experimentally determined structure [27]. ELASPIC, developed by Berliner et al. [23], overcomes this limitation by using homology models, instead of crystal structures, to evaluate the structural impact of mutations. ELASPIC still achieves relatively high accuracy in predicting the effect of mutations on protein stability and protein-protein interaction affinity, but it has much higher coverage, including the majority of proteins in the human proteome and hundreds of thousands of protein-protein interactions.

In this protocol, we describe how to set up and run ELASPIC on a local machine. We describe how precalculated homology models and other data can be downloaded and installed in order to greatly reduce the time taken by ELASPIC to evaluate new mutations. Finally, we show how to use ELASPIC to perform alanine scanning of a protein-protein interaction interface and how to evaluate the structural effect of several thousand mutations that have been implicated in cancer.

2 Materials

This tutorial requires basic knowledge of the Linux command line environment. While we made every effort to make the installation of ELASPIC as simple as possible, the process remains reasonably involved and may take several hours. If you do not wish to make changes to the ELASPIC source code and are planning to run under a few thousand mutations, using the ELASPIC web server [28], available at http://elaspic.kimlab.org, is encouraged. The web server may also be used to verify the results obtained using a local installation of ELASPIC.

The source code for ELASPIC is available at https://gitlab.com/kimlab/elaspic/ and is provided under an MIT license. The documentation for ELASPIC is available at https://kimlab.gitlab.io/elaspic/. ELASPIC should work on any Linux distribution with a version of glibc ≥ 2.14 (e.g., CentOS 6 or newer, Ubuntu 12.04 or newer). At the moment, it does not work on Windows or MacOS (although *see* **Notes 1** and **2**).

ELASPIC can be run using two different "pipelines," the *database pipeline* and the *local pipeline*, as shown in Fig. 1. The *database pipeline* allows us to evaluate the thermodynamic impact of mutations on a proteome-wide scale, without having to specify a structural template for each protein. This pipeline takes as input the UniProt ID of the protein being mutated and one or more mutations affecting that protein. At each decision node, the pipeline queries the database to check whether or not the required information has already been calculated. If the required data has not been calculated, the pipeline executes the appropriate code and stores the results in the database for later retrieval. The pipeline proceeds until homology models of all domains in the protein, and all domain-domain interactions involving the protein, have been calculated and the $\Delta\Delta G$ has been predicted for every specified mutation. The *local pipeline* can be used without downloading and installing a local copy of the ELASPIC databases but requires a PDB structure or template to be provided for every protein. The output from this pipeline is saved as JSON files inside the working directory, rather than being uploaded to the database, as in the case of the database pipeline. Both pipelines use the same internal libraries to perform the majority of the computation.

The ELASPIC database, required by the *database pipeline*, includes many external datasets, which are listed in Table 1. The use of the external datasets is made transparent to the ELASPIC user, who simply has to load the data from the ELASPIC download page (http://elaspic.kimlab.org/static/download) into their local ELASPIC database using the elaspic database load-basic or elaspic database load-complete commands. The only exception is the BLAST nr database, which is required by both the *database pipeline* and the *local pipeline* and has to be downloaded separately from the NCBI website (although *see* **Note 4**). This is described in detail in Subheading 3.

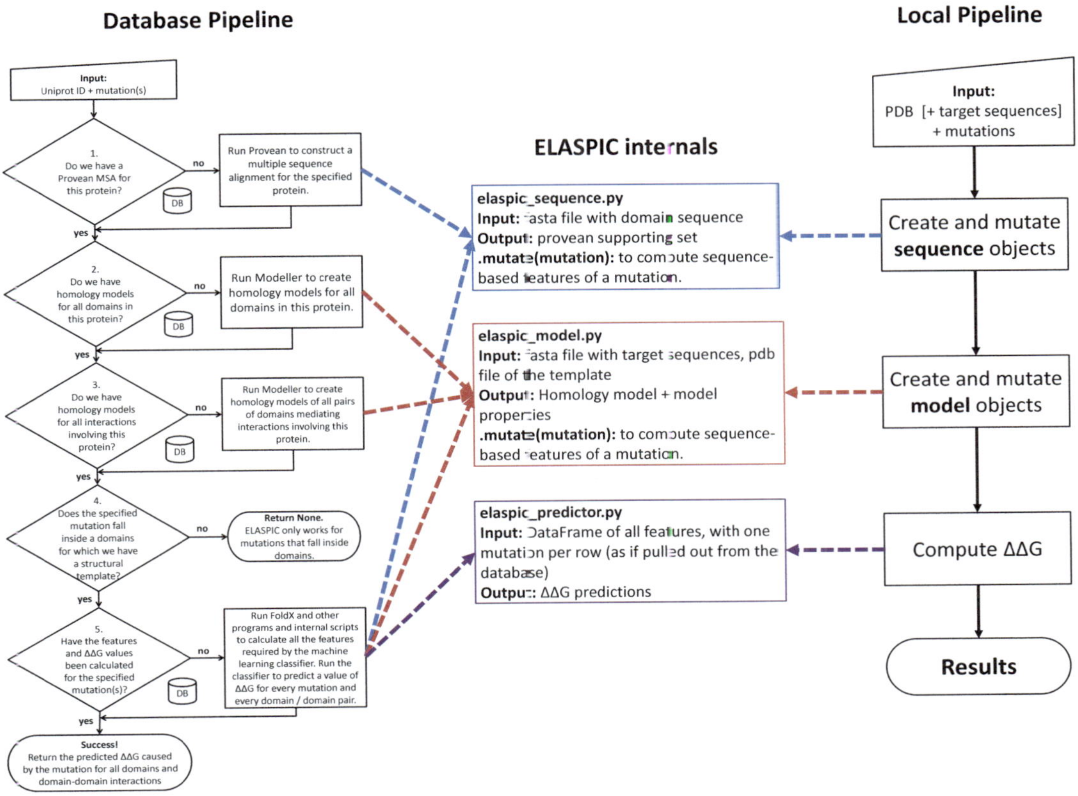

Fig. 1 Schematic providing a general outline of ELASPIC. ELASPIC provides two different pipelines: a *database pipeline* and a *local pipeline*. The *database pipeline* takes as input the UniProt ID of a protein and a mutation and constructs homology models of the domains and domain-domain interactions involving the protein automatically. The *local pipeline* takes as input the structure of a protein, or the sequence or a protein and a structural template, and a mutation. It requires no precalculated data and can run in the absence of the ELASPIC database. Both pipelines use the same code to perform the majority of the calculation

ELASPIC also uses many external programs, which are listed in Table 2. All external dependencies, except for FoldX, are installed automatically when ELASPIC is installed using the conda package manager. Due to licensing restrictions, FoldX has to be downloaded and installed manually, as described in Subheading 3.

3 Methods

3.1 Installing ELASPIC

1. First, we should set the environment variables, which are required for installing and using ELASPIC, in our ~/.bashrc file. This way, those environment variables will be set whenever we start a new bash shell. The required environment variables,

Table 1
External databases that were used in the construction of the ELASPIC database

Database	Description	URL
UniProt [33]	UniProt is a comprehensive collection of protein sequences and their annotations. ELASPIC uses UniProt names to identify all proteins. Homology models constructed by ELASPIC use the UniProt canonical protein sequence	http://www.uniprot.org/
Mentha [34]	Mentha is a database of protein-protein interactions. Mentha combines data from several databases in the IMEx consortium and converts gene names to UniProt identifiers	http://mentha.uniroma2.it/
Profs [28]	Profs is a database of protein domain definitions, which are obtained by integrating data from CATH and PFam (*see* **Note 3**)	https://bitbucket.org/afgiraldofo/profs
BLAST nr database [35]	BLAST nr database includes a collection of unique protein sequences from a variety of sources, including GenPept, Swissprot, PDB, PRF, PIR, and RefSeq (*see* **Note 4**)	ftp://ftp.ncbi.nlm.nih.gov/blast/db/

with reasonable default values, are shown below. The KEY_MODELLER environment variable should contain our Modeller license key. If we do not already have a Modeller license, we should register for a license on the Modeller website: https://salilab.org/modeller/registration.html. Modeller is free for noncommercial use.

```
# Add the following lines to your ~/.bashrc file
export CONDA_DIR="${HOME}/miniconda3"
export LOCAL_BIN_DIR="${HOME}/.local/bin"
export PATH="${CONDA_DIR}/bin:${LOCAL_BIN_DIR}:${PATH}"
export KEY_MODELLER="Put our modeller license here!"
export BLAST_DB_DIR="${HOME}/blast"
export ELASPIC_DB_STRING="sqlite:///${HOME}/elaspic.db"
export ELASPIC_ARCHIVE_DIR="${HOME}/elaspic"
$ source ~/.bashrc
```

2. Download and extract the foldx executable into a folder that has been added to the PATH environment variable. In order to download FoldX, we first need to create an account and register for an academic license at http://foldxsuite.crg.eu/academic-license-info. After registering, we will receive an email with the link to the FoldX download page. We can either download FoldX using a web browser or using wget. If we use wget, we first need to obtain a cookie that will permit us to download the FoldX archive.

Table 2
External software that ELASPIC uses to construct homology models and to calculate some of the sequential and structural features

Software	Description	URL	License
MODELLER [36]	MODELLER is a popular tool for generating homology models of proteins. It requires, as input, the sequence of the protein that we wish to model and the structure of a protein that will be used as a template (*see* **Notes 5** and **6**)	https://salilab.org/modeller/	Closed source. Free for academic use only
Provean [37]	Provean uses a residue conservation score to predict whether a mutation is likely to be deleterious. One advantage of Provean over similar tools is its use of a small "supporting set" of diverse proteins to compute the conservation score. This supporting set can be precalculated for each protein, and evaluating a mutation takes under a second when the supporting set is available	http://provean.jcvi.org/index.php	GPL
FoldX [38]	FoldX is a tool for predicting the thermodynamic impact of mutations on protein folding and protein-protein interaction. ELASPIC uses a number of structural features that are calculated by FoldX	http://foldxsuite.crg.eu	Closed source. Free for academic use only
MSMS [39]	Maximal Speed Molecular Surface (MSMS) is a tool for calculating the solvent-accessible surface area of proteins. It also can be used to quickly calculate the surface mesh of a protein (*see* **Note 7**)	http://mgltools.scripps.edu/packages/MSMS	Closed source. Free for academic use only
STRIDE [40]	STRIDE is a tool for predicting the secondary structure of amino acids in a protein (*see* **Note 7**)	http://webclu.bio.wzw.tum.de/stride/	Open source. Free for academic use only

```
$ mkdir -p "${LOCAL_BIN_DIR}" && cd "${LOCAL_BIN_DIR}"
$ wget --save-cookies cookies.txt --keep-session-cookies --delete-after
http://foldxsuite.crg.eu/node/732/download/{unique_token_from_email}
$ wget --load-cookies cookies.txt
http://foldxsuite.crg.eu/system/files/foldxLinux64.tar__0.gz
$ tar xf foldxLinux64.tar__0.gz
```

3. Download and install either Miniconda or the Anaconda Python distribution, if we do not have them installed already.

```
$ wget https://repo.continuum.io/miniconda/Miniconda3-latest-Linux-x86_64.sh
$ bash Miniconda3-latest-Linux-x86_64.sh -b -p "${CONDA_DIR}"
```

4. Add the conda channels required for installing ELASPIC to our conda settings.

```
$ conda config --add channels conda-forge
$ conda config --append channels bioconda
$ conda config --append channels salilab
$ conda config --append channels kimlab
```

5. Install ELASPIC, including all its dependencies, into a new conda environment. Activate the new environment, and check that elaspic is available by running elaspic --help.

```
$ conda create -n elaspic 'elaspic=0.1' parallel ipython p7zip
$ source activate elaspic
$ elaspic --help
usage: elaspic [-h] {run,database,train} ...
```

optional arguments:

```
-h, --help show this help message and exit
```

command:

```
{run,database,train}
 run Run ELASPIC
 database Perform database maintenance tasks
 train Train the ELASPIC classifiers
```

6. Download the BLAST nonredundant database, check downloaded files for consistency, and uncompress files into our BLAST_DB_DIR folder.

```
$ mkdir -p "${BLAST_DB_DIR}" && cd "${BLAST_DB_DIR}"
$ wget 'ftp://ftp.ncbi.nlm.nih.gov/blast/db/nr*'
$ md5sum -c *.md5
$ for file in *.gz ; do echo "Uncompressing ${file}..." ; tar xf ${file} ; done
```

7. **(Optional)** Create an ELASPIC database, which will contain information about protein domains and domain interactions, structural templates, homology models, and mutation results, and an ELASPIC archive folder, which will store the Provean supporting sets and homology model files. Keeping previously calculated supporting sets and homology models drastically speeds up the evaluation of all subsequent mutations in the same protein.

```
$ elaspic database --connection-string ${ELASPIC_DB_STRING} create
$ mkdir -p "${ELASPIC_ARCHIVE_DIR}"
```

8. **(Optional)** Load precalculated data into the ELASPIC database.
   (Option A) If we do not wish to download precalculated Provean supporting sets and homology models, for example, if we are planning to run ELASPIC for only a handful of human proteins, we should load only the basic dataset.

```
$ elaspic database --connection-string ${ELASPIC_DB_STRING} load-basic \
 http://elaspic.kimlab.org/static/download/latest/homo_sapiens/
```

   (Option B) If we would like to use the Provean supporting sets and homology models that have been calculated previously for the human proteome, we should load the complete dataset into our ELASPIC database, and we should also download and extract a file containing the supporting sets and homology models into our ELASPIC archive folder.

```
$ elaspic database --connection-string ${ELASPIC_DB_STRING} load-complete \
 http://elaspic.kimlab.org/static/download/latest/homo_sapiens/
$ cd ${ELASPIC_ARCHIVE_DIR}
$ wget http://elaspic.kimlab.org/static/download/latest/homo_sapiens/archive.7z
$ 7z x archive.7z
```

Steps 7 and **8** above are required if we want to use ELASPIC to evaluate the effect of mutations on homology models constructed for proteins specified using a UniProt ID (i.e., *database pipeline* in Fig. 1). Those steps can be skipped if we will only use ELASPIC to evaluate the effect of mutations on provided protein structures (*local pipeline* in Fig. 1).

ELASPIC can access data stored in a SQLite, MySQL, or PostgreSQL database. SQLite is recommended if ELASPIC will only be used on a single machine. MySQL or PostgreSQL is recommended if ELASPIC will be used on a cluster of machines, with a single database acting as a centralized store of information. In the steps outlined above, the database can be changed from SQLite to MySQL or PostgreSQL by changing the ELASPIC_DB_STRING environment variable from sqlite:///${HOME}/elaspic.db (which stores the data in a single SQLite database file called elaspic.db, located in our home directory) to {mysql|postgresql}://{username}:{password}@{database_ip}:{database_port}/elaspic, where the words inside curly brackets are replaced by database-specific values.

Protein domain definitions and structural template data are available for a number of organisms from the ELASPIC downloads page: http://elaspic.kimlab.org/static/download/latest/. Note, however, that protein-protein interaction data and precalculated Provean supporting sets and homology models are only available for the human proteome. In the steps outlined above, we can change the organism for which the data is downloaded by changing all occurrences of the string homo_sapiens with the string for the desired organism. For more information on how this data was calculated, *see* the supplemental material in the ELASPIC web server paper [28].

3.2 Running ELASPIC

3.2.1 Evaluating the Effect of Mutations on a Single Protein (Local Pipeline)

The first use case for ELASPIC is to predict the thermodynamic effect of mutations on a protein or a protein-protein interaction for which a crystal structure is available (*local pipeline* in Fig. 1). In this case, the crystal structure of the protein can be provided to ELASPIC directly, and no homology model needs to be created. In the following example, we will show how to use ELASPIC to perform alanine scanning of the dimerization interface of glutathione *S*-transferase.

1. Make sure that the environment variables that we set in **step 1** of the ELASPIC installation are available.

```
$ [[ -z ${BLAST_DB_DIR} ]] && source ~/.bashrc
```

2. Download a structure of glutathione *S*-transferase epsilon, which we will be using for this example.

```
$ wget https://files.rcsb.org/download/3zml.pdb
```

3. Run ELASPIC to calculate Provean supporting sets and optimized structural models for the protein of interest. We can see a description of every argument accepted by the elaspic run command by running elaspic run --help.

```
$ source activate elaspic
$ elaspic run -p 3zml.pdb -t sequence.model
```

4. Run ELASPIC to evaluate the structural impact of each individual mutation. We should specify mutations using the {chain_id}_{residue_wt}{resnum}{residue_mut} format and separate different mutations using a comma or a colon.

```
$ elaspic run -p 3zml.pdb -m
A_M47A:A_P51A:A_Q52A:A_H53A:A_T63A:A_I65A:A_T66A:A_E67A:A_H69A:A_I73A:A_Y74A:
A_T77A:A_Y86A:A_P90A:A_V91A:A_Q93A:A_N97A:A_L100A:A_H101A:A_F102A:A_S104A:A_G105A:
A_R110A:A_R112A:A_F113A:A_E116A:A_R117A:A_Y121A:A_D129A:A_R130A:A_Y133A:A_K136A:
A_L140A:A_D143A:A_T144A -n 1 -vvv
```

Alternatively, we can use GNU parallel to process multiple mutations in parallel.

```
$ parallel --res mutation_logs --joblog mutation_logs.txt \
 elaspic run -p 3zml.pdb -m {1} -t mutation -n 1 -vvv ::: A_M47A A_P51A A_Q52A A_H53A
A_T63A A_I65A A_T66A A_E67A A_H69A A_I73A A_Y74A A_T77A A_Y86A A_P90A A_V91A A_Q93A
A_N97A A_L100A A_H101A A_F102A A_S104A A_G105A A_R110A A_R112A A_F113A A_E116A A_R117A
A_Y121A A_D129A A_R130A A_Y133A A_K136A A_L140A A_D143A A_T144A
```

5. Once the above commands have finished running, we can read ELASPIC results from the .elaspic/sequence.json, .elaspic/model.json, and .elaspic/mutation_{mutation}.json files using the pandas.load_json command in Python (here {mutation} should be replaced with the mutation of interest). All other files generated by ELASPIC, including Provean supporting sets and structures containing each mutation, are also stored inside the .elaspic folder. The idxs column contains the indexes of the chains that make up the interface for which the ΔΔG was calculated; it is NaN if the ΔΔG is calculated for protein folding. Note that ELASPIC renumbers the PDB resnum field to start from 1, and therefore the mutations listed in the JSON files have a different residue number than what was provided as input.

```
$ ipython>>>
import pandas as pd
>>> mutations = ['A_M47A', 'A_P51A', 'A_Q52A', 'A_H53A', 'A_T63A', 'A_I65A',
'A_T66A', 'A_E67A', 'A_H69A', 'A_I73A', 'A_Y74A', 'A_T77A', 'A_Y86A', 'A_P90A',
'A_V91A', 'A_Q93A', 'A_N97A', 'A_L100A', 'A_H101A', 'A_F102A', 'A_S104A',
'A_G105A', 'A_R110A', 'A_R112A', 'A_F113A', 'A_E116A', 'A_R117A', 'A_Y121A',
'A_D129A', 'A_R130A', 'A_Y133A', 'A_K136A', 'A_L140A', 'A_D143A', 'A_T144A']
>>> results = []
>>> for mutation in mutations:
     results.append(
      pd.read_json('.elaspic/mutation_{}.json'.format(mutation)))
>>> mutation_df = pd.concat(results)
>>> mutation_df[['chain_modeller', 'mutation', 'ddg', 'idxs']].head()
  chain_modeller mutation ddg idxs
0   A M44A 0.862108 NaN
1   A M44A 1.475214 [0, 1]
0   A P48A 0.936352 NaN
1   A P48A 2.066441 [0, 1]
0   A Q49A 0.353452 NaN
```

6. We can compare the results that we obtained by running ELASPIC locally with the results that are obtained using the ELASPIC web server by going to http://elaspic.kimlab.org/result/3zml00/.

3.2.2 Evaluating the Effect of Mutations Proteome Wide (Database Pipeline)

A second use case for ELASPIC is to evaluate the effect of mutations in a large number of proteins and protein-protein interactions for which a crystal structure may not be available (*database pipeline* in Fig. 1). In the following example, we will show how to use ELASPIC to predict the effect of missense mutations found in the OncoKB database [29] on protein stability and protein-protein interaction affinity. OncoKB is a database of mutations in known cancer genes with well-established clinical ramifications.

1. Make sure that the environment variables that we set in **step 1** of the ELASPIC installation are available.

```
$ [[ -z ${BLAST_DB_DIR} || -z ${ELASPIC_DB_STRING} || -z
${ELASPIC_ARCHIVE_DIR} ]] && source ~/.bashrc
```

2. Download oncokb.tsv file from the ELASPIC downloads page. This file is derived from the allAnnotatedVariants.txt file obtained from OncoKB [29]. We processed this file to convert HGNC gene identifiers to UniProt accession numbers and to exclude non-missense mutations.

```
$ wget http://elaspic.kimlab.org/static/download/protocol/oncokb.tsv
```

3. Process oncokb.tsv to create a file containing only unique UniProt protein identifiers (uniprot_ids.txt) and a file containing only UniProt identifiers and mutations (uniprot_ids_and_-mutations.txt).

```
$ tail -n +2 oncokb.tsv | awk '{print $2}' | sort -u > uniprot_ids.txt
$ tail -n +2 oncokb.tsv | awk '{print $2 "\t" $3}' | sort -u >
uniprot_ids_and_mutations.txt
```

4. Run ELASPIC to calculate Provean supporting sets and homology models for each mutated protein. We can see a description of every argument accepted by the elaspic run command by running elaspic run --help. Note that calculating Provean supporting sets and homology models is not necessary if we downloaded precalculated data for homo_sapiens from the ELASPIC downloads page (*see* **step 8** in Installing ELASPIC).

```
$ source activate elaspic
$ while read uniprot_id ; do
  echo $uniport_id
  elaspic run -u $uniprot_id -t sequence.model -vvv
 done < uniprot_ids.txt
```

Alternatively, we can use GNU parallel to process multiple proteins in parallel.

```
$ parallel --res log_sequence_model --joblog log_sequence_model.txt \
  elaspic run -u {1} -t sequence.model -vvv :::: uniprot_ids.txt
```

5. Run ELASPIC to evaluate each individual mutation.

```
$ while read uniprot_id mutation ; do
  echo $uniport_id $mutation
  elaspic run -u $uniprot_id -t sequence.model -vvv
  done < uniprot_ids_and_mutations.txt
```

Alternatively, we can use GNU parallel to process multiple mutations in parallel.

```
$ parallel --colsep '\t' --res log_mutation --joblog log_mutation.txt \
  elaspic run -u {1} -m {2} -t mutation -vvv :::: uniprot_ids_and_mutations.
txt
```

6. The results are stored in the uniprot_domain_mutation and uniprot_domain_pair_mutation tables inside the ELASPIC database, which is specified with the ELASPIC_DB_STRING environment variable. We can extract all calculated mutations using the following script.

```
$ ipython
>>> import os
>>> import pandas as pd
>>> import sqlalchemy as sa
>>> engine = sa.create_engine(os.environ['ELASPIC_CONNECTION_STRING'])
>>> # Show all core mutations
>>> core_df = pd.read_sql_query("""\
   SELECT uniprot_id, mutation, ddg core_ddg
   FROM uniprot_domain_mutation
   """, engine)
>>> core_df.head()
 uniprot_id mutation core_ddg
0 P22681 C396R 0.282069
1 P22681 D390Y -0.848413
2 P22681 H398Q 1.543390
3 P22681 H398Y 2.162050
4 P22681 K382E -0.856646
>>> # Show all interface mutations
>>> interface_df = pd.read_sql_query("""\
SELECT uniprot_id, mutation, CASE WHEN uniprot_id = uniprot_id_1 THEN
```

```
uniprot_id_2 ELSE uniprot_id_1 END partner_uniprot_id, ddg interface_ddg
FROM uniprot_domain_pair udp
JOIN uniprot_domain_pair_mutation udpm USING (uniprot_domain_pair_id);
""", engine)
>>> interface_df.head()
uniprot_id mutation partner_uniprot_id interface_ddg
0 P22681 H398Q P60604 1.718820
1 P22681 H398Y P60604 0.992272
2 P22681 K382E P62253 0.496983
3 P22681 K382E P60604 0.169106
4 P22681 K382E P51668 1.109610
```

7. We can compare the results that we obtained by running ELASPIC locally with the results that are obtained using the ELASPIC web server by going to http://elaspic.kimlab.org/result/oncokb/.

4 Notes

1. It is likely that ELASPIC would work on Windows 10 subsystem for Linux (which is based on Ubuntu 14.04), although this has not been tested.
2. ELASPIC should work on MacOS if the external dependencies, such as MSMS and Stride, are recompiled on the MacOS platform.
3. Profs domain definitions are no longer being updated and may be replaced with Gene3D domain definitions in a future release of ELASPIC.
4. A Provean supporting set includes a FASTA file with the sequence of every protein in that supporting set. However, due to a bug in Provean (up to, and including, version 1.1.5), even if a previously calculated supporting set is available, Provean still requires the BLAST nr database to obtain the sequences of the proteins in that supporting set. If this bug were fixed, downloading the BLAST nr database would no longer be required when the Provean supporting sets are available for all proteins that are mutated.
5. If MODELLER finds a gap in the alignment of the protein sequence to the structural template, it will fill this gap by constructing a loop. However, the loop modeling capabilities of MODELLER are limited. If more plausible loops are desired, Rosetta loop modeling [30] may be used instead.

6. I-TASSER can be used instead of Modeller to construct homology models. According to recent CASP competitions [31], I-TASSER is able to construct homology models with the best accuracy relative to the reference crystal structures.
7. MSMS and Stride can be replaced with MDTraj [32], which is distributed under a more permissive LGPL license, and also can calculate secondary structure and solvent-accessible surface area.
8. ELASPIC has a comprehensive test suite, which is run using GitLab continuous integration for every commit that is pushed to the repository (*see* https://gitlab.com/kimlab/elaspic/pipelines). The integration tests test-standalone and test-database are similar to the protocols described above (although they are run on a test database that is much smaller than the full ELASPIC database).

Acknowledgments

Funding: P.M.K. acknowledges support from a NSERC Discovery Grant (RGPIN-2017-064).

References

1. Rockah-Shmuel L, Tóth-Petróczy Á, Tawfik DS (2015) Systematic mapping of protein mutational space by prolonged drift reveals the deleterious effects of seemingly neutral mutations. PLoS Comput Biol 11:e1004421
2. Huber CD, Kim BY, Marsden CD, Lohmueller KE (2017) Determining the factors driving selective effects of new nonsynonymous mutations. Proc Natl Acad Sci U S A 114:4465–4470
3. Brender JR, Zhang Y (2015) Predicting the effect of mutations on protein-protein binding interactions through structure-based interface profiles. PLoS Comput Biol 11:e1004494
4. Albanaz ATS, Rodrigues CHM, Pires DEV, Ascher DB (2017) Combating mutations in genetic disease and drug resistance: understanding molecular mechanisms to guide drug design. Expert Opin Drug Discov 12:553–563
5. Jelesarov I, Bosshard HR (1999) Isothermal titration calorimetry and differential scanning calorimetry as complementary tools to investigate the energetics of biomolecular recognition. J Mol Recognit 12:3–18
6. Sahni N, Yi S, Taipale M et al (2015) Widespread macromolecular interaction perturbations in human genetic disorders. Cell 161:647–660
7. Sun MGF, Seo M-H, Nim S et al (2016) Protein engineering by highly parallel screening of computationally designed variants. Sci Adv 2: e1600692
8. Weile J, Sun S, Cote AG, et al (2017) Expanding the atlas of functional missense variation for human genes. BioRxiv 166595
9. Ng PC, Henikoff S (2003) SIFT: predicting amino acid changes that affect protein function. Nucleic Acids Res 31:3812–3814
10. Adzhubei I, Jordan DM, Sunyaev SR (2013) Predicting functional effect of human missense mutations using PolyPhen-2. Curr Protoc Hum Genet Chapter 7: Unit 7.20
11. Li B, Krishnan VG, Mort ME et al (2009) Automated inference of molecular mechanisms

of disease from amino acid substitutions. Bioinformatics 25:2744–2750

12. Kircher M, Witten DM, Jain P et al (2014) A general framework for estimating the relative pathogenicity of human genetic variants. Nat Genet 46:310–315
13. Shihab HA, Gough J, Mort M et al (2014) Ranking non-synonymous single nucleotide polymorphisms based on disease concepts. Hum Genomics 8:11
14. Choi Y, Sims GE, Murphy S et al (2012) Predicting the functional effect of amino acid substitutions and indels. PLoS One 7:e46688
15. Dorfman R, Nalpathamkalam T, Taylor C et al (2010) Do common in silico tools predict the clinical consequences of amino-acid substitutions in the CFTR gene? Clin Genet 77:464–473
16. Shirts M, Mobley D (2013) An introduction to best practices in free energy calculations. In: Monticelli L, Salonen E (eds) Biomolecular simulations, Methods in molecular biology. Humana Press, Totowa, NJ, pp 271–311
17. Benedix A, Becker CM, de Groot BL et al (2009) Predicting free energy changes using structural ensembles. Nat Methods 6:3–4
18. Pires DEV, Ascher DB, Blundell TL (2014) mCSM: predicting the effects of mutations in proteins using graph-based signatures. Bioinformatics 30:335–342
19. Laimer J, Hofer H, Fritz M et al (2015) MAESTRO - multi agent stability prediction upon point mutations. BMC Bioinformatics 16:116
20. Petukh M, Li M, Alexov E (2015) Predicting binding free energy change caused by point mutations with knowledge-modified MM/PBSA method. PLoS Comput Biol 11: e1004276
21. Dehouck Y, Grosfils A, Folch B et al (2009) Fast and accurate predictions of protein stability changes upon mutations using statistical potentials and neural networks: PoPMuSiC-2.0. Bioinformatics 25:2537–2543
22. Baugh EH, Simmons-Edler R, Müller CL et al (2016) Robust classification of protein variation using structural modelling and large-scale data integration. Nucleic Acids Res 44:2501–2513
23. Berliner N, Teyra J, Çolak R et al (2014) Combining structural modeling with ensemble machine learning to accurately predict protein fold stability and binding affinity effects upon mutation. PLoS One 9:e107353
24. Li M, Simonetti FL, Goncearenco A, Panchenko AR (2016) MutaBind estimates and interprets the effects of sequence variants on protein-protein interactions. Nucleic Acids Res 44:W494–W501
25. Kumar MDS, Bava KA, Gromiha MM et al (2006) ProTherm and ProNIT: thermodynamic databases for proteins and protein–nucleic acid interactions. Nucleic Acids Res 34: D204–D206
26. Moal IH, Fernández-Recio J (2012) SKEMPI: a structural kinetic and energetic database of mutant protein interactions and its use in empirical models. Bioinformatics 28:2600–2607
27. Rose PW, Prlić A, Altunkaya A et al (2017) The RCSB protein data bank: integrative view of protein, gene and 3D structural information. Nucleic Acids Res 45:D271–D281
28. Witvliet DK, Strokach A, Giraldo-Forero AF et al (2016) ELASPIC web-server: proteome-wide structure-based prediction of mutation effects on protein stability and binding affinity. Bioinformatics 32:1589–1591
29. Chakravarty D, Gao J, Phillips SM et al (2017) OncoKB: a precision oncology knowledge base. JCO Precis Oncol 2017. https://doi.org/10.1200/PO.17.00011
30. Das R, Baker D (2008) Macromolecular modeling with rosetta. Annu Rev Biochem 77:363–382
31. Moult J, Fidelis K, Kryshtafovych A et al (2014) Critical assessment of methods of protein structure prediction (CASP)--round x. Proteins 82(Suppl 2):1–6
32. McGibbon RT, Beauchamp KA, Harrigan MP et al (2015) MDTraj: a modern open library for the analysis of molecular dynamics trajectories. Biophys J 109:1528–1532
33. Consortium TU (2015) UniProt: a hub for protein information. Nucleic Acids Res 43: D204–D212
34. Calderone A, Castagnoli L, Cesareni G (2013) mentha: a resource for browsing integrated protein-interaction networks. Nat Methods 10:690–691
35. McGinnis S, Madden TL (2004) BLAST: at the core of a powerful and diverse set of sequence analysis tools. Nucleic Acids Res 32: W20–W25
36. Webb B, Sali A (2016) Comparative protein structure modeling using MODELLER. Curr Protoc Bioinformatics 54:5.6.1–5.6.37

37. Choi Y (2012) A fast computation of pairwise sequence alignment scores between a protein and a set of single-locus variants of another protein. In: Proceedings of the ACM Conference on Bioinformatics, Computational Biology and Biomedicine - BCB '12. ACM, New York, NY.
38. Schymkowitz J, Borg J, Stricher F et al (2005) The FoldX web server: an online force field. Nucleic Acids Res 33:W382–W388
39. Sanner MF, Olson AJ, Spehner J (1996) Reduced surface: an efficient way to compute molecular surfaces. Biopolymers 38:305–320
40. Heinig M, Frishman D (2004) STRIDE: a web server for secondary structure assignment from known atomic coordinates of proteins. Nucleic Acids Res 32:W500–W502

Chapter 2

Accurate Calculation of Free Energy Changes upon Amino Acid Mutation

Matteo Aldeghi, Bert L. de Groot, and Vytautas Gapsys

Abstract

Molecular dynamics based free energy calculations allow for a robust and accurate evaluation of free energy changes upon amino acid mutation in proteins. In this chapter we cover the basic theoretical concepts important for the use of calculations utilizing the non-equilibrium alchemical switching methodology. We further provide a detailed step-by-step protocol for estimating the effect of a single amino acid mutation on protein thermostability. In addition, the potential caveats and solutions to some frequently encountered issues concerning the non-equilibrium alchemical free energy calculations are discussed. The protocol comprises details for the hybrid structure/topology generation required for alchemical transitions, equilibrium simulation setup, and description of the fast non-equilibrium switching. Subsequently, the analysis of the obtained results is described. The steps in the protocol are complemented with an illustrative practical application: a destabilizing mutation in the Trp cage mini protein. The concepts that are described are generally applicable. The shown example makes use of the pmx software package for the free energy calculations using Gromacs as a molecular dynamics engine. Finally, we discuss how the current protocol can readily be adapted to carry out charge-changing or multiple mutations at once, as well as large-scale mutational scans.

Key words Molecular dynamics, free energy calculations, alchemistry, amino acid mutation, pmx, hybrid structure, hybrid topology, non-equilibrium transitions

1 Introduction

Due to the central role of the free energy in thermodynamics and kinetics, the accurate prediction of free energy changes upon amino acid mutation is one of the central goals in computer-aided molecular design, with potential applications ranging from the engineering of thermostable proteins [1] to that of biosensors [2, 3], sequestrants [4], and protein–protein interactions [5–7]. Predicting mutation effects allows understanding the causes of drug resistance

Electronic supplementary material: The online version of this chapter (https://doi.org/10.1007/978-1-4939-8736-8_2) contains supplementary material, which is available to authorized users.

Tobias Sikosek (ed.), *Computational Methods in Protein Evolution*, Methods in Molecular Biology, vol. 1851,
https://doi.org/10.1007/978-1-4939-8736-8_2, © Springer Science+Business Media, LLC, part of Springer Nature 2019

[8, 9]. Engineered stable proteins with high affinity and specificity toward their binding targets may also serve as biopharmaceuticals [10, 11]. Accurate and robust estimation of the free energy differences between protein sequence variants, thus, is crucial to the successful design of proteins with the desired thermodynamic features.

Different approaches have thus been developed that can return an estimate of free energy changes that relate to the different stabilities or binding affinities of wild-type and mutant proteins. These include fast scoring methods [12–16], implicit-solvent approaches based on the post-processing of molecular dynamics (MD) simulations [17–19], and the computationally more expensive but theoretically rigorous (from a statistical mechanics viewpoint) *alchemical* free energy methods [1, 20]. In this chapter, we focus on the latter category of calculations, which are based on all-atom computer simulations that correctly sample the Boltzmann distribution of microstates and inherently take into account entropic and discrete solvent effects.

In alchemical free energy calculations, an amino acid can be transformed into another one via a non-physical path, hence the name that is reminiscent of the ancient practice that aimed at the transmutation of lead into gold. The amino acid transformation can be carried out reversibly, in what are referred to as *equilibrium* free energy calculations, or irreversibly, in *non-equilibrium* calculations [21]. In both cases, the amount of work needed for the transformation and free energy difference between the initial and final states can be recovered. However, the setup of the calculations differs. In this chapter, we discuss non-equilibrium approaches that carry out this transformation irreversibly and describe protocols that can be used for the accurate estimation of free energy changes upon amino acid mutation. In the text, we use the prediction of protein stability changes upon an amino acid mutation as an example application. The methodology and protocol presented here are of generic character and can be applied to study other biophysical processes, assuming a suitable thermodynamic cycle can be built, e.g., changes in protein–protein, protein–DNA, or protein–ligand binding affinities.

In this chapter, we first provide some background concepts that are at the foundation of the non-equilibrium alchemical free energy method; for a more detailed description we give references to more specialized literature sources. Further, we concentrate on the description of the practical steps involved in preparing and subsequently carrying out the free energy calculations following a general protocol. As an example, we use a Trp cage mini protein [22] that provides a real case on which we illustrate setting up and running alchemical free energy calculations of protein mutation. We assume the reader is familiar with the general principles of molecular dynamics simulations. Throughout this chapter, we discuss the potential caveats and solutions for some of the frequently encountered issues. In the last section of the chapter, we describe how the protocol can be easily modified and expanded to perform large-scale mutational scans or to calculate other free energy changes of interest, such as changes in protein–protein or protein–ligand affinities upon protein mutation. Finally, in the Notes section, we provide a few technical remarks that may prove helpful when setting up alchemical free energy calculations using Gromacs 2016 [23] and the pmx python library with the specialized set of scripts [24].

2 Theory

In this section, we briefly review some of the central concepts that allow the estimation of free energy differences from physics-based computer simulations, like Monte Carlo or molecular dynamics (MD) simulations. We place particular focus on the theoretical foundations of non-equilibrium work (NEW) calculations and how they can be used to estimate free energy differences along alchemical (i.e., non-physical) paths. The interested reader can find a broader appraisal of theoretical aspects, also including equilibrium free energy calculations and geometrical transformations, in the numerous excellent reviews that have been written on the subject, [21, 25–29] as well as in the publications by Jarzinski [30, 31], Crooks [32–34], and Hummer [35–37].

2.1 Definition of Free Energy and Irreversible Work

The free energy surface of a system determines its thermodynamic and kinetic properties and, as such, it provides access to understanding biophysical processes, including protein folding, ligand binding, protein–protein association, etc. For instance, a polypeptide chain in solution may be found in many disordered conformations, or in ordered conformations with well-defined secondary and tertiary structure. We can define the set of disordered conformations as the unfolded state of the system (state A), and the set of ordered conformations as the folded state (state B). It is rarely possible to sample the whole phase space of a protein, which would require observing all the folded and unfolded conformations multiple times. However, in practice free energy differences rather than free energies are typically of interest. The difference between the free energy of state A and B alone will give the relative equilibrium probability of finding the protein in its unfolded form with respect to the folded form; i.e., the free energy difference ΔG is proportional to the ratio of probabilities of finding the system in state A or B:

$$\frac{p_A}{p_B} = \frac{e^{-\beta G_A}}{e^{-\beta G_B}} = e^{-\beta(G_A - G_B)} \quad (1)$$

$$\Delta G = G_A - G_B = -k_B T \ln \frac{p_A}{p_B} \tag{2}$$

where G is the free energy of the whole phase space of the system for an ensemble with a fixed number of particles, constant pressure and temperature (T), i.e., isothermal–isobaric conditions. G_A is the free energy of the unfolded state, G_B is the free energy of the folded state, and $\beta = 1/k_B T$, with k_B is the Boltzmann constant with T denoting the absolute temperature.

This free energy difference also determines the maximum amount of work that can be extracted from the closed system during a thermodynamic process, which can only be achieved in the limit of reversibility. During a reversible process, the system is always in thermodynamic equilibrium, which implies that only infinitesimal changes are applied to it and the transformation is infinitely slow. However, for any finite time interval τ, the system will be driven out of equilibrium, resulting in heat dissipation and hysteresis effects, so that the process will be irreversible. In fact, in accordance to the second law of thermodynamics, the work done during a process is on average equal or larger, due to dissipative work, than the free energy difference between the initial and final state:

$$\langle W(\tau) \rangle \geq \Delta G \tag{3}$$

The equality holds only in the limiting case of a reversible process where ($\tau \to \infty$), whereas for finite τ, the difference between $\langle W(\tau) \rangle$ and ΔG is caused by dissipative work and its magnitude will also depend on the chosen thermodynamic path.

If we use a parameter λ to drive a non-equilibrium process along a certain path, such that the process is started at $\lambda = 0$ and it is concluded at $\lambda = 1$, with λ being constantly modified at each time step, one can calculate the work performed on the system by integrating the energetic cost required to modify it:

$$W(\tau) = \int_{\lambda=0}^{\lambda=1} \frac{\partial \mathcal{H}(\boldsymbol{x}, \boldsymbol{v}, \lambda)}{\partial \lambda} d\lambda \tag{4}$$

where $\mathcal{H}$ is the Hamiltonian of the system, which depends on the phase space coordinates $\boldsymbol{x}$ and velocities $\boldsymbol{v}$ of the system and the coupling parameter λ.

2.2 Estimating Free Energy Differences from Non-equilibrium Simulations

From the considerations above, it is possible to derive estimators that allow calculating free energy differences from equilibrium and non-equilibrium simulations. Both, the Zwanzig's formula [38], which lies at the basis of free energy perturbation (FEP) approaches, and thermodynamic integration (TI) [39] make use of ensemble averages obtained from equilibrium simulations for the estimation of free energy differences. More recently, Jarzynski has shown how one can derive an identity from the inequality in Eq. 3, such that a free energy difference can also be obtained from an ensemble of non-equilibrium simulations in which the system is driven irreversibly from one state to another [30, 31]. In fact, it is possible to show that both FEP and TI are limit cases of Jarzynski's equality, in which the non-equilibrium transformation is performed instantaneously (infinitely fast: $\tau \rightarrow 0$) or reversibly (infinitely slowly: $\tau \rightarrow \infty$), respectively [21]. The Crooks Fluctuation Theorem (CFT) [32–34] has further generalized the Jarzynski's equality by relating the equilibrium free energy difference to the ratio of non-equilibrium work distributions collected by performing the process in the forward and reverse directions. In the following, we focus on non-equilibrium work (NEW) approaches. More specifically, we review the free energy estimators based on the Jarzynski's equality and Crooks Fluctuation Theorem (Crooks Gaussian Intersection and Bennet's Acceptance Ratio).

2.2.1 Jarzynski's Equality

The equality derived by Jarzynski in 1997 [30, 40] relates the uni-directional non-equilibrium work average to the equilibrium free energy difference:

$$\langle e^{-\beta W(\tau)} \rangle = e^{-\beta \Delta G} \tag{5}$$

The work W depends on the chosen path connecting the initial ($\lambda = 0$) and final ($\lambda = 1$) states. The parameter λ controls the time evolution of a system with a time-dependent Hamiltonian. The average on the left-hand side of the equation is an ensemble over both equilibrium initial conditions and non-equilibrium transformations. In fact, the above equality requires the non-equilibrium transitions to be started from an equilibrium ensemble; on the other hand, there is no such requirement for the final state of the system at the end of the transition [21, 30]. The non-equilibrium trajectories are then weighted with the Boltzmann factor of the external work done on the system. The work W can be calculated from Eq. 4 by numerical integration; note how instantaneous, rather than ensemble average (as done in TI), $\partial \mathcal{H}/\partial \lambda$ values are evaluated. In the limit of an infinitely fast ($\tau \rightarrow 0$) or slow ($\tau \rightarrow \infty$) transformation, Eq. 5 reduces to the Zwanzig equation and TI, respectively [21]. In fact, if the system is brought from $\lambda = 0$ to $\lambda = 1$ instantaneously, its configurations at both end states are the same and W simply corresponds to the change in Hamiltonian (which, for transformations that conserve the kinetic energy of the system, corresponds to the change in potential energy). On the other hand, for an infinitely slow transformation, the system is always in equilibrium so that $\langle W \rangle = \Delta G$.

From Eq. 5 one can directly estimate the free energy difference as follows, with N being the number of non-equilibrium trajectories sampled:

$$\widehat{\Delta G} = -k_B T \ln\left[\frac{1}{N}\sum_{i}^{N} e^{-\beta W_i}\right] \quad (6)$$

However, in practice, this exponential estimator is affected by statistical and systematic errors. In fact, due to the exponential weight, the average will mostly depend on values at the tail of the work distribution. This means that rare events where little work is dissipated will dominate the estimate; consequently, the free energy will converge slowly to the true value given that rare events are most likely poorly sampled. Furthermore, it has been shown that this estimator is biased [41, 42], i.e., it introduces a systematic error in the free energy estimate for finite numbers of N.

2.2.2 Crooks Fluctuation Theorem

Jarzynski's equality considers the transitions in one direction only, e.g., from $\lambda = 0$ to $\lambda = 1$. The Crooks Fluctuation Theorem (CFT) takes into account the work values obtained from performing the process in both forward (λ: 0 $\rightarrow$ 1) and reverse (λ: 1 $\rightarrow$ 0) directions. According to the CFT, the forward and reverse work distributions relate to the free energy difference as follows:

$$\frac{P_f(W)}{P_r(-W)} = e^{\beta(W-\Delta G)} \quad (7)$$

where $P_f(W)$ and $P_r(-W)$ are the normalized probability distributions of work values obtained from the forward and reverse transformation paths. Note that Jarzynski's equality can be derived from Eq. 7 by integration over W [21]. With enough overlap between the forward and reverse work distributions, the free energy difference can be estimated directly from Eq. 7 as follows:

$$\widehat{\Delta G} = W + k_B T \ln \frac{P_f(W)}{P_r(-W)} \quad (8)$$

with $\widehat{\Delta G} = W$ at the intersection of the work distributions. However, this approach has known limitations: firstly, for certain paths it might be difficult to obtain substantial overlap between $P_f(W)$ and $P_r(-W)$. Secondly, mainly the tails of the distributions, which are defined by rare events of low work dissipation, will contribute to the free energy difference.

To partly alleviate these problems, one can approximate the work distributions with an analytical function [43]. One such strategy, which leads to accurate free energy estimates, was proposed by Goette and Grubmüller [29]. By using a Gaussian approximation, a Crooks Gaussian Intersection (CGI) estimator was derived:

$$\widehat{\Delta G} = \frac{\frac{\langle W_f \rangle}{\sigma_f^2} - \frac{\langle W_r \rangle}{\sigma_r^2} = \sqrt{\frac{1}{\sigma_f^2 \sigma_r^2}\left(\langle W_f \rangle + \langle W_f \rangle\right)^2 + 2\left(\frac{1}{\sigma_f^2} - \frac{1}{\sigma_r^2}\right)\ln\frac{\sigma_r}{\sigma_f}}}{\frac{1}{\sigma_f^2} - \frac{1}{\sigma_r^2}} \quad (9)$$

where σ_f and σ_r are the variances of the forward and reverse work distributions. Note that the accuracy of this estimator relies on the Gaussian approximation. Thus, it is advised to check this assumption by, for instance, using a statistical test like the Kolmogorov–Smirnov test [44]. The CGI estimator does not have an analytical error estimate, but the error can be estimated by the bootstrap approach [45].

Another ΔG estimator, termed BAR (Bennet's Acceptance Ratio), does not require an analytical approximation for the work distributions. Originally, the BAR relation was derived in 1976 by Bennet for a system sampling two states at equilibrium and performing instantaneous transformations between the states. Bennet showed that the information from the forward and reverse distributions of the potential energy difference (ΔU) could be combined in order to obtain an optimal estimate of the free energy difference [46]. For a non-equilibrium process carried out during a finite amount of time, the same derivation holds by substituting ΔU with the non-equilibrium work W. In 2003, Shirts and coworkers showed how the same estimator can be derived starting from the Crooks Fluctuation Theorem using maximum-likelihood arguments [47]. The BAR estimates the free energy difference by satisfying the following relation:

$$\sum_{i=1}^{N_f} \frac{1}{1 + \frac{N_f}{N_r} e^{\beta(W_i - \widehat{\Delta G})}} = \sum_{j=1}^{N_r} \frac{1}{1 + \frac{N_r}{N_f} e^{-\beta(W_j - \widehat{\Delta G})}} \quad (10)$$

where N_f and N_r are the number of forward and reverse trajectories. The BAR equation needs to be solved numerically, for instance, by using a Newton–Raphson or Nelder–Mead solver [48]. This estimator is asymptotically unbiased and an analytical expression for its variance is available [47]. Furthermore, a convergence criterion for BAR has been proposed [49].

The main assumption in the BAR derivation is that the work values are statistically independent [46, 47]. It is thus important to bear this in mind, because if initial configurations are selected from an equilibrium simulation with high frequency, the resulting work values may be correlated [21].

2.3 Free Energy Differences Upon Protein Mutation: The Alchemical Path

To calculate a free energy difference, firstly we need to define the initial and final states of interest, and secondly the path connecting them. If we consider the folding example already used, then the initial state would be the unfolded protein and the final state would be the folded protein, with the free energy difference we want to calculate being the protein folding free energy. If the structure of

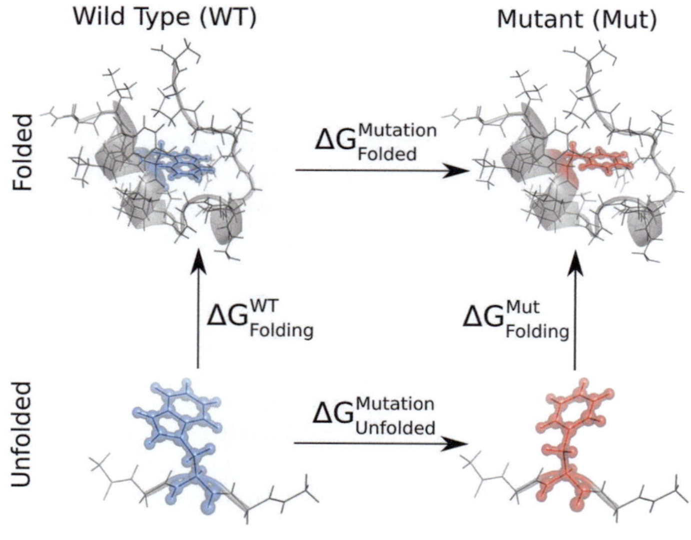

$$\Delta\Delta G^{\text{Mutation}}_{\text{Folding}} = \Delta G^{\text{Mut}}_{\text{Folding}} - \Delta G^{\text{WT}}_{\text{Folding}} = \Delta G^{\text{Mutation}}_{\text{Folded}} - \Delta G^{\text{Mutation}}_{\text{Unfolded}}$$

Fig. 1 Schematic representation of a thermodynamic cycle to calculate changes in protein folding free energy upon mutation ($\Delta\Delta G^{\text{Mutation}}_{\text{Folding}}$). The left column shows the folding process of a wild-type protein, with the associated folding free energy $\Delta G^{\text{WT}}_{\text{Folding}}$; the right column shows the same folding reaction but for a mutated protein, resulting in the folding free energy $\Delta G^{\text{Mut}}_{\text{Folding}}$. The process depicted in the bottom row corresponds to the alchemical transformation of the wild-type unfolded protein into the mutant with the associated free energy difference $\Delta G^{\text{Mutation}}_{\text{Unfolded}}$. The reaction in the top row corresponds to the same alchemical transformation but done on the folded protein, so that the free energy difference between the two mutants is $\Delta G^{\text{Mutation}}_{\text{Folded}}$. The free energy differences for the vertical processes are computationally demanding to compute, but those for the horizontal transformations are more accessible. Thus, $\Delta\Delta G^{\text{Mutation}}_{\text{Folding}}$ can be calculated from the difference between $\Delta G^{\text{Mutation}}_{\text{Folded}}$ and $\Delta G^{\text{Mutation}}_{\text{Unfolded}}$

the folded protein is known, we can then transform state B into state A (via a reversible or irreversible process) by, for instance, pulling the N- and C-termini apart and measuring the work needed to unfold the protein. Although this is in principle possible, such a large perturbation of the system will likely require a lot of computation in order to achieve convergence. However, if the interest is in evaluating *changes* in folding free energy upon protein mutation ($\Delta\Delta G_{\text{folding}}$), it is possible to build a thermodynamic cycle (Fig. 1) that allows to calculate this quantity via alchemical (i.e., non-physical) paths that introduce smaller perturbations in the system, and which are easier to converge. Thus, thanks to the fact that computationally we have control over the topology and potential energy function describing the system, we can take full advantage of the better convergence properties of the non-physical transformation over physical ones; i.e., it is easier to obtain accurate results by alchemically mutating a wild-type protein into its mutant, rather than (un)folding both of them. As the free energy is a state variable, obtained free energy changes are path-independent. It is therefore unproblematic to choose unphysical pathways.

2.3.1 The Thermodynamic Cycle

As shown in Fig. 1, one can define a cycle where for both the initial (unfolded) and final (folded) states the wild-type protein is transformed into a mutant of interest via a non-physical path. The free energy difference of protein folding upon an amino acid mutation ($\Delta\Delta G_{\text{Folding}}^{\text{Mutation}}$) can be recovered by following both, the physical paths of folding the WT and mutant protein ($\Delta G_{\text{Folding}}^{\text{Mut}} - \Delta G_{\text{Folding}}^{\text{WT}}$), and the alchemical paths of morphing the amino acids in the folded and unfolded states ($\Delta G_{\text{Folded}}^{\text{Mutation}} - \Delta G_{\text{Unfolded}}^{\text{Mutation}}$).

From the thermodynamic cycle in Fig. 1 it is clear that in order to calculate $\Delta\Delta G_{\text{Folding}}^{\text{Mutation}}$ we need to be able to simulate the protein's unfolded state. However, the unfolded state of the full-length protein is by its nature poorly defined and would be challenging to simulate [50, 51]. Therefore, short protein fragments have been typically used [52–54]. In particular, it has been observed that capped sequence context independent tripeptides (GXG, where X is the mutated residue) serve as a good approximation of the unfolded state for estimating changes in protein thermostability [20]. In practice, the context independent are convenient to use, as they allow to systematically precompute all possible residue mutations. In such a way, one only needs to calculate $\Delta G_{\text{Folded}}^{\text{Mutation}}$, while $\Delta G_{\text{Unfolded}}^{\text{Mutation}}$ can be found in a precomputed table.

Although here we take protein folding as an example, the same alchemical approach can easily be used to build other thermodynamic cycles by changing the end states; for instance, differences in ligand–protein, protein–protein, or protein–DNA/RNA binding free energy can be calculated by using the apo protein as the initial state and the complex as the final one. Note that while the $\Delta G_{\text{Mutation}}$ values refer to non-physical transformations, the final $\Delta\Delta G$ value obtained from such cycles is that of a physical process (e.g., folding, association, etc.) and can be directly compared to experimental values that measure the same free energy differences.

2.3.2 Single and Dual Topology

We have described how alchemical transformations can be used to build thermodynamic cycles that allow one to calculate changes in free energy differences upon an amino acid mutation. However, how can one alchemically mutate one residue into another during a simulation? Given the separate Hamiltonians at the two end states, it is necessary to define a hybrid topology that contains both physical states. In the specific case of mutating an amino acid into another one, the residue being mutated must be able to represent both the wild-type and mutant residue. This is typically achieved using the single or dual topology approach [55–57].

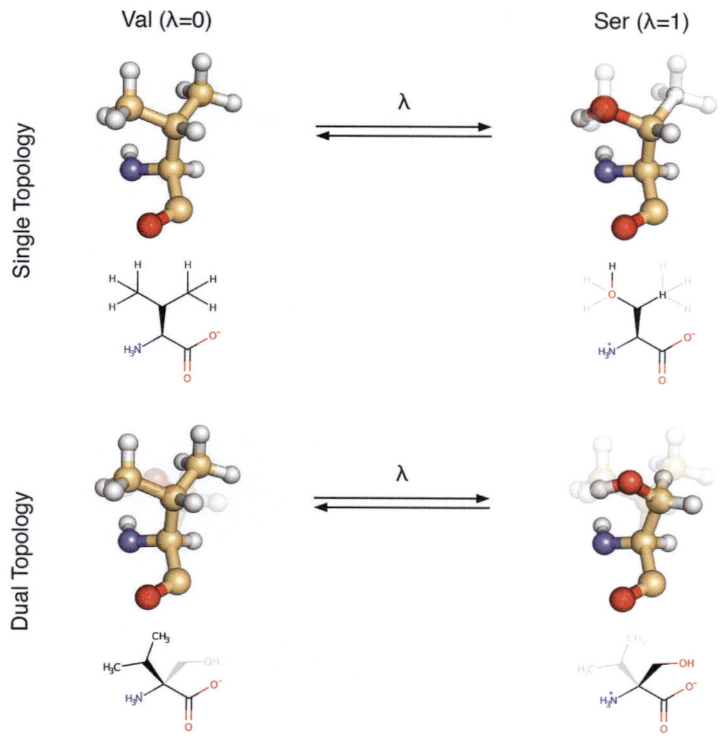

Fig. 2 Example of the single and dual topology setup for the mutation of valine into serine. Dummy atoms in the three-dimensional rendering are shown as transparent balls and sticks, whereas in the chemical structure drawings they are shown in grey. In the single topology approach, a methyl part of valine's side chain is transformed into serine's hydroxyl group, with a carbon becoming an oxygen, while two hydrogens are turned into non-interacting dummy particles; all hydrogens of the second methyl are decoupled as well, while the carbon becomes a C_β hydrogen. In the dual topology approach, no element mutation occurs, because both valine and serine side chains are present in both states, where, however, only one of the two is coupled to the system, with the other one being non-interacting

In the single topology approach (Fig. 2), a number of atoms of state A is mapped onto the atoms of state B. This means that not only a particle's partial charge, but also its chemical element (i.e., atom type) can change according to the λ parameter. For instance, in the example shown in Fig. 2 for a valine being mutated into a serine, one of valine's carbon atoms at $\lambda = 0$ becomes a serine's oxygen at $\lambda = 1$. Effectively, this means that along the alchemical path controlled by λ, the Lennard-Jones and bonded parameters for that particle are interpolated between those of a carbon atom and those of an oxygen atom. Note that such change in chemical identity implies that also the associated equilibrium bond lengths will be modified (e.g., a C–H bond will be shorter than a C–C bond) [55, 58, 59]. Often, the number of atoms in the two end states is not equal, thus not all atoms of the states A and B can be matched. Therefore, non-interacting particles are used either in state A or B. These dummy atoms do not have electrostatic and van der Waals (vdW) interactions with the system; however, they maintain their bonded interactions, so that they effectively are in a vacuum-like state. In the example in Fig. 2, five of valine's hydrogen atoms are turned into dummy atoms.

In the dual topology approach, atoms that are different between the two end states are not morphed directly, but rather transformed into dummy particles [26, 56, 57]. For amino acids, this effectively means that the side chains of both residues are present at the same time. However, at $\lambda = 0$ the side chain of the initial state is interacting with the system and the side chain of the final state is present as non-interacting particles. On the other hand, at $\lambda = 1$ the side chain of the final state is interacting and that of the initial state is turned into non-interacting dummy atoms. This can be seen in Fig. 2: in the initial state, the methanol side chain of serine is decoupled, whereas in the final state it is the propyl side chain of valine being turned off.

In practice, there does not need to be a clear separation between a single and dual topology setup. While some atoms may be morphed between the states following a single topology approach, other atoms in the same system may be turned into dummies according to a dual topology approach.

It is important to bear in mind that the free energy change (ΔG) of the mutation differs depending on whether the single or dual topology approach is used. This is due to the fact that the end states are effectively different due to different dummy atom constructions. In addition, in the single topology approach there is a contribution to the free energy difference from the change in bond lengths. However, the contributions to the free energy difference resulting from the details of the atom mapping between the end states cancel out in a thermodynamic cycle like the one in Fig. 1, such that the final $\Delta\Delta G$ value is independent of how the hybrid topology is implemented [57, 59].

Using dummy particles in alchemical transitions requires introduction and annihilation of particles into the system. Such transformations impose a large perturbation, e.g., creating a particle interacting with the environment in a place of a non-interacting dummy atom results in strong van der Waals repulsions and Coulombic interactions. In turn, large forces are exerted on the atoms which leads to instabilities in dynamics and integration artifacts. To circumvent these issues, it is a common practice to modify, "soften," the non-bonded interactions during the alchemical transformations. A number of functional forms and parameter sets to such soft-cored interactions have been proposed [60–64]. Altering the non-bonded interactions along the alchemical pathway does not affect the final free energy estimates, because the physical end states are still described by the correct unmodified Hamiltonian. The official release of Gromacs 2016 implements a soft-core variant [60] allowing to modify both the van der Waals and Coulombic interactions (*see* **Note 1**).

3 Alchemical Amino Acid Mutations

In this section we use a Trp cage mini protein [22] as a model system to illustrate the process of performing a single amino acid mutation. The pmx [20, 26] software package will be used to introduce a point mutation in this 20 amino acid peptide. pmx provides a single topology-based setup of the alchemical calculations allowing for an automated generation of hybrid amino acid structures and topologies compatible with the Gromacs [23] MD simulation engine.

In this example we will describe in detail the steps needed to prepare the alchemical simulations (Fig. 3) and calculate the free energy difference upon a tryptophan, W6, to phenylalanine mutation (W6F) in the Trp cage protein. W6 is the key residue in the hydrophobic core of this mini protein providing stability to its fold.

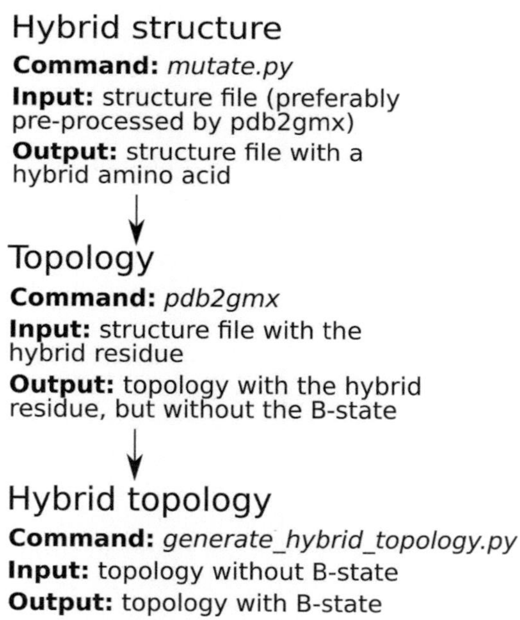

Fig. 3 A schematic depiction of the main steps in generating hybrid structures and topologies for the alchemical simulations using pmx. Firstly, pmx is used to introduce a mutation into the protein. Afterwards, the Gromacs tool *pdb2gmx* generates a topology for the protein with the hybrid residue in a user chosen molecular mechanics force field. In the last step, pmx is used again to add the *B*-state parameters to the topology file

We will assess the change in the thermodynamic stability by calculating the double free energy difference ($\Delta\Delta G$) for the W6F mutation in the folded Trp cage and its unfolded variant approximated by a capped tripeptide (Fig. 1).

For the results of more alchemical mutations in the Trp cage protein, *see* [65].

3.1 Setting Up pmx

`pmx` is a python library that allows the convenient manipulation of biomolecular structure and topology files. Within the framework of `pmx`, a number of scripts have been developed and specifically designed to prepare and analyze alchemical free energy calculations. `pmx` generates topology files that are compatible with the Gromacs simulation engine.

Mutations in a number of contemporary molecular mechanics force fields are supported. This is achieved by means of pre-generated mutation libraries compatible with the Gromacs force field organization. After installing Gromacs and `pmx`, the GMXLIB environmental variable needs to be set to specify the path to the mutation libraries that come with the `pmx` package (*see* **Note 2**).

3.2 Hybrid Structure

The first step in the setup comprises the generation of the hybrid structure for the amino acid to be mutated (Fig. 3). The only file required for this step is the protein structure in .pdb or .gro format. The protein structure needs to be complete, i.e. all heavy and hydrogen atoms need to be present. In order to add missing heavy atoms, external software needs to be used, e.g., Rosetta [15], Modeller [66], or PyMol [67]. Furthermore, given that structures resolved by means of X-ray crystallography usually contain no hydrogen atoms, these need to be added as well. Various software packages, like WhatIf [68] or Rosetta, offer assignment of hydrogen coordinates for protein structures. The Gromacs tool *pdb2gmx* can do this too. In fact, it is convenient to pre-process a .pdb file with *pdb2gmx* because it produces a structure file with atom names already compatible with the Gromacs internal atom naming given the selected force field. *pdb2gmx* also identifies whether any heavy atoms in a protein are missing, so that the tool can be used to identify incomplete residues. While *pdb2gmx* will not model missing heavy atoms, it will inform about such deficiencies. Note that *pdb2gmx* will fail if the input structure contains molecules that are not readily recognized by Gromacs. Therefore, molecules that are not present in the force field file have to be removed from the structure at this stage and processed independently.

For the Trp cage model system we use an NMR structure (PDB-ID 1L2Y) [22] that was deposited with 38 conformers. After manually extracting conformer #2, we pre-process the structure by running it through *pdb2gmx*:

```
gmx pdb2gmx -f 1l2y_conf2.pdb -o 1l2y_conf2_pdb2gmx.pdb
-ff amber99sb-star-ildn-mut -water none -ignh
```

In this example we have selected an updated version of the Amber99sb*ILDN force field [69–71] for which the mutation library has been pre-generated. No water model needs to be chosen at this stage, because with this step we only want to obtain a pre-processed structure file with added hydrogens and Gromacs compatible atom names. The "-ignh" flag ignores the hydrogen atoms already present in the structure and adds them again using the *pdb2gmx* logic, ensuring the names of the hydrogen atoms are compatible with Gromacs and the selected force field (*see* **Note 3**).

The output structure file obtained as described above is then used as an input for the pmx script *mutate.py*:

```
python mutate.py -f 1l2y_conf2_pdb2gmx.pdb -o mut.pdb -ff amber99sb-
star-ildn-mut
```

Upon execution, the command prompts for an interactive selection of a residue to mutate (W6) and a target amino acid (Phe or F). When developing a workflow for a large-scale mutation scan, it may be convenient to provide the information about the amino acid mutations as a text file. For this purpose a "-script" flag in *mutate.py* is available: this option expects a text file with an amino acid number and the name of the residue to mutate into. In the case of the Trp cage example: `6 Phe`.

3.3 Topology

At this point we use the hybrid structure from the previous step ("mut.pdb") as an input to *pdb2gmx* (Fig. 3). This time we want to obtain the topology file containing all the information needed by Gromacs to run the simulations. The topology file will also include the description of the hybrid mutated residue, however, parameters only for one physical state (state *A*) are defined in the output topology file. It is also important to note that at this step the "-ignh" flag should not be set, since the hydrogen atoms have already been added in the previous step.

```
gmx pdb2gmx -f mut.pdb -o mut_pdb2gmx.pdb -ff amber99sb-star-ildn-mut
-water tip3p -p topol.top
```

If one wants to include a ligand that has been parameterized separately, this can be added to the structure ("mut_pdb2gmx.pdb") and topology file ("topol.top") at this stage.

3.4 Hybrid Topology

The generated topology file ("topol.top") has the hybrid residue W2F incorporated. However, it is a non-standard hybrid amino acid with two physical states (*A* and *B*). While state *A* is included in the topology, state *B* still needs to be included explicitly. The required topology parameters for state B can be added by the `pmx` script *generate_hybrid_topology.py* (Fig. 3):

```
python generate_hybrid_topology.py -p topol.top -o hybrid.top -ff
amber99sb-star-ildn-mut
```

3.5 Webserver

The procedure detailed above (and summarized in Fig. 3) can also be executed via a webserver interface: http://pmx.mpibpc.mpg.de. Provided with a protein structure file, the pmx webserver will perform a user-selected mutation in one of the supported molecular mechanics force fields.

The webserver runs a number of additional structure pre-processing steps that simplify the setup procedure. While broken or incomplete proteins will not be repaired, a number of other useful modifications are applied: residue and atom names are matched to the force field nomenclature, terminal residues are dealt with, and if needed hydrogen atoms may be added via *pdb2gmx*. Optionally, the structure may be checked before the mutation is performed, so that the user is informed about any potential deficiencies in the input file. In addition, the setup offered by the webserver is not limited to single amino acid mutations, but also allows to prepare files for mutation scans over selected protein chains.

3.6 Alchemical Simulations

The hybrid structures and topologies we just obtained can readily be used for MD simulations and to calculate free energy differences. Numerous protocols for relative alchemical free energy calculations are currently available: equilibrium approaches (TI, FEP) as well as non-equilibrium methods. Here, we employ non-equilibrium calculations based on the Crooks Fluctuation Theorem.

3.6.1 System Preparation

Firstly, the hybrid structure and topology are used in preparing the system for molecular dynamics simulations following a standard procedure. The protein needs to be placed in a simulation box and solvated. Then ions need to be added to neutralize the system and, optionally, reach a desired salt concentration. These are conventional steps used to prepare an ordinary MD simulation: for a more detailed description of this procedure in Gromacs we refer the reader to a specialized protocol [72].

3.6.2 Equilibrium Simulations

Next, we set up two equilibrium simulations: one for the WT Trp cage (W6, state A, $\lambda = 0$) and another for the mutated protein (F6, state B, $\lambda = 1$) (Fig. 4). We start with an energy minimization performed on both states separately. The parameters for the energy minimization (.mdp) are the same as those used in non-alchemical simulations, with the exception of two flags. The `free-energy` flag has to be set to `yes`. This indicates that the free energy code in

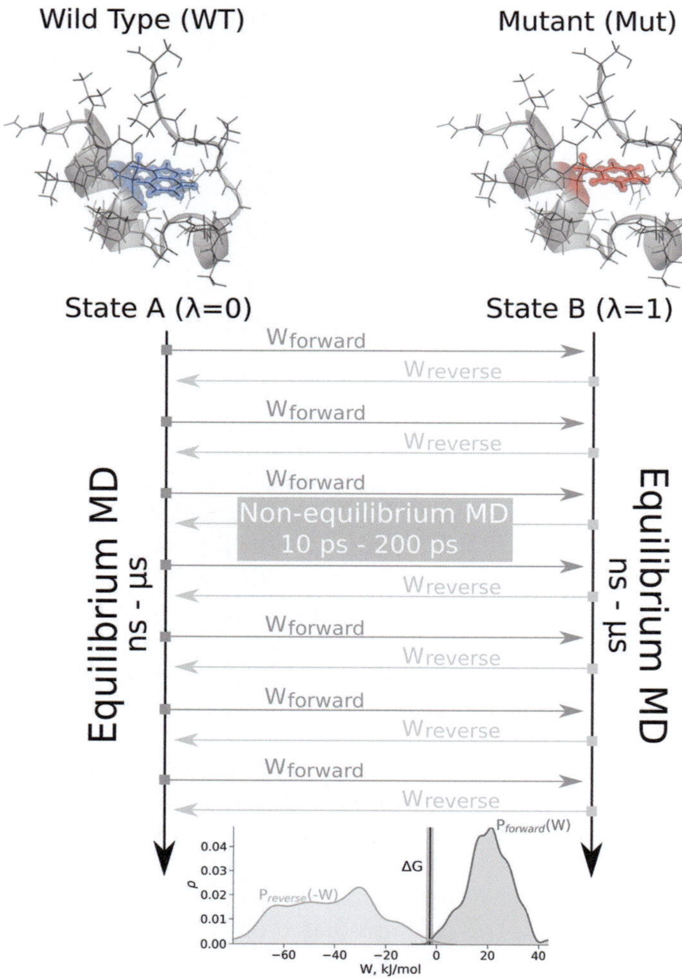

Fig. 4 The procedure of non-equilibrium alchemical simulations for one leg of the thermodynamic cycle: mutation in the folded state of a protein. Two independent equilibrium simulations are performed by keeping the system in its physical states: WT ($\lambda = 0$) and mutant ($\lambda = 1$). These simulations need to sufficiently sample the end state ensembles, as the accuracy of the free energy estimate will depend on the convergence of the equilibrium sampling. Typically, the equilibrium simulations are in the nanosecond to microsecond time range. From the generated trajectories, snapshots are selected to start fast (typically 10–200 ps) transitions driving the system in the forward (λ: 0 $\rightarrow$ 1) and reverse (λ: 1 $\rightarrow$ 0) directions. The work values required to perform these transitions are collected and the Crooks Fluctuation Theorem is used to calculate the free energy difference between the two states

Gromacs will be activated for those interactions that have two sets of parameters (states *A* and *B*) in the topology file. In addition, the `init-lambda` flag has to be set to `0` for the simulation in state *A* (WT Trp cage) and to `1` for state *B* (mutated Trp cage).

After the energy minimization runs for the A and B states are complete, MD simulations can be started from the energy minimized conformations. Similarly to the energy minimization, the simulation parameters are identical to the conventional MD runs, except for setting the `free-energy` and `init-lambda` flags for the simulations in state *A* and *B*, respectively (Fig. 4). These equilibrium runs are used to sample the relevant phase space volumes, i.e., the conformational changes in the WT and mutated variant of Trp cage. Therefore, the ensembles generated during the equilibrium runs will define how accurately the free energy difference will be estimated. This consideration dictates the sampling time: the simulation time should be sufficient to sample the transitions that are considered to be relevant. For example, if a protein is known to undergo large-scale conformational changes and the introduced mutation may be affecting the populations of these conformers, the simulation time has to be long enough to properly sample such transitions. Equilibrium simulations in this case could require microseconds or longer to converge. On the other hand, it is often important to estimate the free energy difference for a structure that would remain close to its experimentally resolved structure. In this scenario, it is sufficient to sample smaller changes in rotameric states of the side chains and minor backbone motions. In previous large-scale amino acid scans investigating protein thermodynamic stabilities, we have observed good agreement with experimental data when using 10–20 ns of equilibrium sampling [1, 20].

Another issue to consider when choosing the sampling time is the definition of states for which the free energy difference will be calculated. In the Trp cage example, we are aiming to estimate the mutation-induced free energy difference in folding free energy. This implies that one of the end states that we need to simulate needs to be the *folded* state, while the other needs to be the *unfolded* state. If we were to introduce a destabilizing mutation (in fact W6F has been shown to strongly destabilize Trp cage [73, 74]), over a longer simulation time the protein would unfold. Thus, the definition of the *folded* state used in the free energy calculation would be violated, rendering the calculated free energy differences inaccurate. For the Trp cage W6F mutation example, we will use equilibrium simulations of 10 ns: short enough such that no spontaneous unfolding occurs.

3.6.3 Non-equilibrium Transitions

Once the equilibrium simulations are completed, we can proceed to the non-equilibrium part of the simulation protocol. Fast non-equilibrium transitions serve the purpose of connecting the two physical states (*A* and *B*) and allow obtaining the free energy difference between them. These transitions are started from snapshots extracted from the two equilibrium trajectories. From each equilibrium trajectory, we discard the first 2 ns as equilibration time. Then we use the last 8 ns to extract 100 frames equidistant in time (1 frame every 80 ps) representing an equilibrium ensemble of starting conformations for the non-equilibrium transitions. These frames can be conveniently extracted using the Gromacs tool *gmx trjconv* (*see* **Note 4**). Note that the number of non-equilibrium transitions performed will influence the accuracy of the free energy estimate. More transitions allow acquiring more work values, which consequentially allow for a more accurate ΔG estimate. In previous investigations we have observed that 50–100 transitions are generally sufficient to obtain accurate estimates of folding free energy changes upon protein mutation [1, 20].

Another parameter influencing the accuracy of the ΔG estimate is the transition time. Over the course of a shorter transition, the system is driven further away from equilibrium, and more work is dissipated along the alchemical path. In turn, the free energy estimate becomes less accurate. The optimal transition time depends on the size of the perturbation and the nature of the system, e.g., replacing a small residue with a large one represents a larger perturbation than a small-to-small residue mutation. Larger perturbations may require slower transitions to obtain free energies at the desired level of accuracy. Transition times ranging from tens to hundreds of picoseconds are usually enough to return accurate results [1, 20].

For the Trp cage example considered here, we use non-equilibrium transitions of 50 ps. Using a 2 fs time step, this means running 25,000 integration steps. Therefore, the λ value will be changed at a speed of 1/25,000=4e−5. For the transition simulations in the forward (state A to B) direction, we set the following parameters in the .mdp file:

```
nsteps        = 25000
nstcalcenergy = 1
nstdhdl       = 1
free-energy   = yes
init-lambda   = 0
delta-lambda  = 4e-5
sc-alpha      = 0.3
sc-sigma      = 0.25
sc-power      = 1
sc-coul       = yes
```

`nsteps` defines the number of steps. The system starts at `init-lambda=0` and is morphed into the $\lambda = 1$ state over the course of a transition in `nsteps` number of steps. The energy and $\partial H/\partial \lambda$ are calculated at every integration step (`nstcalcenergy` and `nstdhdl` set to 1). The parameters starting with a prefix `sc-` control the soft-core interactions. In this non-equilibrium protocol we soften both the van der Waals and Coulombic interactions: `sc-coul=yes` (*see* **Note 1**).

For the transitions in the reverse direction (state B to A), the .mdp parameters are the same, with the exception of the starting state and the direction of the transition:

```
init-lambda    = 1
delta-lambda   = -4e-5
```

For each of the 100 transitions in both directions, the data we need to collect are the $\partial H/\partial\lambda$ values, which are stored in the ".dhdl.xvg" files. Integration over these values gives the work performed during the non-equilibrium transitions in the forward and reverse directions (Fig. 4).

3.7 Analysis

The integration over the $\partial H/\partial\lambda$ curves and the free energy difference estimation can be performed with the `pmx` script *analyze_dhdl.py*:

```
python analyze_dhdl.py -fA stateA/dhdl*.xvg -fB stateB/dhdl*.xvg
```

The script will output the summary of results in a text file containing the estimate of the free energy difference using three estimators: Crooks Gaussian Intersection (CGI), Bennet's Acceptance Ratio (BAR), and Jarzynski's equality. While CGI and BAR use the work distributions generated in both, forward and reverse, directions, Jarzynski's estimator is one-directional. We recommend using the BAR estimation for the ΔG value, as it utilizes all the available work values from both directions and makes no assumptions about the shape of the work distributions. Conveniently, the script also generates plots of the work values over time and of their distributions (Fig. 5), which are useful to detect potential sampling or lack of the work distribution overlap issues.

The convergence of the results can be assessed in various ways. Firstly, if a systematic drift of the work values over time is observed, it usually indicates lack of convergence during the equilibrium sampling stage. The work values are likely to drift due to a conformational change and it may be important to thoroughly sample the significant conformational motions in the protein. Lack of convergence may also be deduced from the error values provided together with the free energy estimates. The uncertainties of the CGI and BAR estimators are sensitive to the lack of the overlap between the forward and reverse work distributions (*see* **Note 5**). A large uncertainty in the ΔG estimate indicates that the overlap between the work distributions might be insufficient. Slower transitions keep the system closer to equilibrium, so that less work is dissipated along the path and the overlap between work distributions generally increases. Running more non-equilibrium transitions increases the probability of observing work values with low dissipation, which also contributes toward good overlap of the work distributions.

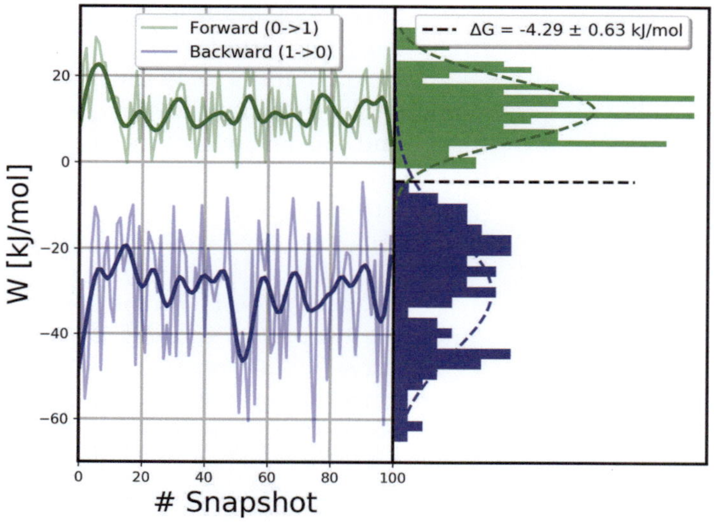

Fig. 5 A standard output generated by the `pmx` *analyze_dhdl.py* script. On the left, the work values for the forward and reverse transitions are depicted for every starting structure. On the right, the distributions of these work values are shown as histograms, the intersection of which allows obtaining the free energy difference. In the current case, the BAR estimator yields a ΔG value of -4.29 ± 0.63 kJ/mol for the W6F mutation in the folded state of the Trp cage protein

The most reliable way to assess the precision of the free energy estimates obtained is to repeat the whole procedure, including equilibrium and non-equilibrium simulations, multiple times. The calculated ΔG values and their spread obtained from multiple independent calculations more accurately capture under-sampling issues. For the Trp cage W6F mutation, we have obtained a ΔG value of -4.29 ± 0.63 kJ/mol (Fig. 5) from a single calculation. Then, we repeated the whole calculation five times, from the system preparation to the equilibrium and non-equilibrium simulations. The average free energy value we obtained was of -3.73 kJ/mol with a standard error of 0.88 kJ/mol. This result confirms that the ΔG estimate obtained can be considered to be reliable.

3.8 Double Free Energy Difference

So far we have calculated the free energy difference for one leg of the thermodynamic cycle (Fig. 1): mutation in a folded protein. To obtain the final double free energy difference the same procedure needs to be performed for the unfolded Trp cage peptide. It has been demonstrated that in the context of the alchemical free energy calculations the unfolded state can be approximated by a capped tripeptide with the residue of interest surrounded by two glycines [20].

Given the tripeptide approximation, the ΔG values can be pre-calculated for every amino acid combination of interest. The tabulated tripeptide ΔG values can be found on the pmx webserver: http://pmx.mpibpc.mpg.de (*see* **Note 6**). The tripeptide's W2F mutation in the Amber99sb*ILDN force field has an associated free energy change ($\Delta G^{\mathrm{Mutation}}_{\mathrm{Unfolded}}$) of -17.96 ± 0.32 kJ/mol. The $\Delta\Delta G$ of interest can thus be calculated as $\Delta G^{\mathrm{Mutation}}_{\mathrm{Folded}} - \Delta G^{\mathrm{Mutation}}_{\mathrm{Unfolded}}$. Thus, in our Trp cage example, the $\Delta\Delta G$ of folding for the W6F mutation is estimated to be 13.67 ± 0.71 kJ/mol. This calculated estimate closely matches the experimentally measured destabilization of 12.5 ± 0.6 kJ/mol [73, 74]. A previous large-scale study compared calculated and experimental $\Delta\Delta G$ values for protein thermostability changes upon mutation for the proteins barnase and Staphylococcal nuclease [1]. It was found that the mean unsigned error in the predictions was of approximately 4 kJ/mol, with the uncertainty due to finite sampling, the force field, and the experimental error equally contributing to the discrepancy between calculated and experimental $\Delta\Delta G$ values. Therefore, the calculated $\Delta\Delta G$ value for the Trp cage W6F mutation falls well within the range of the expected accuracy.

4 Miscellaneous Applications of the Protocol

In the previous section we have outlined a general protocol for the calculation of free energy changes upon an amino acid mutation. However, the protocol can and in some cases should be expanded or modified in order to fit the specific needs of the problem at hand. In particular, charge-changing mutations (e.g., Ala to Asp, Trp to Arg, etc.) need special care as artifacts that affect the accuracy of the calculations are otherwise introduced when applying Ewald summation for long range electrostatics. In addition, one is not limited to a single amino acid mutation per calculation. However, when mutating more residues at the same time, more attention should be paid to possible convergence and work overlap issues. Finally, for protein design applications, large mutational scans need to be set up and run, and with a few precautions the protocol can be executed in a more efficient and convenient manner. All these additional points are discussed in the following paragraphs.

4.1 Charge-Changing Mutations

It is often of interest to calculate free energy differences upon amino acid mutations that cause a net charge change. In principle, there is no fundamental difference whether an alchemical mutation is to be charge-changing or charge-conserving. The issue, however, is of a technical nature and has to do with the treatment of long range electrostatic interactions in molecular dynamics simulations. The state-of-the-art long range Coulombic interaction calculations utilize Particle Mesh Ewald (PME) summation [75, 76]. Due to the specifics of the PME algorithm, for a non-neutral system any extra charge is neutralized by the implicit introduction of a uniform background charge. Taking into account the contributions of this effect to the free energy difference requires additional, and technically involved, correction schemes [77, 78]. If neglected, significant artifacts may occur [79]. Therefore, conserving a system's charge during the alchemical transition is preferred.

To accomplish this, we suggest using a double-system/single-box setup [24, 26]. In this approach both legs of a thermodynamic cycle, e.g., mutation in the folded and unfolded states in Fig. 1, are placed in the same simulation box. The systems of the separate legs in the thermodynamic cycle are set in different physical states: if the folded state protein has a WT residue at $\lambda = 0$, then the unfolded state must be in its mutated form at $\lambda = 0$. During the alchemical transition, the system goes from $\lambda = 0$ to 1, and the folded state is transformed into the mutant while the unfolded (mutant) state is simultaneously transformed into the WT. In this way, the charge of the system will be conserved during the transformation, and the free energy difference calculated already refers to the $\Delta\Delta G$ across the thermodynamic cycle of interest.

The assumption underlying the double-system/single-box approach is that both legs of the thermodynamic cycle are independent when placed in a single simulation box. Therefore, it is important to place the systems (e.g., the folded protein and unfolded peptide) sufficiently far apart. The distance between the molecules needs to be larger than the short range electrostatic cutoff. We have obtained reliable free energy estimates by setting the distance to be at least 3 nm between any atoms of the molecules from the separate legs of the cycle [1, 24]. To ensure that the proteins do not diffuse and come closer to one another during the course of simulations, position restraints on a single backbone atom close to the center of mass of each protein ought to be used. These position restraints affect only the translational degrees of freedom of the proteins. The contribution of such position restraints to the translational partition function will be the same in both legs of the thermodynamic cycle and will cancel from the $\Delta\Delta G$ estimate.

4.2 More Than One Mutation at Once

The protocol in this chapter described an example of a single amino acid mutation in a protein. `pmx`, however, also allows introducing multiple mutations at once as well. This can be done either by interactively selecting more than one mutation to be applied or by providing an external file with every mutation defined in a new line of a text file. The `pmx` webserver also provides the option to introduce multiple mutations.

The caveat of performing an alchemical transformation for several amino acid mutations at once is a slower convergence of the free energy estimate. Having more mutations imposes a larger perturbation to the system. Hence, more work will be dissipated along the path and the free energy estimate will become less accurate. In such a case, performing the non-equilibrium transitions slower may be necessary.

Another way to calculate the effect of multiple mutations is to perform the mutations sequentially. For example, the free energy difference of introducing the mutations X and Y at once is equal to the combined ΔG of performing the mutation X first and in a separate setup calculating ΔG for the Y mutation in a system where the X mutation is already present. In fact, since free energy is a state function, the sequence of introducing the mutations does not influence the final ΔG estimate, thus the mutation Y can be performed first and then the mutation X can follow. The free energy differences calculated in all three scenarios (X and Y at once, first X then Y, first Y then X) ought to yield the same estimate. Therefore, the spread of these three ΔG values could serve as an indicator of the uncertainty in the calculations.

4.3 Mutation Scan

In protein design studies or large-scale mutation investigations, it is common to perform mutation scans by replacing every amino acid in a protein sequence with another residue; e.g., alanine scans are often employed. The command line scripts described in the current protocol make it easy to build workflows allowing for any number of mutations to be introduced. The `pmx` webserver also allows for an automated generation of the hybrid structures and topologies for a mutation scan with the user-selected amino acids.

When a protein needs to be mutated multiple times, the end state representing the wild-type does not need to be simulated at equilibrium multiple times. The same WT equilibrium simulation can be reused for all mutants to increase the efficiency of the protocol. However, in order to be able to reuse the same equilibrium run, the generation of the hybrid structure and topology files has to be postponed to the step after the equilibrium simulations. Thus, effectively, the steps described in Subheadings 3.2–3.4 need to be performed after the equilibrium simulations have been carried out (Subheading 3.6.2). In this scenario, the equilibrium simulations would be performed without invoking the free energy code and using standard non-hybrid structures and topologies for both the WT and mutant proteins. The hybrid topology can be generated only once, using one of the many frames extracted from the equilibrium simulations. On the other hand, the hybrid coordinates need to be built by the `mutate.py` script for all the frames extracted from the equilibrium trajectories. For example, a custom bash script with a *for* loop, calling `mutate.py` for each of the extracted frames, would add the atoms for state B to all the snapshots extracted. In such a way, the same trajectory can be reused for all of the mutations of interest; for each mutation, one needs to create the topology with the hybrid residues of interest, and generate the corresponding hybrid structures.

5 Summary

In summary, we have presented a step-by-step protocol for the calculation of free energy changes upon amino acid mutation. As an example, we have shown how these calculations can be used to estimate the destabilizing effect of the W6F mutation on the fold of the Trp cage protein using the `pmx` software. Throughout the text, we have also pointed out common issues that may be encountered in alchemical non-equilibrium free energy calculations, as well as their solutions. Furthermore, we discussed how the protocol can be automated and scaled up in order to better fit the requirements of applications that involve large mutational scans, such as protein design. Finally, we remind the reader that while here we focussed on amino acid mutations, `pmx` also allows to set up, run, and analyze alchemical free energy calculations that involve the mutation of nucleic acids and development is in progress to support arbitrary organic molecules (ligands). Thus, overall, `pmx` and the free energy calculation protocol presented here are flexible tools that can find broad application in various fields of computational biophysics and chemistry.

6 Notes

1. When using equilibrium alchemical free energy calculation protocols (equilibrium TI or FEP) it is usually recommended to perform transformations of the van der Waals interactions after turning off the charges on the morphed atoms. In this scenario, only the van der Waals interactions need to be soft-cored, while the Coulombic interactions may be calculated using the unmodified Hamiltonian. In principle, the same procedure could be applied for the non-equilibrium transitions as well, however, it is more convenient to perform the fast alchemical transitions by morphing the Lennard-Jones parameters and charges simultaneously. In this case, both the van der Waals and Coulombic interactions have to be modified using the soft-core potential. If the default Gromacs 2016 soft-core implementation leads to an erratic behavior of the $\partial H/\partial \lambda$ curves (e.g., unreproducible spikes orders of magnitude larger than the average values), an alternative soft-core implementation can be found on the `pmx` webserver's download section, which is more suitable for the non-equilibrium protocol described in this chapter [63].
   Depending on the software version, Gromacs may issue a warning during *grompp* execution regarding the use of the soft-core interactions when van der Waals interactions are not decoupled. In the context of the non-equilibrium free energy calculations it is safe to ignore this warning: flag `-maxwarn`.

2. In order to be able to use the hybrid/alchemical force fields available in `pmx`, the environment variable `$GMXLIB` needs to be set. This is required as Gromacs uses the path specified in `$GMXLIB` to locate additional force field libraries. In `pmx`, all available hybrid force fields can be found in `$PMXHOME/data/mutff45`, where `$PMXHOME` is the absolute path to the `pmx` source folder. Thus, to allow Gromacs to find the `pmx` force fields, you should run the following command (in bash shell):

```
export GMXLIB=$PMXHOME/data/mutff45
```

3. The "-ignh" flag tells *pdb2gmx* to ignore the hydrogen atoms present in the input structure. In this way, the tool adds the hydrogen atoms again using its own logic. This can be useful when there are hydrogen atoms in the input structure with atom names that are not recognized by Gromacs and/or not present in the force field of choice. If the flag is not set, *pdb2gmx* will keep the hydrogen atoms present in the structure, which can be useful if external programs were used to determine the protonation states of the protein's residues, or if it is preferred to keep the protonation states determined experimentally (e.g., via neutron diffraction). However, in this case one needs to make sure the names of the hydrogen atoms conform to the naming used in the selected force field, otherwise *pdb2gmx* will raise an error. An alternative is to expand the `aminoacids.arn` file in the force field library of interest to introduce a mapping between the hydrogen atom names in the input structure and in the force field.
4. The Gromacs command `trjconv` allows the user to convert and manipulate trajectory files, and comes handy when one wants to extract single frames to be used as starting points for the non-equilibrium simulations. In particular, the flag `-b` allows to choose the frames to discard before a certain time defined in picoseconds. The flag `-sep` tells the program to write each snapshot as a separate indexed coordinate file. The flag `-skip` tells the program to extract only every n-th frame. The flags `-ur` and `-pbc` keeps molecules intact across the periodic boundaries. For instance, in the example with Trp cage, we ran the following `trjconv` command:

```
gmx trjconv -f equilibrium_sim.xtc -s equilibrium
_sim.tpr -o frame_.pdb -sep -b 2001 -skip 1 -ur com
pact -pbc mol
```

In this way, as we saved coordinates to the trajectory file every 80 ps, we obtained 100.pdb files called `frame_`*n*`.pdb`, with n from 0 to 99.

Here, we used an .xtc file to store the trajectory data and .pdb files to extract the snapshots. These file formats contain only the atom coordinates, but no velocities, therefore, when generating non-equilibrium runs, the flag gen-vel=yes in the .mdp file needs to be set, together with a reference temperature, to generate velocities from a Maxwell distribution. Another option is to use the .trr files for storing the trajectory data from equilibrium simulations. The .trr files can also store the velocities along with the coordinates. If the .trr files are used, the starting snapshots should be extracted as .gro files instead of .pdb, as the .gro file format allows storing both coordinates and velocities. Non-equilibrium runs started from initial structures generated in this way will use velocities as obtained from the equilibrium sampling; thus, the `gen-vel` flag in the .mdp file can be set to `no`.

5. The analytical error for the Bennet's Acceptance Ratio estimator grows very rapidly even for a minor lack of the overlap between the work distributions. The rate of growth for the analytical error often does not match the bootstrapped error estimate, which warrants further investigation into BAR uncertainty estimators. Nevertheless, a large value of the analytical estimator may serve as a good indicator for the lack of convergence during the non-equilibrium transitions.

6. The current implementation of the `pmx` mutation libraries follow the single topology formalism and the bond lengths are allowed to change between the two end states. When bond length constraints are used during the simulations, the contribution of the constraints (upon changes in bond lengths) to $\delta H/\delta\lambda$ is taken into account by Gromacs. Therefore, TI-based approaches and the non-equilibrium free energy calculations in Gromacs properly account for the changes in the bond length. However, Gromacs currently does not incorporate this contribution into the data used by FEP approaches to estimate free energy differences. This means that while equilibrium FEP and non-equilibrium approaches should return the same ΔG values for the same mutations in theory, in practice this is not the case. Since the mutation libraries have been generated using the non-equilibrium free energy protocol, the tabulated values for the tripeptide mutations should be used only in combination with free energy calculations that make use of the $\delta H/\delta\lambda$ curve integration. For FEP-based approaches, the tripeptide mutations need to be calculated separately.

References

1. Gapsys V, Michielssens S, Seeliger D, de Groot BL (2016) Accurate and rigorous prediction of the changes in protein free energies in a large-scale mutation scan. Angew Chem Int Ed Engl 55(26):7364–7368
2. Griss R, Schena A, Reymond L, Patiny L, Werner D, Tinberg CE, Baker D, Johnsson K (2014) Bioluminescent sensor proteins for point-of-care therapeutic drug monitoring. Nat Chem Biol 10(7):598–603
3. Feng J, Jester BW, Tinberg CE, Mandell DJ, Antunes MS, Chari R, Morey KJ, Rios X, Medford JI, Church GM, Fields S, Baker D (2015) A general strategy to construct small

molecule biosensors in eukaryotes. eLife 4:323–329

4. Zhou L, Bosscher M, Zhang C, Özçubukçu S, Zhang L, Zhang W, Li CJ, Liu J, Jensen MP, Lai L, He C (2014) A protein engineered to bind uranyl selectively and with femtomolar affinity. Nat Chem 6(3):236–241
5. Correia BE, Bates JT, Loomis RJ, Baneyx G, Carrico C, Jardine JG, Rupert P, Correnti C, Kalyuzhniy O, Vittal V, Connell MJ, Stevens E, Schroeter A, Chen M, MacPherson S, Serra AM, Adachi Y, Holmes MA, Li Y, Klevit RE, Graham BS, Wyatt RT, Baker D, Strong RK, Crowe JE, Johnson PR, Schief WR (2014) Proof of principle for epitope-focused vaccine design. Nature 507(7491):201–206
6. Koday MT, Nelson J, Chevalier A, Koday M, Kalinoski H, Stewart L, Carter L, Nieusma T, Lee PS, Ward AB, Wilson IA, Dagley A, Smee DF, Baker D, Fuller DH (2016) A computationally designed hemagglutinin stem-binding protein provides in vivo protection from influenza independent of a host immune response. PLoS Pathog 12(2):e1005409
7. Clark AJ, Gindin T, Zhang B, Wang L, Abel R, Murret CS, Xu F, Bao A, Lu NJ, Zhou T, Kwong PD, Shapiro L, Honig B, Friesner RA (2017) Free energy perturbation calculation of relative binding free energy between broadly neutralizing antibodies and the gp120 glycoprotein of HIV-1. J Mol Biol 429(7):930–947
8. Fowler PW, Cole K, Gordon NC, Kearns AM, Llewelyn MJ, Peto TEA, Crook DW, Walker AS (2018) Robust prediction of resistance to trimethoprim in Staphylococcus aureus. Cell Chem Biol 25:339–349
9. Hauser K, Negron C, Albanese SK, Ray S, Steinbrecher T, Abel R, Chodera JD, Wang L (2018) Predicting resistance of clinical Abl mutations to targeted kinase inhibitors using alchemical free-energy calculations. Commun Biol 1:70
10. Tinberg CE, Khare SD, Dou J, Doyle L, Nelson JW, Schena A, Jankowski W, Kalodimos CG, Johnsson K, Stoddard BL, Baker D (2013) Computational design of ligand-binding proteins with high affinity and selectivity. Nature 501(7466):212
11. Yang W, Lai L (2017) Computational design of ligand-binding proteins. Curr Opin Struct Biol 45:67–73
12. Brender JR, Zhang Y (2015) Predicting the effect of mutations on protein-protein binding interactions through structure-based interface profiles. PLoS Comput Biol 11(10):e1004494
13. Pires DEV, Ascher DB, Blundell TL (2014) mCSM: predicting the effects of mutations in proteins using graph-based signatures. Bioinformatics 30(3):335–342
14. Schymkowitz J, Borg J, Stricher F, Nys R, Rousseau F, Serrano L (2005) The FoldX web server: an online force field. Nucleic Acids Res 33(Suppl 2):W382–W388
15. Kortemme T, Baker D (2002) A simple physical model for binding energy hot spots in protein-protein complexes. Proc Natl Acad Sci USA 99(22):14116–14121
16. Leaver-Fay A, Tyka M, Lewis SM, Lange OF, Thompson J, Jacak R, Kaufman K, Renfrew PD, Smith CA, Sheffler W, Davis IW, Cooper S, Treuille A, Mandell DJ, Richter F, Ban YEA, Fleishman SJ, Corn JE, Kim DE, Lyskov S, Berrondo M, Mentzer S, Popović Z, Havranek JJ, Karanicolas J, Das R, Meiler J, Kortemme T, Gray JJ, Kuhlman B, Baker D, Bradley P (2011) Rosetta3: an object-oriented software suite for the simulation and design of macromolecules. Methods Enzymol 487 (C):545–574
17. Petukh M, Li M, Alexov E (2015) Predicting binding free energy change caused by point mutations with knowledge-modified MM/PBSA method. PLoS Comput Biol 11 (7):e1004276
18. Beard H, Cholleti A, Pearlman D, Sherman W, Loving KA (2013) Applying physics-based scoring to calculate free energies of binding for single amino acid mutations in protein-protein complexes. PLoS ONE 8(12):e82849
19. Moreira IS, Fernandes PA, Ramos MJ (2007) Computational alanine scanning mutagenesis - An improved methodological approach. J Comput Chem 28(3):644–654
20. Seeliger D, de Groot BL (2010) Protein thermostability calculations using alchemical free energy simulations. Biophys J 98 (10):2309–2316
21. Chipot C, Pohorille A (eds) (2007) Free energy calculations: theory and applications in chemistry and biology, vol 86. Springer, Berlin
22. Neidigh JW, Fesinmeyer RM, Andersen NH (2002) Designing a 20-residue protein. Nat Struct Mol Biol 9(6):425–430
23. Abraham MJ, Murtola T, Schulz R, Páll S, Smith JC, Hess B, Lindahl E (2015) GROMACS: high performance molecular simulations through multi-level parallelism from laptops to supercomputers. SoftwareX 2:1–7
24. Gapsys V, Michielssens S, Seeliger D, de Groot BL (2015) pmx automated protein structure and topology generation for alchemical perturbations. J Comput Chem 36(5):348–354

25. Chipot C (2014) Frontiers in free-energy calculations of biological systems. Wiley Interdiscip Rev Comput Mol Sci 4(1):71–89
26. Gapsys V, Michielssens S, Peters JH, de Groot BL, Leonov H (2015) Molecular modeling of proteins, vol 1215. Humana Press, New York
27. Pohorille A, Jarzynski C, Chipot C (2010) Good practices in free-energy calculations. J Phys Chem B 114(32):10235–10253
28. Hansen N, van Gunsteren WF (2014) Practical aspects of free-energy calculations: a review. J Chem Theory Comput 10(7):2632–2647
29. Goette M, Grubmüller H (2009) Accuracy and convergence of free energy differences calculated from nonequilibrium switching processes. J Comput Chem 30(3):447–456
30. Jarzynski C (1997) Nonequilibrium equality for free energy differences. Phys Rev Lett 78 (14):2690–2693
31. Jarzynski C (1997) Equilibrium free-energy differences from nonequilibrium measurements: A master-equation approach. Phys Rev E 56:5018–5035
32. Crooks GE (1998) Nonequilibrium measurements of free energy differences for microscopically reversible Markovian systems. J Stat Phys 90(5/6):1481–1487
33. Crooks GE (1999) Entropy production fluctuation theorem and the nonequilibrium work relation for free energy differences. Phys Rev E 60(3):2721–2726
34. Crooks GE (2000) Path-ensemble averages in systems driven far from equilibrium. Phys Rev E 61(3):2361–2366
35. Hummer G, Szabo A (2001) Free energy reconstruction from nonequilibrium single-molecule pulling experiments. Proc Natl Acad Sci USA 98(7):3658–3661
36. Hummer G (2001) Fast-growth thermodynamic integration: error and efficiency analysis. J Chem Phys 114(17):7330–7337
37. Hummer G, Szabo A (2005) Free energy surfaces from single-molecule force spectroscopy. Acc Chem Res 38(7):504–513
38. Zwanzig RW (1954) High-temperature equation of state by a perturbation method. I. nonpolar gases. J Chem Phys 22 (8):1420–1426
39. Kirkwood JG (1935) Statistical mechanics of fluid mixtures. J Chem Phys 3(5):300–313
40. Cuendet MA (2006) The Jarzynski identity derived from general Hamiltonian or non-Hamiltonian dynamics reproducing NVT or NPT ensembles. J Chem Phys 125 (14):144109
41. Wood RH, Mühlbauer WCF, Thompson PT (1991) Systematic errors in free energy perturbation calculations due to a finite sample of configuration space: sample-size hysteresis. J Phys Chem 95(17):6670–6675
42. Gore J, Ritort F, Bustamante C (2003) Bias and error in estimates of equilibrium free-energy differences from nonequilibrium measurements. Proc Natl Acad Sci USA 100 (22):12564–12569
43. Nanda H, Lu N, Woolf TB (2005) Using non-Gaussian density functional fits to improve relative free energy calculations. J Chem Phys 122(13):134110
44. Massey FJ Jr (1951) Kolmogorov-Smirnov test for goodness of fit. Test 46(253):68– 78
45. Efron B, Tibshirani RJ (1994) An introduction to the bootstrap, vol 5, 1st edn. Chapman and Hall/CRC, London/West Palm Beach
46. Bennett CH (1976) Efficient estimation of free energy differences from Monte Carlo data. J Comput Phys 22(2):245–268
47. Shirts MR, Bair E, Hooker G, Pande VS (2003) Equilibrium free energies from non-equilibrium measurements using maximum-likelihood methods. Phys Rev Lett 91 (14):140601
48. Nelder JA, Mead R (1964) A simplex method for function minimization. Comput J 7 (4):308–313
49. Hahn AM, Then H (2010) Measuring the convergence of Monte Carlo free-energy calculations. Phys Rev E Stat Nonlinear Soft Matter Phys 81(4):041117
50. Lindorff-Larsen K, Trbovic N, Maragakis P, Piana S, Shaw DE (2012) Structure and dynamics of an unfolded protein examined by molecular dynamics simulation. J Am Chem Soc 134(8):3787–3791
51. Rauscher S, Gapsys V, Gajda MJ, Zweckstetter M, de Groot BL, Grubmüller H (2015) Structural ensembles of intrinsically disordered proteins depend strongly on force field: a comparison to experiment. J Chem Theory Comput 11(11):5513–5524
52. Prevost M, Wodak SJ, Tidor B, Karplus M (1991) Contribution of the hydrophobic effect to protein stability: analysis based on simulations of the Ile-96 → Ala mutation in barnase. Proc Natl Acad Sci USA 88(23):10880–10884
53. Sneddon SF, Tobias DJ (1992) The role of packing interactions in stabilizing folded proteins. Biochemistry 31(10):2842–2846
54. Pitera JW, Kollman PA (2000) Exhaustive mutagenesis in silico: multicoordinate free

energy calculations on proteins and peptides. Proteins Struct Funct Bioinf 41(3):385–397
55. Pearlman DA, Kollman PA (1991) The overlooked bond-stretching contribution in free energy perturbation calculations. J Chem Phys 94(6):4532
56. Pearlman DA (1994) A comparison of alternative approaches to free energy calculations. J Phys Chem 98(5):1487–1493
57. Boresch S, Karplus M (1999) The role of bonded terms in free energy simulations: 1. Theoretical analysis. J Phys Chem A 103 (1):103–118
58. Boresch S, Karplus M (1996) The Jacobian factor in free energy simulations. J Chem Phys 105(12):5145–5154
59. Boresch S, Karplus M (1999) The role of bonded terms in free energy simulations. 2. Calculation of their influence on free energy differences of solvation. J Phys Chem A 103 (1):119–136
60. Beutler TC, Mark AE, van Schaik RC, Gerber PR, van Gunsteren WF (1994) Avoiding singularities and numerical instabilities in free energy calculations based on molecular simulations. Chem Phys Lett 222(6):529–539
61. Zacharias M, Straatsma TP, McCammon JA (1994) Separation-shifted scaling, a new scaling method for Lennard-Jones interactions in thermodynamic integration. J Chem Phys 100:9025–9031
62. Pham TT, Shirts MR (2011) Identifying low variance pathways for free energy calculations of molecular transformations in solution phase. J Chem Phys 135(3):034114
63. Gapsys V, Seeliger D, de Groot BL (2012) New soft-core potential function for molecular dynamics based alchemical free energy calculations. J Chem Theory Comput 8 (7):2373–2382
64. Buelens FP, Grubmüller H (2012) Linear-scaling soft-core scheme for alchemical free energy calculations. J Comput Chem 33 (1):25–33
65. Gapsys V, de Groot BL (2017) pmx Webserver: a user friendly interface for alchemistry. J Chem Inf Model 57(2):109–114
66. Šali A, Blundell TL (1993) Comparative protein modelling by satisfaction of spatial restraints. J Mol Biol 234(3):779–815
67. Schrödinger, LLC (2015) The PyMOL molecular graphics system, version 1.8, November 2015
68. Vriend G (1990) WHAT IF: a molecular modeling and drug design program. J Mol Graph 8 (1):52–56
69. Hornak V, Abel R, Okur A, Strockbine B, Roitberg A, Simmerling C (2006) Comparison of multiple amber force fields and development of improved protein backbone parameters. Proteins Struct Funct Bioinf 65 (3):712–725
70. Best RB, Hummer G (2009) Optimized molecular dynamics force fields applied to the helix-coil transition of polypeptides. J Phys Chem B 113(26):9004–9015
71. Lindorff-Larsen K, Piana S, Palmo K, Maragakis P, Klepeis JL, Dror RO, Shaw DE (2010) Improved side-chain torsion potentials for the Amber ff99SB protein force field. Proteins Struct Funct Bioinf 78(8):1950–1958
72. Lindahl E (2015) Molecular dynamics simulations. In: Molecular modeling of proteins. Springer, Berlin, pp 3–26
73. Barua B, Andersen NH (2001) Determinants of miniprotein stability: can anything replace a buried H-bonded Trp sidechain? Lett Pept Sci 8(3–5):221–226
74. Barua B, Lin JC, Williams VD, Kummler P, Neidigh JW, Andersen NH (2008) The Trp-cage: optimizing the stability of a globular miniprotein. Protein Eng Des Sel 21 (3):171–185
75. Darden T, York D, Pedersen L (1993) Particle mesh Ewald: an Nlog(N) method for Ewald sums in large systems. J Chem Phys 98 (12):10089–10092
76. Essmann U, Perera L, Berkowitz ML, Darden T, Lee H, Pedersen LG (1995) A smooth particle mesh Ewald method. J Chem Phys 103 (19):8577–8593
77. Rocklin GJ, Mobley DL, Dill KA, Hünenberger PH (2013) Calculating the binding free energies of charged species based on explicit-solvent simulations employing lattice-sum methods: an accurate correction scheme for electrostatic finite-size effects. J Chem Phys 139(18):184103
78. Lin Y-L, Aleksandrov A, Simonson T, Roux B (2014) An overview of electrostatic free energy computations for solutions and proteins. J Chem Theory Comput 10(7):2690–2709
79. Hub JS, de Groot BL, Grubmüller H, Groenhof G (2014) Quantifying artifacts in Ewald simulations of inhomogeneous systems with a net charge. J Chem Theory Comput 10 (1):381–390

Chapter 3

Protocols for the Molecular Evolutionary Analysis of Membrane Protein Gene Duplicates

Laurel R. Yohe, Liang Liu, Liliana M. Dávalos, and David A. Liberles

Abstract

Gene duplication is an important process in the evolution of gene content in eukaryotic genomes. Understanding when gene duplicates contribute new molecular functions to genomes through molecular adaptation is one important goal in comparative genomics. In large gene families, however, characterizing adaptation and neofunctionalization across species is challenging, as models have traditionally quantified the timing of duplications without considering underlying gene trees. This protocol combines multiple approaches to detect adaptation in protein duplicates at a phylogenetic scale. We include a description of models for gene tree-species tree reconciliation that enable different types of inference, as well as a practical guide to their use. Although simulation-based approaches successfully detect shifts in the rate of duplication/retention, the conflation between the duplication and retention processes, the distinct trajectories of duplicates under non-, sub-, and neofunctionalization, as well as dosage effects offer hitherto unexplored analytical avenues. We introduce mathematical descriptions of these probabilities and offer a road map to computational implementation whose starting point is parsimony reconciliation. Sequence evolution information based on the ratio of nonsynonymous to synonymous nucleotide substitution rates (dN/dS) can be combined with duplicate survival probabilities to better predict the emergence of new molecular functions in retained duplicates. Together, these methods enable characterization of potentially adaptive candidate duplicates whose neofunctionalization may contribute to phenotypic divergence across species.

Key words Gene duplication, Gene tree, Birth-death models, Molecular evolution, dN/dS

1 Introduction

1.1 Gene Duplication and Membrane Proteins

The evolutionary mechanisms for generating novelty are key to understanding variation in phenotypic and taxonomic diversity across the Tree of Life. Identifying the genetic mechanisms behind the origin and maintenance of phenotypic diversity is therefore a fundamental objective of evolutionary genetics. While base pair substitutions provide a means for understanding the novel function of existing genes, the duplication of entire genes and genomes offers a source of new variation for functional diversification. Duplications are primary sources of innovation, from large-scale whole-

Tobias Sikosek (ed.), *Computational Methods in Protein Evolution*, Methods in Molecular Biology, vol. 1851,
https://doi.org/10.1007/978-1-4939-8736-8_3,

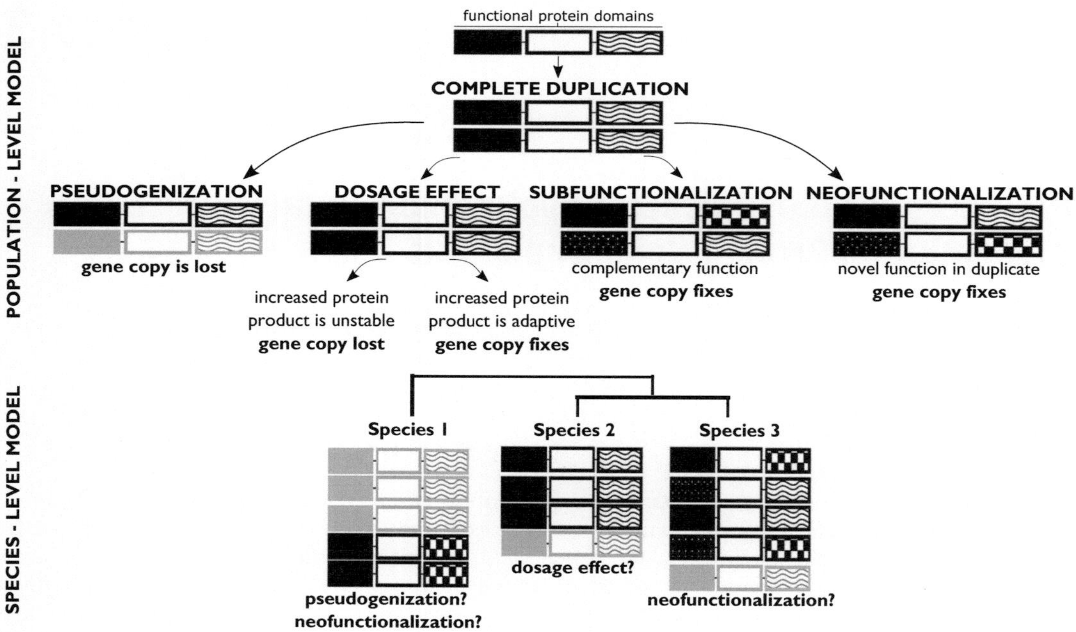

Fig. 1 Theoretical model of single-copy gene duplication and mechanisms for how a duplicate is fixed or lost in a population (top). Different patterns indicate different fixed amino acid differences. Grayed genes indicate loss of function. Note changes can also happen in regulatory regions, but are not shown here. The species-level model (bottom) is a cartoon of hypothetical scenarios that may be observed across species and their potential mechanisms. Figure adapted from [11]

genome duplications that may prompt speciation, seen in notable examples of teleost fish [1–3] or extraordinary polyploidy observed across plants [4–6], to duplications of a single gene, such as the expansion of multiple ion channels associated with the evolution of neural system complexity [7].

Just as new species evolve from ancestral lineages, new genes can evolve from those already present in the genome, and gene duplication is a primary molecular mechanism for the evolution of novel genes [8]. However, testing whether gene duplication is adaptive remains an unresolved challenge in evolutionary biology. In this chapter, we present an overview of the current methods used for studying gene duplication across species, and we describe a theoretical approach that integrates across several methodologies. We focus specifically on detecting adaptation in small-scale duplications from a single gene. Our primary emphasis is on membrane proteins, as many of these proteins are encoded by genes that evolve through a birth-death process that is a central mechanism to the model we propose.

A new gene may follow one of several trajectories after duplication (Fig. 1). Most probably, the duplicate is deleterious or neutral, does not fix in the population, and is lost [9–12]. It may also be retained, either because it is adaptive or because of drift. The adaptive scenario may occur by either taking on a novel coding sequence or expression function or maintaining identical coding and expression domain functions as its ancestor, but increasing the expression of the gene product from redundant gene copies—a phenomenon known as dosage effect [8–10, 13]. In a nonadaptive scenario, the copy may fix but will likely pseudogenize after many generations unless subfunctionalization occurs. In each of these outcomes, the probability of gene retention and loss can be modeled as a function of time. The rates of amino acid-changing and silent substitutions that occur in each of these outcomes differs and can be informative in determining the fate of a gene duplicate. The overarching objective of this chapter is to quantify these distinct processes, present methods of simulation for different models, and synthesize the outcomes into biologically relevant interpretations of adaptation and loss.

The domain of a protein is the coding sequence that encodes the amino acid residues, and proteins can be composed of a single domain or several. These domains are the evolutionary unit of a protein, as part or all of the domain may undergo duplication or recombination or accumulate mutations that may affect protein function [14]. Membrane proteins are critical to several indispensable cellular functions including signal recognition, signal transduction, and transportation of materials into and out of the cell. In addition to these functions, membrane proteins are constrained to maintaining domains that enable the insertion, and that preserve the orientation, of the protein in the lipid bilayer of the cell membrane [15]. Membrane proteins also show a preference for positively charged residues that interact with the cytoplasmic side of the membrane [16]. With these constraints in mind, membrane proteins that respond to extracellular signals from the environment must also have binding sites for their respective ligands. Chemosensory receptors and immune-related membrane proteins involved in pathogen recognition encounter natural selection to detect ever-changing environmental cues. Many genes that encode these proteins evolve in a concerted birth-death fashion, in which genes duplicate, and duplicates may evolve a new function or pseudogenize [17]. This mechanism leads to a pattern of many closely related genes with similar and divergent function that can be classified as a multigene family.

2 Methods

2.1 Approaches and Limitations to Studying Gene Duplication

There are two major approaches to investigating the evolutionary process of gene duplication among species: birth-death models fit to a species tree and gene tree-species tree reconciliation. Several methodologies have been published using gene tree-species tree reconciliation [18–22]. This approach allows detection of branches in which duplications and losses of particular gene copies occur, modeling the history of gene copies as a function of speciation events. However, currently available methods are either parsimony-based or do not estimate rates of gene retention [18, 23]. Importantly, any computed rate of loss is a homogeneous function of time along branches of the species tree, instead of a function relating loss to the age of the duplicate. This is a problem because the loss rate should not be constant through time. Instead, the probability of gene retention decays with duplicate age, making the loss rate a function of the time since duplication. Current inter-specific models also conflate mutation and fixation, overlooking the time between these events. Future work could include the development of mutation-selection style models for gene duplication.

The second approach estimates rates of birth (duplication) and death (pseudogenization/loss) and tests if there are increased rates of either in different parts of the tree [24–26]. These methods calculate the likelihood of gene family data based on a birth and death rate while also considering branch lengths of species divergence times [24, 25]. This framework allows for explicit hypothesis testing of different birth-death rates in different parts of the tree but is subject to several assumptions, discussed below.

We provide an overview of these methods used for studying adaptation of gene duplication. Our examples provide a conceptual framework on how to define biologically meaningful questions in gene duplication analyses in a way that enables quantitative tests. Our examples also demonstrate strong caveats and ever-present assumptions in gene duplication analyses at the phylogenetic scale. First, we demonstrate a gene tree-species tree reconciliation method using parsimony. Second, we show how to test if the number of inferred duplications and losses is significantly higher or lower than expected under a null birth-death process through simulations. Third, we present the theory for developing a more integrated approach to characterize the different fates of gene duplicates.

2.1.1 Parsimony-Based Reconciliation

One early and common approach to gene tree-species tree reconciliation is to use the principle of parsimony to minimize either the duplication or the loss cost associated with mapping lineages of gene trees to branches of the species tree. This approach provides a valuable preliminary analysis for identifying discordance between the gene tree topology and the species tree (when the species tree relationship is not recovered within the gene family). Early approaches required the gene and species trees to be fully resolved with binary nodes, but subsequent approaches relaxed this assumption (*see* [27] for a review). As in parsimony-based tree reconstruction, the insensitivity of parsimony to duplication rates on branches with different lengths is a potential problem. A previous study has evaluated the relationship of different costs of accounting for gene tree discordance to each other in a parsimony context, which represents a starting point for comparing these with model-based reconciliations under different models [28].

Here we provide an example of the amino acid transport protein gene family known as the amino acid-polyamine-organocation (APC) transporters in the sap-feeding insect suborder Sternorrhyncha. These insects have evolved a tight symbiotic relationship with gut bacteria that provides essential amino acids to supplement a nutrient-poor diet of phloem. Amino acid transport proteins facilitate the exchange of amino acids between the symbiont and its host across the bacteriocytes. It was known that some species of sap-feeding insects had multiple gene copies of APC transporters [29], but whether these duplications occurred prior to the radiation of sap-feeding insects was unclear. If an expansion of the number of APC transport proteins had occurred within this clade, it might be related to the increased reliance on nutrient supplements from gut symbionts. To answer this question, a published study implemented several reconciliation and birth-death methods to model the evolutionary history of the gene tree [30]. We first present the reconciliation of the APC gene tree with the Hemiptera species tree to demonstrate parsimony inference of duplications and losses (Fig. 2). Parsimony reconciliation was inferred using Notung [18]. Reconciliation can also be performed using a likelihood-based method (in this case, DupliPHY-ML [31]) that yields similar results (Fig. 3a).

Figure 2a shows that several lineages within Sternorrhyncha have experienced an expansion in the number of copies of APC transporters, as well as an expansion at the base of the group. However, in addition to the statistical inconsistency of parsimony inference when many changes accumulate, there is no hypothesis testing involved in describing whether any of these duplications or losses differ than from what is expected under a null evolutionary model of birth-death.

2.1.2 Birth-Death Models of Gene Duplication

Early models for gene duplication were traditional birth and death models. In these, the number of duplicate copies evolves through a stochastic birth-death process in which retention and loss are modeled with an exponential distribution [32]. Key parameters estimated in birth-death models are the birth and death rates of the genes, as well as the number of gene copies at each internal node. These models set up a statistical framework that describes how rates of gene duplication and loss may vary in different parts of the tree.

In the context of our example with the APC transporters in hemipteran insects, the parsimony inference suggests there may be an increased rate of gene duplication in Sternorrhyncha compared to other insects in the order. Likelihood-based birth-death models

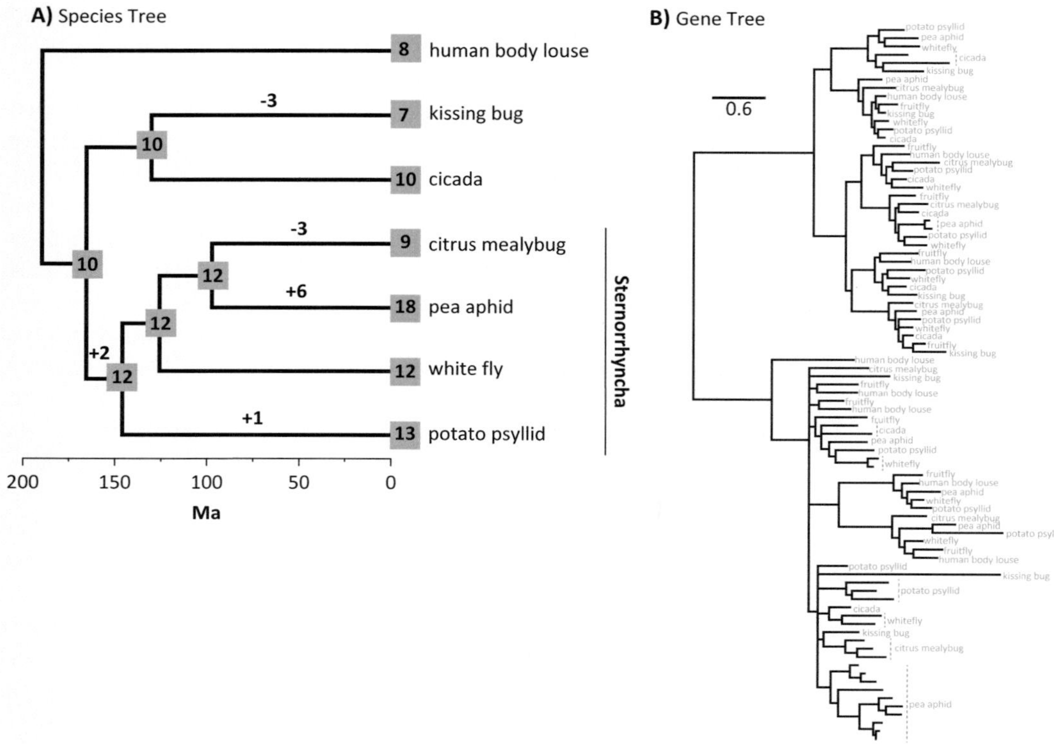

Fig. 2 (**a**) Species tree for Hemiptera insect order, denoted with the Sternorrhyncha sap-feeding insect suborder. The human body louse is an outgroup. The fruit fly (*Drosophila melanogaster*) was omitted from the species tree for clarity. Gray boxes indicate the number of gene copies inferred for each species and at each ancestral node. Branch labels indicate the number of duplications (+) or losses (−) inferred to have occurred at each respective branch as inferred using parsimony. (**b**) Gene tree of the APC amino acid transporter family. Each tip is a unique gene copy belonging to the species labeled at the tip

explicitly test whether multiple birth rates in different parts of the tree (in this case Sternorrhyncha v. background branches) better fit the data than a single birth rate for the entire phylogeny. A previous study estimated the birth rate (b) for different parts of the tree and found that a model with a single b for the entire phylogeny was a better fit than a model with a separate estimate of b for the Sternorrhyncha suborder (Table 1) [30]. Thus, from this approach, evidence does *not* support increased rates of duplication in sap-feeding insects.

While this approach can identify the species tree branches in which increased rates of duplication events occurred, it ignores the gene tree. Unlike reconciliation approaches, phyletic birth-death models simply fit parameters to numbers of gene copies, instead of actually considering if particular orthologs or paralogs are observed across species. Simulations of gene trees under similar birth and death rates estimated from one's data can provide a more thorough understanding of a null model of birth and death rate estimates

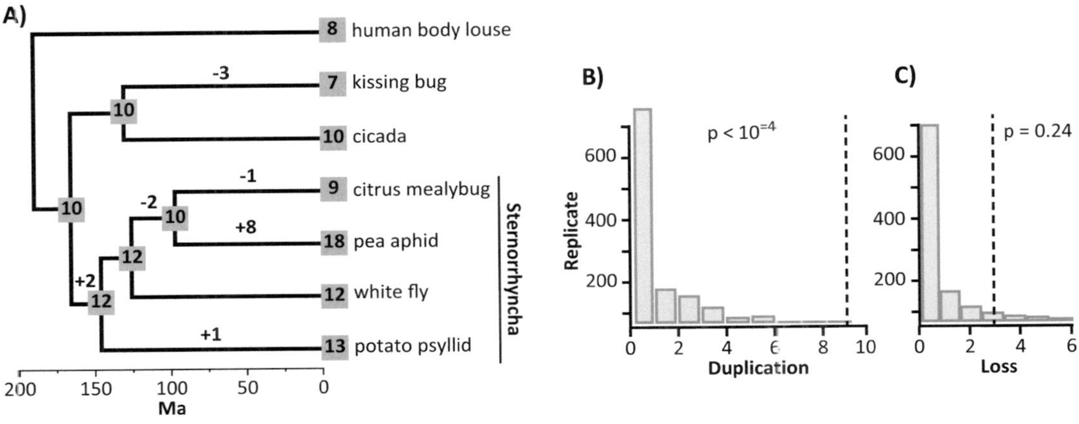

Fig. 3 Likelihood-based reconciliation of the APC transport proteins in Hemiptera. (**a**) Duplications and losses labeled on branches were inferred from reconciliation analyses in DupliPHY-ML v. 1.2 [31]. Gray boxes are number of APC transporter gene copies in each species or inferred at the ancestral node. (**b**) Simulation of expected number of duplications for Sternorrhyncha under a null birth-death process. The dotted line is the cumulative number of duplications observed from the DupliPHY-ML results. (**c**) Simulation of expected number of losses for Sternorrhyncha under a null birth-death process. The dotted line is the cumulative number of losses observed from the DupliPHY-ML results. P-values test whether the observed value is significantly different than the null distribution. Simulations were performed using GenPhyloData within the JPrIME v. 0.3.6 software [21]. Code for simulations is available in the supplementary material of [30]

Table 1
Hemiptera APC transporter gene family parameter estimates of likelihood-based birth-death model and likelihood ratio test results of model comparisons between a null model of a single birth rate (*b*) for the entire tree or two rates of *b*, one for the background branches and one for sternorrhynchans

Model	$b_{background}$	$b_{Sternorrhyncha}$	ML	np	LR	*p*-value
Single *b*	1.22×10^3	–	34.3	1	–	–
Multiple *b*	0.73×10^3	2.50×10^3	33.5	2	1.52	0.20

ML is the log-likelihood. np is the number of parameters. LR is the likelihood ratio. Inferences were made using CAFE v. 3.1 [40]. This model assumed the rate of birth to be equal to the rate of death. Analysis derived from [30].

under a neutral process. If the number of observed fixed duplicates or losses differs significantly from what is estimated from simulated data, the probability of fixation might be higher or lower than is expected by the null birth-death process. In our example with the APC transporter genes, the study used 1000 birth-death simulations based upon a birth rate estimated from the single *b* model in Table 1 [30]. From these gene trees, the expected number of duplications and losses could be estimated for each node of the tree. The study compared the observed values from the likelihood-based reconciliation (Fig. 3a) and found that sternorrhynchan insects did indeed have a significantly higher number of duplications (but no difference in losses) compared to what was expected under a null birth-death scenario (Fig. 3b, c). While this approach is still subject to assumptions made by birth-death models, simulation experiments can provide a useful insight into null expectations for the underlying evolutionary process.

We argue, however, that these methods may be testing the wrong question. All models discussed so far conflate an increased rate of birth, which is a Poisson process similar to mutation events, with an increased rate of gene retention. In other words, instead of testing for an increased "birth rate," which should be intrinsically stochastic and homogeneous throughout long time scales, it would be ideal to measure an increased rate of gene retention. In the case of increased rates of gene retention, duplicates may be subject to selection and may indicate adaptation. Different processes lead to gene retention (Fig. 1), and these processes can be modeled. We propose an integrated framework to quantitatively differentiate among different gene retention scenarios that may lead to more biologically meaningful interpretations of adaptation that result from gene duplication.

2.2 Modeling Different Fates of Gene Duplicates: Integrating Reconciliation and Birth-Death

Several biological models have been proposed to depict the mechanisms that lead to different evolutionary fates for a gene duplicate (Fig. 1), including pseudogenization, neofunctionalization, subfunctionalization, or dosage effect. These mechanisms give rise to quite different retention dynamics that can lead to a time-dependent loss rate of gene duplicates, expressed as a function $\lambda(t)$. For nonfunctionalization, the loss rate is constant over time. In contrast, the loss rates of neofunctionalization and subfunctionalization decline over time and have been described with a Weibull hazard function [8]. For dosage effect, the rate of loss increases over time unless dosage effects are combined with subsequent neofunctionalization or subfunctionalization [33]. Alternative formulations with very similar dynamics have also been proposed [13]. Figure 4 depicts the shapes of these hazard functions under different scenarios.

From Reconciliation Probabilities to Birth-Death Models

In most birth-death model frameworks, the time-dependent loss rates have been incorporated in a generalized birth-death process to model the fate of gene duplicates. This means the evolution of the gene copies in a gene family is modeled as a pure birth process with a time-dependent birth rate, which is a function of the loss and birth rates in the original birth-death process. Since the loss rate characterizes the underlying retention mechanisms, the inference of the loss rates can identify either nonfunctionalization, subfunctionalization, dosage, or neofunctionalization as responsible for the observed site patterns of gene family data. However, an important caveat of all time-dependent models is that any rate of loss that is computed is a function of time along branches of the

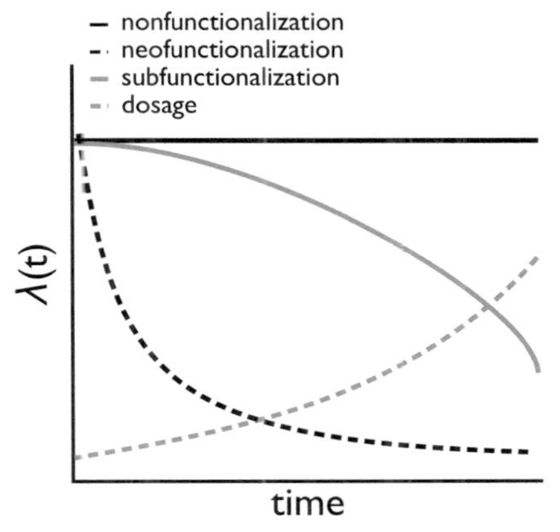

Fig. 4 Shape of the hazard function through time representing the rate of gene loss under the four different gene retention scenarios. Figure modified from [39]

species tree, instead of relating to the age of the gene duplicate. This is a problem because the loss rate should not be constant through time but instead be a function of the time since duplication, as the probability of gene retention decays with duplicate age.

Hence, it is more realistic to treat the loss rate as a function of the ages of gene copies. We propose a theoretical solution. Let $\lambda(t^*)$ be the loss rate of a gene copy at age t^*. The age-dependent model assumes the number of gene copies increases or decreases by 1 or remains the same during an infinitesimal interval $(t, t + \Delta t)$ with probabilities described as follows [12]:
the probability of a gene duplication

$$P(n_{t-\Delta t} = n_t + 1) = n_t b \Delta t + o(\Delta t),$$

the probability of a gene loss

$$P(n_{t+\Delta t} = n_t - 1) = \sum_{i=1}^{n_t} \lambda(t_i^*) \Delta t + o(\Delta t),$$

and the probability that the number of copies stays the same

$$P(n_{t+\Delta t} = n_t) = 1 - \left(n_t b + \sum_{i=1}^{n_t} \lambda(t_i^*) \right) \Delta t + o(\Delta t).$$

The parameter b is the birth rate; n is the number of gene copies at the present time; $\lambda(t_i^*)$ is the loss rate of gene copy i at age t_i^*. The three equations lead to a stochastic differential equation characterizing the age-dependent birth-death process. When the loss rate is constant (nonfunctionalization), the age-dependent birth-death model is identical to the time-dependent birth-death model derived from the reconstructed process (*see* [34] for derivation). For neofunctionalization and subfunctionalization, it has been demonstrated by simulation that the likelihood function of the age-dependent model differs from that of the time-dependent model [12], and presumably for dosage as well. However, at the present time, there is no analytic solution to the stochastic differential equation when subfunctionalization, neofunctionalization, and dosage are the underlying mechanisms governing the age-dependent birth and loss rates. Research on the age-dependent model will provide indispensable insights on the evolution of gene duplicates.

The model we propose differs from existing approaches, as it constrains the inference of duplication events with speciation events while also calculating an age-dependent survival probability of gene copies. If a speciation event occurs at t_i, the probability of gene copy retention is a survival probability E_j calculated from the hazard function $\lambda(t)$, which represents the instantaneous loss at time t. Instead of modeling the time associated with retention or loss as constant through time, it will actually be calculated from the moment the duplication occurred, which can be denoted as t^*, reflecting the age-related duplicate notation described in the equations above [8, 9, 13].

We present a simple example to demonstrate how these probabilities may be calculated and how these probabilities can then be integrated with a gene tree-species tree reconciliation framework. Figure 5 shows an example gene tree with one specific reconciliation solution that may have occurred throughout the history of the gene family and species phylogeny. The solution is shown based on parsimony. In this scenario, two duplications and one loss have occurred in the phylogeny. The probability of retention is the product of all survival probabilities of different events in Table 2. The hazard function $\lambda(t)$ and its corresponding survival function are different for each outcome in Fig. 5 [8], including nonfunctionalization, neofunctionalization, subfunctionalization, and dosage effect. The product of all survival probabilities occurring for each event (e.g., Table 2) will reflect the survival probability of all

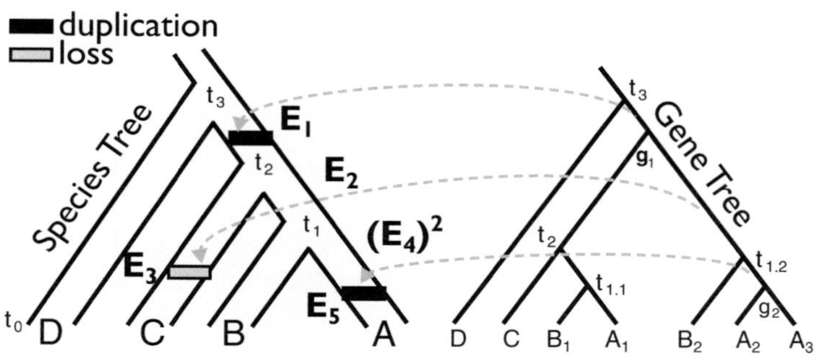

Fig. 5 Cartoon of species tree-gene tree reconciliation. Speciation times (t_i) and gene divergence times (g_i) are noted on nodes. E_4 is squared because it is counting both branches from time t_1. Event probabilities are listed in Table 2

Table 2
Events and probabilities of Fig. 5

Event	Description	Probability
E_1	Duplication and retention	$e^{-\int_0^{g_1-t_2} \lambda(t)dt}$
E_2	Retain duplicate	$e^{-\int_{g_1-t_2}^{g_1-t_1} \lambda(t)dt}$
E_3	Lose duplicate	$1 - e^{-\int_{g_1-t_2}^{g_1} \lambda(t)dt}$
E_4	Retain duplicate	$e^{-\int_{g_1-t_1}^{g_1} \lambda(t)dt}$
E_5	Duplication and retention	$e^{-\int_0^{g_2} \lambda(t)dt}$

The probability of the reconciled tree in Fig. 5 is the product of all event probabilities. Gray arrows indicate probabilities that do not include a speciation event. The branch length-dependent birth rate can also be incorporated, when relevant.

duplicates in the gene tree. The best-fit hazard function model can be determined by model selection using the Akaike or Bayesian Information Criterion.

It should be emphasized that this example only accounts for a single set of events for one proposed reconciliation solution, as opposed to multiple hidden events that may have also occurred. Integrating over all possible reconciliation histories is, in theory, the only way to account for all possible hidden events. However, this is not a feasible solution given the possible number of hidden events that may have occurred. A more tractable solution is to begin with a parsimonious reconciliation and iteratively consider hidden events and alternative reconciliations according to a branch and bound-style approach. In this regard, a finite set of events (such as those shown in Table 2) for each reconciliation history can be compared with one another, and the most probable solution among this finite set of specific histories can be calculated.

2.3 Combining Survival Probabilities with dN/dS

For each outcome in Fig. 1, there is an expected behavior of the ratio rates of nonsynonymous (dN) to synonymous (dS) substitutions (dN/dS or ω) for the gene copy (Fig. 6). The behaviors of this ratio can reveal biologically meaningful interpretations relevant to molecular adaptation. For example, analyses of mammalian olfactory receptors, a hyperdiverse gene family that encodes G-protein-coupled chemosensory receptors, have shown that some particular orthologous gene groups have undergone rapid expansions and have high dN/dS relative to the median, suggesting functional diversification of these receptor types [35]. However, dN/dS is not currently modeled in any methodology used to study gene duplication, despite predictable functions under different gene retention scenarios. When genes are initially redundant following

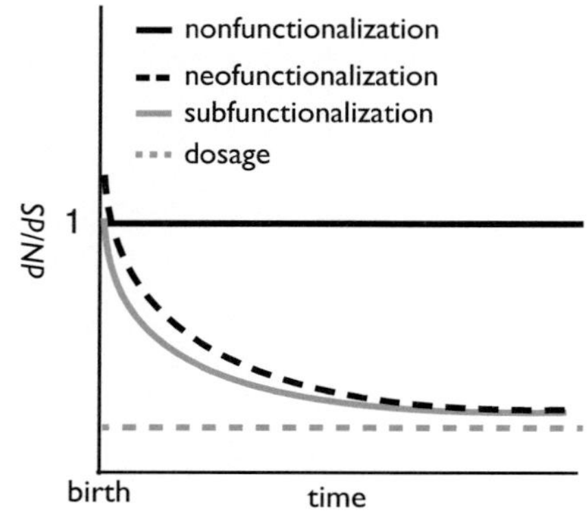

Fig. 6 Expected *dN/dS* of duplicated copy after gene duplication

duplication, they are expected to show neutral evolution or at least relaxation from purifying selection. Genes that nonfunctionalize should continue to evolve with $dN/dS = 1$, whereas duplicates that are retained through either the neofunctionalization or subfunctionalization process should see dN/dS decay toward a rate consistent with non-duplicated genes as an asymptote (Fig. 6). Indeed, it has been empirically shown that accelerated rates of dN/dS occur after duplication and then subsequently decline [36]. There may be little information to differentiate between neofunctionalization and subfunctionalization with these data, although it might be anticipated that neofunctionalizing genes at some early point have $dN/dS > 1$ (depending upon several factors, including the starting value of purifying selection and the strength of positive selection), something not expected for subfunctionalization. For subfunctionalization, dN/dS may not initially be as high as with neofunctionalization, as part of the gene is still under strong purifying selection to maintain ancestral function. In the case of selection for increased dosage, strong purifying selection is expected from the moment of duplication, as duplicates are functionally the same ($dN/dS << 1$ and constant).

One previously used approach is to approximate the age of the duplication event by building a histogram of pairwise dN/dS values of duplicates related to dS values [9]. Across collections of genes from a genome, each empirical frequency distribution is a sample of an underlying duplication process. When a gene family is known, an alternative is to examine branch-specific changes in dN/dS in lineages downstream from a duplication event. In this scenario, the onset of selection post-duplication in individual lineages can be evaluated.

The dN/dS statistic is one of the most commonly used approaches to measure the strength of selection among species, but it can be susceptible to false positives if there is purifying selection on synonymous mutations [37]. Meaningful dN/dS estimates may also be problematic for recent duplicates in closely related lineages [38]. Mutation-selection models can offer a complementary set of tools to estimate the strength of selection and should also be considered in this framework [37].

3 Concluding Thoughts

Gene duplication is a fundamental mechanism underlying novel protein function. However, the fate of a gene duplicate is complex, and it can be challenging to determine whether or not gene duplication events are adaptive at phylogenetic time scales. Reconciling the evolutionary history of the gene family with the species tree and estimating rates of duplication and loss are the two most common approaches to analyzing gene duplication, but current methods are prone to assumptions that hinder a meaningful biological interpretation of parameter estimates. We proposed an approach that integrates both reconciliation and birth-death models to estimate the probabilities of different gene retention scenarios. Future research on the implementation of such an approach will bridge theory to practical application for a more comprehensive understanding of adaptive gene duplication, a key process in protein evolution.

Acknowledgements

This research was supported in part by DEB-1442142 to L.M.D., DEB-1701414 to L.M.D., D.A L., and L.R.Y., and DBI-1222940 to D.A.L. and L.L.

References

1. Hoegg S, Brinkmann H, Taylor JS et al (2004) Phylogenetic timing of the fish-specific genome duplication correlates with the diversification of teleost fish. J Mol Evol 59:190–203
2. Jaillon O, Aury J-M, Brunet F et al (2004) Genome duplication in the teleost fish *Tetraodon nigroviridis* reveals the early vertebrate proto-karyotype. Nature 431:946–957
3. Lien S, Koop BF, Sandve SR et al (2016) The Atlantic salmon genome provides insights into rediploidization. Nature 533:200–205
4. The Arabidopsis Genome Initiative (2000) Analysis of the genome sequence of the flowering plant *Arabidopsis thaliana*. Nature 408:796–815
5. De Bodt S, Maere S, Van De Peer Y (2005) Genome duplication and the origin of angiosperms. Trends Ecol Evol 20:591–597
6. Hollister JD (2015) Polyploidy: adaptation to the genomic environment. New Phytol 205:1034–1039
7. Liebeskind BJ, Hillis DM, Zakon HH (2015) Convergence of ion channel genome content in early animal evolution. Proc Natl Acad Sci U S A 112:E846–E851
8. Konrad A, Teufel AI, Grahnen JA et al (2011) Toward a general model for the evolutionary dynamics of gene duplicates. Genome Biol Evol 3:1197–1209
9. Hughes T, Liberles DA (2007) The pattern of evolution of smaller-scale gene duplicates in mammalian genomes is more consistent with neo- than subfunctionalisation. J Mol Evol 65:574–588
10. Hahn MW (2009) Distinguishing among evolutionary models for the maintenance of gene duplicates. J Hered 100:605–617

11. Sikosek T, Bornberg-Bauer E (2010) Evolution after and before gene duplication? In: Dittmar K, Liberles D (eds) Evolution after gene duplication. Wiley-Blackwell, Hoboken, NJ, pp 105–131
12. Zhao J, Teufel AI, Liberles DA et al (2015) A generalized birth and death process for modeling the fates of gene duplication. BMC Evol Biol 15:275
13. Teufel A, Zhao J, O'Reilly M et al (2014) On mechanistic modeling of gene content evolution: Birth-death models and mechanisms of gene birth and gene retention. Computation 2:112–130
14. Chothia C, Gough J, Vogel C et al (2003) Evolution of the protein repertoire. Science 300:1701–1703
15. von Heijne G (2006) Membrane-protein topology. Nat Rev Mol Cell Biol 7:909–918
16. Poolman B, Geertsma ER, Slotboom D-J (2007) A missing link in membrane protein evolution. Science 315:1229–1231
17. Nei M, Rooney AP (2005) Concerted and birth-and-death evolution of multigene families. Annu Rev Genet 39:121–152
18. Chen K, Durand D, Farach-colton M (2000) NOTUNG: a program for dating gene duplications. J Comput Biol 7:429–447
19. Berglund-Sonnhammer AC, Steffansson P, Betts MJ et al (2006) Optimal gene trees from sequences and species trees using a soft interpretation of parsimony. J Mol Evol 63:240–250
20. Doyon JP, Ranwez V, Daubin V et al (2011) Models, algorithms and programs for phylogeny reconciliation. Brief Bioinform 12:392–400
21. Sjöstrand J, Sennblad B, Arvestad L et al (2012) DLRS: gene tree evolution in light of a species tree. Bioinformatics 28:2994–2995
22. Hermansen RA, Hvidsten TR, Sandve SR et al (2016) Extracting functional trends from whole genome duplication events using comparative genomics. Biol Proced Online 18:11
23. Bielawski JP, Yang Z (2003) Maximum likelihood methods for detecting adaptive evolution after gene duplication. J Struct Funct Genom 3:201–212
24. Hahn MW, De Bie T, Stajich JE et al (2005) Estimating the tempo and mode of gene family evolution from comparative genomic data. Genome Res 15:1153–1160
25. Liu L, Yu L, Kalavacharla V et al (2011) A Bayesian model for gene family evolution. BMC Bioinformatics 12:426
26. Han MV, Thomas GWC, Lugo-Martinez J et al (2013) Estimating gene gain and loss rates in the presence of error in genome assembly and annotation using CAFE 3. Mol Biol Evol 30:1987–1997
27. Eulenstein O, Huzurbazar S, Liberles DA (2010) Reconciling phylogenetic trees. In: Dittmar K, Liberles D (eds) Evolution after gene duplication. Wiley-Blackwell, Hoboken, NJ, pp 185–206
28. Górecki P, Eulenstein O (2014) Refining discordant gene trees. BMC Bioinformatics 15:S3
29. Duncan RP, Husnik F, Van LJT et al (2014) Dynamic recruitment of amino acid transporters to the insect/symbiont interface. Mol Ecol 23:1608–1623
30. Dahan RA, Duncan RP, Wilson AC et al (2015) Amino acid transporter expansions associated with the evolution of obligate endosymbiosis in sap-feeding insects (Hemiptera: Sternorrhyncha). BMC Evol Biol 15:52
31. Ames RM, Money D, Ghatge VP et al (2012) Determining the evolutionary history of gene families. Bioinformatics 28:48–55
32. Arvestad L, Lagergren J, Sennblad B (2009) The gene evolution model and computing its associated probabilities. J ACM 56(7):44
33. Teufel AI, Liu L, Liberles DA (2016) Models for gene duplication when dosage balance works as a transition state to subsequent neo-or sub-functionalization. BMC Evol Biol 16:45
34. Nee S, May RM, Harvey PH (1994) The reconstructed evolutionary process. Philos Trans R Soc Lond Ser B Biol Sci 344:305–311
35. Niimura Y, Matsui A, Touhara K (2014) Extreme expansion of the olfactory receptor gene repertoire in African elephants and evolutionary dynamics of orthologous gene groups in 13 placental mammals. Genome Res 24:1485–1496
36. Pegueroles C, Laurie S, Albà MM (2013) Accelerated evolution after gene duplication: a time-dependent process affecting just one copy. Mol Biol Evol 30:1830–1842
37. Spielman SJ, Wilke CO (2015) The relationship between *dN/dS* and scaled selection coefficients. Mol Biol Evol 32:1097–1108
38. Mugal CF, Wolf JBW, Kaj I (2014) Why time matters: codon evolution and the temporal dynamics of *dN/dS*. Mol Biol Evol 31:212–231
39. Liberles DA, Teufel AI, Liu L et al (2013) On the need for mechanistic models in computational genomics and metagenomics. Genome Biol Evol 5:2008–2018
40. De Bie T, Cristianini N, Demuth JP et al (2006) CAFE: A computational tool for the study of gene family evolution. Bioinformatics 22:1269–1271

Chapter 4

Computational Prediction of De Novo Emerged Protein-Coding Genes

Nikolaos Vakirlis and Aoife McLysaght

Abstract

De novo genes, that is, protein-coding genes originating from previously noncoding sequence, have gone from being considered impossibly unlikely to being recognized as an important source of genetic novelty in eukaryotic genomes. It is clear that de novo gene evolution is a rare but consistent feature of eukaryotic genomes, being detected in every genome studied. However, different studies often use different computational methods, and the numbers and identities of the detected genes vary greatly. Here we present a coherent protocol for the computational identification of de novo genes by comparative genomics. The method described uses homology searches, identification of syntenic regions, and ancestral sequence reconstruction to produce high-confidence candidates with robust evidence of de novo emergence. It is designed to be easily applicable given the basic knowledge of bioinformatic tools and scalable so that it can be applied on large and small datasets.

Key words De novo genes, Gene birth, New gene evolution, Novel genes, ORF formation, Protein-coding genes, Genome-wide detection, Genome evolution

1 Introduction

New genes and protein functions are essential to the evolution of novel phenotypes, to the adaptation to new environments, and to the process of speciation [1]. Novel genes arise by reuse and recombination of pre-existing ones but can also originate from genomic sequences that were previously noncoding [2]. In the latter scenario, a new gene, either in its entirety or in part, emerges "de novo" (*see* [3] for detailed definitions). De novo gene emergence can be thought of as true "gene birth" and has the greatest potential to result in an entirely novel protein function, since the novel protein will be free of constraints present in pre-existing, already functional sequences [4]. Once considered so improbable as to be impossible, the origin of new protein-coding genes de novo has now been demonstrated in every eukaryotic lineage studied, and these new genes have been shown to integrate into central

Tobias Sikosek (ed.), *Computational Methods in Protein Evolution*, Methods in Molecular Biology, vol. 1851, https://doi.org/10.1007/978-1-4939-8736-8_4,

cellular functions (*see* [5] for a complete review). More and more, researchers are coming to recognize de novo emergence as a universal evolutionary phenomenon and to appreciate its potential as a mechanism of rapid phenotypic innovation [6] and as a genome-shaping force [7]. As the interest around de novo genes grows, so does the need for their accurate identification. This, however, is not a trivial task. De novo genes are a subset of "orphan genes" also known as "ORFans" or species-specific genes. These are genes that are found only in a single genome (or in a closely related group of genomes, in which case the term taxonomically restricted gene is used) and lack homologues in any other organism. Disentangling the evolutionary origins of orphan genes can be challenging [8], and the results highly depend upon the employed methodology. The initial de novo gene studies necessarily followed a stringent, painstaking approach involving a substantial amount of manual curation and multiple lines of evidence [9–12]. The goal was to provide solid proof that a functional species-specific gene had emerged from an ancestrally noncoding or nonfunctional region. Since then, multiple studies have adopted a different type of approach, with more relaxed criteria, but its advantages and pitfalls are still a matter of debate [13–17].

In this chapter, we will present what can be considered as a stringent best practice for the **identification of all protein-coding genes that have emerged entirely de novo in a single genome**. Conceptually this is the same as identifying genes that have originated de novo on a particular branch of a tree with the only difference being that the novel gene will be present in the organisms descended from that branch, and not in any outgroups. The methodology described here can be easily adapted to that type of study. The evolution of a novel gene requires, at the very minimum, the origin of an open reading frame (ORF) and regulatory signals for transcription and translation. Here we will mostly focus on the emergence of the ORF. Starting with the complete set of annotated protein-coding genes, we remove the ones with significant similarity to genes in other genomes, resulting in a set of species-specific genes. This set is then further reduced to the ones with identifiable sequence similarity to their orthologous noncoding regions in closely related outgroup genomes, from which an ancestral sequence can be inferred. Finally, the ones for which the inferred ancestral sequence can be shown to lack coding potential are retained as de novo gene candidates (*see* Fig. 1 for a complete outline). The approach is designed to (1) err on the side of caution (i.e., we endeavor to avoid false positives), (2) be applicable as widely and easily as possible, and (3) be scalable so that it can work for large and small datasets. It is for this reason that the method we are describing here is command-line oriented and specific commands are provided. Nevertheless, the choice of parameters that one has to set throughout the application of this

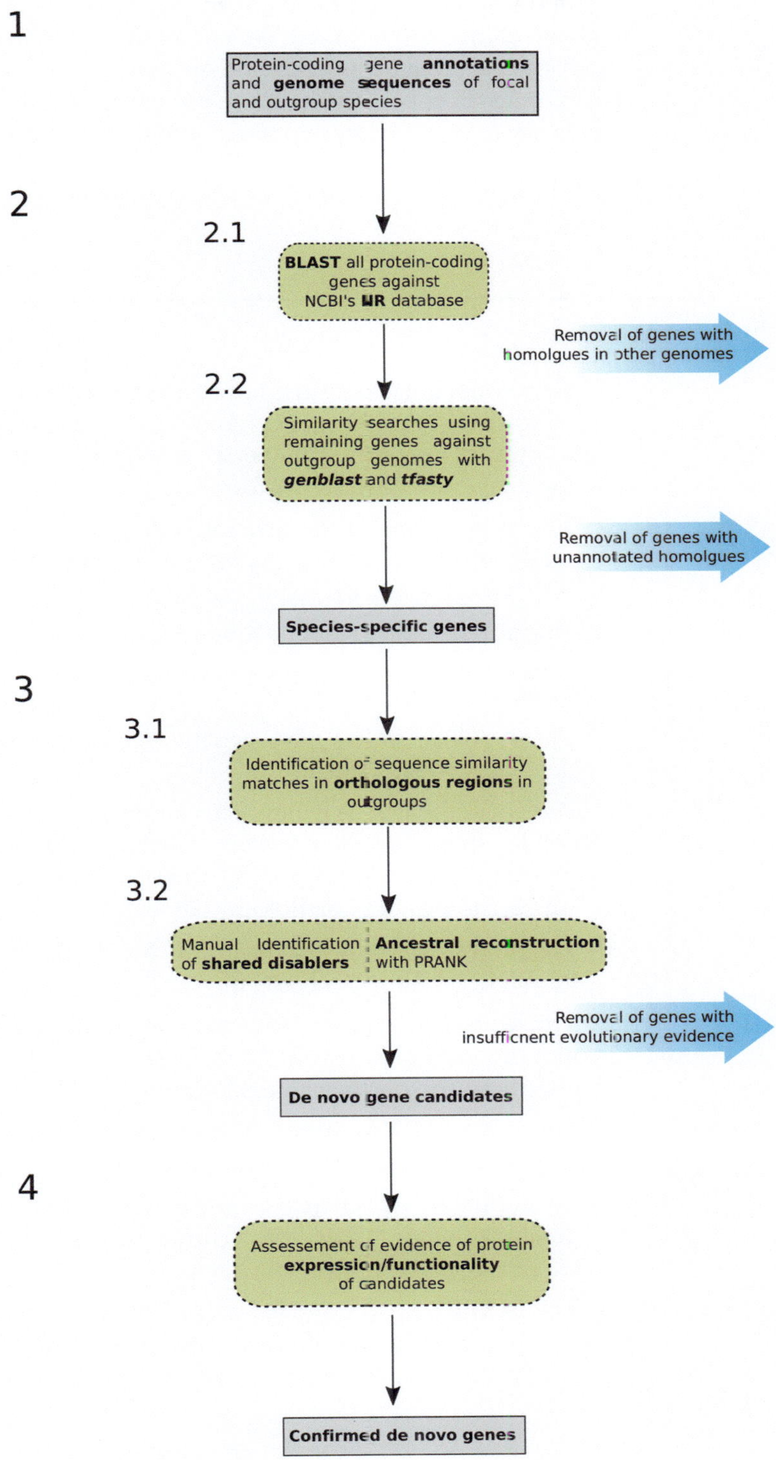

Fig. 1 An outline of the different steps of the methodology described here

method will depend on the context, the biological questions asked, and the choice of genomes. We encourage the reader to carefully study the different parameters and options of the tools before using them, regardless of whether these are mentioned in the examples. Application of the computational procedure requires familiarity with the command-line of UNIX-based systems (Bash) and some basic knowledge of a high-level scripting language such as Python or Perl.

2 Materials

1. Genome sequence for the focal genome (within which we want to identify de novo genes) and *at least two but ideally more* outgroup genomes (the most phylogenetically close to the focal genome), in FASTA format.
2. The annotations of the protein-coding genes in the focal and outgroup genomes in any of the commonly used formats (GENBANK, EMBL, GFF, etc.).
3. Nucleotide and protein sequences of all annotated protein-coding genes of the focal and outgroup genomes, in FASTA format.
4. The topology of the phylogenetic tree of the species in question, in Newick format.
5. Pairs of orthologous genes between the focal genome and each of the neighboring genomes. Orthology resources can be found at the Quest for Orthologs website (https://questfororthologs.org/orthology_databases).
6. The preformatted files of the NCBI NR database available from ftp.ncbi.nlm.nih.gov/blast/db/.
7. The BLAST [18] stand-alone programs available from ftp.ncbi.nlm.nih.gov/blast/executables/blast+/2.6.0/ (choose according to your machine's OS).
8. The *fasta* [19] programs available from http://fasta.bioch.virginia.edu/fasta_www2/fasta_down.shtml.
9. An alignment viewer such as *Jalview* (http://www.jalview.org/download) or *SeaView* (http://doua.prabi.fr/software/seaview).
10. The MAFFT [20] sequence alignment program available from http://mafft.cbrc.jp/alignment/software/ and the PRANK [21] program available from http://wasabiapp.org/software/prank/.
11. The *genblast* [22] program available from http://genome.sfu.ca/genblast/download.html.

12. The GNU *parallel* command line tool available from https://www.gnu.org/software/parallel/.
13. The *faSomeRecords* and *faSize* tools available from http://hgdownload.cse.ucsc.edu/admin/exe/.
14. The SAMtools suite of programs available from http://samtools.sourceforge.net/.
15. The EMBOSS suite of programs available from http://emboss.sourceforge.net/download/.
16. The *phyml* [23] phylogenetic reconstruction program available from http://www.atgc-montpellier.fr/phyml/binaries.php.
17. The *tantan* [24] tool for low-complexity masking in biological sequences.

3 Methods

Note: The choice of genomes to be studied is an initial point to be taken into account. As we will see later on, in order to conclusively show that a gene has emerged de novo, we need to be comparing genomes that have not diverged too much so that detectable similarity exists between intergenic regions and that some degree of synteny is conserved. It is also important to note that the pipeline described here is based on existing annotations, and the results will thus be heavily dependent upon the annotation methodology. High-quality annotation performed according to similar criteria for each genome under comparison is crucial for the robustness of the outcome. Homology to known, already annotated genes in other genomes is frequently used in annotation pipelines thereby biasing the annotation away from detection of species-specific genes [8]. Furthermore, short (usually <300 nt) open reading frames (ORFs) will often be ignored during annotation, and since young de novo genes are predicted to be short, this can lead to further underestimation of their true numbers. A more complete but significantly more time-consuming approach can be to include all ORFs above a certain length threshold, regardless of the annotation.

3.1 Retrieve the Data

The first step is to download the necessary data to a local machine, where all the subsequent computations will take place. To start the analysis, we need the genomic sequences in FASTA format, annotation files in one of the commonly used formats (GenBank, EMBL, GFF), and the amino-acid and coding DNA sequences (CDS) for all annotated protein-coding genes for the focal genome and the outgroup genomes. There exist multiple sources for genome data, and the choice depends on the genome being investigated. NCBI's Genome Resource (https://www.ncbi.nlm.nih.

gov/genome/) is one of the most rich ones. Ensembl is useful when studying vertebrate species (http://www.ensembl.org/index.html). Other organism-specific resources are sometimes simpler to navigate such as FlyBase (http://www.flybase.org/) or Saccharomyces Genome Database (https://www.yeastgenome.org/). It is a good practice to always work on the most recent releases of genomes and annotations. In order to download the actual files, one may have to access an ftp server for which clients such as FileZilla are particularly useful (https://filezilla-project.org/).

Let us now assume that we have downloaded the necessary files for the focal species and four phylogenetically closest outgroup species. We have created a main working directory, and we have moved the following files there:

1. The genome sequences of the outgroup species. Here we will assume that we are dealing with a chromosome-level assembly (one FASTA file per genome containing multiple records, being one record per chromosome): outgen1_chrom.fsa, outgen2_chrom.fsa, outgen3_chrom.fsa, and outgen4_chrom.fsa.
2. The annotation files in GFF format: focal.gff, outgen1.gff, outgen2.gff, outgen3.gff, and outgen4.gff.
3. The CDS files: focal_cds.fsa, outgen1_cds.fsa, outgen2_cds.fsa, outgen3_cds.fsa, and outgen4_cds.fsa.
4. The amino-acid sequence files: focal_aa.fsa, outgen1_aa.fsa, outgen2_aa.fsa, outgen3_aa.fsa, and outgen4_aa.fsa.

For practical purposes, it is a good idea to move the different types of files into separate directories. Let's assume that we have changed directory (*cd*) to where we have just placed our files. To create the sub-directories and move the files, we can execute the following commands:

```
$ mkdir ./chrom/ ./annot/ ./prot/ ./cds/
$ mv *_chrom.fsa ./chrom/ ; mv *gff ./annot/ ; mv *_cds.fsa ./cds/ ;
mv *_aa.fsa ./prot/
```

We are now ready to start the analysis.

Note: Throughout the chapter, we will assume that the headers of the FASTA files are formatted as follows:

>SEQUENCE_ID[SPACE]OTHER_INFORMATION

and that the SEQUENCE_ID part of the header is formatted as follows

[SPECIES][underscore][GENE_NAME]

For example : Hsap_SOMEGENENAME

We invite the reader to adjust the individual commands provided to accommodate the formatting of their FASTA headers.

3.2 Identify All Species-Specific Genes

By definition, de novo genes are derived from previously noncoding sequence. If we are considering recently evolved de novo genes, then they will be a subset of species-specific genes, having no homologs outside the focal genome. Thus the first step is to identify all species-specific genes in our focal genome. De novo genes can be categorized according to whether or not they contain any genetic material that is copied or descended from a pre-existing gene [3]. The most intuitive cases are type I de novo genes which are completely derived from noncoding sequence. However, depending on the purpose of the study, one may also be interested in de novo genes with small or large portions derived from sequences previously under selection (type II and type III de novo genes, respectively), and the similarly search criteria can be adjusted accordingly.

3.2.1 Similarity Search in NCBI's NR Database

We will first perform a similarity search of all the protein-coding genes in the focal genome against NCBI's NR database using the *blastp* executable from the BLAST suite of programs. A commonly used E-value threshold is 10^{-3}, generally accepted to result in a good trade-off between sensitivity and specificity. Using more permissive thresholds than 10^{-3} is very likely to produce a lot of false hits. The NR database is relatively large; the total size of the uncompressed files of the preformatted version, as of this writing, is 106 GB. To speed up the search, we can use BLAST's multithreading option with an appropriate number of threads (-num_threads X) to parallelize the alignment step, as well as the GNU *parallel* command line tool to further accelerate the search. Let's assume that we have downloaded the NR database preformatted files along with the taxonomy information file (taxdb.tar.gz) and we have uncompressed them and placed them in the NR_DIRECTORY directory. The command to execute is the following:

```
$ cat focal_aa.fsa | parallel --GNU --block 100k --recstart '>' --pipe 'blastp
-query - -db [NR_DIRECTORY]/nr -outfmt "6 std slen qlen stitle staxids sscinames"
-max_target_seqs 500 -num_threads [NUM_OF_CPUS] -evalue 0.001' > focal_nr_out.txt
```

The time that this command will take to complete will depend on the number and size of the query protein sequences. The "-max_target_seqs 500" argument is used to limit the number of target sequences reported, since we are only interested in the general presence or absence thereof, and not in the sequences themselves.

We then need to parse the *blastp* output and store in a file a list of all the genes without hits, notwithstanding self-hits. This list will then be compared to the full list of genes to extract a list of genes without any BLAST hit. We then select the protein sequences of these genes and store them in a separate file; the same is also done for the nucleotide sequences. Before proceeding, make sure that the FASTA files do not contain duplicate records.

First, we select all the lines of the output file that do not represent a match to a protein sequence of the focal genome itself. If, for example, the focal species' scientific name is "*Saccharomyces cerevisiae*," we execute the following:

```
$ grep -v "Saccharomyces cerevisiae" focal_nr_out.txt | cut -f 1 |
sort -u > focal_found_genes.txt
```

All the genes with at least one significant match outside of the focal taxon are now stored in the file "focal_found_genes.txt." Then we remove them from the list of all the genes:

```
$ grep ">" focal_prot.aa | cut -f 1 -d ' ' | tr -d '>' | sort -u > all_focal_genes.txt
$ comm -3 all_focal_genes.txt focal_found_genes.txt > focal_ss_names.txt
```

Our species-specific genes are now stored in the file "focal_ss_names.txt." We now need to extract their sequences from the initial FASTA files; this can be done using the utility *faSomeRecords*:

```
$ faSomeRecords focal_aa.fsa focal_ss_names.txt focal_ss_aa.fsa
$ faSomeRecords focal_cds.fsa focal_ss_names.txt focal_ss_cds.fsa
```

3.2.2 Similarity Search in Outgroup Genomes

Using the species-specific genes, we will perform a similarity search in the outgroup genomes' sequences. This is needed to ensure that no homologous, unannotated genes exist in the outgroup species and is also required for other downstream steps. First, it is a good idea to mask low-complexity segments on the chromosome sequences so that we avoid spurious matches. This can be achieved with the *tantan* program:

```
$ ls ./chrom/*chrom.fsa | parallel --GNU 'tantan -x N {} > {}.masked'
```

Next, we use the *genblast* program to identify gene models based on homology searches. First we need to set a variable to let the program know where to look for the auxiliary legacy BLAST programs, installed during *genblast* installation. Replace the part after the "=" by the path of the install directory of the *genblast* program in your machine.

```
$ export GBLAST_PATH="/users/User/Documents/tools/genBlast_v138_mac_os_X/"
```

We also need to copy the auxiliary file *alignscore.txt* from the *genblast* directory to our working directory.

Next, we execute the program once for every species that we want to search in.

```
$ ls ./chrom/*chrom.fsa.masked | parallel --GNU 'genblasta -p genblastg -q focal_s-
s_aa.fsa -t {} -e 0.001 -c 0.5 -cdna -gff -pid -o {}.gb'
```

We are next going to parse the ".gff" files generated by *genblast* and apply identity percentage (60%) and coverage percentage (50%) thresholds (also possible for coverage via the -c option of *genblast*, in bold above). You can adjust these thresholds, highlighted in bold in the following command, according to the parameters of your study:

```
$ for i in ./chrom/*gff ; do grep -v "^#" $i | grep "transcript" | cut -f 9 | tr ';'
'\t' | cut -f 2-4 | sed 's/[A-z]*\=//g' | awk '{ if ($2>60 && $3>50) { print } }' | cut
-f 1 ; done | sort -u > ss_missing_homologs.txt
```

The names of the species-specific genes for which a homologous gene model can be retrieved are now stored in file "ss_missing_homologs.txt." We next need to remove these sequences from the FASTA files and from our list:

```
$ faSomeRecords -exclude focal_ss_aa.fsa ss_missing_homologs.txt focal_ss_aa_final.
fsa
$ faSomeRecords -exclude focal_ss_cds.fsa ss_missing_homologs.txt focal_ss_cds_f-
inal.fsa
$ comm -3 focal_ss_names.txt ss_missing_homologs > focal_ss_names_final.txt
```

At this stage, we will also use the *tfasty* executable from the *fasta* suite of programs, to do a similarity search using as query the protein sequences of the species-specific genes and the masked chromosome sequences from the four outgroup species as subject. The command is run twice to get two different output formats (controlled by the "-m" argument): the tabular one which is useful for parsing and the detailed one which is useful for visual inspection and manual curation (*see* **Note 1**):

```
$ ls ./chrom/*chrom.fsa.masked | parallel --GNU 'tfasty36 -E 0.00001 -m 9C -p -s BP62
focal_ss_aa_final.fsa {} > {}.against.trgs.detailed.txt'
$ ls ./chrom/*chrom.fsa.masked | parallel --GNU 'tfasty36 -E 0.00001 -m 8 -p -s BP62
focal_ss_aa_final.fsa {} > {}.against.trgs.tabular.txt'
```

Next, we need to filter out low identity and low coverage hits. These thresholds can vary, but here we will apply a percentage identity threshold of 50% and a protein coverage threshold of 50%.

To apply a coverage percentage threshold, we first need to use the *faSize* utility to calculate the length of each species-specific sequence:

```
$ faSize -detailed focal_ss_aa_final.fsa | sort -k1 > focal_ss_aa_final_lengths.txt
```

Then, we need to append the corresponding sequence length to each line of our *tfasty* tabular output files:

```
$ for i in ./chrom/*against.trgs.tabular.txt ; do sort -k1 -o $i $i ; join -1 1 -2 1 $i
focal_ss_aa_final_lengths.txt | tr ' '  '\t' >  ${i%\.*}_with_lengths.txt ; done
```

Finally, we filter the files:

```
$ for i in ./chrom/*with_lengths.txt ; do awk '{ if ($3 > 50.0 && sqrt(($8-$7)*
($8-$7))/$13 > 0.5) { print } }' $i > ${i%.*}_filtered.txt ; done
```

The parameters highlighted in bold are, in that order, the percentage identity and query coverage and can be adjusted at will.

3.3 Showing that the Ancestral Sequence Lacked Protein-Coding Potential

The most robust evidence that a gene emerged de novo is provided when its ancestral sequence can be shown to have lacked protein-coding potential. In order to achieve that, one needs to detect the candidate gene's orthologous genomic sequences in at least two closely related outgroup species. At this point it is crucial to note that while this step is relatively straightforward for single-exon genes and genes with simple gene structures, it necessitates significantly more manual curation and involves more uncertainty in the case of complex gene structures. For simplicity and because in the majority of cases young genes are short and have very simple gene structures, we will consider only the single-exon gene case. To extend to multiple exons, each exon would have to be searched separately during the *tfasty* step described in Subheading 3.2.2. That would then allow the manual "stitching together" of the orthologous region of each exon into a single putative CDS which can then be aligned and inspected as described in the following subsections (*see* **Note 2**).

3.3.1 Retrieving the Orthologous Regions in Outgroup Genomes

Check if orthology or synteny information between the focal genome and the neighboring genomes exists in comparative genomic resource databases, a list of which can be found at the Quest for Orthologs website https://questfororthologs.org/orthology_databases. The goal at this step is to locate, if possible, the orthologous region of the candidate genes in closely related outgroup genomes (*see* Fig. 2), and for this we will need lists of orthologous pairs of genes, which can be extracted from one of the aforementioned databases. If orthology information is not available, the orthologous pairs need to be computed from scratch using a dedicated tool.

By combining the results of Subheading 3.2.2 and the orthology information, we can build multiple alignments of each candidate gene and its orthologous sequences. These MSAs will be used in the next step.

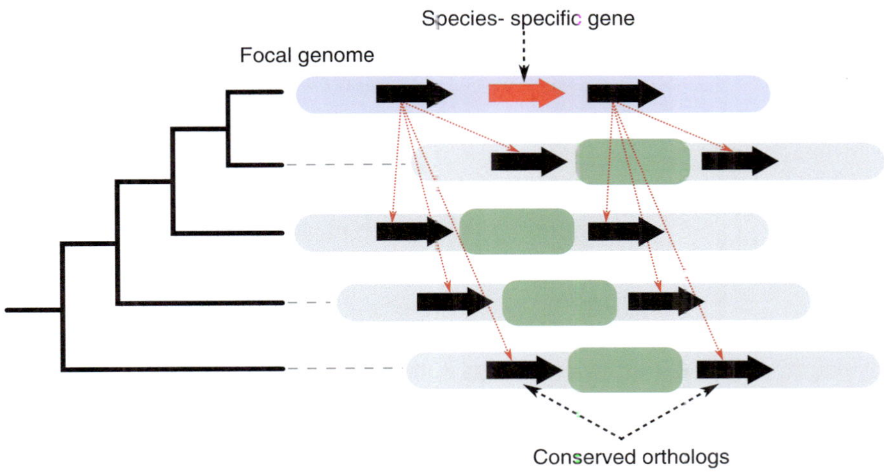

Fig. 2 Graphical representation of the configuration described in Subheading 3.1, in a scenario with four closely related outgroup species. The regions of interest are highlighted in green. Note that in actual cases, neighboring genes and regions might overlap, and so the region of interest might not be as clearly defined as in the example here

If the number of candidates is not excessive, one can opt to do it manually by grouping each candidate with all of its matches in the outgroup genomes and then removing the ones that do not fall in the predicted orthologous region. Here's a short bash script that accomplishes the initial, grouping part:

```
#!/bin/shcat focal_ss_names_final.txt |\
while read f_line ;
    do
        echo $f_line > faIn_temp ;
        faSomeRecords focal_ss_cds_final.fsa faIn_temp ${f_line}_ortho.fsa ;
        grep $f_line chrom/*tabular_with_lengths_filtered.txt |\
        while read line ;
        do
            START=$(echo $line | cut -f 9 -d ' ') ;
            END=$(echo $line | cut -f 10 -d ' ') ;
            REV_FLAG=0 ;
            if [ $START -gt $END ] ;
            then
                read START END <<< "$END $START" ;
                REV_FLAG=1 ;
            fi ;
            FILE_NAME=$(echo $line | cut -f 1 -d ':') ;
            CHROM_NAME=$(echo $line | cut -f 2 -d ' ') ;
            samtools faidx
            chrom/${FILE_NAME%%\.*}.chrom.fsa.masked ${CHROM_NAME}:$START-$END
            | tr ':' '_' > temp ;
```

```
            if [ $REV_FLAG -gt 0 ] ;
            then
               revseq -tag -sequence temp -outseq temp_rev ;
               mv temp_rev temp;
            fi ;
            cat temp | sed 's/^>\(.*\)/>\1 OUTGR/' | tr -d ':' >> ${f_line}_
            ortho.fsa ;
      done ;
done
```

To execute, after replacing with the correct file names, copy and paste this code into a file called *parse_candidate_gen_hits.sh*. Then, permit that file to be executed by running the following:

```
$ chmod +x parse_candidate_gen_hits.sh
```

Then execute the script. By doing so, we will have generated a file for each candidate (with the extension "_ortho.fas"), containing its sequence and all its genomic hits. Then, by manually opening and inspecting each file, we can cross-check the coordinates of each hit with the orthologous regions that have been inferred previously. At the same time, we can also verify that no sequencing gaps exist in the orthologous regions (*see* **Note 3**).

3.3.2 Reconstruction of the Ancestral Sequence

Next, we infer the state of the ancestral sequence. Almost always in the literature, this step would be performed manually, by "walking along" the alignment and trying to identify "shared disablers." Simply put, this consists of manually inspecting the alignments and identifying common ORF-disrupting mutations in the outgroup orthologous sequences. The ancestral state of these positions can then be parsimoniously inferred, revealing whether the ancestral sequence had an intact ORF. We consider the ORF as ancestrally not "intact" when it's shorter than 70% of the de novo candidate gene's length.

The first step is to align the sequences. The alignments can be generated using the *linsi* executable of MAFFT, as follows (as before, replace [CPU_NUM] by the number of cores in your machine):

```
$ ls *_ortho.fas | parallel --GNU 'linsi --quiet --thread
[CPU_NUM] {} > {.}.aln'
```

The aligned sequences are now stored in the files with the extension "_ort'ho.aln." By opening the alignment files with an alignment viewer, we can detect frameshift mutations (*see* Fig. 3a) and stop codons (Fig. 3b) in the orthologous sequences. It is important that we follow any change of reading frame that occurs

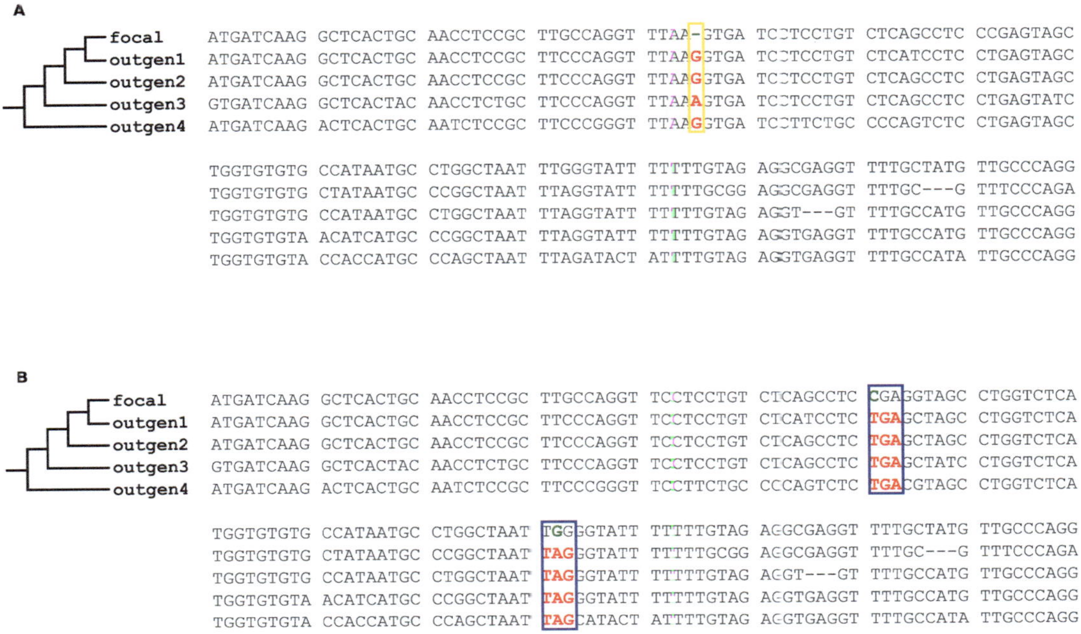

Fig. 3 Two hypothetical examples of shared disablers. (**a**) A single-nucleotide deletion (highlighted in yellow) that occurred along the terminal branch of the focal genome results in a frameshift, making available a different potential translation of the sequence that avoids the TGA stop codon that is in frame for the potential ORF in other species. (**b**) Two base substitutions in the focal genome lineage lead to the removal of two stop codons (nonsense-to-sense mutations, highlighted in blue) leading to the formation of a longer ORF

in the alignment, either by getting the correct reading frame translation from our viewer or by searching the match of the specific region in the *tfasty* detailed output files that we generated in Subheading 3.2.2. Note that in the literature, even if no specific "shared disablers" are detected, de novo emergence may still be assumed if all outgroup orthologous sequences have at least one mutation disrupting the ORF at least 70% of its length.

Cases of de novo emergence inferred manually can be corroborated by performing true ancestral sequence reconstruction. This is especially useful when the alignment contains combinations of multiple frameshifts and nonsense-to-sense mutations. It can be achieved using the PRANK multiple sequence alignment program, but before it can be done, we need to infer phylogenetic trees to use as guide trees in the reconstruction.

First, we must convert our FASTA multiple alignment files in PHYLIP format using the *seqret* tool from the EMBOSS package:

```
$ ls *_ortho.aln | parallel --GNU 'seqret -sequence FASTA::{} -outseq PHYLIP::{.}.
phylip'
```

Then, we need to save the topology of the species tree (assumed to be known and robust) in Newick format in a file, here called "species_tree.nwk." For our hypothetical example, that tree would be the following:

```
$ echo "(relgen4,(relgen3,(relgen2,(relgen1,focal))));" > species_tree.nwk
```

Then, we can build trees, one for each alignment, that follow the species topology by executing the following command:

```
$ ls *phylip | parallel --GNU 'phyml -i {} -d nt -v e -o lr -c 4 -a e -b 0 -f e -u species_tree.nwk'
```

Finally, we can run PRANK with the following command (*see* **Note 4**):

```
$ ls *_ortho.fas | parallel --GNU 'prank -d={} -showanc -showevents -F -once -t={.}.phylip_phyml_tree.txt -o={.}'
```

Once PRANK has successfully finished running, we can inspect the ancestral sequences for intact ORFs. The ancestral sequences can be found aligned to the extant ones in the files with the extension "anc.fas" and have numbers as identifiers. You can see which ancestor corresponds to which branch in the Newick tree files ending with ".anc.dnd," which you can open with a phylogenetic tree viewer such as *SeaView*. *See* Fig. 4 for the reconstruction of a toy example.

When the number of candidates is prohibitive for one-by-one manual investigation, we can automatize the previous step. To do this, we first need to "de-align" the ancestral sequences that are found in the files produced by PRANK ending with "anc.fas":

```
$ for i in *anc.fas ; do cat $i | sed "/^[^>]/s/-//g" > $i.daln ; done
```

Next, we need to check that no intact ORF existed ancestrally. In the PRANK output files, the names of the ancestors follow the format #NUMBER# (#1#, #2#, etc.). The ORFs we are interested in are in the reading frame of the candidate, so all we need to do is translate the ancestral sequences and look for stop codons in the relevant ancestors, meaning the ones that are part of the focal genome's lineage. In the example of Fig. 3, that would be all of them (1,2,3,4), but in other cases, some ancestors might be irrelevant (i.e., not in the lineage of interest).

First we translate the sequences using *transeq* from EMBOSS:

```
$ for i in *anc.fas.daln ; do transeq -sequence $i -outseq $i.prt ; done
```

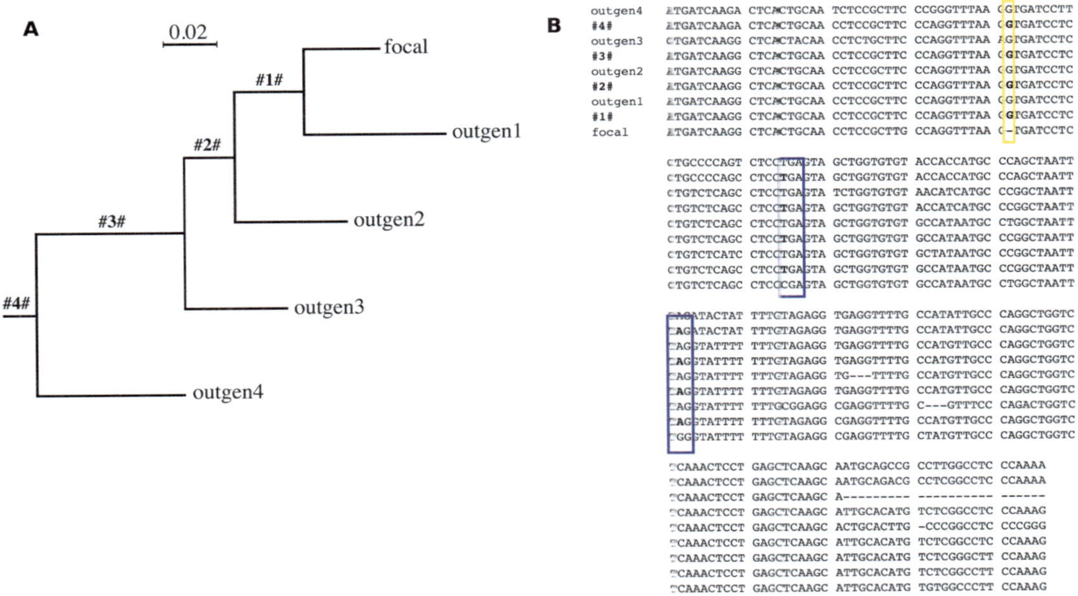

C

```
#4#:  MIKTHCNLRFPGLR*SSCPSLLSSWCVPPCPANLDTIL*R*GFAILPRLVSNS*AQAMQTPRPPKX
#3#:  MIKAHCNLRFPGLR*SSCLSLLSSWCVPSCPANLGIFL*R*GFAMLPRLVSNS*AQALHMSRPPKX
#2#:  MIKAHCNLRFPGLR*SSCLSLLSSWCVP*CPANLGIFL*R*GFAMLPRLVSNS*AQALHMSRPPKX
#1#:  MIKAHCNLRFPGLR*SSCLSLLSSWCVP*CPANLGIFL*RRGFAMLPRLVSNS*AQALHMSRPSKX
focal:MIKAHCNLRLPGLSDPPVSASRVAGVCHNAWLIWVFFCRGEVLLCCPGWSQTPELKHCTCVALPK
```

Fig. 4 Inferring de novo emergence for a hypothetical example alignment combining both frameshifts and nonsense-to-sense mutations of Fig. 3, using ancestral reconstruction. (**a**) The phylogenetic tree generated by PRANK, the same as the input guide tree but with the assigned ancestor identifiers. (**b**) The PRANK alignment containing the extant and ancestral sequences. The positions of interest are highlighted as in Fig. 3. The ancestral states at these positions confirm the results of the manual inference. (**c**) The "de-aligned," translated ancestral proteins and the focal extant one (*see* below for relevant commands), allowing to verify that no intact ORF existed before the focal leaf of the tree

Then, we extract the ancestors' sequences, measure the size of the ORF, and apply the filter (*see* **Note 5**):

```
$ for i in *prt ;do CAND_LEN=$(grep ${i%_*} focal_ss_aa_final_lengths.txt | cut -f 2) ; cat $i | tr '#' '_' > temp ; mv temp $i ; echo -n ${i%_*}$'\t' ; RESULT=$(grep ">_[1234]_" $i | tr -d '>' | while read line ; do echo -n $CAND_LEN$'\t' ; echo -n $line | tee temp_names.txt ; faSomeRecords $i temp_names.txt temp_seq.fasta ; cat temp_seq.fasta | grep -v ">" | tr -d '\n' | egrep -o "^.*?\*|^[^\*]*$" | head -1 | wc -c ; done | awk '{if ($3/$1 > 0.7) {print}}') ; if [ -z "$RESULT" ] ; then echo "Keep" ; else echo "Discard" ; fi ; done > results.txt
```

The final results are stored in the file *results.txt*. The candidates tagged with "keep" are the ones that do not have an intact ORF larger than **70%** of the length of the candidate, in any of the ancestors **1**, **2**, **3**, and **4**. To change the length threshold, change the number in bold (0.7) in the previous command line. Adjust the relevant *grep* matching pattern accordingly to get the desired ancestors.

3.4 Showing that the Candidate Genes Are Protein-Coding/Functional

By default, young de novo genes are only present in a single genome. They therefore lack one of the main lines of evidence that is put forth to prove that a piece of DNA is indeed protein-coding or functional, namely, conservation due to purifying selection. It is thus necessary to provide some evidence that the putative de novo gene expresses a functional protein and is not simply a spurious result. This is a difficult issue which depends on the very definition of a functional protein-coding gene, complicated further by pervasive transcription [25] and pervasive translation [26]. Consequently, what constitutes sufficient evidence of "coding-ness" and functionality will depend on the context and the assumptions of the study.

In the absence of specific functional annotation or identification of the protein by other means, experimental evidence for its expression can be provided if proteomics or ribosome profiling data are available. Major repositories of results from mass spectrometry proteomic experiments include PRIDE (https://www.ebi.ac.uk/pride/archive/) and PeptideAtlas (http://www.peptideatlas.org/) where one can check whether peptides matching a de novo candidate have been experimentally detected (*see* [27] for additional information on mass spectrometry proteomic resources). Ribosome profiling result databases include GWIPS (http://gwips.ucc.ie/) and RFPDB (http://sysbio.sysu.edu.cn/rpfdb/index.html) (*see* [28] for a more complete list of resources). Alternatively, one can calculate what are referred to as "coding scores" based on the intrinsic sequence composition of the candidates [17, 29]. This is sometimes done as part of the initial genome annotation, as is the case, for example, in the *Saccharomyces* [30]. One possible solution is the CPAT tool [31] which has been developed to be applied on entire transcripts but can work on single ORFs as well. It involves training the model on data of known coding and noncoding sequences first (unless your genome is one of human, fly, mouse, or zebrafish, already available at http://lilab.research.bcm.edu/cpat/) and so will not be covered in detail here. At any rate, a sequence annotated as coding remains at the very least a candidate, even if no other evidence exists of its functionality.

4 Notes

1. If this command crashes, replace *parallel* with a normal *for* loop.
2. Here we will describe, in a general fashion, how to look for the aforementioned evidence. However, some of this work is already done for some genomes, especially the ones from model organisms. For example, in human, we can directly extract the orthologous genomic sequences in vertebrate genomes for any part of the genome. To do this, all we need is to download the multiple alignments of mammalian genomes to human, available from http://hgdownload.soe.ucsc.edu/goldenPath/hg38/multiz20way/ in .MAF files (according to our needs, we can choose files with more or less genomes; *see* http://hgdownload.soe.ucsc.edu/downloads.html). Then by converting the human genome coordinates of our species-specific genes in BED format and feeding them to "*mafsInRegion*" (downloadable from http://hgdownload.cse.ucsc.edu/admin/exe/), we can directly generate the desired alignments. A similar shortcut is possible in genomes like *S. cerevisiae*, *D. melanogaster*, and other model organisms.
3. If the number of candidates is high and/or manual curation is not desirable, we will need to write a script to perform the task. The resulting files are assumed to be the same. Here's what a straightforward algorithm might look like in pseudocode, but any effective solution is acceptable:

```
for candidate in species_specific_gene_list :
    find gene_neighbors immediately downstream (candidate_-1) and upstream (candi-
    date_+1)
    for genome in outgroup_species_list :
        find orthologs for candidate_-1 (ortho_candidate_-1) and candidate_+1
        (ortho_candidate_+1)
        if both ortho_candidate_-1 and ortho_candidate_+1 exist :
            Extract their coordinates, calculate the interval
            if they are contiguous on their chromosome :
                for hit in genomic_hits_of_in_genome :
                    if hit within the interval :
                        Extract the genomic sequence
                        Store in file
```

4. In order for the last two commands to work, the names at the leaves of the tree files and the names of the sequences in the FASTA files must match. That means we first must remove the part of the header following the species name in the de novo candidate's record in the FASTA file and accordingly remove any extra information from the header of the orthologous matches.

Moreover, if, for example, we have a candidate for which orthologous matches are found in only three of the four outgroup species but we still want to proceed with the reconstruction, we must adjust the file "*species_tree.nwk*" by removing the extra species.

5. *faSomeRecords* has trouble with sequence identifiers that start with "#," so we replace "#" by "_."

References

1. Long M, Betrán E, Thornton K et al (2003) The origin of new genes: glimpses from the young and old. Nat Rev Genet 4:865–875
2. Andersson DI, Jerlström-Hultqvist J, Näsvall J (2015) Evolution of new functions de novo and from preexisting genes. Cold Spring Harb Perspect Biol 7:a017996
3. McLysaght A, Hurst LD (2016) Open questions in the study of de novo genes: what, how and why. Nat Rev Genet 17:567–578
4. Schlötterer C (2015) Genes from scratch—the evolutionary fate of de novo genes. Trends Genet 31:215–219
5. McLysaght A, Guerzoni D (2015) New genes from non-coding sequence: the role of de novo protein-coding genes in eukaryotic evolutionary innovation. Philos Trans R Soc Lond B Biol Sci 370:20140332
6. Li D, Dong Y, Jiang Y et al (2010) A de novo originated gene depresses budding yeast mating pathway and is repressed by the protein encoded by its antisense strand. Cell Res 20:408–420
7. Vakirlis N, Sarilar V, Drillon G et al (2016) Reconstruction of ancestral chromosome architecture and gene repertoire reveals principles of genome evolution in a model yeast genus. Genome Res 26:918–932
8. Tautz D, Domazet-Lošo T (2011) The evolutionary origin of orphan genes. Nat Rev Genet 12:692–702
9. Cai J, Zhao R, Jiang H et al (2008) De novo origination of a new protein-coding gene in Saccharomyces cerevisiae. Genetics 179:487–496
10. Heinen TJAJ, Staubach F, Häming D et al (2009) Emergence of a new gene from an intergenic region. Curr Biol 19:1527–1531
11. Knowles DG, McLysaght A (2009) Recent de novo origin of human protein-coding genes. Genome Res 9:1752–1759
12. Levine MT, Jones CD, Kern AD et al (2006) Novel genes derived from noncoding DNA in Drosophila melanogaster are frequently X-linked and exhibit testis-biased expression. Proc Natl Acad Sci 103:9935–9939
13. Carvunis A-R, Rolland T, Wapinski I et al (2012) Proto-genes and de novo gene birth. Nature 487:370–374
14. Domazet-Lošo T, Carvunis A-R, Albà MM et al (2017) No evidence for phylostratigraphic bias impacting inferences on patterns of gene emergence and evolution. Mol Biol Evol 34:843–856
15. Moyers BA, Zhang J (2014) Phylostratigraphic bias creates spurious patterns of genome evolution. Mol Biol Evol 32:258–267
16. Moyers BA, Zhang J (2016) Evaluating phylostratigraphic evidence for widespread de novo gene birth in genome evolution. Mol Biol Evol 33:1245–1256
17. Vakirlis N, Hebert AS, Opulente DA et al (2018) A molecular portrait of de novo genes in yeast. Mol Biol Evol 35:631–645
18. Altschul SF, Madden TL, Schäffer AA et al (1997) Gapped BLAST and PSI-BLAST: a new generation of protein database search programs. Nucleic Acids Res 25:3389–3402
19. Pearson WR, Wood T, Zhang Z et al (1997) Comparison of DNA sequences with protein sequences. Genomics 46:24–36
20. Katoh K, Standley DM (2013) MAFFT multiple sequence alignment software version 7: improvements in performance and usability. Mol Biol Evol 30:772–780
21. Löytynoja A, Goldman N (2008) Phylogeny-aware gap placement prevents errors in sequence alignment and evolutionary analysis. Science 320:1632–1635
22. She R, Chu JS-C, Wang K et al (2009) GenBlastA: enabling BLAST to identify homologous gene sequences. Genome Res 19:143–149
23. Guindon S, Delsuc F, Dufayard J-F et al (2009) Estimating maximum likelihood phylogenies with PhyML. Methods Mol Biol 537:113–137

24. Frith MC (2011) A new repeat-masking method enables specific detection of homologous sequences. Nucleic Acids Res 39:e23–e23
25. Clark MB, Amaral PP, Schlesinger FJ et al (2011) The reality of pervasive transcription. PLoS Biol 9:e1000625
26. Ingolia NT, Lareau LF, Weissman JS (2011) Ribosome profiling of mouse embryonic stem cells reveals the complexity and dynamics of mammalian proteomes. Cell 147:789–802
27. Chen T, Zhao J, Ma J et al (2015) Web resources for mass spectrometry-based proteomics. Genomics Proteomics Bioinformatics 13:36–39
28. Wang H, Wang Y, Xie Z (2017) Computational resources for ribosome profiling: from database to Web server and software. Brief Bioinform. https://doi.org/10.1093/bib/bbx093
29. Ruiz-Orera J, Messeguer X, Subirana JA et al (2014) Long non-coding RNAs as a source of new peptides. Elife 3:e03523
30. Scannell DR, Zill OA, Rokas A et al (2011) The awesome power of yeast evolutionary genetics: new genome sequences and strain resources for the Saccharomyces sensu stricto genus. G3 (Bethesda) 1:11–25
31. Wang L, Park HJ, Dasari S et al (2013) CPAT: coding-potential assessment tool using an alignment-free logistic regression model. Nucleic Acids Res 41:e74

Chapter 5

Coevolutionary Signals and Structure-Based Models for the Prediction of Protein Native Conformations

Ricardo Nascimento dos Santos, Xianli Jiang, Leandro Martínez, and Faruck Morcos

Abstract

The analysis of coevolutionary signals from families of evolutionarily related sequences is a recent conceptual framework that provides valuable information about unique intramolecular interactions and, therefore, can assist in the elucidation of biomolecular conformations. It is based on the idea that compensatory mutations at specific residue positions in a sequence help preserve stability of protein architecture and function and leave a statistical signature related to residue-residue interactions in the 3D structure of the protein. Consequently, statistical analysis of these correlated mutations in subsets of protein sequence alignments can be used to predict which residue pairs should be in spatial proximity in the native functional protein fold. These predicted signals can be then used to guide molecular dynamics (MD) simulations to predict the three-dimensional coordinates of a functional amino acid chain. In this chapter, we introduce a general and efficient methodology to perform coevolutionary analysis on protein sequences and to use this information in combination with computational physical models to predict the native 3D conformation of functional polypeptides. We present a step-by-step methodology that includes the description and application of software tools and databases required to infer tertiary structures of a protein fold. The general pipeline includes instructions on (1) how to obtain direct amino acid couplings from protein sequences using direct coupling analysis (DCA), (2) how to incorporate such signals as interaction potentials in C_α structure-based models (SBMs) to drive protein-folding MD simulations, (3) a procedure to estimate secondary structure and how to include such estimates in the topology files required in the MD simulations, and (4) how to build full atomic models based on the top C_α candidates selected in the pipeline. The information presented in this chapter is self-contained and sufficient to allow a computational scientist to predict structures of proteins using publicly available algorithms and databases.

Key words Coevolution, Structure-based model, Energy landscapes, Molecular dynamics, Protein Folding, Structure prediction

1 Introduction

The knowledge of the three-dimensional fold of proteins is fundamental for the goal of understanding biological function. Therefore, a major goal of structural biology is the determination of the tertiary protein structures, either by experimental or computational

Tobias Sikosek (ed.), *Computational Methods in Protein Evolution*, Methods in Molecular Biology, vol. 1851, https://doi.org/10.1007/978-1-4939-8736-8_5, © Springer Science+Business Media, LLC, part of Springer Nature 2019

methods. Computational methods have been for years able to confidently predict protein folds for structures for which experimental models of similar proteins are available. More recently, the development of the theory and software to elucidate contact information from a collection of related amino acid sequences has enhanced the prediction of protein folds to a more prominent level [1, 2]. One idea behind coevolutionary coupled amino acid sites is that during evolution proteins experience mutations which are only partially deleterious for function or retain marginal stability and, thus, are propagated through generations. However, compensatory mutations can appear in the population of individuals, restoring or even enhancing the fitness of the protein, and the associated pair of mutations might become the dominant genotype. Many of these compensatory mutations have structural origins, that is, they counteract the loss of favorable interactions between amino acid side chains which are structurally in contact [3]. The emergence and propagation of compensatory mutations between structurally related residues can be studied through statistical methods to identify which are the pairs of residues that coevolve and provide clues about physical residue contacts from collections of evolutionary-related sequences [4].

The use of coevolutionary-derived contact information has been successful in improving the quality of 3D-fold predictions to a precision which was not attainable for proteins for which no similar experimental structure was available [1, 5, 6]. The identification of potential contacts between pairs of residues contributes to the prediction of protein structure [7–16] and protein interactions [8, 17–19] and even assists the study of protein folding [20, 21], conformational changes [22, 23], and complex-forming mechanisms [24, 25].

Here we present a pratical guide on how to predict protein fold structures using contact information obtained from coevolutionary signals at amino acid positions in combination with coarse-grained physical models of proteins. Initially, we describe how to infer residues that are physically interacting from sequence alignments along a protein family using direct coupling analysis (DCA) [1]. DCA is a global probabilistic model used to calculate an estimate of the amino acid pairs that are directly coupled in sequence families. We provide the instructions and references for the implementation of DCA, starting from data collection to the final computation of amino acid couplings. Then we show how to combine secondary structure prediction with residue contacts predicted by DCA to define a force field to simulate folding using structure-based models (SBMs) [26, 27]. Finally, we describe how to visualize the simulated model and evaluate the structural similarity between the predicted and experimental coordinates. The procedures described here are expected to facilitate and generalize research focused on the prediction of tertiary and quaternary molecular structures of proteins.

The combination of coevolutionary information for amino acids with structure-based models exhibits reproducible model prediction with fold-level accuracy. The prediction performance with raw input secondary structure for simulation suggests that the method is versatile for any kind of secondary structure. This methodology and the tools we provide are convenient and are aimed to enable the scientific community to solve problems in structural bioinformatics.

2 Materials

In this section we describe all computational tools and web-based resources that will be necessary to generate inputs, run molecular simulations for folding, as well as evaluate and visualize results.

2.1 UniProt Server

UniProt is a comprehensive genomic sequence and analysis database containing a large dataset of protein sequences, accompanied by diverse biological annotations for biological function, domain composition, subcellular location, and possible molecular interactions [28]. This database is freely accessible at www.uniprot.org.

2.2 Pfam Server

Pfam is a public database of protein families (groups of evolutionary-related proteins) that features thousands of annotated entries and is constantly being curated and updated [29, 30]. This database provides multiple sequence alignments (MSAs) generated using Hidden Markov Models, a statistical modeling paradigm based on dynamic Bayesian networks. Entries for identified families and respective MSAs can be accessed at http://pfam.xfam.org.

2.3 Direct Coupling Analysis (DCA) Server

Correlations in amino acid mutations can be identified by applying statistical inference in MSAs [31]. Several algorithms and refinements to perform this analysis have been developed by our group and others, with good performance [1, 4, 11]. Direct coupling analysis is an efficient technique to compute coevolutionary signals that is able to disentangle indirect and direct correlations that are hard to distinguish by usual correlation analysis such as mutual information [1, 5]. Identification of direct couplings is especially important in structure predictions, since they can be interpreted as regions that are physically interacting in the functional state of macromolecules. A web server for estimation of direct correlations in residue pairs using the DCA approach is available at http://dca.rice.edu/ and http://morcoslab.org. Moreover, a standalone version of DCA is also available in the same web page. A more detailed description of DCA usage can be found in another publication [5].

2.4 HMMER Profiling

HMMER is a software framework developed for the identification of homologous sequences in biological databases and to perform efficient multiple sequence alignments. Its methodology employs probabilistic Hidden Markov Models for pattern recognition. HMMER is comprised of several tools such as *hmmscan* for query sequence against homologs databases and *hmmbuild*, to generate a MSA. This software is freely accessible and can be downloaded at http://hmmer.org/.

2.5 SBM Generation Script

Structure-based models (SBMs) are simplified representations of molecular interactions that allow to accelerate the conformational search of biomolecular systems in molecular dynamic simulations [25, 26, 32]. They are based on the idea of minimally frustrated energy landscapes, which describe realistic protein-folding mechanisms [33]. Due to the SBM formulation, interaction data such as coevolutionary couplings can be added to SBMs to drive conformational search [6, 19]. We provide an efficient script to automatically generate coordinates and SBM models based solely on the primary sequence and predicted secondary structure of a protein. This script incorporates coevolution data as energy potentials with optimized parameters for protein-folding simulations. This procedure provides an initial unfolded model and the topology and force-field files necessary to run folding simulations. The scripts can be downloaded at the following link, http://morcoslab.org, under the tab *Research -> Software Tools*. Python scripts and parameter files to assist the process of generation and visualization of protein models from coevolution data are also provided.

2.6 Gnuplot

Gnuplot is an open source and portable software for graphical visualization of data and mathematical functions. It is very intuitive and native to most Linux distributions, in addition to running in all major operating systems. To check if Gnuplot is already included in your operating system, type *gnuplot* in an operating system's terminal. Details about download and usage can be found at http://gnuplot.sourceforge.net/.

2.7 Jpred Server

While recognition of tertiary structure is still challenging and methodologies are under development, tools for estimation of local secondary motifs are very mature and display good accuracy [34, 35]. One of the most recent implementations of secondary structure prediction methods, which was selected for this protocol, is Jpred version 4 [36]. This approach uses multilayered neural networks that are trained to identify secondary motif patterns from primary sequence as input. Sequence queries can be submitted to the Jpred server at http://www.compbio.dundee.ac.uk/jpred/index.html. Alternatively, other prediction methodologies can also be used, such as PSIPRED, Spider2, and PredictProtein [37–39].

2.8 GROMACS with Support to Gaussian Potentials

GROningen MAchine for Chemical Simulations (GROMACS) is a molecular dynamic platform designed for computational simulations of biomolecular systems, such as proteins, nucleic acids, lipids, and small organic molecules [40, 41]. It is very fast and compatible to many force-field representations and supports parallelization through central and graphical processing units (CPUs and GPUs, respectively). As an open-source package, versions with diverse additional supports can be independently developed. In the protocol described herein, a GROMACS version with support for SBMs containing Gaussian potentials for representation of distance constraints will be used. This modified version can be downloaded at http://smog-server.org/extension/gromacs-4.5.4_sbm1.0.tar.gz. Details about GROMACS installation procedure for a diverse set of operating systems can be found at http://www.gromacs.org/Documentation/Installation_Instructions_4.5. This protocol will assume a single precision GROMACS installation.

2.9 Protein Data Bank

Protein Data Bank (PDB) is a structural database containing the 3D relative coordinates of thousands of biological macromolecules obtained from experimental techniques such as X-ray and electron crystallography, nuclear magnetic resonance (NMR), and cryo-electron microscopy (cryo-EM). Most of the database comprises protein structures, but also nucleic acid structures are found [42, 43]. This source of structural genomic information is in continuous growth, with more than 130, 000 entries by August 2017. It can be accessed at https://www.rcsb.org/pdb/home/home.do.

2.10 UCSF Chimera

UCSF Chimera is a free-of-charge software for visualization and analysis of molecular structures. It is very flexible on supporting a wide range of file formats and provides many tools for data analysis, such as sequence alignment, charge distribution and energy optimizations, interpolation of structure conformations, and solvation. Moreover, Chimera also provides a platform to generate high-quality scientific images and animations. It can be downloaded at https://www.cgl.ucsf.edu/chimera/download.html.

2.11 LovoAlign

Structural similarity of two protein conformations can be computed by rigid-body structural alignment. Here we use LovoAlign, a free and fast package for structural alignment [44]. It supports several well-established similarity measurements, such as the root-mean-square deviation (RMSD), template modeling score (TM-score), and global distance test (GDT). Directions for downloading the latest version of LovoAlign and installation can be found at http://www.ime.unicamp.br/~martinez/lovoalign.

2.12 REMO Server

REconstruct atomic MOdel (REMO) is a program that generates full atomic coordinates of proteins from C_α models. This reconstruction process employs an algorithm to optimize the network of hydrogen bonding in backbone and side chains [45]. A server and a stand-alone version of REMO are available at https://zhanglab.ccmb.med.umich.edu/REMO/. Moreover, alternative approaches to reconstruct all-atom models can also be used in this practice [46, 47].

3 Methods

We describe a step-by-step protocol to run folding simulations driven by coevolutionary data. The protocol will be divided into several main steps, which should be executed in succession (*see* **Note 1**). We will assume the reader is working on a Unix-like environment or operating system. Further details about software compatibility and installation in specific versions of operating systems can be found in the list of links provided in *Materials* section.

3.1 Protein Sequence and Family

As a first example of protein folding prediction, we will work with the human transmembrane protein aquaporin-1. This macromolecule is part of a large family of transport proteins that controls the flow of water through the cell [48–50]. The presence of water channels allows cells to increase and decrease intracellular water content at a faster rate than diffusion through membranes, and functional defects are related to diseases [48]. Moreover, structural characterization of transmembrane proteins such as aquaporin-1 is particularly challenging for experimental studies, due to limitations of available techniques. Therefore, computational methodologies like the one presented here in are particularly useful. Also, while many full-atom ab initio methods have been successful in predicting the conformation of small proteins (<100 residues), the study of larger systems such as aquaporin-1 is still intangible even when using high-performance computation. Together, these attributes justify aquaporin-1 as an excellent example for folding studies that is compatible to the complexity of real-case research problems.

In order to infer coevolutionary signals for a given system, we need the primary amino acid sequence of the target. To perform this task, we will access the UniProt server (Subheading 2.1) and type *human aquaporin 1* in the main query field at the top of the page, with the option UniProtKB selected. A list of entries for aquaporins organized by relevance will be provided. We should select the top relevant entry that corresponds to our target of study (entry P29972, AQP1_HUMAN). By clicking in the link of this entry, another page will display all information already annotated about this molecule (such as biological function and molecular interactions). In order to perform coevolutionary analysis, we need the amino acid sequence and the Pfam family of this protein. In section *Sequences* of the aquaporin-1 UniProt page, we can download its amino acid sequence corresponding to isoform 1 using the *Fasta* button (next to entry code P29972-1, 269 residues, click with the right button, and go to *Save link as...*). Open the downloaded *.fasta* file with any text processing program, and check the entry code and length.

Next we should identify to which family aquaporin-1 belongs and obtain a curated MSA containing many representative proteins of this family. Still at the page resulting from our query, we can identify in the section *Family and Domains* the aquaporin-1 families annotated for distinct databases. For the Pfam annotation, we observe a unique hit for a family named MIP (major intrinsic proteins), with Pfam accession code PF00230. By clicking in this link, we are redirected to Pfam page for this MIP family (Subheading 2.2). An initial page provides a summary of functional details about the members of this family. In the *Alignments* tag, we see all MSAs available considering distinct proteome databases (*see* **Note 2**). We will use the full MSA already curated by Pfam with the maximum number of sequences. In the section *Format an alignment* of this tab, select the options Alignment = Full, format = FASTA, order = Tree, Sequence = Inserts lower case, and gaps = mixed (Gaps as "." or "-"), and download option. Save the MSA file by clicking on *Generate* button. Figure 1 depicts the settings used to generate this MSA file.

The MSA file generated by Pfam contains many position inserts that are not important for the analysis of coevolution, but in some cases, it can result in intractable file sizes. To optimize the size of these working files keeping the same amount of information, a python script named *msa_clean.py* is provided (Subheading 2.5). Use this script with the following command in the terminal to generate a new MSA with reduced file size.

```
python msa_clean.py PF00230_full.txt
```

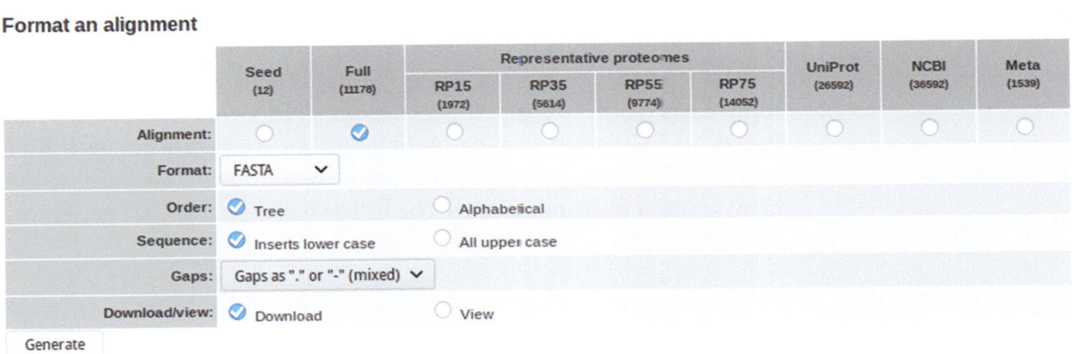

Fig. 1 Parameters to download MSA for a specific family in Pfam server. The generated format is compatible with DCA

This step will remove insertions from the original MSA file *PF00230_full.txt* downloaded from Pfam and rewrite this information in a file named *PF00230_full_clean*. Furthermore, in order to identify the region and specific residues from the human aquaporin-1 sequence that are part of MIP family, we also need to know the profile used to generate the MSA (Pfam employs Hidden Markov Models for multiple alignments). We can download the model named *MIP.hmm* at the *Curation & model* tab using the download link provided at the bottom.

3.2 Coevolutionary Information from Direct Coupling Analysis

From the MSA of the MIP family, we can now perform a statistical analysis over many sequences using direct coupling analysis (DCA). This analysis will provide a quantitative measure of the direct couplings of each possible pair of positions in the MSA, which can be used to infer physical contacts in aquaporin-1. To perform this step, go to the DCA server website (*see* Subheading 2.3), and use the full MSA for MIP family obtained (*PF00230_full_clean*) as input. Use the DCA button on the *Workbench* tab. Choose a job name in the first blank field, and upload the MSA file in *FASTA_IN* option. Set the *relative_pseudo_count* option to 1.0 and the *homolog_radius* parameter to 0.8. Details about the influence of these parameters to DCA performance are provided in the same submission page as well as in [1]. Finally, run the coevolutionary analysis with the button *Start Job*.

After finishing the calculations, the DCA server will provide a preliminary heatmap with an overview of highly ranked correlation distribution in MSA. Furthermore, a file containing the list of direct correlations for each pair of position in the family can be downloaded from the link named *DI_values.DI*. Save this file to your computer, and open it using any text editor of choice. Check that three columns are provided for each line, corresponding to the pair positions in MSA and to the direct correlation level named as direct information (DI), respectively. Notice that lines in the DCA list are sorted by pair numbering starting at 0, but we want to distinguish only the pairs with maximum DI values (third column). Moreover, since adjacent positions in MSA usually correspond to neighboring residues in the backbone, then we expect an intrinsic and trivial high correlation among those residues. In order to sort this list based on DI values and to filter neighboring positions with local correlations, we will open a terminal in the folder containing the downloaded files and execute the following command to generate a new list or pairs (*see* **Note 3**):

```
awk '{if($2-$1>4)print $1+1,$2+1,$3}' DI_values.DI | sort -g -k 3 -r > PF00230_full_
ranked.DI
```

Open the file just created (*PF00230_full_ranked.DI*), and check if this new list presents the desired properties (descending

DI values for nonadjacent pairs). It should be noticed that aquaporin-1 is only a representative member of MIP family. Now that we have the top coevolving pairs in MIP family, we need to identify the matching pairs of MSA in the respective residue pairs of our target aquaporin-1. In other words, we need to map the position of MIP family in human aquaporin-1 sequence. For this step, we will use HMMER software (Subheading 2.4). Still in the same terminal in the folder where the files are located, use the *hmmpress* tool from HMMER to compile the MIP HMM profile:

```
hmmpress MIP.hmm
```

Next, use the *hmmscan* option to search for MIP domain position in aquaporin-1 sequence:

```
hmmscan -o P29972_MIP_scan --notextw MIP.hmm P29972.fasta
```

Check the file *P29972_MIP_scan* using a text editor to verify the correspondence between MSA positions and the residues in the protein sequence. Now, we should use the generated mapping to rewrite top correlated MSA pairs as residue pairs in aquaporin-1 sequence. In order to complete this task, copy to the current folder the python script *map_dca.py* provided in the link of Subheading 2.5, and run the following command:

```
python map_dca.py P29972_MIP_scan PF00230_full_ranked.DI
```

This script will generate two files: (i) a new list of top coevolutionary couplings corresponding to residue pairs in the target protein sequence and (ii) a reference table for each matching position in the MSA and the protein sequence (*see* **Note 4**). You can check the creation of these files by typing the command "*ls*" in the terminal. Open the first file generated (named as *P29972_MIP_scan_ranked_matched.DI*), and compare with the previous MSA list obtained from DCA (*PF00230_full_ranked.DI*). Next, check the generated reference file for mapping (*P29972_MIP_scan_reference.txt*), and observe that some insertions can occur in the map list (represented by "-") that are not in the original MSA of assigned family and vice versa. These instances demonstrate why a detailed pair matching is necessary instead of only knowing the absolute beginning and end of the location of a family in a protein (*see* **Note 5**).

Moreover, to get an idea about the level of interaction diversity in our predicted coevolutionary data, we can visualize the top-ranked couplings using residue-residue plots. To do this, we should open a Gnuplot terminal (Subheading 2.6) by typing *gnuplot* in the working shell terminal and plot the first 200 obtained

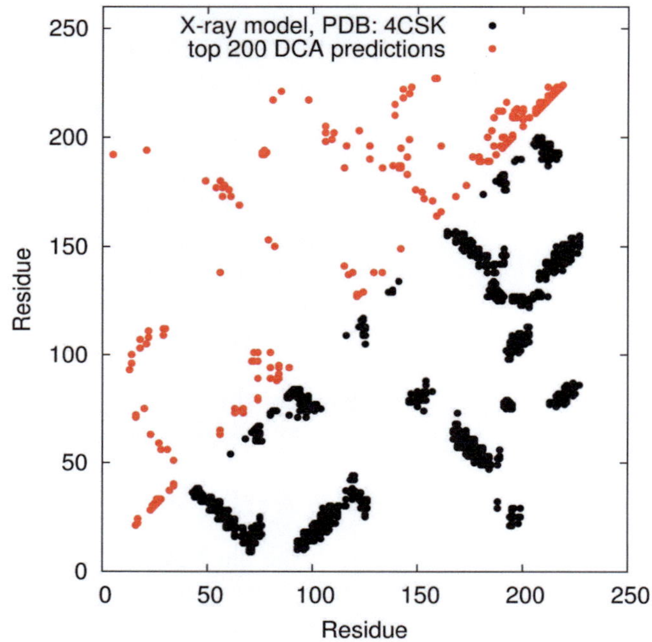

Fig. 2 Representation of top 200 DCA contacts for human aquaporin-1 (UniProt code: P29972). The lower right distribution corresponds to the native contacts from the reference structure PDB: 4CSK

residue couplings with top DI values using the following command in the new terminal:

```
plot 'P29972_MIP_scan_ranked_matched.DI' every::::200 pt 7 notitle
```

This step will generate a residue-residue map pattern similar (but probably not identical, since Pfam MSAs are constantly being updated) to the one shown in Fig. 2. If we have abundant information in the MSA, this pattern of highly coevolving pairs should ideally recover the physical contact map of the target protein at its functional state [1, 5, 51]. As known in statistics, the accuracy of inference is directly proportional to the level of nonredundant information available [52, 53]. Therefore, the larger the number of distinct but still homologous sequence representatives that are available in our MSA, the more accurate the DCA predictions tend to be. In order to evaluate the variation of coevolution signal distribution, try to plot different numbers of top-ranked pairs replacing the *200* in the last command executed in Gnuplot, and see how it changes in comparison to the top DCA pairs shown in Fig. 2. The lower triangular part of Fig. 2 shows the native contacts from the reference structure PDB: 4CSK.

DCA uses a statistical estimation method to compare the strength of direct couplings among positions in MSA. Since this estimation depends on intrinsic features of the MSA (length, evolutionary pressure, sequence variation, and number of instances), the estimated DI value of a residue pair is relative to its counterparts, rather than an absolute value. As a consequence, the accuracy of DCA predictions (rate of true positive interactions) follows a continuum descending profile from the top predicted coupling that allows internal differentiation. Furthermore, interaction couplings from distinct phenomena can contribute and be mixed in DCA signals, such as signals from folding, oligomerization, and allosteric movements. Therefore, to estimate the ideal set of coevolutionary information is not trivial. An initial guess suggested here, based on practical folding examples, is to select the top-ranked pairs proportional to the length of the MSA of the protein family (i.e., the number of DCA pairs equals the MSA length). If we check the file generated by *hmmscan* for MIP family search in aquaporin-1 (*P29972_MIP_scan*), we can observe a match of 226 positions of this family to aquaporin-1 sequence (from position 2 until 227 in MSA of MIP family). Therefore, we will select the top 226 DCA pairs for folding prediction. This can be done using the following command in the working terminal:

```
head -n 226 P29972_MIP_scan_ranked_matched.DI > DCA_top226
```

where the file *DCA_top226* has the top 226 coupled residue pairs separated by more than 4 residues in the amino acid chain.

3.3 Secondary Structure Prediction

After quantifying coevolutionary couplings for our protein of study aquaporin-1, we can now use this information for folding prediction. But first, we need to describe how neighbor residues are locally organized and packed. Clues about local information (secondary structure) can also be inferred from coevolution analysis; however, several mature and more accurate methodologies are available. These approaches use statistical, knowledge-based, or machine-learning techniques (or combinations of them) to predict packing patterns. One robust tool for secondary structure prediction that was selected for this methodology is Jpred (Subheading 2.7).

In order to predict the secondary structure of aquaporin-1, go to the Jpred server website (Subheading 2.7), delete the sample sequence in the query field, paste the amino acid sequence of the protein obtained in Subheading 3.1 and saved in a *.fasta* file (copy the text inside *P29972.fasta* and paste in query field). Run the analysis with the button *Make prediction*. In the following page returned after job submission, a message should appear showing validated structures matching this query sequence. Since we are performing this practice as a validation example for the method, ignore this page by selecting the *continue* button. Notice this option is not true in cases where the architecture of a protein of study is unknown.

Finally, after the job is complete, a new page displaying an overview of the predicted secondary structure will be provided. Select the option *View simple results in HTML*, copy the entire two lines corresponding to the sequence and respective secondary structure, and save this information in a new text file as *P29972_Jpred.txt*.

3.4 Structure-Based Models from Coevolution

Once we have obtained sources of data for secondary structure and coevolutionary couplings as proxy for secondary- and tertiary-fold levels, respectively, we can merge these data and use it as input for structure prediction. First, in order to run folding with molecular dynamic simulations using structure-based models, we need to generate: (i) an initial unfolded model for aquaporin-1 and (ii) a topology file containing all details about the physical properties of the system, such as the mass of atoms, covalent bonds, interaction potentials between specific pairs obtained from DCA, and energies involved in conformational movements (variations in bond lengths, angles, and dihedrals). Further general information about physical models and the approach of molecular dynamic simulations can be found elsewhere [26, 54–56]. To generate these files, we should use a python script provided at the link described in Subheading 2.5 (file named as *dcasbm.tar.gz*). Extract the file provided in a folder inside the working directory using the following command:

```
tar -xzvf DCASBM.tar.gz
```

Now, run the script to generate the protein model and topology files including DCA signals using the following command:

```
python dcasbm/dcasbm.py P29972_Jpred.txt DCA_top226
```

A value for maximum force factor of DCA interactions will be asked. This value will control the force of pairwise potentials that will drive folding and should be set based on the number of coevolutionary restrictions. For a small dataset of restrictions, stronger force factors should be chosen in order to be powerful enough to ensure high conformational convergence and vice versa. A good estimate for the strength of coevolutionary interactions (suggestion based on empirical data) is to consider the force factor equivalent to the ratio of the family length (L) over the number of DCA pairs (factor = L/|DCA|). In our case of aquaporin-1, both values are already the same (226), and therefore the force factor should be set to 1. After setting this parameter, running the script will generate two new files (*P29972_Jpred_calpha.gro* and *P29972_Jpred_calpha.top*). Open these new files with any text editor, and check their organization. The first one (*P29972_Jpred_-calpha.gro*) corresponds to the coordinates of the unfolded model of aquaporin-1 that will be used as an initial conformation for folding simulation. The second file (*P29972_Jpred_calpha.top*) comprises the physical parameters and DCA interactions that will be used to drive conformational search.

3.5 Folding Simulations

With the initial coordinates and topological description files for aquaporin-1 at hand, we can now run folding simulations. This procedure will drive the association of protein residue pairs identified as highly coevolving by the application of a combination of repulsive and attractive energy potentials (SBMs) in molecular simulations [19, 25, 26, 32]. In this process, the system temperature is gradually reduced, in an approach known as simulated annealing, until the total conformational energy of the system achieves a minimum where most predicted interactions are satisfied yielding to a native-like folded conformation [6, 25, 33].

A GROMACS version with support for Gaussian potentials (*see* Subheading 2.8) is required. We should use the files created in the last section as input for the simulation using the generated topology and coordinate files for aquaporin-1 and the file with the simulation parameters downloaded as part of the package provided in Subheading 2.5 (file named *sbm_calpha_SA.mdp*) (*see* **Note 6**). This procedure can be done using the following script:

```
grompp -f sbm_calpha_SA.mdp -c P29972_Jpred_calpha.gro -p
P29972_Jpred_calpha.top -o run.tpr
```

Completion of this step will generate a binary compiled file (run.tpr) that can be used for simulation in GROMACS. Next, a folding simulation can be carried out in GROMACS using the following command:

```
mdrun -deffnm run -pd
```

This simulation procedure will take near a couple of hours to be completed on a Linux-based PC desktop (tested in a AMD FX-8370E cpu, running Ubuntu version 16.04). Furthermore, when multi-core CPUs are available, computational time to finish the process can be reduced using parallelization. Having knowledge about the architecture of processors available, simply add the tag "*-nt x*" in the end of last command, where *x* is the number of available computer threads. Finally, after finishing the folding simulation, a few new files will be generated. Among those we can find the conformation with the final folded protein coordinates (*run.gro*).

3.6 Analysis of Predicted Models

After developing a folding simulation for aquaporin-1, we can now check the predicted conformation and compare with respective experimental models available in the literature. We can look for experimental models for aquaporin-1 in the Protein Data Bank (PDB, Subheading 2.9). At the PDB web page, type the code 4CSK into the search field. You should be redirected to an entry for a human aquaporin-1 X-ray structure model. At the initial *Summary* tab for this entry, many details such as quality measurements of experimental data and related literature can be observed [57]. In order to verify that this protein corresponds exactly to the same aquaporin-1 we have been working with, go to the tab *Sequence*, and look for the annotated UniProt reference code (UniProtKB P29972) that corresponds exactly to the UniProt code of the sequence for our system. Having this information verified, we can now download the coordinates of the X-ray model for human aquaporin-1 by selecting the button *Download Files* located on the right of the page and the option *PDB file*. Save the generated file in pdb format (*4csk.pdb*) to the current working directory of the shell terminal. For an initial visual analysis, open this experimental model together with the final folding conformation for aquaporin-1 using UCSF Chimera package (Subheading 2.10) with the following command:

```
chimera run.gro 4csk.pdb
```

Alternatively, this step can be completed by opening Chimera and selecting the files for both models with the option *File* > *Open* inside the Chimera environment. Set a ribbon representation for both models selecting the following option in Chimera panel: *Presets* > *Publication 4 (depth-cued, licorice)*. Next, we can align the orientation of both structures using the tool *MatchMaker* from Chimera. To do that, go to *Tools* > *Structure Comparison* > *MatchMaker*, and select the experimental and predicted models as reference and structure to match, respectively, on the new window that should be displayed (*see* Fig. 3). Keep the default options of all secondary parameters and align both structures by clicking on the *OK* button.

As a last step in our study, we will evaluate the structural similarity of predicted and experimental aquaporin-1 models using quantitative parameters. The selection of a robust method for structural comparison is crucial to evaluate the performance and identify the correct predictions from molecular modeling methods. For example, although the root-mean square deviation (RMSD) is a well-established measure to estimate relative coordinate differences in molecules, it is not the best similarity measure to compare protein folds. Since RMSD computes the absolute differences on atom coordinates between models, this approach fails dramatically to identify equivalent folding patterns when the system displays some intrinsic flexibility such as those observed in hinges and shearing mechanisms [58]; in other words, RMSD is too sensitive to discrepancies in subsets of the structures. A more appropriate

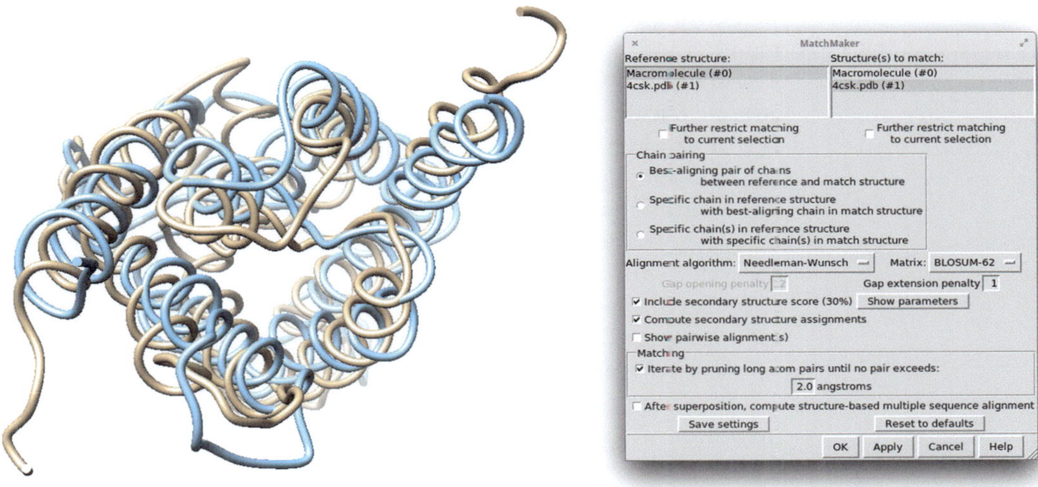

Fig. 3 Visual comparison of predicted and experimental conformations of human aquaporin-1 using UCSF chimera

measure of the similarity between conformations of protein models is the template modeling score (TM-score) [59]. This parameter indicates the similarity between two protein models by a value ranging between 0 and 1. As a general rule, values below 0.2 represent uncorrelated structures, which could be obtained by a random search. TM-score values greater than 0.5 indicate that the models share the same overall fold, while a value of 1.0 is achieved when both models converge to the very same 3D structure [59].

In this chapter, we will use the software LovoAlign (Subheading 2.11) to calculate TM-score values [44]. Since LovoAlign requires input models to be represented in the standard *.pdb* format, we first need to transform our predicted model generated by GROMACS from *.gro* to *.pdb* format. This step can be completed using a GROMACS tool called editconf, with the command:

```
editconf -f run.gro -o run.pdb
```

Now, having LovoAlign installed, one can compute the TM-score between the full-length predicted and experimental models of aquaporin-1 (run.pdb and 4csk.pdb, respectively) by simply using the following command on the working terminal:

```
lovoalign -p1 4csk.pdb -p2 run.pdb -seqnum
```

The *-p1* and *-p2* options define the two structures to be aligned, and the *-seqnum* option assures that the sequence alignment will match exactly to the corresponding residue numbering in the PDB files, which is the case here since the two proteins share the same sequence except for possible deletions. After running this command, a log message will be displayed on terminal, providing the TM-score between models (in the field named *FINAL SCORE:*) and specific details about the calculation (sequence matching, number of residues considered, and alignment cycles). Moreover, this log data for alignment can be saved by adding the extra "> *name.log*" identifier at the end of the command (*name* can be chosen as desired). If the level of coevolutionary information obtained is enough (which is true in our case) and the folding simulation converged to a configuration where DCA geometrical restrictions are maximized, the predicted and experimental models should be very similar and result in a TM-score greater than 0.5 (*see* **Note 7**).

In addition, we can restrict the comparison of folding conformations only to the protein regions that are covered by the family and have assigned coevolutionary restraints. To calculate the TM-score of a specific region of protein sequence, first we need to provide to LovoAlign the fragment that should be analyzed in each model. In order to generate the aquaporin-1 models containing only the region corresponding to MIP family, use the python script *getregion.py* provided in Subheading 2.5 and the information obtained from hmmscan analysis (Subheading 3.2) by running the following commands:

```
python getregion.py P29972_MIP_scan run.pdb

python getregion.py P29972_MIP_scan 4csk.pdb
```

This procedure will generate two new pruned models containing only the family region of the protein. Finally, we can compute the TM-score along the conserved family region of aquaporin-1 with the following command:

```
lovoalign -p1 4csk_pruned.pdb -p2 run_pruned.pdb -seqnum
```

Since we are restricting the analysis only to the sequence region where coevolutionary information exists, the final TM-score should be higher than the one observed when considering the full model. Divergences in TM-scores when comparing full models and their corresponding fragments restricted to family region should be consistent to the relative size of fragments outside family coverage. In this case of aquaporin-1, since the domain encompass almost the full sequence of aquaporin-1, the best TM-score obtained in our tests considering only the domain is very close to the one for the full model (TM-scores of 0.70 and 0.68, respectively).

3.7 Rebuilding All-Atom Protein Structures

In this step, we will generate an all-atom protein structure from our predicted C_α folded model. To perform this step, we will make use of the REMO server [45]. Go to the web page of REMO

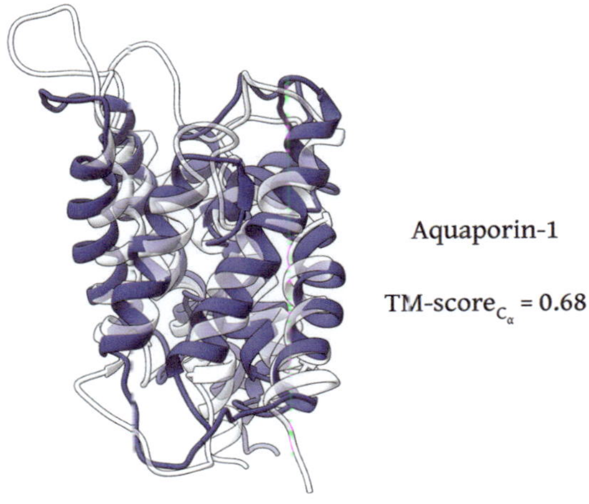

Fig. 4 All-atom model for aquaporin-1 generated by REMO using the coarse-grained predicted fold model. The full protein TM-score between predicted (purple) and experimental (white) models is 0.68, considering C_α carbons

(Subheading 2.12), and upload the final folded model obtained for aquaporin-1 converted to pdb format (run.pdb) using the *Browse...* button. Fill the e-mail form, and submit the process using the button *run REMO* in the bottom of the page. After a couple of minutes, you should receive the link with REMO results in the e-mail that you provided. Download the generated model and visualize it using Chimera (Fig. 4).

3.8 Additional Examples

For additional test cases, we suggest the reader to try the same protocol for folding prediction using other interesting biological systems. Some suggestions are provided:

1. The human small G protein RAP2A. UniProt code P10114 and PDB code 1KAO.
2. The bacterial ABL transporter, a larger transmembrane protein. UniProt code P06609 and PDB code 1L7V.
3. The receiver domain of DesR from *Bacillus subtilis*, a transcriptional regulatory protein. UniProt code O34723 and PDB code 4LE1.

In this chapter we discussed a convenient methodology to predict 3D coordinates of folded protein structures based on coevolutionary information and molecular dynamics. We provide resources and software tools that are free to access and instructions on how to use them to elucidate structures with high TM-scores. The information contained in this chapter is general and can be used to study and infer structures of many proteins for which no structural information is available. With the advent of sequencing technologies, we expect that the applicability of these techniques will become more relevant and will have a positive impact in the quest to accurately characterize and uncover conformations and functional properties of biomolecules.

4 Notes

1. A folder containing all input files necessary to run the protocols described here can be found at the link provided in Subheading 2.5 under the name *example_aqp1.tar.gz*.
2. The described protocol is limited to multiple sequence alignments (MSAs) with available Pfam annotations. Nevertheless, sequences with no assigned Pfam domains can also be used to generate customized MSAs to infer coevolution MSAs. In this case, MSAs can be built using the Jackhmmer search tool inside the HMMER server.
3. If direct coupling analysis is performed using the local Matlab implementation also available in the DCA server, please keep in mind that the output DI file has indexing starting at 1 instead of 0 as in the web server. The Matlab implementation also gives mutual information (MI) values as output; therefore the DI values are shown in the 4th column as opposed to the third as in the DCA server.
4. The python script provided for sequence mapping will only consider the first family found by *hmmscan* in the target protein sequence. For systems with multiple families, mapping for consecutive families can be done manually or using the same procedure described in Subheading 3.2 by deleting the data of previously found families in the output of hmmscan analysis.
5. Mapping an MSA to protein sequence is not only a process of getting the beginning and the end position of domain in protein. Since insertions in domains and proteins can occur, the correspondence of each MSA position in a sequence residue should be considered.
6. When using GROMACS in a double-precision installation version, the same protocol for running folding simulations (Subheading 3.5) can be done by replacing the tag *grompp* by *grompp_d* in commands.
7. When comparing a predicted protein structure with its respective experimental model, make sure that the amino acid sequences used for prediction and the experimental structure use the same residue numbering to avoid indexing errors.

Acknowledgments

The authors thank financial support from the São Paulo Research Foundation (FAPESP) (Grants 2015/13667-9, 2010/16947-9, 2013/05475-7, and 2013/08293-7) and funding from the University of Texas at Dallas.

References

1. Morcos F, Pagnani A, Lunt B et al (2011) Direct-coupling analysis of residue coevolution captures native contacts across many protein families. Proc Natl Acad Sci U S A 108: E1293–E1301
2. Hamilton N, Burrage K, Ragan MA, Huber T (2004) Protein contact prediction using patterns of correlation. Proteins 56:679–684
3. Ivankov DN, Finkelstein AV, Kondrashov FA (2014) A structural perspective of compensatory evolution. Curr Opin Struct Biol 26:104–112
4. de Juan D, Pazos F, Valencia A (2013) Emerging methods in protein co-evolution. Nat Rev Genet 14:249–261
5. Morcos F, Hwa T, Onuchic JN, Weigt M (2014) Direct coupling analysis for protein contact prediction. Methods Mol Biol 1137:55–70
6. Sulkowska JI, Morcos F, Weigt M et al (2012) Genomics-aided structure prediction. Proc Natl Acad Sci 109:10340–10345
7. Hopf TA, Colwell LJ, Sheridan R et al (2012) Three-dimensional structures of membrane proteins from genomic sequencing. Cell 149:1607–1621
8. Ovchinnikov S, Kamisetty H, Baker D (2014) Robust and accurate prediction of residue-residue interactions across protein interfaces using evolutionary information. Elife 3: e02030
9. Kamisetty H, Ovchinnikov S, Baker D (2013) Assessing the utility of coevolution-based residue-residue contact predictions in a sequence- and structure-rich era. Proc Natl Acad Sci U S A 110:15674–15679
10. Skwark MJ, Abdel-Rehim A, Elofsson A (2013) PconsC: combination of direct information methods and alignments improves contact prediction. Bioinformatics 29:1815–1816
11. Ekeberg M, Lövkvist C, Lan Y et al (2013) Improved contact prediction in proteins: using pseudolikelihoods to infer Potts models. Phys Rev E Stat Nonlinear Soft Matter Phys 87:012707
12. Hayat S, Sander C, Marks DS, Elofsson A (2015) All-atom 3D structure prediction of transmembrane β-barrel proteins from sequences. Proc Natl Acad Sci U S A 112:5413–5418
13. Marks DS, Hopf TA, Sander C (2012) Protein structure prediction from sequence variation. Nat Biotechnol 30:1072–1080
14. Jones DT, Singh T, Kosciolek T, Tetchner S (2015) MetaPSICOV: combining coevolution methods for accurate prediction of contacts and long range hydrogen bonding in proteins. Bioinformatics 31:999–1006
15. Sadowski MI, Taylor WR (2013) Prediction of protein contacts from correlated sequence substitutions. Sci Prog 96:33–42
16. Hopf TA, Morinaga S, Ihara S et al (2015) Amino acid coevolution reveals three-dimensional structure and functional domains of insect odorant receptors. Nat Commun 6:6077
17. Schug A, Weigt M, Onuchic JN et al (2009) High-resolution protein complexes from integrating genomic information with molecular simulation. Proc Natl Acad Sci U S A 106:22124–22129
18. Tamir S, Rotem-Bamberger S, Katz C et al (2014) Integrated strategy reveals the protein interface between cancer targets Bcl-2 and NAF-1. Proc Natl Acad Sci U S A 111:5177–5182
19. dos Santos RN, Morcos F, Jana B et al (2015) Dimeric interactions and complex formation using direct coevolutionary couplings. Sci Rep 5:13652
20. Morcos F, Schafer NP, Cheng RR et al (2014) Coevolutionary information, protein folding landscapes, and the thermodynamics of natural selection. Proc Natl Acad Sci U S A 111:12408–12413
21. Mallik S, Kundu S (2015) Co-evolutionary constraints of globular proteins correlate with their folding rates. FEBS Lett 589:2179–2185
22. Morcos F, Jana B, Hwa T, Onuchic JN (2013) Coevolutionary signals across protein lineages

help capture multiple protein conformations. Proc Natl Acad Sci U S A 110:20533–20538
23. Sfriso P, Duran-Frigola M, Mosca R et al (2016) Residues coevolution guides the systematic identification of alternative functional conformations in proteins. Structure 24:116–126
24. Cheng RR, Morcos F, Levine H, Onuchic JN (2014) Toward rationally redesigning bacterial two-component signaling systems using coevolutionary information. Proc Natl Acad Sci U S A 111:E563–E571
25. Jana B, Morcos F, Onuchic JN (2014) From structure to function: the convergence of structure based models and co-evolutionary information. Phys Chem Chem Phys 16:6496–6507
26. Noel JK, Levi M, Raghunathan M et al (2016) SMOG 2: a versatile software package for generating structure-based models. PLoS Comput Biol 12:e1004794
27. Noel JK, Whitford PC, Sanbonmatsu KY, Onuchic JN (2010) SMOG@ctbp: simplified deployment of structure-based models in GROMACS. Nucleic Acids Res 38: W657–W661
28. UniProt Consortium (2015) UniProt: a hub for protein information. Nucleic Acids Res 43: D204–D212
29. Bateman A (2000) The Pfam protein families database. Nucleic Acids Res 28:263–266
30. Finn RD, Coggill P, Eberhardt RY et al (2016) The Pfam protein families database: towards a more sustainable future. Nucleic Acids Res 44: D279–D285
31. Göbel U, Sander C, Schneider R, Valencia A (1994) Correlated mutations and residue contacts in proteins. Proteins Struct Funct Genet 18:309–317
32. Lammert H, Schug A, Onuchic JN (2009) Robustness and generalization of structure-based models for protein folding and function. Proteins 77:881–891
33. Onuchic JN, Luthey-Schulten Z, Wolynes PG (1997) Theory of protein folding: the energy landscape perspective. Annu Rev Phys Chem 48:545–600
34. Pirovano W, Heringa J (2010) Protein secondary structure prediction. Methods Mol Biol 609:327–348
35. Yang Y, Gao J, Wang J et al (2018) Sixty-five years of the long march in protein secondary structure prediction: the final stretch? Brief Bioinform 19:482–494. https://doi.org/10.1093/bib/bbw129
36. Drozdetskiy A, Cole C, Procter J, Barton GJ (2015) JPred4: a protein secondary structure prediction server. Nucleic Acids Res 43: W389–W394
37. Yachdav G, Kloppmann E, Kajan L et al (2014) PredictProtein—an open resource for online prediction of protein structural and functional features. Nucleic Acids Res 42:W337–W343
38. Buchan DWA, Minneci F, Nugent TCO et al (2013) Scalable web services for the PSIPRED Protein Analysis Workbench. Nucleic Acids Res 41:W349–W357
39. Heffernan R, Paliwal K, Lyons J et al (2015) Improving prediction of secondary structure, local backbone angles, and solvent accessible surface area of proteins by iterative deep learning. Sci Rep 5:11476
40. Pronk S, Páll S, Schulz R et al (2013) GROMACS 4.5: a high-throughput and highly parallel open source molecular simulation toolkit. Bioinformatics 29:845–854
41. Kutzner C, Páll S, Fechner M et al (2015) Best bang for your buck: GPU nodes for GROMACS biomolecular simulations. J Comput Chem 36:1990–2008
42. Meyer EE (1997) The first years of the Protein Data Bank. Protein Sci 6:1591–1597
43. Young J, RCSB PDBj PDBe Protein Data Bank (2009) Annotation and curation of the Protein Data Bank. Nat Preced. https://doi.org/10.1038/npre.2009.3379.1
44. Martínez L, Andreani R, Martínez JM (2007) Convergent algorithms for protein structural alignment. BMC Bioinformatics 8:306
45. Li Y, Zhang Y (2009) REMO: a new protocol to refine full atomic protein models from C-alpha traces by optimizing hydrogen-bonding networks. Proteins 76:665–676
46. Maupetit J, Gautier R, Tufféry P (2006) SABBAC: online Structural Alphabet-based protein BackBone reconstruction from Alpha-Carbon trace. Nucleic Acids Res 34:W147–W151
47. Rotkiewicz P, Skolnick J (2008) Fast procedure for reconstruction of full-atom protein models from reduced representations. J Comput Chem 29:1460–1465
48. Agre P (2006) The aquaporin water channels. Proc Am Thorac Soc 3:5–13
49. Ishibashi K, Sasaki S (1997) Aquaporin water channels in mammals. Clin Exp Nephrol 1:247–253
50. Agre P, Kozono D (2003) Aquaporin water channels: molecular mechanisms for human diseases1. FEBS Lett 555:72–78
51. Marks DS, Colwell LJ, Sheridan R et al (2011) Protein 3D structure computed from evolutionary sequence variation. PLoS One 6: e28766

52. Ash RB (2012) Information theory. Courier Corporation, Dover Publications Inc, Mineola, NY
53. Freedman D, Pisani R, Purves R (2007) Statistics: fourth international student edition. W. W. Norton & Company, New York, NY
54. Rapaport DC (2004) The art of molecular dynamics simulation. Cambridge University Press, New York, NY
55. Karplus M, Kuriyan J (2005) Molecular dynamics and protein function. Proc Natl Acad Sci U S A 102:6679–6685
56. Scheraga HA, Khalili M, Liwo A (2007) Protein-folding dynamics: overview of molecular simulation techniques. Annu Rev Phys Chem 58:57–83
57. Ruiz Carrillo D, To Yiu Ying J, Darwis D et al (2014) Crystallization and preliminary crystallographic analysis of human aquaporin 1 at a resolution of 3.28 Å. Acta Crystallogr F Struct Biol Commun 70:1657–1663
58. Subbiah S (1996) Protein motions. Springer, Berlin
59. Zhang Y, Skolnick J (2004) Scoring function for automated assessment of protein structure template quality. Proteins 57:702–710

Chapter 6

Detecting Amino Acid Coevolution with Bayesian Graphical Models

Mariano Avino and Art F. Y. Poon

Abstract

The comparative study of homologous proteins can provide abundant information about the functional and structural constraints on protein evolution. For example, an amino acid substitution that is deleterious may become permissive in the presence of another substitution at a second site of the protein. A popular approach for detecting coevolving residues is by looking for correlated substitution events on branches of the molecular phylogeny relating the protein-coding sequences. Here we describe a machine learning method (Bayesian graphical models) implemented in the open-source phylogenetic software package *HyPhy*, http://hyphy.org, for extracting a network of coevolving residues from a sequence alignment.

Key words amino acid coevolution, Bayesian graphical model, hepatitis C virus, *HyPhy*, epistasis

1 Introduction

Genomes encode an enormous number of components that need to interact in order to function properly. At a low level, for instance, nucleotides within a codon triplet define each other's context: whether a substitution from A to G at the second codon position alters the encoded amino acid depends on what nucleotides occupy the other positions, given the organism's repertoire of transfer RNAs that determine the genetic code. Different amino acids in the same protein may interact through direct contacts or long-range interactions through energetic, structural, or allosteric mechanisms [1]. Interactions between different proteins can also be modulated by amino acid sequence variation [2]. Thus, we cannot understand the biological significance of a genetic polymorphism without accounting for at least some of its genomic context. Conversely, identifying interactions between amino acids may confer information about the higher-order structure and function of a protein and its relation to other components of the genome.

Tobias Sikosek (ed.), *Computational Methods in Protein Evolution*, Methods in Molecular Biology, vol. 1851, https://doi.org/10.1007/978-1-4939-8736-8_6, © Springer Science+Business Media, LLC, part of Springer Nature 2019

There is an enormous literature on detecting interactions between residues within a protein from genetic sequence variation. Although there have been many efforts to review this literature, it is difficult to produce a comprehensive overview because similar methods have been developed repeatedly and independently in diverse disciplines such as biochemistry, computer science, and evolutionary biology [3–6]. A common motivation for the comparative study of amino acid sequences is that covariation among residues may reveal information about the structure of a folded protein [7]. In the medical sciences, sequence covariation provides information about the structural and functional constraints of proteins of viral or bacterial pathogens, and can thereby inform the development of vaccines [8]. We tend to attribute missense point mutations to disease-associated phenotypes, and in many cases there is strong evidence that such mutations are independently sufficient to cause disease [9]. However, the effect of these mutations can also be modulated through their interactions with mutations at other specific sites in the genome [10]. Finally, characterizing the interactions among amino acids can quantify the type and prevalence of epistasis and how it shapes the evolutionary trajectories of populations adapting to new environments [11, 12].

1.1 Correlated Substitutions

The simplest approach to detect interactions between amino acids from the comparative study of protein sequences is to look for correlations between different positions in the protein with respect to the biochemical properties of residues [13], empirical substitution rates [7, 14] or the occurrence of specific amino acids [15]. Correlations are often measured using mutual information [16–18] or extensions of this approach that incorporate other information [19, 20]. One of the major confounding factors affecting the comparative study of protein sequences is that the amino acids in different sequences are not independent observations; instead, they are copies that descend from a common ancestor such that shared genotypes may be due to *identity by descent*. In the worst case scenario, a comparative method may predict a false interaction between residues at different positions of a protein because of two ancestral substitution events that have been propagated to the observed descendants with no further evolution [21]. This confounding due to common ancestry has been independently recognized across fields and a large number of techniques have been proposed to resolve it, e.g., [22–25]. One common approach is to change the focus from the observed characteristics of residues at different sites to the amino acid substitutions that have accumulated in the evolutionary history of the protein sequences [22]. Hence, we are looking for substitutions at different sites to occur on the same branch of a phylogenetic tree, which implies causality by proximity in time. This transformation of the data can substantially reduce the number of false positives caused by identity by descent [26]. However, it can also result in a loss of statistical power to detect true associations because the number of inferred substitutions tends to be much smaller than the number of differences [27], and this approach may also be biased by variation in rates of evolution [28]. It is likely that none of these methods will ever attain a high level of prediction accuracy [29]. Even so, the comparative study of protein variation is a cost-effective means to identify putative interactions for further empirical study.

1.2 Bayesian Graphical Models

The challenge of detecting interactions among residues in a protein is highly susceptible to the "curse of dimensionality": the number of possible interaction networks grows far more rapidly than the amount of data. Many of the previously described methods use false discovery rate methods to account for multiple comparisons. One of the problems of this approach is that each interaction is evaluated from a narrow subset (i.e., pairs) of variables while excluding the rest of the data. Consequently, the end result for this type of analysis is a list of potential pairwise interactions with no framework for assembling them into a coherent whole. Instead, we have previously proposed [27] to examine the joint distribution of all potential interactions, borrowing a class of methods from the field of artificial intelligence known as Bayesian graphical models (BGMs) or Bayesian networks [30].

Each node in a BGM represents a variable, such as an amino acid position in a protein alignment. A directed edge (arrow) is drawn from node A to another node B to indicate that B is conditionally dependent on A. Thus, a BGM concisely and visually expresses the structure of a joint probability distribution for a set of variables as a graph. For example, $P(A, B, C) = P(B|A, C)\, P(A)\, P(C)$ is represented by the graph $A \rightarrow B \leftarrow C$. The challenge is to learn the structure of this distribution from the data. The number of structures increases faster than exponentially with the number of variables. Hence, we employed a Markov chain Monte Carlo sampling algorithm proposed by Daphne Koller and Nir Friedman [31] that collapses the enormous space of all possible graph structures into a more manageable space. This dimensionality reduction results in a severe loss of granularity—instead of sampling over individual graph structures, we are sampling permutations that define hierarchies of nodes (whether A can be a parent of B, or vice versa). Even so, the posterior probability of a specific edge can be calculated from the number of structures consistent with a given node hierarchy that contain this edge.

2 Program Usage

HyPhy is an open-source software package for phylogenetic sequence analysis [32] that is written in C++ and compiled into binary distributions for Mac OS, Windows, and Linux.[1] It features both a command-line and graphical user interface (GUI) and a custom batch language for writing scripts that can be executed from the command line. Most of the standard analyses in *HyPhy* are implemented in the batch language. In addition, many of these standard analyses are made available as web applications on the datamonkey.org server [33]. One of these methods is a BGM analysis of coevolving sites in codon sequences, which we dubbed *Spidermonkey* [34]. Because the server is a shared public resource, every analysis imposes a set of restrictions on the number and length of sequences that can be processed. *Spidermonkey* limits users to alignments of no more than 500 sequences, and the number of codons to no more than 1000. Furthermore, it limits the number of conditional dependencies (parents) to one or two per node and sets a hard limit on the number of steps to run the Markov chain Monte Carlo sampler (1.1×10^5 steps with 10,000 steps discarded as "burn-in"). Researchers will often want to analyze more than 500 sequences and/or to customize various aspects of the Bayesian network analysis, such as focusing on a subset of codon sites or analyzing a protein sequence alignment. Hence, our purpose here is to provide the resources and methods for running *Spidermonkey* as a standalone analysis pipeline that can be customized to a specific problem.

2.1 Obtaining HyPhy and Scripts

HyPhy binaries can be downloaded for free at http://hyphy.org. If you want to compile the software package, the source code can be obtained at http://github.com/veg/hyphy. Alternatively, a POSIX-threaded *HyPhy* binary (hyphymp) for Linux can be obtained with a package manager; for example, in Ubuntu: `sudo apt install hyphy-pt`. The scripts and data used here are available at http://github.com/PoonLab/comet-prot. If you are running *HyPhy* from the command line, then all commands should specify the path to your local installation, e.g.: `HYPHYMPBASEPATH=/usr/local/lib/hyphy <path to script>`.

[1] The scripts in this chapter were tested with *HyPhy* version 2.220170201beta and release 2.2.7. *HyPhy* is a large and complex software package that is constantly undergoing development by a small team of researchers and programmers, and some of the more specialized features such as BGMs may temporarily break as newer versions are released. If you compiled *HyPhy* from source, make sure that you are using a single-threaded (HYPHYSP) or multiprocessing-enabled (HYPHYMP) build and not a message passing interface (MPI)-enabled (HYPHYMPI) build; at the time of writing, there were residual issues in the source code related to MPI processing. If you encounter any other problems, please submit an issue at https://github.com/veg/hyphy/issues and we will attend to it as soon as possible.

Otherwise, you can run the scripts through the graphical user interface by opening the file through the file selection dialog (⌘-O on macOS, Ctrl-E on Windows, or *File > Open > Open Batch File...*).

2.2 Preparing Input Data

To run this analysis, you need to have a codon sequence alignment and a phylogenetic tree relating these sequences. A codon sequence has a single reading frame, excluding any frameshifts or stop codons. In other words, the first three bases should map to a codon, and so on. It does not have to cover the entire gene. Any stop codons need to be replaced with gaps (interpreted as missing data); otherwise, the entire codon site will be stripped from the alignment, throwing out useful data and making it difficult to interpret the end result of the analysis. *HyPhy* also has strict requirements on sequence names, which cannot contain any characters other than the alphanumeric characters and the underscore character "_". This name restriction also applies to tip labels in the tree, so it is often more convenient to reconstruct a tree after the following step.

A convenient tool for adjusting sequence names and simultaneously replacing stop codons with gap characters is provided in the *HyPhy* standard library. In the GUI, you can open the Standard Analysis menu by pressing ⌘-E (macOS), Ctrl-E (Windows), or selecting (*Analysis > Standard Analyses...*), expanding the "Data File Tools" tab, and then selecting `CleanStopCodons.bf`. From the command line, you can launch an interactive menu by calling the *HyPhy* executable (`HYPHYMP` or `hyphymp` if you used a package manager) and then select the options `(4) Data File Tools` and then `(6)` to run the same script. You will be prompted to specify a genetic code and codon data file (see below). The last query is whether to discard duplicate sequences and/or codon sites that are entirely gaps. Duplicate sequences cannot be separated in a phylogeny, so unless you will be using a tree relating these sequences based on additional information, there is no reason to retain all copies for the analysis. Similarly, entirely gapped codon positions are not phylogenetically informative and may be dropped unless they are needed to preserve the coordinate system of the alignment.

A phylogenetic tree can be reconstructed from the sequence alignment using any standard maximum likelihood program such as RAxML [35] (https://github.com/stamatak/standard-RAxML) or PhyML [36] (https://github.com/stephaneguindon/phyml).[2]

[2] For this type of analysis, we prefer using maximum likelihood (ML) methods to reconstruct trees. If it is not feasible to use ML methods due to excessive numbers of sequence and/or sequence lengths, we suggest using the approximate ML program FastTree 2 [37], which can be orders of magnitude faster than the standard ML programs. Neighbor-joining (NJ) methods also scale favorably with larger alignments, but tend to be less accurate for reconstructing branch lengths. While there are NJ and ML tree reconstruction methods implemented in *HyPhy*, they are not as efficient as these specialized programs and we do not recommend using them for larger data sets.

In addition, the tree should not contain bootstrap support values [38]. *HyPhy* interprets these values as internal node labels and does not allow duplicate labels.[3]

2.3 Fit Codon Model

The first step in our analysis pipeline is to fit a codon substitution model [39] to the sequence alignment by running the script `fit_codon_model.bf` (which depends on the utility file `fit_codon_model.ibf`).[4] Although there are standard methods for this task in the default *HyPhy* menu, we implemented a customized method that constrains the branch lengths in the input tree to be rescaled by a global factor.[5] This confers a significant savings in computing time, since we don't need to re-estimate the length of every branch in the tree.

2.3.1 Choose a Genetic Code

In most cases, we will select option `1`, the universal genetic code. However, there is a large selection of genetic codes available in *HyPhy*, and selecting an appropriate code is important for this analysis because it will determine how nucleotide substitutions are interpreted as missense, nonsense, or silent mutations.

2.3.2 Specify a Codon Data File

Enter a relative or absolute path[6] to the file containing the cleaned sequence alignment, or if using the GUI, use the filesystem dialog to navigate to the file. Again, we assume that this alignment comprises codon sequences with a consistent reading frame. The presence of frameshifts due to alignment errors or actual sequence insertions/deletions will prevent *HyPhy* from correctly reconstructing non-synonymous and synonymous substitutions.

2.3.3 Model Options

This option determines how the model parameters are distributed across branches in the tree. The "Local" option assigns an instance of each parameter to every branch in the tree. For example, if we are fitting a model with a transition/transversion bias parameter, then this bias will be estimated independently for every branch. While this results in a more flexible model, there is a greater danger of

---

[3] A bootstrap support value is an empirical measure of confidence in a specific clade given the data. Most phylogeny reconstruction programs should have an option to omit these values. If you already have a Newick tree file and you just need to remove the support values, you can use the following UNIX command: `sed -E s/)[0-9.]+:/):/g [input] > [output]`.

[4] From this point onward, we assume that you are using the command-line interface. Unfortunately, this script may not work properly with the GUI because of how *HyPhy* handles file paths. Even on the command line, this is not straight-forward. For example, we used the following invocation in the macOS Terminal: `HYPHYMP BASEPATH=/usr/local/lib/hyphy/ pwd/fit_codon_model.bf` If you want to take advantage of a multi-core CPU, you can add the argument `CPU=[number of cores]` immediately after `HYPHYMP`. Note that not all steps in this analysis are able to utilize multiple threads.

[5] If you want to examine this scaling factor, you can find it in the serialized likelihood function generated by this script by searching for the parameter name `scalingB`.

[6] If you're using an operating system with a desktop environment, it's often easier to drag the icon representing your file into the terminal window instead of typing out the corresponding path. This works when running *HyPhy* on the command line, but you need to use backspace to remove the space that is automatically appended to end of the path. *HyPhy* won't be able to locate the file otherwise.

overfitting your data and we seldom use this approach. The "Global" option means that each model parameter is estimated using the information from all branches in the tree. This is the other extreme that results in a simpler but less flexible model. The remaining two options represent a compromise between these two extremes by allowing rates of evolution to vary across sites in the alignment.[7] The "Global w/variation" option models this rate variation using one of many parametric distributions, such as the prototypical gamma distribution that was first proposed by Ziheng Yang [40]. The "Global w/variation+HM" option uses a Hidden Markov model to smooth the assignment of rate categories along the length of the sequence alignment [41], such that adjacent sites will tend to be assigned to the same rate category.

2.3.4 Nucleotide Model

The codon substitution model implemented in these scripts has a nested model of nucleotide substitution[8] that needs to be specified by the user. This step uses the 6-digit PAUP*-style model specification string [42], which defines equality constraints for the six symmetric substitution rates in alphabetical order: A $\leftrightarrow$ C, A $\leftrightarrow$ G, A $\leftrightarrow$ T, C $\leftrightarrow$ G, C $\leftrightarrow$ T, and G $\leftrightarrow$ T. For example, the Tamura-Nei model [43] is specified by the string `010020`—all the nucleotide transversions share a single rate identifier (`0`). The most appropriate nucleotide model can be determined using a model selection method such as ModelTest [44].

2.3.5 Specify a Tree File

At the prompt, enter a relative or absolute path to the file containing the reconstructed phylogeny in a Newick tree string format. The tip labels in this tree need to correspond one-to-one with the sequence labels in your alignment file.[9]

2.3.6 Fit a Likelihood Function

Finally, you are prompted to specify a relative or absolute path to a file to write a serialized likelihood function, which encodes the data, model, and parameter estimates.[10] After providing a file path, the analysis will run and eventually converge to the maximum likelihood estimates of the model parameters. It is usually a good idea to open the likelihood function output in a text editor and inspect the

[7] Prior to version 2.3.4, the text in *HyPhy* implies that these options allow rates to vary among branches, not sites: "…branch lengths come from a user-chosen distribution." We have revised this help text as of version 2.3.4 to indicate that the distributions are used to model rate variation across sites, not branches.

[8] A standard codon model is described by a 61-by-61 transition rate matrix and a single parameter R that corresponds to the ratio of non-synonymous and synonymous substitution rates. The model assumes that the system moves from one codon to another by single nucleotide substitutions; codon substitutions that require more than one nucleotide change are not allowed.

[9] Some phylogeny reconstruction programs truncate sequence labels and cause an error at this stage—for example, neither RAxML or FastTree2 will read sequence labels beyond a whitespace character. A quick fix in this situation is to replace all whitespace characters with underscores in a text editor or with `sed`.

[10] By convention, we use the file extension `.lf` and keep the same basename as the codon data file. This makes it easier to track files that belong to the same workflow.

parameter estimates. This output is written in a NEXUS format[11] [45]. The parameter estimates can be found at the top of the `HYPHY` block. One useful diagnostic is to examine the estimate of the non-synonymous/synonymous rate ratio *R*. An *R* value of less than 1 indicates that most of the alignment has undergone purifying selection. In addition, we often look for the transversion rate estimates to be less than 1 (where the reference rate of A $\leftrightarrow$ G transitions is fixed to 1).

2.4 Map Substitutions to the Tree

The next step in our pipeline is to reconstruct ancestral sequences in the tree based on the maximum likelihood parameter estimates of the model [46]. If the descendant sequence has a different codon than its ancestor, then we infer that at least one substitution has occurred along the intervening branch in the tree [47]. This step is implemented by the script `MapMutationsToTree.bf`. Upon running the script, the user is prompted to provide a relative or absolute path to the file containing the serialized likelihood function from the previous step.

2.4.1 Select Reconstruction Option

HyPhy implements the fast joint ancestral reconstruction algorithm formulated by Tal Pupko and colleagues [48]. Our script prompts the user to decide whether to sample ancestral sequences from the posterior distributions at each node of the tree. Sampling enables us to accommodate the uncertainty in reconstructing ancestral states, which is exacerbated for ancestral nodes that are further back in time relative to the observed sequences. On the other hand, each sample will comprise a set of ancestral sequences that compounds the number of replicate analyses to be performed further along the pipeline. We recommend using your discretion for this step: if it is likely that the most recent common ancestor is separated from all the observed sequences by excessive amounts of evolutionary time, then it may be important to sample ancestral states for a more robust but time-consuming analysis.

2.4.2 Output Options

This script was designed to generate two different kinds of outputs. The first option is to generate a binary matrix where each row corresponds to a branch of the tree, and each column corresponds to a codon site in the alignment. This matrix is written to the output file in a comma-separated tabular (CSV) format. A `1` indicates that a non-synonymous substitution was mapped to the respective branch and site. This matrix output is the raw material for a BGM analysis, where each codon site is a potential node in the graph and the branches represent independent observations. The second option is to output a tab-delimited tabular file where each

[11] NEXUS is a widespread format with known issues with standardization and usability, and has been implemented in diverse and often incompatible ways by multiple programs.

row corresponds to a inferred non-synonymous substitution, and the columns correspond to the replicate number (if sampling), branch label, site, and the ancestral and derived amino acids.[12]

If the first option is selected, then the user is prompted to specify if they want the CSV to begin with a header row. Each entry of the header row corresponds to the amino acid at each position of the ancestral sequence reconstructed at the root of the tree.[13]

2.5 BGM Analysis

A Bayesian graphical model (BGM) analysis is implemented in the script `bayesgraph.bf`, which depends on the utility ("include") file `bayesgraph.ibf`. These scripts were designed to emulate the workflow provided by the *Spidermonkey* application on our datamonkey.org webserver. We note here, however, that BGM analyses in *HyPhy* are more versatile than our example demonstrates.[14]

2.5.1 Input Data Matrix

First, the user is prompted for a relative or absolute path to the CSV file containing the substitution map matrix that was produced by the `MapMutationsToTree.bf` script. If the CSV does not contain a header row with column labels that indicate what each variable represents, then they will be assigned integer values. It is preferable to use the ancestral residue labels generated by the `MapMutationsToTree.bf` script because we are going to filter out columns based on the number of inferred non-synonymous substitutions.

2.5.2 Filter Sites

Next, the program will prompt you for the minimum number of substitutions for a site to be included in the BGM model. This cutoff cannot be less than `1` because sites without any non-synonymous substitutions contain no information for inferring conditional dependencies. The script automatically determines the maximum cutoff based on the largest number of substitutions mapped at any single codon site. Once the user selects a number in this range, the script will filter sites that do not meet this cutoff from the data set and populate a BGM model with the remaining variables.[15]

[12] We have previously found this list output to be a more convenient format for debugging the script. It's usually a good idea to manually compare entries in this list against your sequence alignment to make sure that things make sense.

[13] Most phylogenetic tree reconstruction methods, such as maximum likelihood or neighbor-joining, will output an unrooted tree. For an unrooted tree, the labels will be generated for the deepest internal node.

[14] For example, you can customize on a node-by-node basis the number of "parental" nodes on which a given node can be conditionally dependent. You can also load a serialized BGM from a XML Bayesian Interchange format file and use this model to simulate additional data sets. For more details, please refer to the file `bayesgraph.ibf` and the batch file `tests/hbltests/BayesianGraphicalModels/TestBGM.bf` in the *HyPhy* source code distribution.

[15] As a general rule of thumb, we try to not build a BGM model that has many more nodes than observations. The number of substitutions provides a meaningful criterion for reducing the dimensionality of our data.

2.5.3 MCMC Settings

There are four settings that the user needs to specify for running a Markov chain Monte Carlo (MCMC) sample. First, the user has to specify the maximum number of parents that will be allowed per node. This determines the complexity of the analysis. An analysis with a one-parent maximum per node will run very fast and scales easily with large numbers of variables, but loses the sensitivity to detect complex interactions among nodes. Conversely, an analysis that allows many more parents per node is far more computationally complex.[16] Second, the user needs to indicate the number of steps to discard as a "burn-in" period. This budgets an amount of time that one estimates it will require for this random walk to travel from its initial point to a "reasonable" area of model space. Third, the user needs to specify the number of steps to run the chain sample following the "burn-in" period. This length sets an upper limit to the effective sample size, which will almost surely be much smaller because of the highly autocorrelated nature of MCMC. Lastly, the user must specify the number of steps to extract from this chain sample. Because of autocorrelation in the chain, there is usually no benefit in retaining every step. To reduce the output file sizes and increase the efficiency of post-processing, it is standard practice to reduce the chain by sub-sampling at regular intervals. The script defaults to a sub-sample of 100 steps, which results in gaps of 1000 steps (*see* Fig. 1). The user should adjust the size of the sub-sample roughly in proportion to the length of the post-burn-in chain sample.

2.5.4 Output Settings

The `bayesgraph.bf` script generally produces three kinds of outputs. The script will prompt the user for only one relative or absolute path for an output file, and paths for the other output files will automatically be generated based on this first path. First, the script will output the marginal posterior probabilities for directed edges as a CSV formatted file. This is the raw material for assembling the consensus BGM. Next, the script will write this consensus BGM using the network visualization language DOT, which can be converted into an image by several programs such as GraphViz [49], Cytoscape [50], and Gephi [51]. Finally, the script will record the posterior probability trace for all steps sub-sampled from the original MCMC sample. This is important information for assessing the convergence of the chain sample to the posterior distribution (e.g., Fig. 1).

[16] This is where the ability to customize the analysis implemented in the `bayesgraph.bf` script can be very useful. If you have prior information that a subset of codon sites are involved in a large number of interactions, the computational complexity of increasing the number of parents can be greatly reduced by modifying this parameter for only these sites.

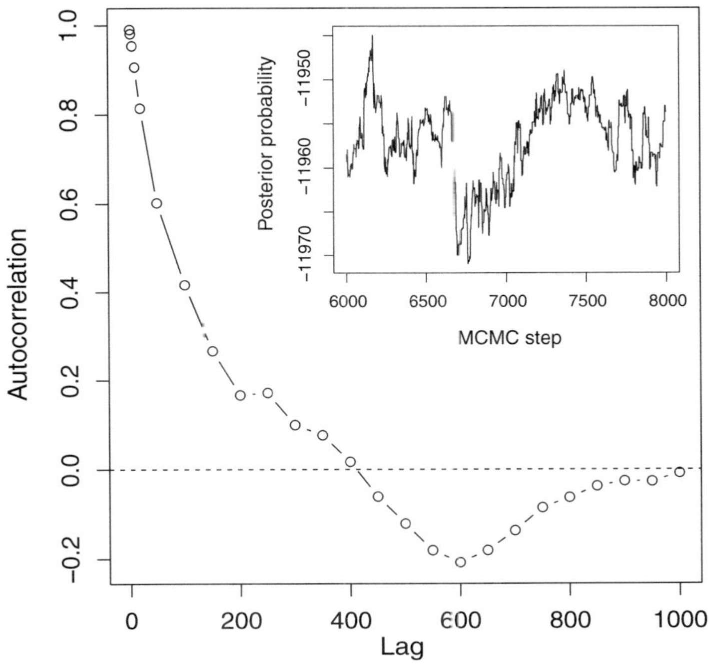

Fig. 1 Autocorrelation in a Markov chain Monte Carlo (MCMC) sample from a BGM analysis. The *x*-axis represents the number of steps separating observations in the sub-sample (lag), and the *y*-axis measures the autocorrelation between adjacent steps in the sub-sampled chain. We observe strong autocorrelation when the lag is short because the MCMC random walk takes short steps in model space. On average, we need to have taken at least 1000 steps until the posterior probability at the current step is not informed by the previous state. The inset figure displays an interval of the MCMC sample, where the autocorrelation between steps is more apparent[18]

3 Example

Hepatitis C virus (HCV) is a rapidly evolving RNA virus that can establish persistent chronic infections in human hosts [52]. Over 70 million people worldwide are estimated to be infected with actively replicating HCV, as diagnosed by the presence of viral RNA in the bloodstream [53]. Comparative methods to detect amino acid coevolution in HCV proteins have been utilized to find potential therapeutic targets [54], associate this variation with treatment outcomes [55], and to explain the global prevalence

[18] (In an MCMC run, we observe autocorrelation when we sample parameter values that are very close in the parameter space and unrepresentative of the true underlying posterior distribution. Therefore, we try to decrease autocorrelation so that the MCMC sample provides a more precise estimate of the posterior sample. One way to accomplish this is by down-sampling to every n-th step).

of drug resistance-associated mutations [56], to name a few applications. In this example, we work through a BGM analysis of amino acid coevolution in HCV nonstructural (NS) protein 5b, an RNA-dependent RNA polymerase that is a major target for the new generation of direct-acting antivirals [57].

We first queried the euHCVdb database [58] (https://euhcvdb.ibcp.fr/euHCVdb/) to retrieve GenBank accession numbers for HCV subtype 1b nucleotide sequences with at least partial coverage of the NS5b gene. Next, we used the Entrez batch interface to retrieve the corresponding FASTA records from GenBank given the list of accession numbers (`HCV1b-NS5b.fasta`).[19] We used a fast pairwise alignment method to extract the NS5b-encoding region from each sequence relative to the H77 reference (`H77-NS5b.txt`), then generated a multiple sequence alignment using MAFFT v7.305b [59] (`HCV1b-NS5b.mafft.fa`) and manually inspected and adjusted the resulting alignment with AliView v1.19-beta-3 [60] (`HCV1b-NS5b.aliview.fa`). The remaining sequences in this data set ($n = 536$) ranged from 1043 to 1776 in nucleotide length. We used the built-in method in *HyPhy* (*see* Subheading 2.2) to clean sequence names and remove stop codons and shortened the FASTA identifiers to facilitate subsequent phylogenetic analyses (`HCV1b-NS5b.cleaned.fa`). To quickly screen for potential alignment errors or subtype misclassifications, a preliminary phylogenetic tree was reconstructed by approximate maximum likelihood with FastTree 2 [37] and examined visually.

Next, a maximum likelihood phylogenetic tree (Fig. 2) was reconstructed using PhyML (v20160207, [36]) with a bootstrap analysis (`HCV1b-NS5b.phyml.nwk`). Using jModeltest v2.1.10, a TPM2 model (PAUP* specification 010212) incorporating invariant sites and a gamma distribution (TPM2+I+G) was selected based on the Akaike information criterion [61, 62]. The likelihood function was constructed and optimized in *HyPhy* using the script `fit_codon_model.bf` with the universal genetic code option. We selected for parameters to be globally constrained and for rate variation across sites to be modeled by the Gamma+Invariant model with 4 rate classes. We then entered the 6 character model designation (010212) corresponding to the TPM2 model. Lastly, we input the file containing the PhyML tree reconstruction and specified the output path to a new file `HCV1b-NS5b.lf`.

To extract the map of non-synonymous substitution events to branches of the phylogenetic tree (`MapMutationsToTree.bf`), we input the serialized likelihood function from the last step and selected maximum likelihood for the ancestral reconstruction option. Next, we selected a binary matrix CSV file output with

[19] We have provided most of the data files in this example on our GitHub repository at https://github.com/PoonLab/comet-prot/tree/master/data.

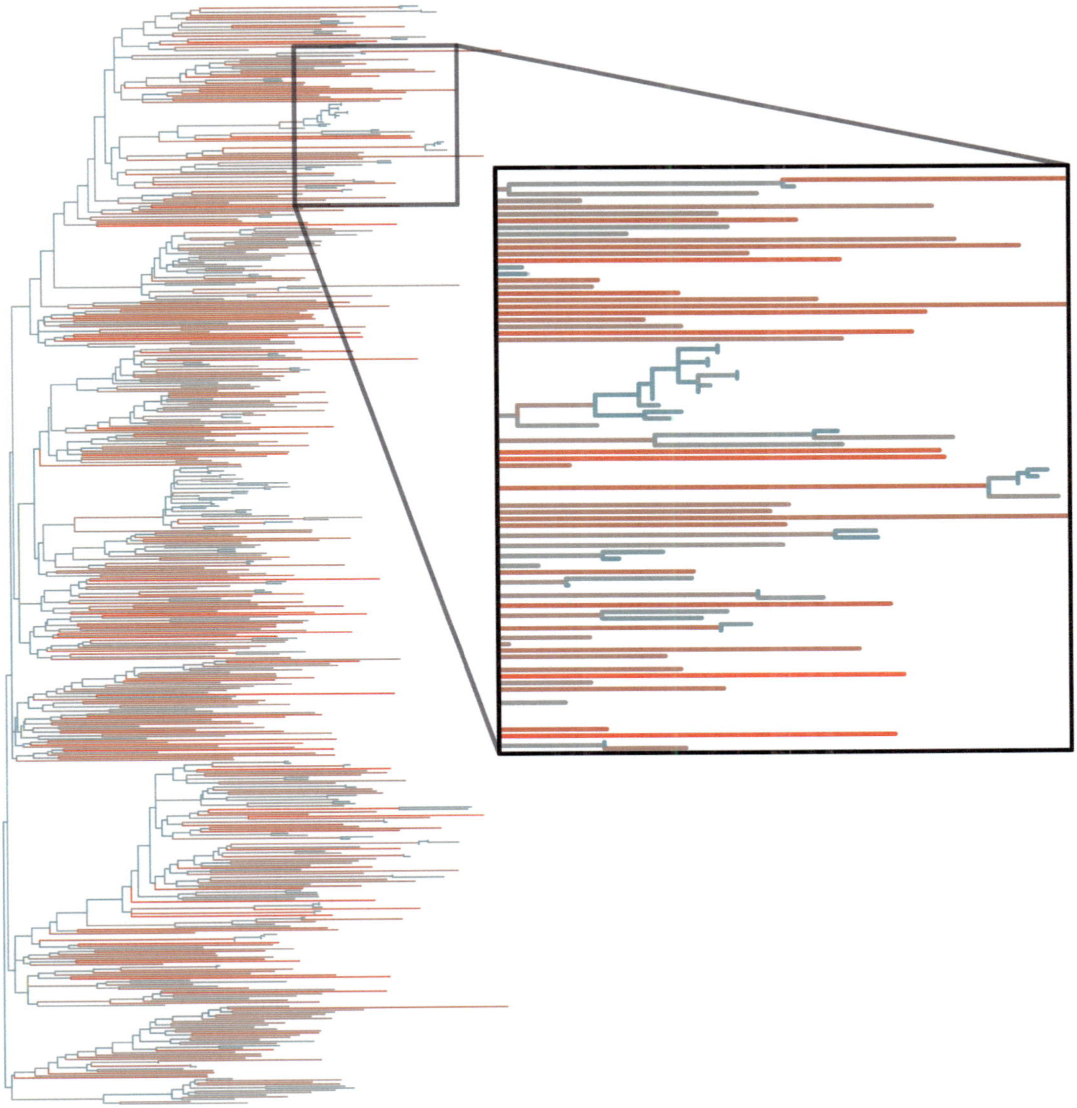

Fig. 2 Excerpt of phylogenetic tree reconstructed from the HCV sequence data using PhyML. The tree was rooted at the midpoint (the halfway point on the longest path separating two tips in the tree). This image was generated using the R package `ggtree` [63]. Branches are colored with respect to the number of non-synonymous substitutions (increasing from blue to red). The shape of the tree is generally consistent with the sequences belonging to a single HCV subtype. However, the tree also contains some clusters (inset) of highly related sequences that may represent multiple sequences from the same individuals, or recent transmission outbreaks of HCV

automatically generated column labels and indicated that this output should be written to the file `HCV1b-NS5b.csv`.

The resulting matrix was used as the input file for our BGM analysis script `bayesgraph.bf`. Since we used the previous script to generate column labels from the reconstructed ancestral amino acid sequence, we specified that the input file contained a header row. To verify that these column labels corresponded to the protein NS5b of HCV subtype 1b, we concatenated these labels into an amino acid sequence using a search-and-replace method[20] and submitted the resulting sequence to the BLASTP database (http://blast.ncbi.nlm.nih.gov/Blast.cgi?PAGE=Proteins). The highest scoring match was a HCV subtype 1b NS5b sequence (GenBank accession number BAW81715) with a 97% (573/591) amino acid identity. <!– COMP: Please set the following "`seq = re.sub([0-9]+,*, , header.strip())`" as in MS. --?>

The `bayesgraph.bf` script indicated that our data matrix contained 1069 cases (rows) and 592 variables (columns). We set the minimum number of non-synonymous substitutions per codon site to 2, which resulted in 231 variables after filtering conserved sites. Next, we set the maximum number of parents per node to 2. We ran two replicate chain samples under the default settings (a 10^4 "burn-in" followed by 10^5 steps thinned to 100). To assess convergence to the posterior distribution, we compared the log-likelihood traces associated with each chain and obtained a potential scale reduction factor of 0.996 (upper 95% confidence limit = 0.998) using the *coda* package in R[21] [64], where factors in excess of 1 imply a lack of convergence [65].

A consensus Bayesian graphical model of the two chains is shown in Fig. 3. This network visualization was generated using the script `make_dotfile.py`. The HCV NS5b structure presents discernible fingers, palm and thumb subdomains, and a biologically significant C-terminal region that lines the RNA binding cleft in the active site. Our analysis shows 32 coevolving pairs associations sites and most of them involve amino acid pairs located in the same subdomain (23 of whom 12 lie on fingers, 4 palm, 3 thumb, and 4 C-terminal). Five of the pairs involve sites belonging to the fingers and palm region while none involved fingers and thumb; this is quite unexpected given the extensive interactions between these regions, which is responsible for the formation of an encircled active site [66]. One pair involves a flanking amino acid of the β-hairpin region (A442) and a site included in the same region (A450); the β-hairpin plays an important role of positioning the

[20] To generate an amino acid sequence from the column labels, we used the regular expression "[0-9]+,*" to replace all instances with an empty string. In Python, this can be achieved with the `re` module: `seq = re.sub([0-9]+,*, , header.strip())`, where `header` is a string variable containing the first line of the CSV file.

[21] This can be accomplished with the following R commands:

```
require(coda)
chain1 <- read.csv("chain1.trace.csv", header=F)
chain2 <- read.csv("chain2.trace.csv", header=F)
chains <- mcmc.list(mcmc(chain1$V1), mcmc(chain2$V1))
gelman.diag(chains, autoburnin=F)
```

where the file names may be different for your run.

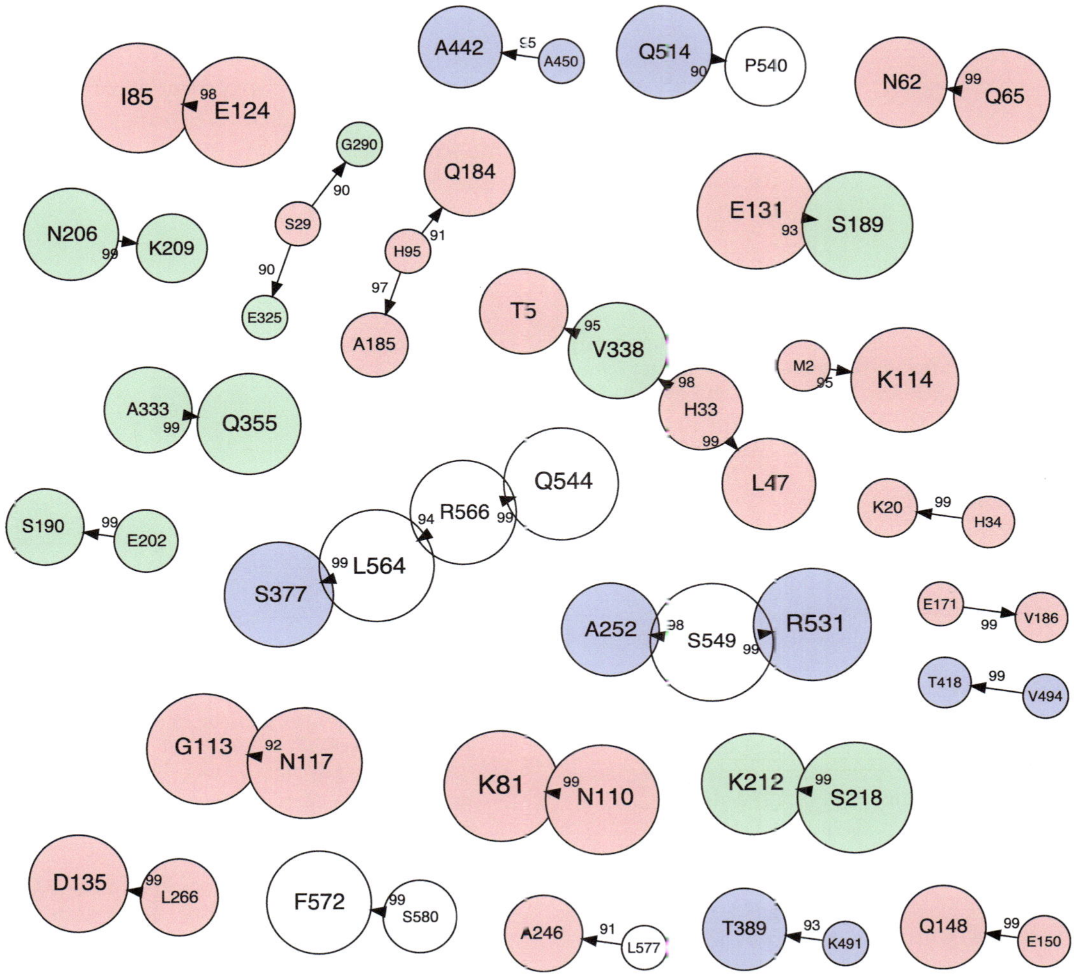

Fig. 3 Visualization of a consensus Bayesian graphical model of residue–residue interactions in HCV1b NS5b proteins. Each node represents a codon site in the NS5b protein, labelled with the ancestral residue and position. The size of the node is scaled to the log-transformed number of non-synonymous substitutions reconstructed at the respective codon sites. Nodes are colored with respect to the NS5b domains: fingers (red), palm (green), and thumb (blue); nodes representing residues in the C-terminal tail are left uncolored. Arrows (edges) are drawn to represent inferred coevolution between the respective codon positions. The edges are annotated with the marginal posterior probabilities (MPP, %). Only edges with an MPP value exceeding 90% were included in this graph

3′ terminus of the viral RNA genome [67]. In general, our results confirmed the general finding that coevolving sites are not restricted to residues that are in physical contact and likely responsible for the structural stability of the proteins. We found a few examples of coevolving pairs of residues involving the β-hairpin and C-terminal regions that may be of functional importance for the functioning NS5b protein, or in proximity of sites that have a biological function.

Acknowledgements

This study was supported in part by the Government of Canada through Genome Canada and the Ontario Genomics Institute (OGI-131), and by grants from the Canadian Institutes of Health Research (PJT-153391 and BOP-149562). AFYP was supported by a CIHR New Investigator Award (FRN-130609).

References

1. Kihara D (2005) The effect of long-range interactions on the secondary structure formation of proteins. Protein Sci 14(8):1955–1963
2. Sprinzak E, Margalit H (2001) Correlated sequence-signatures as markers of protein-protein interaction. J Mol Biol 311 (4):681–692
3. Horner DS, Pirovano W, Pesole G (2007) Correlated substitution analysis and the prediction of amino acid structural contacts. Brief Bioinform 9(1):46–56
4. Taylor WR, Hamilton RS, Sadowski MI (2013) Prediction of contacts from correlated sequence substitutions. Curr Opin Struct Biol 23(3):473–479
5. Marks DS, Hopf TA, Sander C (2012) Protein structure prediction from sequence variation. Nat Biotechnol 30(11):1072–1080
6. De Juan D, Pazos F, Valencia A (2013) Emerging methods in protein co-evolution. Nat Rev Genet 14(4):249
7. Göbel U, Sander C, Schneider R, Valencia A (1994) Correlated mutations and residue contacts in proteins. Proteins Struct Funct Bioinf 18(4):309–317
8. Korber B, Farber RM, Wolpert DH, Lapedes AS (1993) Covariation of mutations in the V3 loop of human immunodeficiency virus type 1 envelope protein: an information theoretic analysis. Proc Natl Acad Sci 90 (15):7176–7180
9. Hirschhorn JN, Lohmueller K, Byrne E, Hirschhorn K (2002) A comprehensive review of genetic association studies. Genet Med 4 (2):45–61
10. Kowarsch A, Fuchs A, Frishman D, Pagel P (2010) Correlated mutations: a hallmark of phenotypic amino acid substitutions. PLoS Comput Biol 6(9):e1000923
11. Weinreich DM, Delaney NF, DePristo MA, Hartl DL (2006) Darwinian evolution can follow only very few mutational paths to fitter proteins. Science 312(5770):111–114
12. Ivankov DN, Finkelstein AV, Kondrashov FA (2014) A structural perspective of compensatory evolution. Curr Opin Struct Biol 26:104–112
13. Neher E (1994) How frequent are correlated changes in families of protein sequences? Proc Natl Acad Sci 91(1):98–102
14. Olmea O, Rost B, Valencia A (1999) Effective use of sequence correlation and conservation in fold recognition. J Mol Biol 293 (5):1221–1239
15. Atchley WR, Wollenberg KR, Fitch WM, Terhalle W, Dress AW (2000) Correlations among amino acid sites in bHLH protein domains: an information theoretic analysis. Mol Biol Evol 17(1):164–178
16. Tillier ER, Lui TW (2003) Using multiple interdependency to separate functional from phylogenetic correlations in protein alignments. Bioinformatics 19(6):750–755
17. Martin L, Gloor GB, Dunn S, Wahl LM (2005) Using information theory to search for co-evolving residues in proteins. Bioinformatics 21(22):4116–4124
18. Gouveia-Oliveira R, Pedersen AG (2007) Finding coevolving amino acid residues using row and column weighting of mutual information and multi-dimensional amino acid representation. Algorithms Mol Biol 2(1):12
19. Fernandes AD, Gloor GB (2010) Mutual information is critically dependent on prior assumptions: would the correct estimate of mutual information please identify itself? Bioinformatics 26(9):1135–1139
20. Jeong CS, Kim D (2012) Reliable and robust detection of coevolving protein residues. Protein Eng Des Sel 25(11):705–713
21. Felsenstein J (1985) Phylogenies and the comparative method. Am Nat 125(1):1–15
22. Shindyalov IN, Kolchanov NA, Sander C (1994) Can three-dimensional contacts in protein structures be predicted by analysis of correlated mutations? Protein Eng 7(3):349–358

23. Wollenberg KR, Atchley WR (2000) Separation of phylogenetic and functional associations in biological sequences by using the parametric bootstrap. Proc Natl Acad Sci 97 (7):3288–3291
24. Gloor GB, Martin LC, Wahl LM, Dunn SD (2005) Mutual information in protein multiple sequence alignments reveals two classes of coevolving positions. Biochemistry 44 (19):7156–7165
25. Pollock DD, Taylor WR, Goldman N (1999) Coevolving protein residues: maximum likelihood identification and relationship to structure. J Mol Biol 287(1):187–198
26. Tuff P, Darlu P (2000) Exploring a phylogenetic approach for the detection of correlated substitutions in proteins. Mol Biol Evol 17 (11):1753–1759
27. Poon AFY, Lewis FI, Pond SLK, Frost SDW (2007) An evolutionary-network model reveals stratified interactions in the V3 loop of the HIV-1 envelope. PLoS Comput Biol 3(11): e231
28. Talavera D, Lovell SC, Whelan S (2015) Covariation is a poor measure of molecular coevolution. Mol Biol Evol 32(9):2456–2468
29. Fodor AA, Aldrich RW (2004) Influence of conservation on calculations of amino acid covariance in multiple sequence alignments. Proteins Struct Funct Bioinf 56(2):211–221
30. Pearl J (1986) Fusion, propagation, and structuring in belief networks. Artif Intell 29 (3):241–288
31. Friedman N, Koller D (2003) Being Bayesian about network structure. A Bayesian approach to structure discovery in Bayesian networks. Mach Learn 50(1–2):95–125
32. Pond SLK, Frost SDW, Muse SV (2005) HyPhy: hypothesis testing using phylogenies. Bioinformatics 21(5):676–679
33. Delport W, Poon AFY, Frost SDW, Kosakovsky Pond SL (2010) Datamonkey 2010: a suite of phylogenetic analysis tools for evolutionary biology. Bioinformatics 26 (19):2455–2457
34. Poon AFY, Lewis FI, Frost SDW, Kosakovsky Pond SL (2008) Spidermonkey: rapid detection of co-evolving sites using Bayesian graphical models. Bioinformatics 24 (17):1949–1950
35. Stamatakis A (2014) RAxML version 8: a tool for phylogenetic analysis and post-analysis of large phylogenies. Bioinformatics 30 (9):1312–1313
36. Guindon S, Dufayard JF, Lefort V, Anisimova M, Hordijk W, Gascuel O (2010) New algorithms and methods to estimate maximum-likelihood phylogenies: assessing the performance of PhyML 3.0. Syst Biol 59 (3):307–321
37. Price MN, Dehal PS, Arkin AP (2010) FastTree 2–approximately maximum-likelihood trees for large alignments. PLoS ONE 5(3): e9490
38. Holmes S (2003) Bootstrapping phylogenetic trees: theory and methods. Stat Sci 18:241–255
39. Muse SV, Gaut BS (1994) A likelihood approach for comparing synonymous and nonsynonymous nucleotide substitution rates, with application to the chloroplast genome. Mol Biol Evol 11(5):715–724
40. Yang Z (1993) Maximum-likelihood estimation of phylogeny from DNA sequences when substitution rates differ over sites. Mol Biol Evol 10(6):1396–1401
41. Felsenstein J, Churchill GA (1996) A hidden Markov model approach to variation among sites in rate of evolution. Mol Biol Evol 13 (1):93–104
42. Swofford D, Begle DP (1993) PAUP: Phylogenetic analysis using parsimony, Version 3.1, March 1993. Center for Biodiversity, Illinois Natural History Survey
43. Tamura K, Nei M (1993) Estimation of the number of nucleotide substitutions in the control region of mitochondrial DNA in humans and chimpanzees. Mol Biol Evol 10 (3):512–526
44. Posada D (2003) Using MODELTEST and PAUP* to select a model of nucleotide substitution. Curr Protoc Bioinformatics 6–5. https://doi.org/10.1002/0471250953.bi0605s00
45. Maddison DR, Swofford DL, Maddison WP (1997) NEXUS: an extensible file format for systematic information. Syst Biol 46 (4):590–621
46. Joy JB, Liang RH, McCloskey RM, Nguyen T, Poon AFY (2016) Ancestral reconstruction. PLoS Comput Biol 12(7):e1004763
47. Nielsen R (2002) Mapping mutations on phylogenies. Syst Biol 51(5):729–739
48. Pupko T, Pe I, Shamir R, Graur D (2000) A fast algorithm for joint reconstruction of ancestral amino acid sequences. Mol Biol Evol 17 (6):890–896
49. Ellson J, Gansner E, Koutsofios L, North SC, Woodhull G (2001) Graphviz—open source graph drawing tools. In: International symposium on graph drawing. Springer, Berlin, pp 483–484
50. Shannon P, Markiel A, Ozier O, Baliga NS, Wang JT, Ramage D, Amin N,

Schwikowski B, Ideker T (2003) Cytoscape: a software environment for integrated models of biomolecular interaction networks. Genome Res 13(11):2498–2504

51. Bastian M, Heymann S, Jacomy M et al (2009) Gephi: an open source software for exploring and manipulating networks. In: Proceedings of the third international ICWSM conference, vol 8, pp 361–362
52. Simmonds P (2004) Genetic diversity and evolution of hepatitis C virus–15 years on. J Gen Virol 85(11):3173–3188
53. Blach S, Zeuzem S, Manns M, Altraif I, Duberg AS, Muljono DH, Waked I, Alavian SM, Lee MH, Negro F et al (2017) Global prevalence and genotype distribution of hepatitis C virus infection in 2015: a modelling study. Lancet Gastroenterol Hepatol 2 (3):161–176
54. Campo D, Dimitrova Z, Mitchell RJ, Lara J, Khudyakov Y (2008) Coordinated evolution of the hepatitis C virus. Proc Natl Acad Sci 105 (28):9685–9690
55. Aurora R, Donlin MJ, Cannon NA, Tavis JE (2009) Genome-wide hepatitis C virus amino acid covariance networks can predict response to antiviral therapy in humans. J Clin Invest 119(1):225–236
56. McCloskey RM, Liang RH, Joy JB, Krajden M, Montaner JS, Harrigan PR, Poon AF (2014) Global origin and transmission of hepatitis C virus nonstructural protein 3 Q80K polymorphism. J Infect Dis 211(8):1288–1295
57. Poveda E, Wyleș DL, Mena Á, Pedreira JD, Castro-Iglesias Á, Cachay E (2014) Update on hepatitis C virus resistance to direct-acting antiviral agents. Antivir Res 108:181–191
58. Combet C, Garnier N, Charavay C, Grando D, Crisan D, Lopez J, Dehne-Garcia A, Geourjon C, Bettler E, Hulo C et al (2006) euHCVdb: the European hepatitis C virus database. Nucleic Acids Res 35(Suppl_1): D363–D366
59. Katoh K, Standley DM (2013) MAFFT multiple sequence alignment software version 7: improvements in performance and usability. Mol Biol Evol 30(4):772–780
60. Larsson A (2014) AliView: a fast and lightweight alignment viewer and editor for large datasets. Bioinformatics 30(22):3276–3278
61. Darriba D, Taboada GL, Doallo R, Posada D (2012) jModelTest 2: more models, new heuristics and parallel computing. Nat Methods 9 (8):772
62. Guindon S, Gascuel O (2003) A simple, fast, and accurate algorithm to estimate large phylogenies by maximum likelihood. Syst Biol 52 (5):696–704
63. Yu G, Smith DK, Zhu H, Guan Y, Lam TTY (2017) ggtree: an R package for visualization and annotation of phylogenetic trees with their covariates and other associated data. Methods Ecol Evol 8(1):28–36
64. Plummer M, Best N, Cowles K, Vines K (2006) CODA: convergence diagnosis and output analysis for MCMC. R News 6(1):7–11
65. Gelman A, Rubin DB (1992) Inference from iterative simulation using multiple sequences. Stat Sci 7:457–472
66. Ranjith-Kumar C, Kao CC (2006) Biochemical activities of the HCV NS5B RNA-dependent RNA polymerase. In: Tan S (ed) Hepatitis C viruses: genomes and molecular biology. Horizon Bioscience, Norfolk, pp 293–310
67. Hong Z, Cameron CE, Walker MP, Castro C, Yao N, Lau JY, Zhong W (2001) A novel mechanism to ensure terminal initiation by hepatitis C virus NS5B polymerase. Virology 285(1):6–11

Chapter 7

Context-Dependent Mutation Effects in Proteins

Frank J. Poelwijk

Abstract

Defining the extent of epistasis—the nonindependence of the effects of mutations—is essential for understanding the relationship of genotype, phenotype, and fitness in biological systems. The applications cover many areas of biological research, including biochemistry, genomics, protein and systems engineering, medicine, and evolutionary biology. However, the quantitative definitions of epistasis vary among fields, and the analysis beyond just pairwise effects can be problematic. Here, we demonstrate the application of a particular mathematical formalism, the weighted Walsh-Hadamard transform, which unifies a number of different definitions of epistasis. We provide a computational implementation of such analysis using a computer-generated higher-order mutational dataset. We discuss general considerations regarding the null hypothesis for independent mutational effects, which then allows a quantitative identification of epistasis in an experimental dataset.

Key words Epistasis, Higher-order epistasis, Context-dependent mutations, Amino acid interactions, Evolutionary biology, Fitness, Combinatorial mutagenesis

1 Introduction

From the lowest to the highest level of biological organization, from biomolecules to ecosystems, the world is shaped by interactions, which make that a biological system as a whole is not simply the sum of its parts. Interactions may lead to unexpected behavior and complex dynamics, which represent both experimental and conceptual challenges for the development of predictive models. To unravel these complexities, a general strategy is to separate the consequences of modifying individual components into direct/independent effects and effects that are dependent on the configuration of other components. In genetics, the context-dependent effects are referred to as "epistasis," a term coined by William

Electronic supplementary material: The online version of this chapter (https://doi.org/10.1007/978-1-4939-8736-8_7) contains supplementary material, which is available to authorized users.

Tobias Sikosek (ed.), *Computational Methods in Protein Evolution*, Methods in Molecular Biology, vol. 1851,
https://doi.org/10.1007/978-1-4939-8736-8_7, © Springer Science+Business Media, LLC, part of Springer Nature 2019

Bateson in 1907 to indicate a genetic interaction in which the state of one allele "masks" the effect of an allele at another locus [1]. This definition was later formalized by Ronald Fisher to refer to a statistical interaction that causes a deviation from additivity [2].

Since then, many alternative measures for epistasis have been developed (*see*, e.g., [3, 4]), which have been used with various levels of quantitativeness. One aspect that often remains underexposed is the establishment of an explicit null model: what does it mean quantitatively for two mutations to act *in*dependently? In most cases independence is equated with additivity or multiplicativity of effects; however, without consideration of the underlying system, such choices are arbitrary. Additionally, empirical datasets usually exhibit overall nonlinearities, for example, due to a limited linear range of the measurement or to a saturating organismal fitness, which, if ignored, lead to an overestimate of the prevalence of epistasis.

In principle, quantitative information on epistasis should help make meaningful descriptions of proteins and, potentially, more complex biological systems by capturing the complexity without an overabundance of parameters. For example, in ref. 5, it was shown how epistatic analysis can identify the cooperative unit in a PDZ-binding domain and a potassium ion channel. Many fields of study should benefit from a firmer grasp of the "typical" level and distribution of epistatic interactions. In molecular evolution, as well as in laboratory evolution experiments, such knowledge could guide our expectation for the repeatability and variability of adaptation, owing to the direct link between epistasis and the accessibility of evolutionary trajectories under selection [6–9]. From the opposite end, knowing the prevalence of epistasis in genes and noncoding sequences may improve phylogenetic reconstruction methods by explicitly incorporating context-dependent effects (*see*, e.g., [10]).

The current protocol illustrates the analysis of a combinatorially complete set of mutations in a protein, which are phenotypically assessed by means of an assay with some inherent nonlinearities. In this dataset, statistically, significant two-way and multi-way interactions between the mutations are identified.

1.1 Walsh-Hadamard Transform and Epistasis

In this protocol we will calculate the epistasis present in a complete combinatorial mutant dataset, i.e., a set that contains all mutants that can be made by recombination of two parental protein sequences that differ at N positions. More precisely, we start with a vector $\bar{y}$ containing the phenotypic measurements for all 2^N combinations of mutations that can be generated at N positions, where each position has two states. Calculating epistasis in such a dataset consists of a linear mapping of vector of 2^N phenotypes $\bar{y}$

onto a vector of 2^N epistatic coefficients $\overline{\omega}$, using an epistasis operator Ω of dimension $2^N \times 2^N$.

$$\overline{\omega} = \Omega\, \bar{y} \tag{1}$$

The specific choice of the matrix elements for operator Ω determines what quantitative definition of epistasis is being calculated [5]. Here we focus on background-averaged epistasis. In this definition, each epistatic coefficient is averaged over all states of the positions not involved in that term. For example, two-way epistasis is calculated as the differential effect of mutating *a* to *A* in the presence of either *b* or *B*, averaged over all states of the remaining positions *C*, *D*, *E*, *F*, . . ., etc. Three-way epistasis between *A*, *B*, and *C* is averaged over all backgrounds involving positions *D*, *E*, *F*,. . ., etc. This definition of epistasis does not require a particular genotype as a reference, unlike the more traditional definition of epistasis where this averaging does not take place [5]. For this reason, the traditional definition of epistasis is also referred to as "local" epistasis (local in sequence space) and background-averaged epistasis as "global." The elements in the operator for global epistasis are defined by the Hadamard matrix [11], *H*, weighted by the entries in a diagonal matrix *V* specifying how many genetic backgrounds a certain epistatic term is averaged over.

$$\overline{\omega} = VH\, \bar{y} \tag{2}$$

The two matrices comprising the operator for background-averaged epistasis can be generated with a recursive definition:

$$V_{n+1} = \begin{pmatrix} \frac{1}{2}V_n & 0 \\ 0 & -V_n \end{pmatrix} \quad \text{and} \quad H_{n+1} = \begin{pmatrix} H_n & H_n \\ H_n & -H_n \end{pmatrix} \tag{3}$$

where $V_0 = 1$, $H_0 = 1$, and $n = \{0, \ldots, N-1\}$. I mention in passing that calculating epistasis using the Hadamard formalism can be seen as a decomposition of the fitness landscape into features of different (genotypic) length scales, analogous to a Fourier decomposition of some temporal signal in its frequency components (*see*, e.g., [12–14]).

The inverse transformation, which reconstructs the data points $\bar{y}$ from complete knowledge of the epistatic coefficients $\overline{\omega}$ in the system, is given by

$$\bar{y} = H^{-1} V^{-1}\, \overline{\omega} \tag{4}$$

In this protocol the inverse will be used at several points, for example, by generating an initial dataset of mutant phenotypes from a computationally generated vector of epistatic coefficients.

Two remarks are in order here. First, throughout this protocol, it is assumed we have a *complete* combinatorial dataset, so that calculating epistasis from phenotypic data (and vice versa) is a one-to-one mapping where no information is lost. In general, since the number of possible combinations grows exponentially with the number of mutable positions, N, the set of observed mutant phenotypes will not be complete. Powerful methods can be applied to estimate epistatic coefficients for incomplete data (*see*, e.g., [15]), and in fact, a main reason for performing epistatic analysis is to be able to predict missing phenotypes from a measured subset of all possibilities. Second, here, every position is assumed to have two possible states, which, for a protein, implies two possible amino acids per position. This assumption is made for simplicity because an extension of the Hadamard formalism to an arbitrary number of states per position is not straightforward. In this way, this protocol can focus on two key parts of epistatic analysis: overall nonlinearities in the dataset and identifying significant epistatic terms.

1.2 Overall Nonlinearities and the Null Hypothesis

Most experimental assays will exhibit overall nonlinearities, meaning that our observation of a quality of interest $\bar{x}$ is "distorted" according to some nonlinear function $f(\bar{x})$ (Fig. 1a, b). This can be specific to the measurement, for example, a limited linear range of fluorescence detection in a flow cytometer or a limited linear concentration range in a binding assay due to non-specific binding. Additionally, it can be inherent to the biological system, for example, a saturating dependence of protein expression on an activator's binding affinity. If such nonlinearities are not taken into account, the empirical dataset may appear more epistatic than it actually is (Fig. 1b). To meaningfully quantify epistasis, an explicit null hypothesis needs to be expressed, defining what it means for mutations to act independently. Note that the null model also addresses the question of whether mutational independence implies additivity or multiplicativity of effects: in fact, additivity in a quantity of interest in φ can appear in the dataset as multiplicativity if the assay measures a quantity with φ in the exponent, for example, when we measure equilibrium dissociation constants but are interested in epistasis with respect to the binding free energy. If we have sufficient knowledge about the system, we can directly choose the applicable nonlinear scaling (in the case of dissociation constants, this would be their logarithm) and define independence as additivity. In general, especially when the system is more complex, we can remove (part of) the overall nonlinearities using a linear-nonlinear optimization (*see* [16, 17] for similar approaches). Here, the vector $\bar{y}$ containing the observables is transformed using a nonlinear function $g(\bar{y})$ that only has a small number of free parameters, after which we attempt to optimize those parameters by maximizing the variance captured with first-order or low-order epistatic

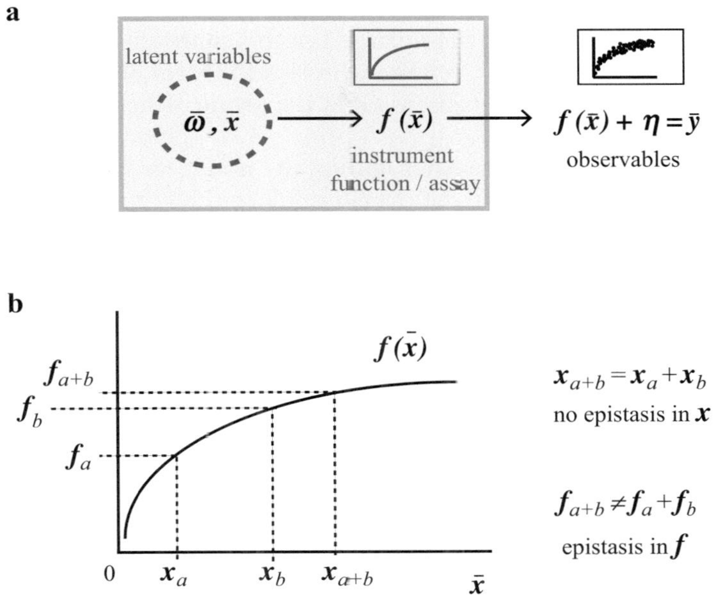

Fig. 1 Latent variables and observables. (**a**) Here the biological system of interest is represented by variable $\bar{x}$ that can be decomposed into its epistatic components $\bar{\omega}$. However, $\bar{x}$ is latent, and we can only observe its effects after some nonlinear transformation $f(\bar{x})$ has occurred, that may indicate saturation in the experimental assay, or an instrument function. Experimental noise is modeled through a random variable η, so that the observed phenotypic data are given by $\bar{y} = f(\bar{x}) + \eta$. (**b**) Calculating epistasis requires an explicit null model. Trivial nonlinearities in the transfer function $f(\bar{x})$ can result in apparent epistasis in the observable $\bar{y}$ without epistasis in the underlying latent variable $\bar{x}$

coefficients. Mathematically, reconstituting a dataset $g(\bar{y})$ from first-order epistatic coefficients is achieved by the transformation

$$g_1(\bar{y}) = H^{-1} V^{-1} S_1 \; VH \; g(\bar{y}) \tag{5}$$

Here, from right to left after the equal sign, first, all epistatic coefficients are calculated by multiplication with VH. Then this vector is multiplied by matrix S_1, the identity matrix with entries $S_{1,\ ii} = 0$ at positions that do not pertain to first-order (linear) terms, so that first-order epistatic terms are kept intact, but everything else is set to zero. Lastly, the inverse transformation $H^{-1}V^{-1}$ reconstructs the data using only the information contained in the first-order terms. The linear-nonlinear optimization procedure now consists of finding the values for the parameters in function g that minimize the quantity

$$h = \frac{\mathrm{var}(g(\bar{y}) - g_1(\bar{y}))}{\mathrm{var}(g(\bar{y}))} \tag{6}$$

which is the sum of squares of the residuals divided by the total sum of squares. For this approach to be successful, there are a number of requirements for the form of the nonlinear function, which will be discussed in the protocol steps and the Notes.

1.3 Error Propagation and Significant Terms

After finding the nonlinear transformation that optimally removes the overall nonlinearities, epistatic coefficients can be determined according to

$$\overline{\omega} = VH\, g(\bar{y}) \tag{7}$$

Since this is a one-to-one mapping from measurements to epistasis, the full set of 2^N epistatic coefficients in $\overline{\omega}$ also captures any measurement noise that is present in the transformed dataset $g(\bar{y})$. As the error for epistatic terms propagates exponentially, with a factor two for each increasing order [5], the effects of noise are more pronounced for higher-order terms. To prevent overfitting, this protocol illustrates a self-consistent approach that determines the noise contributions and establishes a significance threshold for epistatic coefficients of each order. We will show at the end of the protocol that these significant terms allow reconstruction of our original computer-generated model data at high accuracy and with little modeled measurement noise. Not surprisingly, when measurement noise is too large, this approach will break down.

2 Materials

The accompanying computer script is written in MATLAB, but the implementation is kept as simple as possible for straightforward translation to other languages. Most programming languages contain explicit implementations of the Hadamard matrix, but when lacking, the matrices can be easily generated using the recursive definitions provided.

3 Methods

Steps correspond to sections in the accompanying MATLAB script.

3.1 Generate a Combinatorial Mutant Dataset

1. Initialize the parameters and matrices used for the calculation of the epistasis operator (*see* **Note 1**) and its inverse. We use auxiliary variables A and B to indicate which positions are involved in an epistatic term and what the order of that term is.
2. Generate a vector $\overline{\omega}$ of length 2^N containing the epistatic contributions. The entries are generated randomly according to a model of preferential attachment, where higher-order terms are more likely to be non-zero if they involve positions

with non-zero lower-order terms (*see* **Note 2**). The generating function PrefAttach() contains two variables, frac and dExp, setting the fraction of non-zero first-order terms and the decay rate of this fraction for higher-order terms.

3. Generate the phenotypic data by the inverse transform, $\bar{x} = H^{-1} V^{-1} \overline{\omega}$ (*see* **Note 3**). This, in an experimental setting, is the latent variable (Fig. 1a) prior to potential nonlinear scaling due to the measurement instrument or the assay conditions and without measurement noise. This is the quantity for which we aim to map the epistatic contributions.
4. Apply a nonlinear transfer function, mimicking a limited linear range of the measurement instrument or assay, $\bar{x} \to f(\bar{x})$. Here we have chosen a function that saturates for high values of $\bar{x}$:

$$f(\bar{x}) = \frac{c \log(1 + \bar{x})}{1 + \log(1 + \bar{x})}$$

with c as a free parameter. Since the linear-nonlinear optimization procedure below is invariant under scaling and translation ($\bar{x} \to \frac{\bar{x}-a}{b}$), without loss of generality, we can move the original range of values of $\bar{x}$ into an unsaturated part of the transfer function by choosing values for the variables scale and shift (*see also* **Note 4**).
5. Add Gaussian "measurement noise," $f(\bar{x}) \to f(\bar{x}) + \eta$ (*see* **Note 5**). The generated noisy data plays the role of the observables, designated by $\bar{y}$ (Fig. 1a). Plot the observables $\bar{y}$ as a function of the latent variable $\bar{x}$ (Fig. 2a). This completes the generation of the phenotypic dataset.

3.2 Removing the Overall Nonlinearities

1. Decide on the general form of the nonlinear scaling $g(\bar{y})$ to be tested. Ideally, g is the inverse of f, and we have $g(\bar{y}) \approx \bar{x}$, but in general, f is not known accurately. The particular functional form can be chosen based on knowledge of the instrument transfer function or the experimental assay. If we do not have such knowledge about the system, monotonically increasing or decreasing test functions can be tried that have properties consistent with the expected nonlinearities, e.g., saturation for high values of the phenotypic data $\bar{x}$ (*see* **Note 6**). Here, for simplicity, we assume we know the transfer function f but up to some constant c, which we will try to find using the linear-nonlinear optimization. The chosen test function is therefore the inverse of f:

$$g(c, \bar{y}) = e^{\frac{\bar{y}}{c-\bar{y}}} - 1$$

2. Initialize the parameters for the linear-nonlinear optimization. If the transfer function for which we chose the test function has pronounced saturating behavior, a small amount of

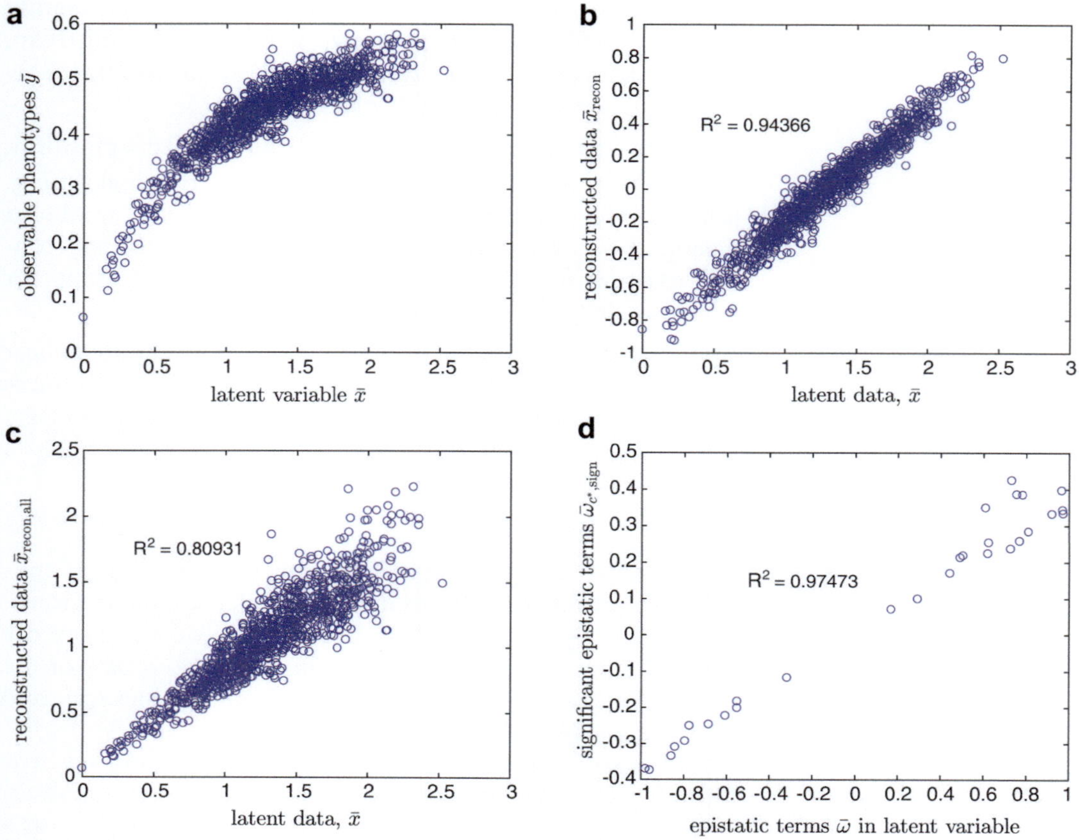

Fig. 2 Example data generated in the protocol. (**a**) Observable mutant phenotypes $\bar{y}$ versus latent variable $\bar{x}$. A clear nonlinear relation is present. (**b**) After removing overall nonlinearities, data is reconstructed using the obtained significant epistatic terms. A linear relation with a high correlation can be observed, provided that the measurement noise is not too large. (**c**) Reconstructed data from all epistatic terms (i.e., the data directly after applying the nonlinear scaling). Large noise contributions are present for data points at larger values of $\bar{x}$. Comparing this to panel **b** shows the noise suppression that can be achieved when only significant epistatic terms are used for reconstruction. (**d**) Comparison of the values of the significant epistatic terms and their counterparts in the latent data

measurement noise will yield extreme variability of the data points in the saturated regime (*see* Fig. 1b). To counteract this, data points in $\bar{y}$ are down-weighted during the optimization according to the slope $g'(c, \bar{y})$ of the test function (*see* **Note 7**). The parameter wExp sets the extent of this down-weighting, and epsilon determines the values for the slope where no down-weighting occurs.

3. Perform the optimization. Here, since there is only a single optimization parameter, we simply scan through its range and record the quantity $h(c) = \frac{\mathrm{var}(g(c,\bar{y}) - g_1(c,\bar{y}))}{\mathrm{var}(g(c,\bar{y}))}$ (where g_1 is defined in Eq. 5). *See* **Note 8**.

4. Find the value c^* that minimizes $h(c)$. With this value, we can perform the nonlinear transformation of the observables $\bar{y} \rightarrow g(c^*, \bar{y})$.

3.3 Epistatic Analysis, Noise Propagation, Significant Terms

1. Calculate the epistatic coefficients of the transformed data: $\overline{\omega}_{c^*} = VH\, g(c^*, \bar{y})$.
2. Make histograms of the epistatic coefficients, per order, with a reasonable number of terms per bin. Fit the histograms using Gaussian distributions and record the widths.
3. Perform a linear fit to the logarithms of widths, for intermediate orders. *See* **Notes 9** and **10**. Calculate the widths for low orders and high orders by extrapolation. *See* **Note 11**.
4. Use the widths to set the significance thresholds, using the Sĭdák correction for multiple testing, i.e., using the numbers of potential epistatic contributions per order.
5. Identify the epistatic coefficients with a value greater than the significance thresholds. The vector containing these terms, $\overline{\omega}_{c^*,\text{sign}}$, is the estimate for the epistasis present in the system.

3.4 Reconstruction of Phenotypes, Corroboration

1. Reconstitute the phenotypic data from significant epistatic contributions, according to $\bar{x}_{\text{recon}} = H^{-1}V^{-1}\,\overline{\omega}_{c^*,\text{sign}}$, and plot the reconstituted data $\bar{x}_{\text{recon}}$ versus the initially generated data $\bar{x}$. *See* **Note 12** and Fig. 2b.
2. Reconstitute the phenotypic data using all epistatic contributions, according to $\bar{x}_{\text{recon,all}} = H^{-1}V^{-1}\,\overline{\omega}_{c^*}$ (where $\overline{\omega}_{c^*}$ was calculated in Subheading 3.3, **step 1**), and plot the reconstituted data $\bar{x}_{\text{recon,all}}$ versus the initially generated data $\bar{x}$. *See* **Note 13** and Fig. 2c.
3. Plot $\overline{\omega}_{c^*,\text{sign}}$ (calculated in Subheading 3.4, **step 1**) versus the actual epistasis $\overline{\omega}$ in the system (Fig. 2d).

4 Notes

1. If the number of mutable positions is large, an explicit Hadamard matrix may impose a large burden on the available memory resources. Fast Hadamard transforms are available in MATLAB and other languages.
2. Preferential attachment is used here for the generation of the initial epistatic vector since it creates a relatively sparse distribution of epistatic terms which seems consistent with the limited amount of data that is available currently [15].
3. For the inverse epistasis operator, we use the simple explicit inverses of H and V: $H^{-1} = \frac{1}{2^N}\,H$ and $V_{ii}^{-1} = 1/V_{ii}$. It is

important for speed and accuracy to not use a general implementation for the inverse, such as inv() in MATLAB.

4. Linear-nonlinear optimization is invariant under transformation of the latent variable according to $\bar{x} \rightarrow \frac{\bar{x}-a}{b}$, because (1) an overall shift ($\frac{a}{b}$) will be absorbed in the zeroth-order epistatic term, which is the average phenotypic value of all data points, and (2) a linear scaling b will be accounted for by the linear (first-order) terms. An example of a shift and linear scaling is, respectively, the arbitrary zero point of a binding free energy ΔG_0, and its expression in either kcal/mol or kJ/mol, both of which physically do not matter to the system. In this protocol, the transformation was used to populate the unsaturated part of the transfer function. The invariance only holds if the transfer function is monotonic and non-singular (*see also* **Note 7**) and, strictly, if the measurement noise is zero. Practically, results are insensitive to low amounts of measurement noise.
5. For increasing levels of measurement noise, the approach presented here will break down, because the propagating noise will dominate the epistatic terms. This can be explored by varying the parameter noise level in the initial generation of $\bar{y}$.
6. In most practical cases, the chosen test function for the removal of nonlinearities $g(\bar{y})$ will not be the exact inverse of the transfer function $f(\bar{x})$. This is not necessarily an issue. The saturation effects present in the transfer function used in the current protocol can be reasonably well captured using other functions that have a monotonously decreasing slope, such as a fractional power $f(\bar{x}) = \bar{x}^{\alpha}$, where $0 < \alpha < 1$, with test function $g(\bar{y}) = \bar{y}^{1/\alpha}$.
7. Apart from the requirement that the modeled transfer function $f(\bar{x})$ is monotonically increasing or decreasing for the linear-nonlinear optimization procedure to work, there are a number of other considerations to keep in mind. Depending on the choice for the test function $g(\bar{y})$, the optimization procedure may encounter some issues if for some range of $\bar{y}$, the slope $g'(\bar{y})$ becomes very large (indicating saturation in $f(\bar{x})$) or if $g(\bar{y})$ contains singular points (indicating an asymptote in $f(\bar{x})$). With a sharply saturating function $f(\bar{x})$, any measurement noise present in $\bar{y}$ will lead to large uncertainties after applying the test function $g(\bar{y})$. In the current protocol, this effect is remediated in two ways: (1) while calculating the minimization quantity $h(c)$, data points are weighted by the inverse of the slope $g'(\bar{y})$, so that data with potentially large uncertainties will be down-weighted, and (2) the calculation of residuals $g(c, \bar{y}) - g_1(c, \bar{y})$ is replaced by a procedure where first a linear fit is made between $g(c, \bar{y})$ and $g_1(c, \bar{y})$ using a robust estimator for the slope (the Theil-Sen estimator [18, 19]). The more

severe case where $f(\bar{x})$ exhibits asymptotic behavior, and thus $g(\bar{y})$ does not exist for certain $\bar{y}$, can be addressed with more sophisticated approaches. In that case some data points have to be ignored, and epistasis has to be calculated based on an incomplete combinatorial dataset (*see*, e.g., [15]).

8. Since there is only one optimization parameter here, we perform a scan through its values, rather than a search procedure. When there are multiple parameters, a nonlinear solver such as fmincon() in MATLAB can perform a constrained minimization.
9. Only widths at intermediate epistatic orders are included in the fit. Low orders may have a large fraction of significant contributions, which would lead to an overestimate of the noise width. Very high orders are noisier because they have fewer epistatic terms to generate the histograms and the terms are also averaged over fewer genetic backgrounds.
10. Since background-averaged epistatic terms of a certain order are differences between two terms of a lower order, but also averaged over half as many genetic backgrounds, the error increases by a factor 2 per order (*see* ref. 5). Therefore the expectation that the width of the Gaussian distributions of epistatic terms increases by a factor 2 per order serves as a consistency check in the analysis.
11. The intersection of the linear fit with the *y*-axis (zeroth-order epistasis) is a measure for the error on the mean for the measurement noise (here the noise in $g(c^*, \bar{y})$) or the per-data point noise divided by a factor $\sqrt{2^N}$.
12. Directly comparing results to $\bar{x}$ and $\bar{\omega}$ is impossible in an experimental setting, because $\bar{x}$ and $\bar{\omega}$ are latent variables (Fig. 1a). In the current protocol, the initial data was computationally generated, allowing for the corroboration in Subheading 3.4. Generally, the effect of the nonlinear scaling can be assessed by verifying that fewer epistatic terms are necessary to accurately reconstitute the scaled data compared to the unscaled data.
13. Comparing the phenotypic data reconstructed using the significant epistatic terms ($\bar{x}_{\text{recon}}$) and using all epistatic terms ($\bar{x}_{\text{recon,all}}$) to the initial data ($\bar{x}$) illustrates two things. First, there is indeed a linear relationship between $\bar{x}_{\text{recon}}$ and $\bar{x}$ (potentially shifted and linearly scaled (*see* Subheading 3.1, **step 4** and **Note 4**)). In fact, in this example, a mere 6% of variance in the dataset is not captured using the significant epistatic terms alone (but *see also* **Note 5**). Second, leaving out the nonsignificant terms has the additional benefit of removing noise contributions that originate from the steep parts of the test function (Fig. 2c).

Acknowledgments

I thank Michael A. Stiffler and DerZen Fan for critical reading of the manuscript.

References

1. Bateson W (1907) Facts limiting the theory of heredity. Science 26:649–660
2. Fisher RA (1918) The correlation between relatives on the supposition of Mendelian inheritance. Trans Roy Soc Edinb 52:399–433
3. Phillips PC (1998) The language of gene interaction. Genetics 149:1167–1171
4. Phillips PC (2008) Epistasis—the essential role of gene interactions in the structure and evolution of genetic systems. Nat Rev Genet 9:855–867
5. Poelwijk FJ, Krishna V, Ranganathan R (2016) The context-dependence of mutations: a linkage of formalisms. PLoS Comput Biol 12: e1004771
6. Weinreich DM, Watson RA, Chao L (2005) Perspective: sign epistasis and genetic constraint on evolutionary trajectories. Evolution 59:1165–1174
7. Weinreich DM, Delaney NF, Depristo MA, Hartl DL (2006) Darwinian evolution can follow only very few mutational paths to fitter proteins. Science 312:111–114
8. Poelwijk FJ, Kiviet DJ, Weinreich DM, Tans SJ (2007) Empirical fitness landscapes reveal accessible evolutionary paths. Nature 445:383–386
9. Poelwijk FJ, T̃nase-Nicola S, Kiviet DJ, Tans SJ (2011) Reciprocal sign epistasis is a necessary condition for multi-peaked fitness landscapes. J Theor Biol 272:141–144
10. Siepel A, Haussler D (2004) Phylogenetic estimation of context-dependent substitution rates by maximum likelihood. Mol Biol Evol 21:468–488
11. Beer T (1981) Walsh transforms. Am J Phys 49:466–472
12. Stoffer DS (1991) Walsh-Fourier analysis and its statistical applications. J Am Stat Assoc 86:461–479
13. Weinberger E (1991) Fourier and Taylor series on fitness landscapes. Biol Cybernetics 65:321–330
14. Stadler PF (2002) Spectral landscape theory. In: Crutchfield JP, Schuster P (eds) Evolutionary dynamics—exploring the interface of selection, accident, and function. Oxford University Press, Oxford, pp 231–272
15. Poelwijk FJ, Socolich M, Ranganathan R (2017) High-order epistasis linking genotype and phenotype in a protein. Submitted
16. Otwinowski J, Nemenman I (2013) Genotype to phenotype mapping and the fitness landscape of the *E. coli* lac promoter. PLoS One 8: e61570
17. Sailer ZR, Harms MJ (2017) Detecting high-order epistasis in nonlinear genotype-phenotype maps. Genetics 205:1079–1088
18. Theil H (1950) A rank-invariant method of linear and polynomial regression analysis. I, II, III, Nederl Akad Wetensch Proc 53: 386–392, 521–525, 1397–1412
19. Sen PK (1968) Estimates of the regression coefficient based on Kendall's tau. J Am Stat Assoc 63:1379–1389

Chapter 8

High-Throughput Reconstruction of Ancestral Protein Sequence, Structure, and Molecular Function

Kelsey Aadland, Charles Pugh, and Bryan Kolaczkowski

Abstract

Ancestral protein sequence reconstruction is a powerful technique for explicitly testing hypotheses about the evolution of molecular function, allowing researchers to meticulously dissect how historical changes in protein sequence impacted functional repertoire by altering the protein's 3D structure. These techniques have provided concrete, experimentally validated insights into ancient evolutionary processes and help illuminate the complex relationship between protein sequence, structure, and function. Inferring the protein family phylogenies on which ancestral sequence reconstruction depends and reconstructing the sequences, themselves, are amenable to high-throughput computational analysis. However, determining the structures of ancestral-reconstructed proteins and characterizing their functions typically rely on time-consuming and expensive laboratory analyses, limiting most current studies to examining a relatively small number of specific hypotheses. For this reason, we have little detailed, unbiased information about how molecular function evolves across large protein family phylogenies. Here we describe a generalized protocol that integrates ancestral sequence reconstruction with structural homology modeling and structure-based molecular affinity prediction to characterize historical changes in protein function across families with thousands of individual sequences. We highlight key steps in the analysis protocol requiring particularly careful attention to avoid introducing potential errors as well as steps for which computationally efficient subroutines can be substituted for more intensive approaches, allowing researchers to scale the analysis up or down, depending on available resources and requirements for reproducibility and scientific rigor. In our view, this approach provides a compelling compliment to more laboratory-intensive procedures, generating important contextual information that can help guide detailed experiments.

Key words Ancestral sequence reconstruction, Structural modeling, Protein function prediction, Affinity prediction, Protein evolution, Molecular evolution

1 Introduction

Computational reconstruction of ancestral protein sequences has been used as a foundation to test molecular-evolutionary hypotheses [1], illuminate how protein structure impacts molecular

---

Electronic supplementary material: The online version of this chapter (https://doi.org/10.1007/978-1-4939-8736-8_8) contains supplementary material, which is available to authorized users.

Tobias Sikosek (ed.), *Computational Methods in Protein Evolution*, Methods in Molecular Biology, vol. 1851,
https://doi.org/10.1007/978-1-4939-8736-8_8, © Springer Science+Business Media, LLC, part of Springer Nature 2019

function [2], and provide functional variation to support protein engineering [3, 4]. Ancestral sequence reconstruction (ASR) is computationally efficient, allowing thousands of ancestral sequences to be inferred in a few seconds using modern computers [5]. However, structural and functional characterization of ancestral sequences can be much more costly and time-consuming, limiting current ASR studies to examining at most a handful of sequences in detail [4, 6–15].

A typical ASR study begins with a sequence alignment of representatives from the protein family of interest and a phylogenetic tree describing the evolutionary relationships among the sequences as well as the relative rates of amino acid substitutions and other evolutionary model parameters needed to adequately describe the molecular-evolutionary process. Using the aligned extant sequences, phylogenetic tree and fully specified evolutionary model, the marginal posterior probability of each amino acid residue at each position in the alignment and each ancestral node on the phylogeny can be calculated [5]. Most evolutionary models do not infer gap states in ancestral sequences, requiring either a secondary reconstruction process to infer insertion-deletion events [16, 17] or use of an explicit insertion-deletion (indel) model as part of the sequence reconstruction [18, 19].

Although genes from extant organisms can be sequenced with little uncertainty, probabilistic reconstruction of ancestral sequences necessarily generates some uncertainty at each position in the sequence. Most ASR studies reconstruct "maximum likelihood" ancestral sequences, choosing the most probable residue at each position in the alignment. Although this approach is expected to minimize errors in the ancestral sequence, even if the evolutionary model is correct and every position is inferred with very high posterior probability, sequence errors can become unavoidable. For example, the probability of at least one sequence error is 1.0 for a 100-residue sequence having every position reconstructed with posterior probability 0.99. Many studies evaluate the impact of ASR error on functional inferences using a "robustness" approach, in which plausible alternative residues (typically non-maximum-likelihood residues reconstructed with posterior probability >0.3) are introduced into the maximum-likelihood sequence to observe the effects on functional characteristics [11, 12, 20–22].

Some researchers have argued that maximum-likelihood ASR might introduce functional biases, particularly when many sequence positions have largely additive contributions to protein function [23]. Maximum-likelihood ASR can also introduce biases in state frequency distributions under some conditions, making them inappropriate for explicitly examining the evolution of state frequencies [24, 25]. Sampling a large number of ancestral sequences directly from the posterior probability distributions at each position has been suggested as a complimentary approach to maximum-likelihood reconstruction [26, 27]. However, functionally characterizing a large ensemble of ancestral sequences requires significantly more laboratory resources. In addition, interpreting the results of such a study is difficult, as many ancestral sequences are expected to include relatively large numbers of potential errors likely degrading protein expression or function.

Although the accuracy of ancestral sequence reconstruction has not been thoroughly evaluated, current studies suggest that the ASR process is likely to be highly accurate, provided a reasonable evolutionary model is used [16, 28]. Interestingly, phylogenetic uncertainty has little impact on ASR accuracy or uncertainty, as the conditions under which the phylogeny becomes less certain also make ancestral sequences increasingly similar across plausible trees [29]. We might expect strong violations of the assumed evolutionary model—such as changes in site-specific evolutionary rates or changes in amino acid substitutability—to reduce ASR accuracy, as model violations have been shown to introduce other phylogenetic errors [30–32]; however, this has not been examined. The impact of alignment error on ASR accuracy has also not been systematically evaluated, although we might expect alignment accuracy to be particularly important for the robustness of ASR.

Once ancestral protein sequences of interest have been identified computationally, genes encoding the ancestral proteins can be synthesized and subjected to nearly any molecular or cellular assay commonly used to examine the functions of extant proteins. Ancestral protein function is typically evaluated using a highly simplified in vitro model system, minimizing the impact of any potential "mismatch" between the ancestral protein and the extant genomic context within which it is being evaluated [4, 6–15, 27, 33–35]. However, some recent studies have begun examining the cellular functions of ancestral proteins [20, 36, 37], with the caveat that an ancestral protein's function within a modern cellular context may not accurately reflect its functional role within the appropriate ancestral cell.

The most detailed ASR studies attempt to identify the specific historical substitutions responsible for changes in a protein's molecular function, as well as the structural mechanisms through which the substitutions impact molecular function. Once a functional difference between two sequential ancestral sequences—typically an ancestor-descendent pair on the phylogeny—has been observed, derived residues can be introduced into the ancestral sequence by site-directed mutagenesis, in order to identify the patterns of substitutions necessary to recapitulate the observed functional shift [4, 6–15, 27, 33–35]. Structural determination of the ancestral proteins can then be used to identify the mechanisms by which changes in sequence impact molecular function [2, 38].

Although ASR studies have provided one of the few experimental means for directly examining historical evolutionary hypotheses, their reliance on rigorous experimental techniques has also severely limited the scope with which these approaches can be applied. Compared to the efficiency of computational reconstruction of ancestral sequences, the synthesis of ancestral genes, expression of ancestral proteins, and functional-structural characterization in the lab require an enormous expenditure of time and money, limiting nearly all current ASR studies to examining only a few functional shifts at key positions in the protein family phylogeny, typically chosen a priori by the investigator based on patterns of gene duplication or speciation. Although undoubtedly useful, by focusing on parts of the tree where we believe function might change, this approach has potentially biased our view of how function evolves across large protein families.

Here we describe a complimentary ASR protocol that combines ancestral sequence reconstruction with high-throughput structural modeling and structure-based affinity prediction to evaluate the evolution of protein function across all ancestral nodes on the phylogeny, potentially providing a more unbiased view of how protein function evolves. This approach can be used to directly evaluate hypotheses about large-scale patterns of functional evolution and guide traditional lab-based functional characterization efforts, expanding both the scope and efficiency of ASR studies.

2 Methods

2.1 Sequence Collection and Protein Family Curation

The first step in any ancestral sequence reconstruction (ASR) study is the collection and curation of protein sequences from the family of interest. In almost every case, the root of the protein family under study is of interest, requiring the collection and curation of "outgroup" sequences. The goal of this step is to collect all available members of the protein family under study—including appropriate outgroup sequences—and no sequences that are not members of the target protein family or outgroup. Given the efficiency of modern phylogenetic analysis software, we see no reason to limit the amount of protein sequence data analyzed, other than elimination of potentially erroneous sequences and redundant sequences. Ideally, a roughly equal number of "ingroup" and outgroup sequences should be included in the analysis.

There are many approaches to collecting sequence data, nearly all of which rely on some form of sequence similarity search to identify members of the protein family of interest. The most common approach starts with a small number of well-annotated protein family members from heavily studied model organisms and collects homologs using some form of protein BLAST search, typically with an e-value cutoff of $1.0e^{-5}$ or an alternative value thought to be appropriate for the family under study. While this approach is probably adequate in most cases, care must be taken to avoid either including spurious homologs based on local hits to common conserved domains or missing distantly related homologs too dissimilar to the well-studied seed sequences to be detected by BLAST or similar approaches.

Here we develop an approach based on a domain architecture definition of the protein family. When available, protein functional domains—available through the Conserved Domain Database [39] or the PFam Database [40]—can provide a concise description of the expected sequence and structural features present in a protein family. Functional domains are typically encoded at the sequence level as high-level statistical models—either hidden Markov models [40] or position-specific scoring matrices [39]—that can be used to search sequence databases for domain-specific matches while avoiding potentially limiting dependencies on specific seed sequences.

In this example, we will use position-specific scoring matrices (PSSMs) from the Conserved Domain Database (CDD) to identify domain matches in NCBI's nr database. This analysis requires a local mirror of the nr database, the BLAST command-line tools, and the CDD, all available through NCBI. Note that some PSSM domain models may need to be rescaled from a scaling factor of 100 to a scaling factor of 1, in order to be used in sequence database searches. The following python script should perform the requisite rescaling (note that this script will overwrite the existing PSSM file):

rescalePSSM.py[1]:

```
#!/usr/bin/env python
import sys, re

fname = sys.argv[1]
f = open(fname, 'r')
content = f.read()
f.close()

start = content.find("scores {")
end = content.find("}", start)
scores = re.sub("([0-9]{2})(?=[\r\n\,])", "", content[start:end])
scores = re.sub("\ (\-)?,", " 0,", scores)
f = open(fname, 'w')
f.write(content[0:start] + scores + content[end:].replace("scalingFactor 100",
"scalingFactor 1"))
f.close()
```

The following BLAST command can then be run at the UNIX command prompt:

[1] The Python scripts described in this chapter are hosted with the online version of the book.

```
psiblast -in_pssm cd00021.smp -db nr -out cd00021.nrhits.csv -outfmt '10 qstart qend
sstart send sacc ssciname stitle evalue qlen sseq' -evalue 0.01 -max_target_seqs
10000000
```

Given the rescaled CDD domain model from cd00021.smp, this command will identify all sequences in the nr database matching the domain model, using an e-value cutoff of 0.01, which is sufficient to separate true domain matches from spurious hits for most complex, globular domains. Matching sequences will be printed to a comma-delimited text file, cd00021.nrhits.csv, which includes information about the matching sequence's accession, the species it is found in, and the location in the full-length protein where the match was identified. One useful approach for eliminating potentially spurious single-domain hits is to eliminate any potential matches of <70–80% of the expected domain length, which can be estimated by examining domain hits from well-annotated model organisms or evaluated based on the length of the PSSM domain model. The following python script can be used to eliminate partial domain hits:

```
removePartialDomainHits.py:
#!/usr/bin/env python
import sys

min_prop = 0.75

handle = open(sys.argv[1],"r")
for line in handle:
    linearr   = line.strip().split(",")
    hitlength = (int(linearr[3]) - int(linearr[2])) + 1
    qlen      = int(linearr[-2])
    if float(hitlength) / float(qlen) >= min_prop:
        sys.stdout.write(line)
handle.close()
```

Most protein families will not consist of a single functional domain, and most functional domains can be found across multiple families. In order to identify only members of the target protein family, it is up to the researcher to develop a "minimal domain architecture" that is believed to capture all members of the target family and not match any other protein families. For each domain in the minimal domain architecture, executing the psiblast command above will identify all protein sequences encoding that domain. Each list of domain hits should then be culled by eliminating partial hits. Finally, individual domain hits must be combined by protein accession to identify all the functional domains in each protein and their ordering. The following Python script should form a reasonable starting point for such an analysis:

```
combineDomainHits.py:
#!/usr/bin/env python
import sys, glob

combined_hits = {}

for f in glob.glob("*.nrhits.csv"):
    domname = f.split(".nrhits.csv")[0]
    handle = open(f, "r")
    for line in handle:
        linearr = line.strip().split(",")
        acc = linearr[4]
        spp = linearr[5]
        beg = int(linearr[2])
        end = int(linearr[3])
        if acc in combined_hits.keys():
            combined_hits[acc].append((beg,end,domname))
        else:
            combined_hits[acc] = [spp,(beg,end,domname)]
    handle.close()
for acc in combined_hits.keys():
    spp  = combined_hits[acc][0]
    doms = combined_hits[acc][1:]
    doms.sort()
    sys.stdout.write("%s,%s" % (acc,spp))
    for (b,e,name) in doms:
        sys.stdout.write(",%s:%d..%d" % (name,b,e))
    sys.stdout.write("\n")
```

Any sequences not matching the defined "minimal domain architecture" should be eliminated. Typically, this will require expert manual curation by a researcher familiar with the protein family under study, its expected diversity in domain architecture, and some familiarity with the quality of genome annotations across the species in which the protein family is present. Common genome annotation errors—such as domain duplications and deletions—can produce variation in domain architecture that is artefactual rather than biological, and it is up to the researcher whether to include or exclude such data. As all subsequent analyses will depend on the initial sequence data set, it is our view that time invested in insuring the initial data collection and curation are both reliable and thorough can pay huge dividends downstream, potentially boosting analysis power and reducing the likelihood of errors.

Once a suitable data set of curated ingroup and outgroup sequences has been identified, full-length protein sequences can be downloaded from the nr database in FASTA format using the blastdbcmd tool:

```
blastdbcmd -db nr -entry NP_004169 -outfmt %f -out unaligned.
fasta
```

Sequences corresponding to individual functional domains can also be downloaded by supplying starting and ending ranges:

```
blastdbcmd -db nr -entry NP_004169 -range 100-200 -outfmt %f
-out unaligned.100-200.fasta
```

Downloading a large number of sequences can be accomplished by entering each accession (and range) on a single line of a text file and using the –entry_batch flag.

2.2 Alignment and Phylogenetic Tree Inference

Multiple sequence alignment forms the basis for inferring the protein family phylogeny and reconstructing ancestral sequences. Conceptually, aligning protein sequences amounts to making residue-level statements of homology: aligned residues from two different sequences are inferred to have arisen from a common ancestor; aligning a residue from one sequence to a gap in another sequence ("-") amounts to inferring that the residue in the first sequence does not have a homologous residue in the second sequence, either due to an insertion in the first or a deletion in the second. Most approaches to phylogenetic inference treat alignment and tree inference as separate problems, first aligning the sequences and then inferring the most likely tree, given that alignment. However, there are approaches that attempt to simultaneously infer the sequence alignment and the phylogeny [19, 41–44].

Unfortunately, different sequence alignment algorithms can produce different residue-level statements of homology, even when overall alignment accuracies are similar [45–47]. Additionally, some regions of the alignment may be easier to infer—e.g., highly conserved functional domains—while other regions may be more error-prone [47]. Any errors in the sequence alignment can potentially impact phylogenetic inference and ancestral sequence reconstruction [48–50]. Ideally, we would like to eliminate alignment errors, but this is typically not possible.

Here we take a "robustness approach" to sequence alignment and phylogenetic inference; we use a variety of alignment strategies to generate a large number of plausible sequence alignments, use each of these alignments to infer the protein family phylogeny, and then combine these inferences—both formally and informally—to identify the protein family phylogeny most "robust" to uncertainty in the sequence alignment.

As an example, we will use the popular (and relatively fast) alignment algorithms from clustalw2 [51, 52], muscle [53], and mafft [54] to align protein sequences. If computational resources permit, other alignment programs such as msaprobs [55], probalign [56], probcons [57], and tcoffee [58] can also be used.

Assuming unaligned sequences have been collected in FASTA format, aligning full-length protein sequences is trivial at the UNIX command line:

```
clustalw2  -output=fasta  -infile=unaligned.fulllength.fasta  -outfile=aligned_clustalw.fasta
muscle     -in unaligned.fulllength.fasta     -out aligned_muscle.fasta
einsi      unaligned.fulllength.fasta        > aligned_einsi.fasta
```

Note that we are using the einsi algorithm in mafft as a general alignment strategy for divergent, multidomain proteins. For convenience, we are standardizing all alignment formats to FASTA and using a file-naming convention of aligned_<algorithm_name>.fasta to facilitate scripting.

In addition to full-length sequence alignments, we recommend extracting the functional domains encoding the protein family's previously defined "minimal domain architecture" from each sequence and aligning them, without including variable domains or intervening sequences, using the same alignment algorithms. Removing variable domains and intervening sequences simplify the alignment process and avoid alignment errors due to highly divergent linker sequences or variation in domain architecture across the sequences under study [47].

Another approach to reducing potential alignment errors is to identify and remove potentially ambiguous alignment columns using an objective alignment-processing methodology [59–62]. The most commonly used approach is Gblocks, a simple and efficient ad hoc alignment-processing algorithm [63]. In our analyses, we typically set the minimum number of sequences for a flank position (-b2) equal to 3/5 of the total number of sequences in the alignment, the maximum number of nonconserved positions (-b3) to 10 and the minimum block length (-b5) to 5. We also typically allow gap positions (-b5 = a). The specific parameter values will probably vary, depending on the protein family under study; ideally, we want to remove potentially unreliable alignment regions while leaving enough data to reconstruct a well-supported phylogeny. Each sequence alignment can be processed at the command line to generate a trimmed alignment with potentially ambiguous regions removed.

```
Gblocks aligned_clustalw.fasta -b2=576 -b3=10 -b4=5 -b5=a
Gblocks aligned_muscle.fasta   -b2=576 -b3=10 -b4=5 -b5=a
Gblocks aligned_einsi.fasta    -b2=576 -b3=10 -b4=5 -b5=a
```

One approach we will use to incorporate alignment uncertainty is an "elision" technique [64], which is similar to a "supermatrix" approach for species tree reconstruction [65]; only in this case are we concatenating different alignments of the same protein sequences, rather than concatenating alignments of different gene families. An "elision" alignment can be constructed from individual sequence alignments using the following python script:

```
makeElision.py:
#!/usr/bin/env python
import sys, glob

def parseFasta(infname):
    alnlen = 0
    alignment = {}
    handle = open(infname, "r")
    line = handle.readline()
    while line:
        if line[0] == ">":
            id = line[1:].strip()
            seq = ""
            line = handle.readline()
            while line and line[0] != ">":
                seq += line.strip()
                line = handle.readline()
            alnlen = len(seq)
            alignment[id] = seq
        else:
            line = handle.readline()
    handle.close()
    return (alnlen, alignment)

allids = []
handle = open("unaligned.fulllength.fasta", "r")
for line in handle:
    if line[0] == ">":
        allids.append(line[1:].strip())
handle.close()
full_aln = {}
for id in allids:
    full_aln[id] = ""

for fname in glob.glob("aligned_*.fasta"):
    (alnlen,aln) = parseFasta(fname)
    for id in allids:
        if id in aln.keys():
            full_aln[id] += aln[id]
        else:
            full_aln[id] += ("-" * alnlen)
```

```
outf = open("aligned_elision.fasta", "w")
for id in allids:
    outf.write(">%s\n%s\n" % (id, full_aln[id]))
outf.close()
```

Each sequence alignment can be used to infer a protein family phylogeny using standard tree inference techniques, such as maximum parsimony, neighbor joining, maximum likelihood, or Bayesian inference. For this example, we will use maximum likelihood phylogenetic inference, as it is considered one of the most accurate approaches [66, 67], and existing software is computationally efficient [68–70].

Statistical phylogenetic inference methods like maximum likelihood and Bayesian inference rely on a probabilistic model of the molecular-evolutionary process, and different models can result in different tree topologies [71, 72]. Although most commonly used models typically have little impact on the main branching pattern of the tree, it is advisable to select the best-fit model for each sequence alignment using a statistical model selection procedure. Perhaps the most widely used approach is to select the best-fit model using either the Akaike information criterion (AIC) or the Bayesian information criterion (BIC), both of which are implemented in ProtTest [73]:

```
prottest3 -i aliged_clustalw.fasta -G -F -all -o modelSelection.clustalw.results.txt
```

Once a suitable evolutionary model has been selected, a maximum likelihood phylogenetic tree can be rapidly inferred using FastTree [68]; for example:

```
FastTree -pseudo -lg aligned_clustalw.fasta > clustalw.fasttree.tre
```

where the –lg parameter specifies the amino acid transition model [74]. FastTree is extremely computationally efficient and can produce highly accurate phylogenetic inferences on its own [69]. Clade support is reported as SH-like aLRT scores, which have been demonstrated to be statistically powerful but typically slightly conservative support estimates [75, 76]. In our analyses, we usually consider SH-like aLRT scores >0.8 to be strongly supported clades. One potential problem with FastTree output is that redundant sequences are collapsed to polytomies in the inferred phylogeny, which can be incompatible with some other analysis programs. This problem can be avoided by removing completely redundant sequences from the analysis prior to alignment and tree inference.

If the best-fit evolutionary model is not available in FastTree, or if the inferred phylogeny is to be refined using what might be considered a more rigorous tree search algorithm, the FastTree phylogeny can be used as a starting tree to perform a maximum-likelihood inference using RAxML [70]:

```
raxml -m PROTCATLGX -f t -s aligned_clustalw.fasta -t clustalw.fasttree.tre -n
clustalwtree
```

RAxML has a large number of parameters that can be set to tune the evolutionary model, the tree search algorithm, and the clade support calculation.

After protein family phylogenies have been generated from each sequence alignment, we recommend manually examining the set of plausible trees to evaluate any strong consistencies or inconsistencies across alignments. Regions of the tree that are consistently recovered with strong clade support across alignments should generally be considered reliable, whereas regions that are weakly supported or that vary across alignments are less robust to alignment uncertainty.

In addition to this informal evaluation, consistency across alignments can be formally evaluated using a "supertree" approach that combines phylogenies inferred from each alignment into a single "consensus." To perform this analysis, use the supertree toolkit (STK) to convert the ensemble of phylogenetic inferences into a clade presence-absence matrix [77]:

```
cat tree1.tre tree2.tre ... treeN.tre > alltrees.tre
stk create_matrix -f nexus alltrees.tre alltrees.nexus
convertMatrix.py alltress.nexus > alltrees.fasta
```

Where convertMatrix.py is the following python script that converts NEXUS to FASTA format:

```
#!/usr/bin/env python
import sys

fname = sys.argv[1]
handle = open(fname, "r")
readmatrix = False
for line in handle:
    linearr = line.split()
    if len(linearr) > 0 and linearr[0] == "matrix":
        readmatrix = True
        continue
    if readmatrix:
        if len(linearr) > 0 and linearr[0] == ";":
            break
        elif len(linearr) > 0:
            id = linearr[0]
```

```
            seq = linearr[1].replace("?","-")
            sys.stdout.write(">%s\n%s\n" % (id,seq))
handle.close()
```

The clade presence-absence matrix can then be used to reconstruct a "consensus supertree" using a binary likelihood model and clade support evaluated as for any other alignment:

```
raxml -m BINCATX -f t -s alltrees.fasta -t startingTree.tre -n supertree
raxml -m BINCATX -f J -s alltrees.fasta -t RAxML_bestTree.supertree -n supertreeSH
```

Note that clade support generated from the concatenated "elision" alignment and the supertree approach are summaries of consistency across different alignments, and do not necessarily reflect the support for a given clade generated by each individual alignment. A weakly supported clade that is consistently recovered across alignment strategies should generally be given less credibility than a clade that is consistently recovered with high support. For this reason, we suggest that researchers investigate both consensus support from the elision alignment and supertree inferences and clade support generated by individual alignments, which can be summarized by providing the maximum, minimum, and mean ± standard deviation support for a given clade across alignment strategies.

2.3 Ancestral Sequence and Insertion-Deletion Reconstruction

Given an alignment and a phylogeny, marginal maximum-likelihood ancestral sequence reconstructions can be computed across all ancestral nodes very efficiently:

```
raxml -f A -m PROTCATLGF -t tree.tre -s alignment.fasta -n
alignment.ancseqs
```

Maximum-likelihood ancestral sequences will be aligned to the input sequence alignment and are available in:

```
RAxML_marginalAncestralStates.alignment.ancseqs
```

RAxML numbers ancestral nodes based on a tree traversal algorithm; a node-labeled tree is provided, so the researcher can map ancestral sequence identifiers to node labels:

```
RAxML_nodeLabelledRootedTree.alignment.ancseqs
```

Finally, the marginal posterior probability distributions across all residues at each site and ancestral node are provided, allowing the researcher to evaluate alternative reconstructions:

```
RAxML_marginalAncestralProbabilities.alignment.ancseqs
```

Although there are some algorithms that attempt to reconstruct ancestral insertion-deletion events (indels) explicitly [17, 19], the "standard" ASR algorithm treats gaps as missing data, so ancestral indels need to be reconstructed separately and then integrated into the sequence reconstruction. Here we perform this task using a simple binary likelihood model applied to the presence-absence sequence alignment. First, convert the amino acid alignment to a presence-absence alignment using the following python script:

```
makePresenceAbssenceMatrix.py:
#!/usr/bin/env python
import sys

handle = open(sys.argv[1], "r")
line = handle.readline()
while line:
    if line[0] == ">":
        id = line[1:].strip()
        seq = ""
        line = handle.readline()
        while line and line[0] != ">":
            seq += line.strip()
            line = handle.readline()
        indelseq = ""
        for c in seq:
            if c == "-":
                indelseq += "0"
            else:
                indelseq += "1"
        sys.stdout.write(">%s\n%s\n" % (id,indelseq))
handle.close()
```

Next, reconstruct ancestral indels using RAxML:

```
raxml -f A -m BINCAT -t tree.tre -s alignment.presenceabsence.fasta -n alignment.
ancindels
```

Finally, combine ancestral sequence reconstructions with ancestral indel reconstructions using a python script:

```
putAncestralIndels.py:
#!/usr/bin/env python
import sys
```

```
SEQF = sys.argv[1]  # RAxML_marginalAncestralStates.alignment.ancseqs
INDF = sys.argv[2]  # RAxML_marginalAncestralStates.alignment.ancindels
ancseqs = {}

handle = open(SEQF,"r")
for line in handle:
    linearr = line.split()
    id = linearr[0]
    seq = linearr[1]
    ancseqs[id] = seq
handle.close()

handle = open(INDF,"r")
for line in handle:
    linearr = line.split()
    id = linearr[0]
    ins = linearr[1]
    seq = ancseqs[id]

    sys.stdout.write(">%s\n" % id)
    for i in range(len(seq)):
        if ins[i] == "0" or seq[i] == "?":
            sys.stdout.write("-")
        else:
            sys.stdout.write(seq[i])
    sys.stdout.write("\n")
handle.close()
```

Although previous studies have suggested that ancestral sequence reconstruction is expected to be highly congruent across plausible tree topologies, it is possible to integrate reconstructions across a set of input trees [29] or to use Bayesian approaches to integrate reconstructions over model parameter values, rather than relying on maximum-likelihood inference [19, 78]. It is up to the researcher to evaluate the robustness of ancestral sequence reconstructions; we recommend adopting the common approach of evaluating alternative reconstructions if they are supported with >0.3 posterior probability. Bayesian sampled ancestral sequences can be generated using a python script (requires SciPy):

```
getBayesASR.py:
#!/usr/bin/env python
import sys
from scipy import stats
import random

if len(sys.argv) < 4:
    sys.stderr.write("randomASR.py N nodeID
```

```
RAxML_marginalAncestralProbabilities.ancseqs
RAxML_marginalAncestralProbabilities.ancindels\n")
      sys.stderr.write("  generates N random ancestral sequence 'draws' from the
marginal probability distributions\n")
    sys.stderr.write("  including indels\n")
    sys.exit(1)

nseqs           = int(sys.argv[1])
nodeid          = sys.argv[2]
ancseqprobfname = sys.argv[3]
ancindprobfname = sys.argv[4]

# read prob distributions for sequences #
seq_labels = ["A","R","N","D","C","E","Q","G","H","I","L","K","M","F","P","S","T",
"W","Y","V"]
seq_probdists = []
handle = open(ancseqprobfname, "r")
line = handle.readline()
while line and line.strip() != nodeid:
    line = handle.readline()
line = handle.readline()
while line:
    linearr = line.split()
    if len(linearr) < 20:
        break
    pdist = []
    idist = []
    i = 0
    for k in linearr:
        p = float(k)
        if p > 0.0:
            pdist.append(p)
            idist.append(i)
        i += 1
    seq_probdists.append(stats.rv_discrete(values=(idist,pdist)))
    line = handle.readline()
handle.close()

# read prob distributions for indels #
ind_labels = [0,1]
ind_probdists = []
handle = open(ancindprobfname, "r")
line = handle.readline()
while line and line.strip() != nodeid:
    line = handle.readline()
line = handle.readline()
while line:
    linearr = line.split()
```

```
    if len(linearr) < 2:
        break
    pdist = []
    idist = []
    i = 0
    for k in linearr:
        p = float(k)
        if p > 0.0:
            pdist.append(p)
            idist.append(i)
        i += 1
    ind_probdists.append(stats.rv_discrete(values=(idist,pdist)))
    line = handle.readline()
handle.close()

# now generate random sequences in FASTA format #
for i in range(nseqs):
    sys.stdout.write(">%s_rep%d\n" % (nodeid,i))
    seq = ""
    for k in range(len(seq_probdists)):
        if ind_probdists[k].rvs():
            seq += seq_labels[seq_probdists[k].rvs()]
        else:
            seq += "-"
    sys.stdout.write("%s\n" % seq)
```

Here we treat the inference of the protein family phylogeny and ancestral sequence reconstruction as separate problems with different objectives. Phylogenetic inference is concerned primarily with determining the most accurate and robust tree topology, whereas the goal of ancestral sequence reconstruction is to identify the most accurate and robust ancestral sequences, given the inferred protein family tree or ensemble of plausible trees. In order to maximize available data and improve statistical power, full-length sequences are typically used to infer the protein family tree. However, in most cases, only a subset of the protein family's functional domains will be used for ancestral reconstruction.

Any of the sequence alignments used for phylogenetic inference can be used to reconstruct ancestral sequences, and examining reconstruction congruence across plausible alignments could be a useful approach for characterizing ASR robustness. However, for this methodology, we assume the researcher has access to 3D structural information for the domain(s) of interest, which will be used to ultimately predict protein-ligand affinities. For cases in which multiple structures of the relevant functional domain(s) are available, we recommend performing a structure-based sequence alignment and using that alignment to reconstruct ancestral sequences. First, align 3D structures using modeler [79, 80], which can be run through a python script.

```
structureAlign.py:
#!/usr/bin/env python
import os
import sys
import glob
from modeller import *
import modeller.salign

if len(sys.argv) < 2:
     sys.stderr.write("usage: structAln.py directory\n")
   sys.stderr.write("  will perform an iterative structural alignment of all\n")
    sys.stderr.write("  the .pdb files in the input directory\n")
    sys.stderr.write("  the resulting alignment is printed to directory_it.pap\n")
    sys.stderr.write("  and directory_it.ali\n")
    sys.exit(1)

thedir = sys.argv[1]
if thedir[-1] == "/":
    thedir = thedir[:-1]

pdbfiles = glob.glob("%s/*.pdb" % thedir)

# set up environment for modeller #
log.verbose()
env = environ()
env.io.atom_files_directory = thedir
aln = alignment(env)

# add all the .pdb files to the modeller alignment object #
for pdb in pdbfiles:
    code = pdb.split("/")[-1].split(".pdb")[0]
    mdl = model(env, file=code)
    aln.append_model(mdl, atom_files=code, align_codes=code)

# perform the iterative structural alignment #
modeller.salign.iterative_structural_align(aln)
aln.write(file='%s_it.pap' % thedir, alignment_format='PAP')
aln.write(file='%s_it.ali' % thedir, alignment_format='PIR')
```

The python script requires modeler to be installed and produces a structure-based alignment of all PDB files in an input directory. The structure-based alignment is written in ALI format, which is very close to FASTA. This structure-based alignment will be used as a "seed" to align homologous domains from the protein family of interest; it is important that only the corresponding functional domains are aligned to the structure-based seed alignment. We will use mafft (ginsi) to align sequences to the seed.

```
ginsi --seed structalign_it.fasta unaligned_domains.fasta > structaligned_domains.
fasta
```

The seed sequences can then be removed prior to ancestral reconstruction. In our view, provided diverse 3D structures of the functional domain of interest are available, structure-based sequence alignment will generally give better results in this application than sequence-based alignment, which can tend to misinterpret highly divergent sequences as insertion/deletion events. Protein structure tends to be much more conserved than amino acid sequence over long evolutionary distances, so structure-based alignments should provide a more appropriate platform for downstream structural modeling and affinity prediction [81, 82].

2.3.1 Structural Modeling and Optimization

It is widely thought that proteins function primarily through their three-dimensional structure, which determines the spatial distribution of biochemical properties and its dynamics [83–85]. Here we will exploit this structural basis for molecular function to provide high-throughput molecular affinity predictions across ancestral and extant protein sequences. We will use structural homology modeling to infer 3D structures of protein sequences for which empirical structures are not available [79]. To facilitate downstream affinity predictions, we will need at least one empirical structure of the functional domain of interest from a protein family member or distantly related homolog in complex with a ligand of interest, which can be a small molecule, DNA/RNA, or another protein. Most often, this will be retrieved from the Protein Data Bank by sequence search [86]. Note that if the 3D structure of the functional domain of interest has not been empirically solved in complex with an appropriate ligand, it will need to be generated before proceeding with this protocol. Generating a starting protein-ligand complex is probably best done using an empirical structure-determination protocol, although de novo structure prediction or protein-ligand docking is an alternative method [87–91].

Once an appropriate protein-ligand complex has been generated, its protein sequence should be aligned to the same alignment used to infer ancestral sequences; this will ensure that all extant and ancestral protein sequences are aligned to the structural template, facilitating high-throughput structural modeling. Given an alignment of a protein sequence to a structural template, modeler can be used to generate 100 structural models and evaluate their accuracy using a number of validation scores [79].

```
generateStructuralModel.py:
#!/usr/bin/env python
from modeller import *
from modeller.automodel import *

#####------------          CONTROL VARIABLES       ------------ #####
##              you should only have to change these.            ##
ALNFILE = 'alignment.ali'       # alignment file
KNOWNS  = 'mystructure'         # name of template (known structure)
SEQ     = 'mysequenceID'        # name of target (sequence of unknown structure)
NMODELS = 100                   # number of models to build
##                                                               ##
#####------------    END CONTROL VARIABLES  ------------ #####

log.verbose()    # request verbose output
env = environ()  # create a new MODELLER environment to build this model in

# directories for input atom (structure) files
env.io.atom_files_directory = ['.','../..','../../..']

a = automodel(env,
       alnfile  = ALNFILE,
       knowns   = KNOWNS,
       sequence = SEQ,
       assess_methods=(assess.DOPE, assess.DOPEHR, assess.normalized_dope,
assess.GA341))
a.starting_model= 1
a.ending_model  = NMODELS
a.make()
```

This script will print a large amount of diagnostic information to the screen, including model assessment scores, which can be used to identify the highest-quality structural model(s) for downstream analysis. The format of the sequence-template alignment file ("alignment.ali" in this example) is modeler-specific and will depend on the particular complex used for homology-based structural modeling. Please consult the documentation for the specific version of modeler used to ensure that the alignment file format is correct. As an example, consider the following alignment.ali file, constructed using PDBID 3ADL:

```
>P1;ANC948
sequence:ANC948::::::0.00: 0.00
----QCDPDNDPSKTPISL-LSQLCEKRN-LCSPEYD------LVSQ-QG---P----
PH---TRTFTMRVTVGD----FV-F-QGT---GRSKKEAKHNAAEKMLDHLRQ-
CPDVPYPT--
/
........../
```

```
..........*

>P1;3ADL
structure:3ADL::A:::::0.00: 0.00
-------------SHEVGA-LQELVVQKG-WRLPEYT------VTQE-SG---P----
AH---RKEFTMTCRVER----FI-E-IGS---GTSKKLAKR-
NAAAKMLLRVHT---------
/
........../
..........*
```

In this case, an ancestral DSRM protein sequence (ANC948) has been aligned to the 3ADL protein structure; the gaps present in both sequences are artifacts created by extracting these two sequences out of a larger multi-sequence alignment and are inconsequential for structural modeling. The forward-slash characters (/) separate structural chains, and the dots (.) indicate non-amino acid molecules, a double-stranded RNA in this case, which is present in the structural template and will be transferred to the homology model.

Automating modeler to generate structural models for a large number of aligned sequences is relatively straightforward but does require a bit of organization. Given an alignment of protein sequence domains and an aligned structural template, the following python script will generate a directory hierarchy including separate directories for each sequence, construct 100 structural models for each sequence, and identify each sequence's best model. Note that this incorporates the previous generateStructuralModel.py script.

```
generateStructuralModels.py:
#!/usr/bin/env python
import sys
import glob
import os

## NOTE: you will need to change the PDBID to the template you are using ##
PDBID = "3ADL"

## the alignment will need to be modified for your system ##
alnstr = """
>P1;%s
sequence:%s:::::::0.00: 0.00
%s
/
........../
..........*

>P1;%s
```

```
structure:%s::A:::::0.00: 0.00
%s
/
........../
..........*
"""

modelerpy="""
#!/usr/bin/env python
from modeller import *
from modeller.automodel import *

#####------------         CONTROL VARIABLES         ------------ #####
##             you should only have to change these.             ##
ALNFILE = 'alignment.ali'         # alignment file
KNOWNS  = '%s'                    # name of template (known structure)
SEQ     = '%s'         # name of target (sequence of unknown structure)
NMODELS = 100                  # number of models to build
##                                                                ##
#####------------       END CONTROL VARIABLES     ------------ #####

log.verbose()    # request verbose output
env = environ()  # create a new MODELLER environment to build this model in

# directories for input atom files
env.io.atom_files_directory = ['.','../..','../../..']

a = automodel(env,
              alnfile  = ALNFILE,
              knowns   = KNOWNS,
              sequence = SEQ,
              assess_methods=(assess.DOPE, assess.DOPEHR))
a.starting_model= 1
a.ending_model  = NMODELS
a.make()
"""

def launchSeq(mydir,myid,myseq,number,mystruct):
    index = number / 100
    topdir = "D%d" % index
    if not os.path.exists("%s/%s" % (mydir,topdir)):
        os.mkdir("%s/%s" % (mydir,topdir))
    if not os.path.exists("%s/%s/%s" % (mydir,topdir,myid)):
        os.mkdir("%s/%s/%s" % (mydir,topdir,myid))
    wrkdir = "%s/%s/%s" % (mydir,topdir,myid)
    #write alignment file#
    handle = open("%s/alignment.ali" % wrkdir, "w")
```

```
    handle.write(alnstr % (myid,myid,myseq,PDBID,PDBID,mystruct))
    handle.close()
    #write modeler run file#
    handle = open("%s/runModeller.py" % wrkdir, "w")
    handle.write(modelerpy % (PDBID,myid))
    handle.close()
    os.system("chmod 775 %s/runModeller.py" % wrkdir)
    os.system("cp parseBestModel.py %s/" % wrkdir)
    os.chdir("%s/" % wrkdir)
    os.system("./runModeller.py > SCORES.txt")
    os.system("./parseBestModel.py SCORES.txt")
    os.chdir("../../../")
    sys.stderr.write("finished: %s %s\n" % (mydir,myid))

num = 0
dr = "models"

alnfilename   = sys.argv[1]     # alignment file, in FASTA
strucfilename = sys.argv[2]     # aligned structural template, in FASTA

# we assume the structural template is aligned to the sequence alignment, is
in FASTA format #
# and has the sequence all on one line (the second line of the file)
#
handle = open(structfilename, "r")
handle.readline()
structuraltemplate = handle.readline().strip()
handle.close()

handle = open(alnfname, "r")
line = handle.readline()
while line:
    if line[0] == ">":
        id = line[1:].strip()
        se = ""
        line = handle.readline()
        while line and line[0] != ">":
            se += line.strip()
            line = handle.readline()
        ## now have id and seq, build directories and launch! ##
        launchSeq(dr,id,se,num,structuraltemplate)
        num += 1
```

This analysis relies on another python script:

```
parseBestModel.py:
#!/usr/bin/env python
import os
import sys
```

```
"""
Calculate mean and standard deviation of data x[]:
    mean = {\sum_i x_i \over n}
    std = sqrt(\sum_i (x_i - mean)^2 \over n-1)
"""
def meanstdev(x):
    from math import sqrt
    n, mean, std = len(x), 0, 0
    for a in x:
        mean = mean + a
    mean = mean / float(n)
    for a in x:
        std = std + (a - mean)**2
    std = sqrt(std / float(n-1))
    return (mean, std)

scorefname = sys.argv[1]

handle = open(scorefname, "r")
line = handle.readline()
# skip over all the model-build output #
while line:
    linearr = line.split()
    if len(linearr)>3 and linearr[0]=="Filename" and linearr
    [1]=="molpdf" and linearr[2]=="DOPE":
        break
    line = handle.readline()

# read model scores #
models = []
molpdf = []
dope   = []
dopehr = []
handle.readline()
line = handle.readline()
while line:
    linearr = line.split()
    if len(linearr) > 3:
        models.append(linearr[0])
        molpdf.append(float(linearr[1]))
        dope.append(float(linearr[2]))
        dopehr.append(float(linearr[3]))
    line = handle.readline()
handle.close()
```

```
# need to scale by 2 * stdev #
(molpdf_mean, molpdf_stdev) = meanstdev(molpdf)
(  dope_mean,   dope_stdev) = meanstdev(dope  )
(dopehr_mean, dopehr_stdev) = meanstdev(dopehr)

newmolpdf  =  [(x-molpdf_mean)/(2.0*molpdf_stdev)  for  x  in
molpdf]
newdope   = [(x-  dope_mean)/(2.0*dope_stdev)   for x in dope  ]
newdopehr  =  [(x-dopehr_mean)/(2.0*dopehr_stdev)  for  x  in  do-
pehr]

# now calculate best model #
# ave of scaled scores     #
bestmodel = ""
bestscore = 100000000.0

print "model [molpdf dope dopehr] score"
printlines = []
for i in range(len(models)):
    score = newmolpdf[i] + newdope[i] + newdopehr[i]
     printlines.append((score,"%s  [%.3f(%.3f)  %.3f(%.3f)  %.3f
(%.3f)]  %.3f" %  (models[i], molpdf[i], newmolpdf[i], dope[i],
newdope[i], dopehr[i], newdopehr[i], score)))
    if score < bestscore:
        bestscore = score
        bestmodel = models[i]

printlines.sort(reverse=True)
for  (s,v)  in printlines:
    print v
print "BEST MODEL (out of %d): %s" % (len(models),bestmodel)

# get top 1 models #
printlines.sort()
NMODELS = 1
for i in range(NMODELS):
    (score,info) = printlines[i]
    model = info.split()[0]
    mname = info.split(".")[0]
    cmd = "cp %s ../%s.BESTMODEL_%d.pdb" % (model,mname,i)
    print cmd
    os.system(cmd)
```

In this case, the "best" structural model is determined by averaging over all model assessment scores, each of which is first scaled to units of standard deviation across the 100 constructed structural models. The researcher can increase the NMODELS variable to evaluate additional models, allowing assessment of robustness to variation in structural modeling. For very large data sets, structural homology modeling should probably be parallelized across multiple processors; the particular approach will depend on the interface available for parallelization on a specific computer system.

Protein-protein complexes are typically modeled appropriately by modeler, but other molecular systems—DNA/RNA or small chemical ligands—may be treated as "block" molecules during structural modeling, which fails to account for the specific biochemical properties of the ligand. In these cases, it is recommended that the resulting structural model be "optimized" to form favorable protein-ligand interactions, prior to affinity prediction. Here we will use pdb2pqr to perform a fast, force-field-based structural optimization [92]; a more computationally expensive alternative would be to optimize the protein-ligand structure using molecular dynamics simulation [93].

```
pdb2pqr --ff amber --chain inmodel.pdb outmodel.pqr
```

This optimization will need to be executed for each structural model used in the analysis.

2.4 Affinity Prediction and Visualization

A variety of structure-based affinity prediction tools are available, most of them tuned for predicting a protein's affinity for small chemical ligands such as drug leads [94–98]. Here we will use generalized linear models developed in our lab to perform structure-based affinity prediction; models are available for predicting protein-small molecule, protein-DNA/RNA, or protein-protein affinities [99, 100]. The software requires the protein and the ligand in separate files, so we will use the following python script to extract specific chains from the complex:

```
extractChains.py:
#!/usr/bin/env python
import sys

pdbfname = sys.argv[1]
chains   = sys.argv[2:]

handle = open(pdbfname, "r")
for line in handle:
    if len(line) > 21:
        chain = line[21]
        if chain in chains:
            sys.stdout.write(line)
handle.close()
```

Assuming the protein domain is chain A, and the ligand is encoded as chains B and C, separate protein and ligand files can be generated using:

```
extractChain.py complex.pdb A   > protein.pdb
extractChain.py complex.pdb B C > ligand.pdb
```

A ligand file in mol2 format is also required, which can be generated using the OpenBabel library [101]:

```
babel -ipdb ligand.pdb -omol2 ligand.mol2
```

Finally, GLM-Score is used to predict the affinity of the protein for its ligand [100]:

```
GLM-Score protein.pdb ligand.pdb ligand.mol2 protein
```

If the ligand is DNA/RNA, the last "protein" should be changed to "DNA," or the ligand type can be identified as "small molecule." Of course, affinity predictions will need to be made for each structural model, which can be scripted by the researcher or parallelized on a large supercomputer.

Affinity predictions are provided as pKds, which are $-\log_{10}$-transformed dissociation constants; larger pKds indicate tighter protein-ligand binding. Results will be available in a "ligand_result.txt" file.

Generating protein-ligand affinity predictions for every ancestral and extant sequence on a large protein family phylogeny can provide valuable information about how one aspect of protein function has evolved, but this information can be difficult to visualize. Here we develop an approach that maps affinity estimates onto the phylogeny used to generate ancestral sequences, coloring branches on a red-blue gradient based on predicted affinity, with high-affinity branches colored red and branches with low affinity colored blue.

First, pKd estimates are collected into a single text file, organized by protein identifier. We are assuming a directory structure in which each protein-ligand complex and resulting affinity prediction is stored in a directory named by the protein ID. Results will be placed in all_pkds.txt.

```
collectPKDs.py:
#!/usr/bin/env python
import sys
import glob

outf = open("all_pkds.txt", "w")
for f in glob.glob("*/*_result.txt"):
```

```
            seqid = f.split("/")[0]
            pkds = []
            handle = open(f,"r")
            for line in handle:
                linearr = line.split()
                if linearr[0] == "predicted" and linearr[1] == "pKD:":
                    pkds.append(float(linearr[2]))
            handle.close()
            pkds.sort(reverse=True)
            outf.write(seqid)
            for k in pkds:
                outf.write("\t%f" % k)
            outf.write("\n")
        outf.close()
```

If multiple affinity estimates were generated for each sequence, the mean affinity estimate could be calculated for display across the tree. The next step is to generate a phylogenetic tree that has both branch lengths and ancestral node identifiers; this will be stored in pkd_base_tree.tre.

```
createPKD_BaseTree.py:
#!/usr/bin/env python
import sys

# need to point to the tree with branch lengths and the labelled tree #
BL_TREE="ASRinput.tre"
LA_TREE="RAxML_nodeLabelledRootedTree.alignment.ancseqs"

handle = open(LA_TREE, "r")
nodetree = ""
for line in handle:
    nodetree += line.strip()
handle.close()

handle = open(BL_TREE, "r")
brlentree = ""
for line in handle:
    brlentree += line.strip()
handle.close()

outf = open("pkd_base_tree.tre", "w")

# parse trees
structurals = ["(",")",",",":",";"]
numericals  = ["0","1","2","3","4","5","6","7","8","9",".","e","E","-"]
i1 = 0
i2 = 0
```

```
        while i1 < len(nodetree):
            if nodetree[i1] in structurals:
                outf.write(nodetree[i1])
                i1 += 1
            else:
                label = ""
                while nodetree[i1] not in structurals:
                    label += nodetree[i1]
                    i1 += 1
                # get branch length #
                while i2 < len(brlentree) and brlentree[i2] != ":":
                    i2 +=1
                brlenstr = ""
                i2 += 1
                while i2 < len(brlentree) and brlentree[i2] in numericals:
                    brlenstr += brlentree[i2]
                    i2 += 1
                # print labelled information #
                if brlenstr == "":
                    brlenstr = "0.0"
                outf.write("%s:%s" % (label,brlenstr))

        outf.write("\n")
        outf.close()
```

Finally, we will convert pKd predictions to colors on a red-blue gradient. The researcher will have to set the lower and upper bounds on pKd values, based on the specific system under study.

```
colorPKDTree.py:
#!/usr/bin/env python
import sys

pkdfname = "all_pkds.txt"
trefname = "pkd_base_tree.tre"

# read pKd values #
pkd_map = {}
handle = open(pkdfname, "r")
handle.readline()
for line in handle:
    linearr = line.split()
    pkd = float(linearr[1])
    pkd_map[linearr[0]] = pkd
handle.close()

colors=["#0025e5","#1926d2","#3327c0","#4c28ae","#66299b",
"#7f2a89","#992b77","#b22c64","#cc2d52",'#e52e40","#ff302e"]
```

```
## the max and min pKd values should be chosen based on the system under study ##
new_max = 9.75
new_min = 4.75

# create break points for color gradient #
breaks = []
totalsize = new_max - new_min
bitsize   = totalsize / (len(colors)-1)
breaks.append(new_min)
for i in range(1,len(colors)-1,1):
    breaks.append(new_min+(bitsize*i))
breaks.append(new_max)

def getColor(nodename):
    if nodename not in pkd_map.keys():
        return "[&!color=#d3d3d3]"    ## missing data gets gray color ##
    else:
        pkd = pkd_map[nodename]
        for i in range(len(breaks)):
            if pkd < breaks[i]:
                return "[&!color=%s]" % colors[i]
    return "[&!color=%s]" % colors[-1]

handle = open(trefname, "r")
nodetree = ""
for line in handle:
    nodetree += line.strip()
handle.close()

sys.stdout.write("#nexus\nbegin trees;\n  tree t1 = [&R] ")

structurals = ["(",")",",",":",";"]
numericals  = ["0","1","2","3","4","5","6","7","8","9",".","e","E","-"]
i1 = 0
while i1 < len(nodetree):
    # parse branch length #
    if nodetree[i1] == ":":
        sys.stdout.write(nodetree[i1])
        i1 += 1
        brlen = ""
        while i1 < len(nodetree) and nodetree[i1] in numericals:
            brlen += nodetree[i1]
            i1 += 1
        sys.stdout.write(brlen)

    # write any tree structural information #
```

```
    elif nodetree[i1] in structurals:
        sys.stdout.write(nodetree[i1])
        i1 += 1

    else:
        label = ""
        while nodetree[i1] not in structurals:
            label += nodetree[i1]
            i1 += 1
        colorlabel = getColor(label)
        sys.stdout.write("%s%s" % (label,colorlabel))
sys.stdout.write("\nend trees;\n")
```

The resulting tree will be in NEXUS format and can be visualized using FigTree (Fig. 1). The observed patterns of changes in predicted protein-ligand affinities can be used to guide the experimental characterization of ancestral protein function.

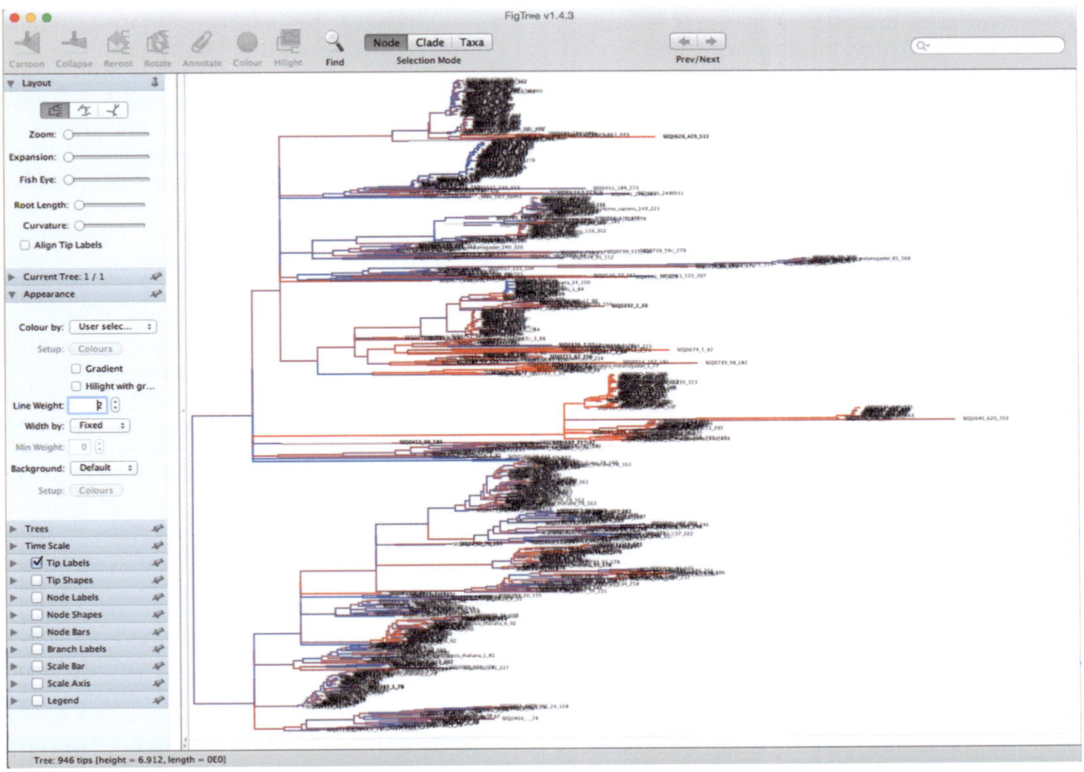

Fig. 1 Visualizing predicted affinities of extant and ancestral-reconstructed double-stranded RNA-binding motif (DSRM) domains for dsRNA targets using FigTree

References

1. Dean AM, Thornton JW (2007) Mechanistic approaches to the study of evolution: the functional synthesis. Nat Rev Genet 8 (9):675–688. https://doi.org/10.1038/nrg2160
2. Harms MJ, Thornton JW (2013) Evolutionary biochemistry: revealing the historical and physical causes of protein properties. Nat Rev Genet 14(8):559–571. https://doi.org/10.1038/nrg3540
3. Cole MF, Gaucher EA (2011) Exploiting models of molecular evolution to efficiently direct protein engineering. J Mol Evol 72 (2):193–203. https://doi.org/10.1007/s00239-010-9415-2
4. Ogawa T, Shirai T (2014) Tracing ancestral specificity of lectins: ancestral sequence reconstruction method as a new approach in protein engineering. Methods Mol Biol 1200:539–551. https://doi.org/10.1007/978-1-4939-1292-6_44
5. Yang Z, Kumar S, Nei M (1995) A new method of inference of ancestral nucleotide and amino acid sequences. Genetics 141 (4):1641–1650
6. Shih P, Malcolm BA, Rosenberg S, Kirsch JF, Wilson AC (1993) Reconstruction and testing of ancestral proteins. Methods Enzymol 224:576–590
7. Zmasek CM, Godzik A (2011) Strong functional patterns in the evolution of eukaryotic genomes revealed by the reconstruction of ancestral protein domain repertoires. Genome Biol 12(1):R4. https://doi.org/10.1186/gb-2011-12-1-r4
8. Whitfield JH, Zhang WH, Herde MK, Clifton BE, Radziejewski J, Janovjak H, Henneberger C, Jackson CJ (2015) Construction of a robust and sensitive arginine biosensor through ancestral protein reconstruction. Protein Sci 24(9):1412–1422. https://doi.org/10.1002/pro.2721
9. Malcolm BA, Wilson KP, Matthews BW, Kirsch JF, Wilson AC (1990) Ancestral lysozymes reconstructed, neutrality tested, and thermostability linked to hydrocarbon packing. Nature 345(6270):86–89. https://doi.org/10.1038/345086a0
10. Clifton BE, Jackson CJ (2016) Ancestral protein reconstruction yields insights into adaptive evolution of binding specificity in solute-binding proteins. Cell Chem Biol 23 (2):236–245. https://doi.org/10.1016/j.chembiol.2015.12.010
11. Bridgham JT, Carroll SM, Thornton JW (2006) Evolution of hormone-receptor complexity by molecular exploitation. Science 312 (5770):97–101. https://doi.org/10.1126/science.1123348
12. Bridgham JT, Ortlund EA, Thornton JW (2009) An epistatic ratchet constrains the direction of glucocorticoid receptor evolution. Nature 461(7263):515–519. https://doi.org/10.1038/nature08249
13. Voordeckers K, Brown CA, Vanneste K, van der Zande E, Voet A, Maere S, Verstrepen KJ (2012) Reconstruction of ancestral metabolic enzymes reveals molecular mechanisms underlying evolutionary innovation through gene duplication. PLoS Biol 10(12): e1001446. https://doi.org/10.1371/journal.pbio.1001446
14. Ugalde JA, Chang BS, Matz MV (2004) Evolution of coral pigments recreated. Science 305(5689):1433. https://doi.org/10.1126/science.1099597
15. van Hazel I, Sabouhanian A, Day L, Endler JA, Chang BS (2013) Functional characterization of spectral tuning mechanisms in the great bowerbird short-wavelength sensitive visual pigment (SWS1), and the origins of UV/violet vision in passerines and parrots. BMC Evol Biol 13:250. https://doi.org/10.1186/1471-2148-13-250
16. Hall BG (2006) Simple and accurate estimation of ancestral protein sequences. Proc Natl Acad Sci U S A 103(14):5431–5436. https://doi.org/10.1073/pnas.0508991103
17. Ashkenazy H, Penn O, Doron-Faigenboim A, Cohen O, Cannarozzi G, Zomer O, Pupko T (2012) FastML: a web server for probabilistic reconstruction of ancestral sequences. Nucleic Acids Res 40(Web Server issue): W580–W584. https://doi.org/10.1093/nar/gks498
18. Redelings BD, Suchard MA (2005) Joint Bayesian estimation of alignment and phylogeny. Syst Biol 54(3):401–418. https://doi.org/10.1080/10635150590947041
19. Suchard MA, Redelings BD (2006) BAli-Phy: simultaneous Bayesian inference of alignment and phylogeny. Bioinformatics 22 (16):2047–2048. https://doi.org/10.1093/bioinformatics/btl175
20. Anderson DP, Whitney DS, Hanson-Smith V, Woznica A, Campodonico-Burnett W, Volkman BF, King N, Thornton JW, Prehoda KE (2016) Evolution of an ancient protein

function involved in organized multicellularity in animals. Elife 5:e10147. https://doi.org/10.7554/eLife.10147

21. Thornton JW (2004) Resurrecting ancient genes: experimental analysis of extinct molecules. Nat Rev Genet 5(5):366–375. https://doi.org/10.1038/nrg1324
22. Chang BS, Jonsson K, Kazmi MA, Donoghue MJ, Sakmar TP (2002) Recreating a functional ancestral archosaur visual pigment. Mol Biol Evol 19(9):1483–1489
23. Williams PD, Pollock DD, Blackburne BP, Goldstein RA (2006) Assessing the accuracy of ancestral protein reconstruction methods. PLoS Comput Biol 2(6):e69. https://doi.org/10.1371/journal.pcbi.0020069
24. Matsumoto T, Akashi H, Yang Z (2015) Evaluation of ancestral sequence reconstruction methods to infer nonstationary patterns of nucleotide substitution. Genetics 200 (3):873–890. https://doi.org/10.1534/genetics.115.177386
25. Susko E, Roger AJ (2013) Problems with estimation of ancestral frequencies under stationary models. Syst Biol 62(2):330–338. https://doi.org/10.1093/sysbio/sys075
26. Pollock DD, Chang BS (2007) Dealing with uncertainty in ancestral sequence reconstruction: sampling from the posterior distribution. In: Liberles DA (ed) Ancestral sequence reconstruction. Oxford University Press, Oxford
27. Dias R, Manny A, Kolaczkowski O, Kolaczkowski B (2017) Convergence of domain architecture, structure, and ligand affinity in animal and plant RNA-binding proteins. Mol Biol Evol 34(6):1429–1444. https://doi.org/10.1093/molbev/msx090
28. Randall RN, Radford CE, Roof KA, Natarajan DK, Gaucher EA (2016) An experimental phylogeny to benchmark ancestral sequence reconstruction. Nat Commun 7:12847. https://doi.org/10.1038/ncomms12847
29. Hanson-Smith V, Kolaczkowski B, Thornton JW (2010) Robustness of ancestral sequence reconstruction to phylogenetic uncertainty. Mol Biol Evol 27(9):1988–1999. https://doi.org/10.1093/molbev/msq081
30. Kolaczkowski B, Thornton JW (2004) Performance of maximum parsimony and likelihood phylogenetics when evolution is heterogeneous. Nature 431(7011):980–984. https://doi.org/10.1038/nature02917
31. Blanquart S, Lartillot N (2006) A Bayesian compound stochastic process for modeling nonstationary and nonhomogeneous sequence evolution. Mol Biol Evol 23 (11):2058–2071. https://doi.org/10.1093/molbev/msl091
32. Blanquart S, Lartillot N (2008) A site- and time-heterogeneous model of amino acid replacement. Mol Biol Evol 25(5):842–858. https://doi.org/10.1093/molbev/msn018
33. Risso VA, Gavira JA, Mejia-Carmona DF, Gaucher EA, Sanchez-Ruiz JM (2013) Hyperstability and substrate promiscuity in laboratory resurrections of Precambrian beta-lactamases. J Am Chem Soc 135 (8):2899–2902. https://doi.org/10.1021/ja311630a
34. Korithoski B, Kolaczkowski O, Mukherjee K, Kola R, Earl C, Kolaczkowski B (2015) Evolution of a novel antiviral immune-signaling interaction by partial-gene duplication. PLoS One 10(9):e0137276. https://doi.org/10.1371/journal.pone.0137276
35. Pugh C, Kolaczkowski O, Manny A, Korithoski B, Kolaczkowski B (2016) Resurrecting ancestral structural dynamics of an antiviral immune receptor: adaptive binding pocket reorganization repeatedly shifts RNA preference. BMC Evol Biol 16(1):241. https://doi.org/10.1186/s12862-016-0818-6
36. Finnigan GC, Hanson-Smith V, Stevens TH, Thornton JW (2012) Evolution of increased complexity in a molecular machine. Nature 481(7381):360–364. https://doi.org/10.1038/nature10724
37. Kratzer JT, Lanaspa MA, Murphy MN, Cicerchi C, Graves CL, Tipton PA, Ortlund EA, Johnson RJ, Gaucher EA (2014) Evolutionary history and metabolic insights of ancient mammalian uricases. Proc Natl Acad Sci U S A 111(10):3763–3768. https://doi.org/10.1073/pnas.1320393111
38. Ortlund EA, Bridgham JT, Redinbo MR, Thornton JW (2007) Crystal structure of an ancient protein: evolution by conformational epistasis. Science 317(5844):1544–1548. https://doi.org/10.1126/science.1142819
39. Marchler-Bauer A, Derbyshire MK, Gonzales NR, Lu S, Chitsaz F, Geer LY, Geer RC, He J, Gwadz M, Hurwitz DI, Lanczycki CJ, Lu F, Marchler GH, Song JS, Thanki N, Wang Z, Yamashita RA, Zhang D, Zheng C, Bryant SH (2015) CDD: NCBI's conserved domain database. Nucleic Acids Res 43(Database issue):D222–D226. https://doi.org/10.1093/nar/gku1221
40. Finn RD, Bateman A, Clements J, Coggill P, Eberhardt RY, Eddy SR, Heger A, Hetherington K, Holm L, Mistry J, Sonnhammer EL, Tate J, Punta M (2014) Pfam: the protein families database. Nucleic Acids

Res 42(Database issue):D222–D230. https://doi.org/10.1093/nar/gkt1223

41. Yue F, Shi J, Tang J (2009) Simultaneous phylogeny reconstruction and multiple sequence alignment. BMC Bioinformatics 10 (Suppl 1):S11. https://doi.org/10.1186/1471-2105-10-S1-S11
42. Fleissner R, Metzler D, von Haeseler A (2005) Simultaneous statistical multiple alignment and phylogeny reconstruction. Syst Biol 54(4):548–561. https://doi.org/10.1080/10635150590950371
43. Herman JL, Challis CJ, Novak A, Hein J, Schmidler SC (2014) Simultaneous Bayesian estimation of alignment and phylogeny under a joint model of protein sequence and structure. Mol Biol Evol 31(9):2251–2266. https://doi.org/10.1093/molbev/msu184
44. Liu K, Warnow TJ, Holder MT, Nelesen SM, Yu J, Stamatakis AP, Linder CR (2012) SATe-II: very fast and accurate simultaneous estimation of multiple sequence alignments and phylogenetic trees. Syst Biol 61(1):90–106. https://doi.org/10.1093/sysbio/syr095
45. Nuin PA, Wang Z, Tillier ER (2006) The accuracy of several multiple sequence alignment programs for proteins. BMC Bioinformatics 7:471. https://doi.org/10.1186/1471-2105-7-471
46. Pervez MT, Babar ME, Nadeem A, Aslam M, Awan AR, Aslam N, Hussain T, Naveed N, Qadri S, Waheed U, Shoaib M (2014) Evaluating the accuracy and efficiency of multiple sequence alignment methods. Evol Bioinformatics Online 10:205–217. https://doi.org/10.4137/EBO.S19199
47. Thompson JD, Linard B, Lecompte O, Poch O (2011) A comprehensive benchmark study of multiple sequence alignment methods: current challenges and future perspectives. PLoS One 6(3):e18093. https://doi.org/10.1371/journal.pone.0018093
48. Ogden TH, Rosenberg MS (2006) Multiple sequence alignment accuracy and phylogenetic inference. Syst Biol 55(2):314–328. https://doi.org/10.1080/10635150500541730
49. Simmons MP, Muller KF, Webb CT (2011) The deterministic effects of alignment bias in phylogenetic inference. Cladistics 27 (4):402–416
50. Wang LS, Leebens-Mack J, Kerr Wall P, Beckmann K, dePamphilis CW, Warnow T (2011) The impact of multiple protein sequence alignment on phylogenetic estimation. IEEE/ACM Trans Comput Biol Bioinform 8(4):1108–1119. https://doi.org/10.1109/TCBB.2009.68
51. Larkin MA, Blackshields G, Brown NP, Chenna R, McGettigan PA, McWilliam H, Valentin F, Wallace IM, Wilm A, Lopez R, Thompson JD, Gibson TJ, Higgins DG (2007) Clustal W and Clustal X version 2.0. Bioinformatics 23(21):2947–2948. https://doi.org/10.1093/bioinformatics/btm404
52. Sievers F, Wilm A, Dineen D, Gibson TJ, Karplus K, Li W, Lopez R, McWilliam H, Remmert M, Soding J, Thompson JD, Higgins DG (2011) Fast, scalable generation of high-quality protein multiple sequence alignments using Clustal Omega. Mol Syst Biol 7:539. https://doi.org/10.1038/msb.2011.75
53. Edgar RC (2004) MUSCLE: multiple sequence alignment with high accuracy and high throughput. Nucleic Acids Res 32 (5):1792–1797. https://doi.org/10.1093/nar/gkh340
54. Katoh K, Standley DM (2013) MAFFT multiple sequence alignment software version 7: improvements in performance and usability. Mol Biol Evol 30(4):772–780. https://doi.org/10.1093/molbev/mst010
55. Liu Y, Schmidt B, Maskell DL (2010) MSAProbs: multiple sequence alignment based on pair hidden Markov models and partition function posterior probabilities. Bioinformatics 26(16):1958–1964. https://doi.org/10.1093/bioinformatics/btq338
56. Roshan U, Livesay DR (2006) Probalign: multiple sequence alignment using partition function posterior probabilities. Bioinformatics 22(22):2715–2721. https://doi.org/10.1093/bioinformatics/btl472
57. Do CB, Mahabhashyam MS, Brudno M, Batzoglou S (2005) ProbCons: probabilistic consistency-based multiple sequence alignment. Genome Res 15(2):330–340. https://doi.org/10.1101/gr.2821705
58. Notredame C, Higgins DG, Heringa J (2000) T-Coffee: a novel method for fast and accurate multiple sequence alignment. J Mol Biol 302(1):205–217. https://doi.org/10.1006/jmbi.2000.4042
59. Talavera G, Castresana J (2007) Improvement of phylogenies after removing divergent and ambiguously aligned blocks from protein sequence alignments. Syst Biol 56 (4):564–577. https://doi.org/10.1080/10635150701472164
60. Gouveia-Oliveira R, Sackett PW, Pedersen AG (2007) MaxAlign: maximizing usable data in an alignment. BMC Bioinformatics 8:312.

https://doi.org/10.1186/1471-2105-8-312

61. Capella-Gutierrez S, Silla-Martinez JM, Gabaldon T (2009) trimAl: a tool for automated alignment trimming in large-scale phylogenetic analyses. Bioinformatics 25 (15):1972–1973. https://doi.org/10.1093/bioinformatics/btp348
62. Wu M, Chatterji S, Eisen JA (2012) Accounting for alignment uncertainty in phylogenomics. PLoS One 7(1):e30288. https://doi.org/10.1371/journal.pone.0030288
63. Castresana J (2000) Selection of conserved blocks from multiple alignments for their use in phylogenetic analysis. Mol Biol Evol 17 (4):540–552
64. Wheeler WC, Gatesy J, DeSalle R (1995) Elision: a method for accommodating multiple molecular sequence alignments with alignment-ambiguous sites. Mol Phylogenet Evol 4(1):1–9. https://doi.org/10.1006/mpev.1995.1001
65. de Queiroz A, Gatesy J (2007) The supermatrix approach to systematics. Trends Ecol Evol 22(1):34–41. https://doi.org/10.1016/j.tree.2006.10.002
66. Mar JC, Harlow TJ, Ragan MA (2005) Bayesian and maximum likelihood phylogenetic analyses of protein sequence data under relative branch-length differences and model violation. BMC Evol Biol 5:8. https://doi.org/10.1186/1471-2148-5-8
67. Kolaczkowski B, Thornton JW (2009) Long-branch attraction bias and inconsistency in Bayesian phylogenetics. PLoS One 4(12):e7891. https://doi.org/10.1371/journal.pone.0007891
68. Price MN, Dehal PS, Arkin AP (2010) FastTree 2--approximately maximum-likelihood trees for large alignments. PLoS One 5(3):e9490. https://doi.org/10.1371/journal.pone.0009490
69. Liu K, Linder CR, Warnow T (2011) RAxML and FastTree: comparing two methods for large-scale maximum likelihood phylogeny estimation. PLoS One 6(11):e27731. https://doi.org/10.1371/journal.pone.0027731
70. Stamatakis A (2014) RAxML version 8: a tool for phylogenetic analysis and post-analysis of large phylogenies. Bioinformatics 30 (9):1312–1313. https://doi.org/10.1093/bioinformatics/btu033
71. Ripplinger J, Sullivan J (2008) Does choice in model selection affect maximum likelihood analysis? Syst Biol 57(1):76–85. https://doi.org/10.1080/10635150801898920
72. Ripplinger J, Sullivan J (2010) Assessment of substitution model adequacy using frequentist and Bayesian methods. Mol Biol Evol 27 (12):2790–2803. https://doi.org/10.1093/molbev/msq168
73. Darriba D, Taboada GL, Doallo R, Posada D (2011) ProtTest 3: fast selection of best-fit models of protein evolution. Bioinformatics 27(8):1164–1165. https://doi.org/10.1093/bioinformatics/btr088
74. Le SQ, Gascuel O (2008) An improved general amino acid replacement matrix. Mol Biol Evol 25(7):1307–1320. https://doi.org/10.1093/molbev/msn067
75. Anisimova M, Gascuel O (2006) Approximate likelihood-ratio test for branches: a fast, accurate, and powerful alternative. Syst Biol 55(4):539–552. https://doi.org/10.1080/10635150600755453
76. Anisimova M, Gil M, Dufayard JF, Dessimoz C, Gascuel O (2011) Survey of branch support methods demonstrates accuracy, power, and robustness of fast likelihood-based approximation schemes. Syst Biol 60 (5):685–699. https://doi.org/10.1093/sysbio/syr041
77. Hill J, Davis KE (2014) The Supertree Toolkit 2: a new and improved software package with a Graphical User Interface for supertree construction. Biodivers Data J 2:e1053. https://doi.org/10.3897/BDJ.2.e1053
78. Pagel M, Meade A, Barker D (2004) Bayesian estimation of ancestral character states on phylogenies. Syst Biol 53(5):673–684. https://doi.org/10.1080/10635150490522232
79. Eswar N, Eramian D, Webb B, Shen MY, Sali A (2008) Protein structure modeling with MODELLER. Methods Mol Biol 426:145–159. https://doi.org/10.1007/978-1-60327-058-8_8
80. Madhusudhan MS, Webb BM, Marti-Renom MA, Eswar N, Sali A (2009) Alignment of multiple protein structures based on sequence and structure features. Protein Eng Des Sel 22 (9):569–574. https://doi.org/10.1093/protein/gzp040
81. Kalaimathy S, Sowdhamini R, Kanagarajadurai K (2011) Critical assessment of structure-based sequence alignment methods at distant relationships. Brief Bioinform 12 (2):163–175. https://doi.org/10.1093/bib/bbq025
82. Kim C, Lee B (2007) Accuracy of structure-based sequence alignment of automatic methods. BMC Bioinformatics 8:355. https://doi.org/10.1186/1471-2105-8-355

83. Ashtawy HM, Mahapatra NR (2012) A comparative assessment of ranking accuracies of conventional and machine-learning-based scoring functions for protein-ligand binding affinity prediction. IEEE/ACM Trans Comput Biol Bioinform 9(5):1301–1313. https://doi.org/10.1109/TCBB.2012.36

84. Ashtawy HM, Mahapatra NR (2015) BgN-Score and BsN-Score: bagging and boosting based ensemble neural networks scoring functions for accurate binding affinity prediction of protein-ligand complexes. BMC Bioinformatics 16(Suppl 4):S8. https://doi.org/10.1186/1471-2105-16-S4-S8

85. Brylinski M (2013) Nonlinear scoring functions for similarity-based ligand docking and binding affinity prediction. J Chem Inf Model 53(11):3097–3112. https://doi.org/10.1021/ci400510e

86. Rose PW, Bi C, Bluhm WF, Christie CH, Dimitropoulos D, Dutta S, Green RK, Goodsell DS, Prlic A, Quesada M, Quinn GB, Ramos AG, Westbrook JD, Young J, Zardecki C, Berman HM, Bourne PE (2013) The RCSB Protein Data Bank: new resources for research and education. Nucleic Acids Res 41(Database issue):D475–D482. https://doi.org/10.1093/nar/gks1200

87. Comeau SR, Gatchell DW, Vajda S, Camacho CJ (2004) ClusPro: an automated docking and discrimination method for the prediction of protein complexes. Bioinformatics 20 (1):45–50

88. Kastritis PL, Bonvin AM (2010) Are scoring functions in protein-protein docking ready to predict interactomes? Clues from a novel binding affinity benchmark. J Proteome Res 9(5):2216–2225. https://doi.org/10.1021/pr9009854

89. Kozakov D, Beglov D, Bohnuud T, Mottarella SE, Xia B, Hall DR, Vajda S (2013) How good is automated protein docking? Proteins 81(12):2159–2166. https://doi.org/10.1002/prot.24403

90. Lensink MF, Wodak SJ (2013) Docking, scoring, and affinity prediction in CAPRI. Proteins 81(12):2082–2095. https://doi.org/10.1002/prot.24428

91. Roberts VA, Thompson EE, Pique ME, Perez MS, Ten Eyck LF (2013) DOT2: macromolecular docking with improved biophysical models. J Comput Chem 34 (20):1743–1758. https://doi.org/10.1002/jcc.23304

92. Dolinsky TJ, Czodrowski P, Li H, Nielsen JE, Jensen JH, Klebe G, Baker NA (2007) PDB2PQR: expanding and upgrading automated preparation of biomolecular structures for molecular simulations. Nucleic Acids Res 35(Web Server issue):W522–W525. https://doi.org/10.1093/nar/gkm276

93. Pronk S, Pall S, Schulz R, Larsson P, Bjelkmar P, Apostolov R, Shirts MR, Smith JC, Kasson PM, van der Spoel D, Hess B, Lindahl E (2013) GROMACS 4.5: a high-throughput and highly parallel open source molecular simulation toolkit. Bioinformatics 29(7):845–854. https://doi.org/10.1093/bioinformatics/btt055

94. Dias R, Timmers LF, Caceres RA, de Azevedo WF Jr (2008) Evaluation of molecular docking using polynomial empirical scoring functions. Curr Drug Targets 9(12):1062–1070

95. De Paris R, Quevedo CV, Ruiz DD, Norberto de Souza O, Barros RC (2015) Clustering molecular dynamics trajectories for optimizing docking experiments. Comput Intell Neurosci 2015:916240. https://doi.org/10.1155/2015/916240

96. Seo MH, Park J, Kim E, Hohng S, Kim HS (2014) Protein conformational dynamics dictate the binding affinity for a ligand. Nat Commun 5:3724. https://doi.org/10.1038/ncomms4724

97. Kruger DM, Ignacio Garzon J, Chacon P, Gohlke H (2014) DrugScorePPI knowledge-based potentials used as scoring and objective function in protein-protein docking. PLoS One 9(2):e89466. https://doi.org/10.1371/journal.pone.0089466

98. Camacho CJ, Zhang C (2005) FastContact: rapid estimate of contact and binding free energies. Bioinformatics 21(10):2534–2536. https://doi.org/10.1093/bioinformatics/bti322

99. Dias R, Kolaczkowski B (2017) Improving the accuracy of high-throughput protein-protein affinity prediction may require better training data. BMC Bioinformatics 18(Suppl 5):102. https://doi.org/10.1186/s12859-017-1533-z

100. Dias R, Kolazckowski B (2015) Different combinations of atomic interactions predict protein-small molecule and protein-DNA/RNA affinities with similar accuracy. Proteins 83(11):2100–2114. https://doi.org/10.1002/prot.24928

101. O'Boyle NM, Banck M, James CA, Morley C, Vandermeersch T, Hutchison GR (2011) Open Babel: an open chemical toolbox. J Cheminform 3:33. https://doi.org/10.1186/1758-2946-3-33

Chapter 9

Ancestral Sequence Reconstruction as a Tool for the Elucidation of a Stepwise Evolutionary Adaptation

Kristina Straub and Rainer Merkl

Abstract

Ancestral sequence reconstruction (ASR) is a powerful tool to infer primordial sequences from contemporary, i.e., extant ones. An essential element of ASR is the computation of a phylogenetic tree whose leaves are the chosen extant sequences. Most often, the reconstructed sequence related to the root of this tree is of greatest interest: It represents the common ancestor (CA) of the sequences under study. If this sequence encodes a protein, one can "resurrect" the CA by means of gene synthesis technology and study biochemical properties of this extinct predecessor with the help of wet-lab experiments.

However, ASR deduces also sequences for all internal nodes of the tree, and the well-considered analysis of these "intermediates" can help to elucidate evolutionary processes. Moreover, one can identify key mutations that alter proteins or protein complexes and are responsible for the differing properties of extant proteins. As an illustrative example, we describe the protocol for the rapid identification of hotspots determining the binding of the two subunits within the heteromeric complex imidazole glycerol phosphate synthase.

Key words Ancestral sequence reconstruction, Vertical analysis, Evolutionary biochemistry, In silico mutagenesis, Protein–protein interaction

1 Introduction

A major goal of life scientists is to understand the function of proteins at the residue level, and often, computational biology contributes a lot to the finding of functionally or structurally important residues; for a review *see* [1] For example, if the 3D structure of a protein is known, one can assess the contribution of individual residues to protein stability [2]; additionally, one can predict catalytic sites [3] and protein interfaces [4] by analyzing cavities or surface residues. Moreover, the comparison of results deduced for homologous proteins allows one to elucidate the evolution of specific protein functions [5]. Similarly, protein sequences can be utilized; however, the predictive power of corresponding algorithms depends on the number of sequences

Tobias Sikosek (ed.), *Computational Methods in Protein Evolution*, Methods in Molecular Biology, vol. 1851, https://doi.org/10.1007/978-1-4939-8736-8_9,

that are at hand. In the post-genomic era, computational protein biology benefits from the enormous number of known orthologs, i.e., sequences from different species that have the same ancestor and encode identical or similar functions. In order to identify residue positions that are crucial for a specific family, it is a common approach to generate a multiple sequence alignment (MSA), which is subsequently utilized to determine for each position in the protein the conservation level of each residue [6].

This and similar approaches are often named "horizontal", because they are based on the analysis of a certain phase of evolution represented by the proteins found in extant species. Due to the enormous number of known sequences, these residue distributions can be determined quite precisely and the horizontal approach allows the identification of residues that are important for all members of a family. However, this method rarely identifies sets of residues that determine specificity in a family of functionally diverse proteins [7]. Thus, to study protein evolution, a more detailed analysis is needed, for example, based on a clustering of sequences by means of neighbor joining [8]. A state-of-the-art method for the study of divergent evolution even in very large protein families is the usage of sequence similarity networks and genome neighborhood networks; for a recent review *see* [9]. Such cluster algorithms are based on a simplified model of protein evolution; due to their computational complexity, models that are more elaborate are not applicable for the analysis of large datasets.

Although only applicable to a relatively small number of sequences, the implementation of highly reliable phylogenetic algorithms has added a further dimension to sequence analysis: It makes possible to trace back the evolution of a fair number of extant orthologs to common ancestors. If functional diversity is known for some of the extant orthologs, this "vertical" approach has great potential, because one can reconstruct the sequences of putative predecessors and identify those mutations that occurred along that branch of the family tree on which functional diversification occurred [7].

The vertical approach is a specific application of ancestral sequence reconstruction (ASR), which became popular during the last decade, especially in combination with "resurrection" experiments; for recent reviews *see* [10–13]. The typical protocol of each ASR consists of two steps: First, the user has to compute a phylogenetic tree tr_{phylo}. In all cases, the extant orthologs chosen by the user constitute the leaves, but the topology of tr_{phylo} is determined by sequence similarity, the selected evolutionary model, and the algorithm used for its computation. In contrast to a classical phylogenetic analysis, ASR requires a subsequent step that deduces for all internal nodes of tr_{phylo} sequences that represent predecessors. The composition of these sequences critically depends on the content of the leaves (extant orthologs) but also on the topology of tr_{phylo}.

This is why tr_{phylo} has to fulfill certain quality criteria to guarantee proper sequence reconstruction. Nowadays, it is straightforward to supplement such an in silico reconstruction with wet-lab experiments: One can recombinantly resurrect proteins with the help of gene synthesis and characterize them with classical biochemical and biophysical methods [11]. Besides their relevance for answering evolutionary problems, resurrected proteins became increasingly important in protein engineering, because one can beneficially exploit their promiscuity [14] to tailor protein function [15].

In addition, the fact that ancestral proteins are frequently "generalists" motivates their usage in vertical approaches. In the following, we detail a protocol for the identification of specificity-determining residues. The general strategy is to select a protein family of interest and a property to be evaluated. Then, one has to infer a phylogenetic tree and choose the branches of the family tree to be analyzed. The selection of branches may depend on in silico or wet-lab experiments aimed at finding branch-determining leaves, i.e., extant proteins with differing functions. The final task is to reconstruct the sequences of predecessors with the help of ASR (Subheading 2.1) and to identify specificity-determining residues by comparing the sequences of ancestral sequences within the chosen branches (Subheading 2.2). Again, the assessment of these residues may comprise in silico and/or wet-lab analyses.

We used this strategy to study the stepwise adaptation of the protein–protein interface (PPI) from the heterodimeric imidazole glycerol phosphate synthase (ImGPS). This enzyme mediates the incorporation of nitrogen into PRFAR by catalyzing the transfer of the amido nitrogen of glutamine to an acceptor substrate [16, 17]. In bacteria and archaea, ImGPS consists of the cyclase subunit HisF and the glutaminase subunit HisH, which assemble with high affinity to a bi-enzyme complex [18]. Despite detailed biochemical and structural studies [19], the specific residue positions responsible for HisF:HisH complex formation were unknown. This is why we identified key residue positions of this PPI by means of a vertical approach [20, 21], which is illustrated in Fig. 1.

2 Method

2.1 Ancestral Sequence Reconstruction

1. Collect a large number of orthologs. Start with a specific sequence of interest and use BLAST [23] to deduce orthologs from the *nr* or *refseq_protein* databases of the NCBI [24] or the EBI database *UniProt* [25]; alternatively select the corresponding *InterPro* family [26] (*see* **Note 1**). Choose a *bona fide* protein as a reference sequence and, if possible, several sequences that can serve as an outgroup. Additionally, include the sequences of those proteins ($prot_i$) that possess differing

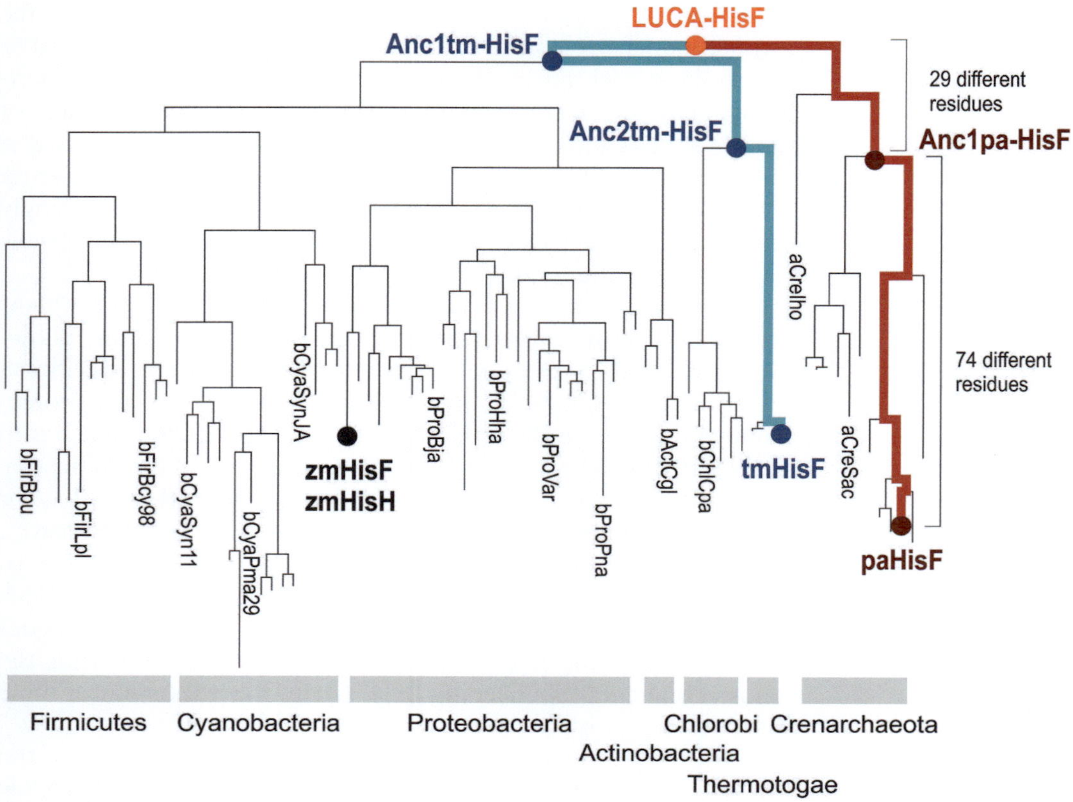

Fig. 1 Identification of specificity-determining residue positions of the HisF:HisH interface by means of a vertical approach. Initial binding studies had shown that subunits from phylogenetically unrelated species are not compatible: The HisF subunit from the Crenarchaeon *Pyrobaculum arsenaticum* (paHisF) did not bind HisH from the Proteobacterium *Zymomonas mobilis* (zmHisH). For the rapid identification of crucial residue positions within the HisF interface, 87 HisF sequences from seven phyla were chosen for a vertical analysis. Thus, we deduced ancestral sequences linking the native interaction partner of zmHisH, namely zmHisF (the leaf of the gray branch) and the distant paHisF (the leaf of the brown branch). Ancestral proteins were resurrected, and their binding to zmHisH was characterized experimentally. HisF corresponding to the last universal common ancestor (LUCA-HisF) bound zmHisH. In contrast, the first intermediate (Anc1pa-HisF) on the branch leading to paHisF that differed markedly from LUCA-HisF did not bind zmHisH. Anc1pa-HisF deviates from LUCA-HisF by not more than 29 residues, but from paHisF by 74 residues. A subsequent in silico analysis focusing on the PPI of HisF allowed us to narrow down the number of putative key residue positions to two. Their role was assessed by experimental binding studies; one was identified as an interface hotspot. To trace the species-specific evolution of PPIs in more detail, the two predecessors (Anc1tm-HisF and Anc2tm-HisF) on the path (shown in blue) leading to HisF from *Thermotoga maritima* (tmHisF) were resurrected as well. Both intermediates bound zmHisH, but tmHisF was a poor binder. The mutual exchange of residues from the latter three sequences at corresponding positions confirmed their hotspot quality; for details *see* [21]. Note that these residues are located at the rim of the PPI and only moderately conserved, which explains why they have not been discovered previously. To avoid overloading the graph, only a few of the extant sequences are shown with their *Key2Ann* [22] annotation indicating the phylogenetic lineage, i.e., the superkingdom (first character), the phylum (following three characters), and the species name (last three characters)

properties, whose determinants shall be elucidated by the subsequent analysis.

2. Create an MSA. According to our experience, *MAFFT* [27] is a highly versatile and robust method that can cope with large sequence sets (*see* **Note 2**).
3. Eliminate redundant sequences and obvious outliers like those that are much shorter or longer than the reference sequence. Additionally, eliminate sequences that induce conspicuously large indels in the MSA (*see* **Note 3**). A versatile tool supporting these tasks is *Jalview* [28] (*see* **Note 4**).
4. Repeat **steps 2** and **3** until the MSA consists of a homogeneous set of sequences.
5. If the protein under study is part of a larger complex, perform MSA generation for each subunit. Afterward, concatenate the sequences in a species-specific manner (*see* **Note 5**), and create an MSA consisting of the concatenated sequences.
6. Optionally, replace the database identifiers with more informative names for the sequences (*see* **Note 6**).
7. Remove less informative residue positions from the MSA. Apply *Gblocks* [29] to eliminate all columns containing more than 50% gaps. Use the resulting MSA for the inference of the phylogenetic tree, but not for the subsequent sequence reconstruction, which is based on the full MSA.
8. Compute a phylogenetic tree tr_{phylo} with a method of choice. We prefer *PhyloBayes* [30] and start eight independent MCMC samplings in parallel with a maximal length of 50,000 samples to guarantee congruence (*see* **Note 7**). If congruence is reached, we deduce the consensus tree computed by *readpb* from the samples following the burn-in phase of the MCMC computation. The number of samples that have to be excluded (burn-in) can be determined with *VMCMC* [31]; often, the first 25% of the samples are considered as burn-in and discarded. Alternatively, use other state-of-the-art probabilistic methods like *MrBayes* [32] or *BEAST* [33] to compute the phylogenetic tree (*see* **Note 8**). For a given MSA of amino acid sequences, one can utilize *ProtTest* [34] to determine the best fitting evolutionary model prior to MCMC sampling.
9. Visualize tr_{phylo} by means of *NJplot* [35] or *FigTree* [36] and assess the length of the individual edges and their posterior probabilities. All edge lengths must indicate mutation rates $<<$ 1 mutation per site and the posterior probabilities of relevant internal nodes must exceed the value of 0.75. Furthermore, make sure that the resulting phylogenetic hierarchy of the chosen sequences (species) is plausible: For example, compare the topology of tr_{phylo} with the relationships of the sequences

(species), determined for the iTOL project [37] or the "nearly universal tree" of life [38]. This comparison allows one to eliminate cases of horizontal gene transfer and to avoid long-branch attraction. If tree topology is not plausible, consider to choose a different set of sequences and repeat the procedure (*see* **Note 9**).

10. If the sequence set does not contain an outgroup, use *NJplot* [35] or an alternative algorithm to root tr_{phylo} for subsequent sequence reconstruction. Positioning the root is critical for the computation of the CA sequence. Choose the location of the root according to a plausible hierarchy to be determined by one of the methods described in **step 9**. If an outgroup was used for rooting, we recommend to eliminate the corresponding sequences during sequence reconstruction to prevent undesired effects on residue composition.
11. Use the rooted tree prepared in the last step and the full MSA to reconstruct the ancestral sequences related to internal nodes. Methods of choice are *PAML* [39] or *FastML* [40], which can handle indels (*see* **Note 10**). If possible, choose the same substitution model as used for tree construction. ASR programs compute for each residue position posterior probabilities for all 20 amino acids. If alternative predictions with relatively high posterior probabilities exist, a near-ancestor sequence ensemble can be calculated for each node; for details *see* [22]. If one sequence per internal node is of interest, select for each position the residue possessing the highest posterior probability.

2.2 Identification of Specificity-Determining Residues by Means of Intermediate Sequences

1. In analogy to Fig. 1, determine the branches of tr_{phylo} that interconnect the two or more recent proteins $prot_i$ under study, i.e., those that possess diversified properties.
2. Compile an initial set *anc_prot*, consisting of ancestral proteins that differ most likely from the extant proteins $prot_i$ and support an efficient characterization. For example, one can pairwise compare all ancestral sequences to choose several intermediates, i.e., ancestral sequences that span the sequence differences between the $prot_i$ in approximately similar proportions. We recommend the usage of *Jalview* for sequence selection (*see* **Note 11**). The finding that primordial proteins are often generalists suggests to add the CA sequence to *anc_prot* and to characterize the corresponding protein with high preference.
3. Optional step: If the 3D structure of a $prot_i$ is known, compute homology models of all *anc_prot* (*see* **Note 12**) and try to minimize further the number of candidate residues to be studied in the following steps. If protein function is of interest, use

the compiled annotations of *PDBsum* (*www.ebi.ac.uk/pdbsum/*) or an alternative database to assess the position of the differing residues with respect to a catalytic center or a binding site. If complex formation is under study, consider a webserver-like *PISA* (*www.ebi.ac.uk/pdbe/pisa/*) that details characteristics of residues located in PPIs. One can also predict the contribution of residues to protein or complex stability by utilizing force fields to calculate differences in free energy (*see* **Note 13**). For the example presented in Fig. 1, we could reduce the number of putative key residue positions to two by combining in silico approaches.

4. Optional step: If experimental characterization is intended, choose protein sequences for the resurrection experiments and design their gene sequences. Produce the proteins recombinantly and characterize them according to the specific problem. The choice of suitable wet-lab experiments depends on the characteristics under assessment and may contain tests of enzyme activity or complex stability. Additionally, it is advisable to confirm proper protein folding by means of far-UV CD spectroscopy.
5. Associate the determined effects with the introduced mutations to deduce the stepwise evolutionary adaptation toward the properties of recent proteins. In case of ambiguous results, repeat **steps 2–4** of the protocol given in Subheading 2.2 and extend the analyses to additional intermediates and/or single point mutations.

3 Notes

1. Compiling an appropriate sequence set for ASR is more an art than an artisanal activity and sequence selection is an iterative process that requires several rounds of user interaction. This is why the initial number of sequences should be as high as possible. Choose sequences that are most likely orthologs and avoid the addition of paralogous sequences by comparing gene duplicates. If a Bayesian approach is used to infer the phylogenetic tree, running time is an issue that currently limits the finally selected number of recent sequences to $\approx$200. Make sure that the chosen sequences originate from phyla needed to deduce the intended set of predecessors. If one wants to represent the last universal common ancestor, the chosen sequences must at least come from several bacterial and archeal clades. Store sequence sets in multi-FASTA file format, which is accepted by most tools required for ASR.
2. Use a *MAFFT* method that is accuracy-oriented, i.e., one of the "*INS*" modes. This selection depends on the size of the MSA;

for details see the *MAFFT* manual. For the initial generation of large MSAs, the option – *auto* is also appropriate.

3. Modeling the history of insertions and deletions on an evolutionary time scale is difficult and requires for most ASR algorithms the manual adjustment of primordial sequences. One can minimize errors by choosing a set of sequences of relatively uniform length.
4. *Jalview* is an excellent tool for the preparation of sequence sets used in ASR. The *Jalview* command *Edit\Remove redundancy* allows the selection of a percentage identity threshold and initiates the subsequent comparison of all sequence pairs. If the similarity of any two sequences exceeds this cutoff, the shorter sequence is discarded. A cutoff of 95% or lower is useful to remove redundant sequences and to avoid highly articulated subtrees. The command *Calculate\Sort by length* makes it possible to identify easily sequences that are much shorter or longer than the reference sequence. These sequences and those introducing strikingly long indels can be erased by clicking their name and the delete button. The command *Web Service\Alignment* offers several alternatives for MSA creation, among them is *MAFFT.*
5. Concatenation helps to deduce a robust tree due to the stronger phylogenetic signal spread over a larger set of residue positions. Make sure that the sequences originate from the same species by using for their linkage the *Tax-Id* assigned by the *taxonomy browser* of the NCBI. Note that concatenation is only valid for sequences that coevolve and share the same evolutionary history for the entire period under study.
6. For the visual inspection of trees, it is helpful to replace the hard to interpret database identifiers with names that indicate the function of the proteins and/or the phylogenetic position of the species contributing the sequences. We use our in-house tool *Key2Ann* [41] to denote the phylogenetic lineage; *see* Fig. 1 for an example.
7. A detailed description of all the programs and their options belonging to the software suite *PhyloBayes* can be found at *www.phylobayes.org*. For the reconstruction of amino acid sequences, we use the *CAT* or *JTT* model and specify a minimal effective sample size of 100. Congruence can be tested by calculating the maximum difference (*maxdiff*) of posterior probabilities of tree bipartitions by using the *PhyloBayes* tool *bpcomp*; the *maxdiff* value should be below 0.3 [29]. Computation time can be reduced by using the multi-core version *PhyloBayes-MPI*. Note that an MCMC calculation may take several weeks, if a large number of recent sequences were chosen.

8. A detailed description of the *BEAST* functionality can be found at *www.beast2.org*. The *BEAST* tool *LogCombiner* can be used to discard the burn-in samples and *Tracer* allows one to determine the effective sample size. *TreeAnnotator* assists the user in summarizing information from a sample of trees onto a consensus tree. Computation time of *BEAST* can be reduced by incorporating the *BEAGLE* library for parallel processing.
9. Long branches (>1.0 mutations per site) and low posterior probabilities (<0.75) prevent the reliable computation of ancestral states. The same is true, if divergence of the sequence set is too small or if the tree is highly articulated. To overcome these problems, the content of the sequence set has to be altered. For example, one can exclude sequence sets amendable to long branches and erase some sequences in highly articulated subtrees.
10. According to our experience with MSAs containing a small number of indels, *FastML* performs well in ASR. If the MSA contains a larger number of indels, one can try several values of the advanced option *probability cutoff to prefer ancestral indel over character* and compare the results. For further processing, choose the sequences computed as a *marginal reconstruction*. Note that *FastML* does not offer all evolutionary models implemented for *PhyloBayes* or *BEAST*. Alternatively, one can use *PRANK* [42] or *Historian* [43] that are based on alternative models of indel evolution. Due to the method used for indel reconstruction, the lengths of reconstructed sequences may deviate from the mean length of extant sequences as N- and C-termini are of higher variability than the rest of the sequence. Thus, it might be necessary to trim the reconstructed sequences.
11. For a set of sequences, the similarity of all pairs can easily be determined by executing the *Jalview* command *Calculate \Pairwise Alignment*.
12. Several alternatives are available to compute homology models of subunits and protein complexes, among them are *YASARA* [44], *I-Tasser* [45], or *HHSearch* [46] in combination with *Modeller* [47]. For ASR experiments, one can expect reliable models, because the sequence similarity between the template (a $prot_i$) and the target (an *anc_prot*) is usually high.
13. The effect of a mutation on protein or complex stability can be assessed in silico by utilizing programs like *FoldX* [48], which is a stand-alone application, but also integrated into *YASARA*. To predict the contribution of point mutations on protein stability, assess the corresponding $\Delta\Delta G$ values. To estimate the effect on complex stability, compute the $\Delta\Delta G$ value indicating the binding energy difference between a "wild-type

complex" and a complex with a mutated PPI. |ΔΔG values| > 2 kcal/mol are considered a significant contribution of one residue to complex stability. For this *FoldX* analysis, three functions have to be executed subsequently, namely *RepairPDB*, *BuildModel*, and *AnalyseComplex*. For this specific application of *FoldX*, the "wild-type complexes" may consist of $prot_i$ or *anc_prot* sequences, which can differ in their length. In order to identify the corresponding residues, create an MSA containing $prot_i$ and *anc_prot* sequences to coordinate their positions.

Acknowledgement

This work was supported by the Deutsche Forschungsgemeinschaft (ME2259/2-1). Calculations were facilitated by using advanced computational infrastructure provided by the Leibniz Supercomputing Center of the Bavarian Academy of Sciences and Humanities (*www.lrz.de*) under grant pr48fu. We thank Samuel Blanquart for continuous support, many helpful hints, and fruitful discussions.

References

1. Lee D, Redfern O, Orengo C (2007) Predicting protein function from sequence and structure. Nat Rev Mol Cell Biol 8(12):995–1005. https://doi.org/10.1038/nrm2281
2. Schymkowitz J, Borg J, Stricher F et al (2005) The FoldX web server: an online force field. Nucleic Acids Res 33(Web Server issue): W382–W388
3. Janda JO, Meier A, Merkl R (2013) CLIPS-4D: a classifier that distinguishes structurally and functionally important residue-positions based on sequence and 3D data. Bioinformatics 29(23):3029–3035. https://doi.org/10.1093/bioinformatics/btt519
4. Zellner H, Staudigel M, Trenner T et al (2012) PresCont: predicting protein-protein interfaces utilizing four residue properties. Proteins 80 (1):154–168. https://doi.org/10.1002/prot.23172
5. Plach MG, Löffler P, Merkl R, Sterner R (2015) Conversion of anthranilate synthase into isochorismate synthase: implications for the evolution of chorismate-utilizing enzymes. Angew Chem Int Ed 54(38):11270–11274. https://doi.org/10.1002/anie.201505063
6. Edgar RC, Batzoglou S (2006) Multiple sequence alignment. Curr Opin Struct Biol 16(3):368–373
7. Harms MJ, Thornton JW (2010) Analyzing protein structure and function using ancestral gene reconstruction. Curr Opin Struct Biol 20 (3):360–366. https://doi.org/10.1016/j.sbi.2010.03.005
8. Saitou N, Nei M (1987) The neighbor-joining method: a new method for reconstructing phylogenetic trees. Mol Biol Evol 4(4):406–425
9. Gerlt JA (2017) Genomic enzymology: web tools for leveraging protein family sequence-function space and genome context to discover novel functions. Biochemistry 56 (33):4293–4308. https://doi.org/10.1021/acs.biochem.7b00614
10. Merkl R, Sterner R (2016) Ancestral protein reconstruction: techniques and applications. Biol Chem 397(1):1–21. https://doi.org/10.1515/hsz-2015-0158
11. Thornton JW (2004) Resurrecting ancient genes: experimental analysis of extinct molecules. Nat Rev Genet 5(5):366–375. https://doi.org/10.1038/nrg1324
12. Liberles DA (2007) Ancestral sequence reconstruction. Oxford University Press, Oxford
13. Hochberg GKA, Thornton JW (2017) Reconstructing ancient proteins to understand the causes of structure and function. Annu Rev Biophys 46:247–269. https://doi.org/10.1146/annurev-biophys-070816-033631

14. Bornscheuer UT, Huisman GW, Kazlauskas RJ et al (2012) Engineering the third wave of biocatalysis. Nature 485(7397):185–194. https://doi.org/10.1038/nature11117
15. Romero-Romero ML, Risso VA, Martinez-Rodriguez S et al (2016) Engineering ancestral protein hyperstability. Biochem J 473 (20):3611–3620. https://doi.org/10.1042/BCJ20160532
16. Massiere F, Badet-Denisot MA (1998) The mechanism of glutamine-dependent amidotransferases. Cell Mol Life Sci 54(3):205–222
17. Zalkin H, Smith JL (1998) Enzymes utilizing glutamine as an amide donor. Adv Enzymol Relat Areas Mol Biol 72:87–144
18. Beismann-Driemeyer S, Sterner R (2001) Imidazole glycerol phosphate synthase from *Thermotoga maritima*. Quaternary structure, steady-state kinetics, and reaction mechanism of the bienzyme complex. J Biol Chem 276 (23):20387–20396
19. List F, Vega MC, Razeto A et al (2012) Catalysis uncoupling in a glutamine amidotransferase bienzyme by unblocking the glutaminase active site. Chem Biol 19(12):1589–1599. https://doi.org/10.1016/j.chembiol.2012.10.012
20. Reisinger B, Sperl J, Holinski A et al (2014) Evidence for the existence of elaborate enzyme complexes in the Paleoarchean era. J Am Chem Soc 136(1):122–129. https://doi.org/10.1021/ja4115677
21. Holinski A, Heyn K, Merkl R, Sterner R (2017) Combining ancestral sequence reconstruction with protein design to identify an interface hotspot in a key metabolic enzyme complex. Proteins 85(2):312–321. https://doi.org/10.1002/prot.25225
22. Bar-Rogovsky H, Stern A, Penn O et al (2015) Assessing the prediction fidelity of ancestral reconstruction by a library approach. Protein Eng Des Sel 28(11):507–518. https://doi.org/10.1093/protein/gzv038
23. Altschul SF, Gish W, Miller W et al (1990) Basic local alignment search tool. J Mol Biol 215(3):403–410
24. Pruitt KD, Tatusova T, Klimke W, Maglott DR (2009) NCBI Reference Sequences: current status, policy and new initiatives. Nucleic Acids Res 37(Database issue):D32–D36. https://doi.org/10.1093/nar/gkn721
25. Apweiler R, Martin M, O'Donovan C et al (2013) Update on activities at the Universal Protein Resource (UniProt) in 2013. Nucleic Acids Res 41(D 1):D43–D47
26. Hunter S, Jones P, Mitchell A et al (2012) InterPro in 2011: new developments in the family and domain prediction database. Nucleic Acids Res 40(Database issue): D306–D312. https://doi.org/10.1093/nar/gkr948
27. Katoh K, Standley DM (2013) MAFFT multiple sequence alignment software version 7: improvements in performance and usability. Mol Biol Evol 30(4):772–780. https://doi.org/10.1093/molbev/mst010
28. Waterhouse AM, Procter JB, Martin DMA, Clamp M, Barton GJ, (2009) Jalview Version 2—a multiple sequence alignment editor and analysis workbench. Bioinformatics 25 (9):1189–1191. https://doi.org/10.1093/bioinformatics/btp033
29. Castresana J (2000) Selection of conserved blocks from multiple alignments for their use in phylogenetic analysis. Mol Biol Evol 17 (4):540–552
30. Lartillot N, Lepage T, Blanquart S (2009) PhyloBayes 3: a Bayesian software package for phylogenetic reconstruction and molecular dating. Bioinformatics 25(17):2286–2288. https://doi.org/10.1093/bioinformatics/btp368
31. Ali RH, Bark M, Miro J et al (2017) VMCMC: a graphical and statistical analysis tool for Markov chain Monte Carlo traces. BMC Bioinformatics 18(1):97. https://doi.org/10.1186/s12859-017-1505-3
32. Ronquist F, Huelsenbeck JP (2003) MrBayes 3: Bayesian phylogenetic inference under mixed models. Bioinformatics 19 (12):1572–1574
33. Bouckaert R, Heled J, Kuhnert D et al (2014) BEAST 2: a software platform for Bayesian evolutionary analysis. PLoS Comput Biol 10 (4):e1003537. https://doi.org/10.1371/journal.pcbi.1003537
34. Abascal F, Zardoya R, Posada D (2005) ProtTest: selection of best-fit models of protein evolution. Bioinformatics 21(9):2104–2105. https://doi.org/10.1093/bioinformatics/bti263
35. Perriere G, Gouy M (1996) WWW-query: an on-line retrieval system for biological sequence banks. Biochimie 78(5):364–369
36. Rambaut A (2012) FigTree v1.4. http://tree.bio.ed.ac.uk/software/figtree/
37. Ciccarelli FD, Doerks T, von Mering C et al (2006) Toward automatic reconstruction of a highly resolved tree of life. Science 311 (5765):1283–1287
38. Puigbo P, Wolf YI, Koonin EV (2009) Search for a 'Tree of Life' in the thicket of the phylogenetic forest. J Biol 8(6):59. https://doi.org/10.1186/jbiol159
39. Yang Z (2007) PAML 4: phylogenetic analysis by maximum likelihood. Mol Biol Evol 24

(8):1586–1591. https://doi.org/10.1093/molbev/msm088

40. Ashkenazy H, Penn O, Doron-Faigenboim A et al (2012) FastML: a web server for probabilistic reconstruction of ancestral sequences. Nucleic Acids Res 40(Web Server issue): W580–W584. https://doi.org/10.1093/nar/gks498
41. Pürzer A, Grassmann F, Birzer D, Merkl R (2011) Key2Ann: a tool to process sequence sets by replacing database identifiers with a human-readable annotation. J Integr Bioinform 8(1):153. https://doi.org/10.2390/biecoll-jib-2011-153
42. Löytynoja A, Goldman N (2008) Phylogeny-aware gap placement prevents errors in sequence alignment and evolutionary analysis. Science 320(5883):1632–1635. https://doi.org/10.1126/science.1158395
43. Holmes IH (2017) Historian: accurate reconstruction of ancestral sequences and evolutionary rates. Bioinformatics 33(8):1227–1229. https://doi.org/10.1093/bioinformatics/btw791
44. Krieger E, Joo K, Lee J et al (2009) Improving physical realism, stereochemistry, and side-chain accuracy in homology modeling: four approaches that performed well in CASP8. Proteins 77 Suppl 9:114–122. https://doi.org/10.1002/prot.22570
45. Zhang Y (2008) I-TASSER server for protein 3D structure prediction. BMC Bioinformatics 9:40. https://doi.org/10.1186/1471-2105-9-40
46. Söding J (2005) Protein homology detection by HMM-HMM comparison. Bioinformatics 21(7):951–960. https://doi.org/10.1093/bioinformatics/bti125
47. Webb B, Sali A (2014) Protein structure modeling with MODELLER. Methods Mol Biol 1137:1–15. https://doi.org/10.1007/978-1-4939-0366-5_1
48. Guerois R, Nielsen JE, Serrano L (2002) Predicting changes in the stability of proteins and protein complexes: a study of more than 1000 mutations. J Mol Biol 320(2):369–387

Chapter 10

Enhancing Statistical Multiple Sequence Alignment and Tree Inference Using Structural Information

Joseph L. Herman

Abstract

For highly divergent sequences, there is often insufficient information to reliably construct alignments and phylogenetic trees. Since protein structure may be strongly conserved despite large divergences in sequence, structural information can be used to help identify homology in such cases.

While there exist well-studied models of sequence evolution, structurally informed alignment methods have typically made use of geometric measures of deviation that do not take into account the underlying mutational processes. In order to integrate structural information into sequence-based evolutionary models, we recently developed a stochastic model of structural evolution on a phylogenetic tree and implemented this as the StructAlign plugin for the StatAlign statistical alignment package.

In this chapter, we will outline the types of analyses that can be carried out using StructAlign, illustrating how the inclusion of structural information can be used to inform joint estimation of alignments and trees. StructAlign can also be used to infer branch-specific rates of structural evolution, and analysis of an example globin dataset highlights strong variation in the inferred rate across the tree. While structure is more highly conserved within clades, the rate of structural divergence as a function of sequence variation is larger between functionally divergent proteins. Allowing for the rate of structural divergence to vary over the tree results in an improved fit to the empirically observed pairwise RMSD values.

Key words Protein structure, Structural alignment, RMSD, Statistical alignment, Alignment uncertainty, Bayesian hierarchical models, MCMC, Parallel tempering, Molecular phylogenetics, Globins

1 Introduction

Homologous sequences may diverge substantially while still preserving function. When comparing such sequences, a large number of alternative alignments may achieve a very similar score or likelihood [1, 2], and selecting any one of these alignments can strongly bias the downstream inference [3–7], especially if a single guide tree is used to construct the alignment [8].

The original version of this chapter was revised. The correction to this chapter is available at https://doi.org/10.1007/978-1-4939-8736-8_23

Tobias Sikosek (ed.), *Computational Methods in Protein Evolution*, Methods in Molecular Biology, vol. 1851, https://doi.org/10.1007/978-1-4939-8736-8_10, © Springer Science+Business Media, LLC, part of Springer Nature 2019

One solution to this issue is to jointly estimate alignments along with other parameters of interest (e.g., tree topologies, branch lengths, or indel rates), as part of a hierarchical probabilistic model [9, 10]. This allows the uncertainty in each of these variables, including the alignment, to be incorporated into the inference of the other parameters, eliminating the bias associated with selecting a single alignment for downstream inference. As well as reducing errors in tree reconstruction, joint estimation of alignments has been found to improve inference of indel rates [11, 12], positive selection parameters [13], and the location of putative regulatory elements [14, 15].

However, for sequences that are highly divergent, sequence similarity alone may be insufficient for identifying homology, giving rise to high uncertainty, and a diffuse posterior distribution over alignments, trees, and other model parameters. While inclusion of additional sequences into the dataset may reduce tree uncertainty given a particular alignment and set of model parameters, the set of plausible alignments increases as more sequences are added, especially in regions of low homology, such that the resulting distribution over trees may become less reliable and more sensitive to the choice of model [16, 17]. Filtering out regions of the alignment that are predicted to be unreliable or highly variable may also appear to reduce uncertainty in such cases [18, 19], although systematic removal of indel-rich regions may introduce additional bias into the inferred trees [20–22].

Since protein structure often needs to be conserved in order to preserve function, structural similarity can be used to identify homologous proteins even in the so-called twilight zone of very low sequence identity [23]. However, while there have been some attempts to construct phylogenetic trees directly from pairwise structural distances [24–26], the lack of well-developed models for structural evolution has made it difficult to translate structural deviation into evolutionary time. Examining the relative rates of sequence and structural divergence over large sets of proteins, there appear to be some global trends across multiple classes [27], although the exact relationship may vary between families [28, 29] as well as across different sites within a protein [30]. In addition, sequences may undergo structurally constrained neutral evolution while still preserving fold [31–36], such that the increase in structural divergence may be a nonlinear [37] or discontinuous [32, 38, 39] function of evolutionary distance.

Challis and Schmidler [40] developed a stochastic model describing the joint evolution of structure and sequence over time, whereby local structural deviations within a particular fold are described using an Ornstein-Uhlenbeck diffusion process. In order to facilitate efficient inference, the authors focused on an independent-sites model of structural evolution, for which the overall metric of structural similarity between two proteins can be expressed as a sum of per-site contributions, such that dynamic programming algorithms can still be used to perform analytical summation over alignments. Herman et al. [41] extended this model to multiple structures related by a tree, with an additional baseline variance term that takes account of the intrinsic structural variability at different sites. This model allows for the rate of structural divergence to vary independently for each branch of the tree, allowing for more complex relationships between sequence and structure divergence, and was incorporated into the StatAlign statistical alignment package [42] as the StructAlign plugin.

2 Materials

2.1 Running StatAlign

StatAlign is written in Java and is run via a JAR file, which can either be obtained from the pre-compiled distribution or built from source. The pre-compiled package can be obtained from GitHub at

https://github.com/statalign/statalign/releases/download/v3.3/StatAlign-v3.3.zip.

Source code can also be downloaded and compiled from the GitHub repository if desired.

The graphical version of StatAlign can be run on Windows, Mac, and Linux by double-clicking on the JAR archive (*see* **Note 1**). Instructions for using this GUI version can be found on the StatAlign website at http://statalign.github.io/doc/user_manual.html.

For longer-running analyses running on multiple cores, we need to make use of the command-line version of StatAlign. In this chapter, we will present the commands required to run the command-line version under Unix-based systems; Windows users can run these commands using a terminal emulator such as Cygwin.

The single-threaded version can be invoked from the JAR via

```
java -jar StatAlign.jar [OPTIONS] DATA_FILE1 [DATA_FILE2 ...]
```

For the MPI-based parallel version, there is a wrapper script that passes the arguments to the JAR in the appropriate form

```
StatAlignParallel NCORES [OPTIONS] DATA_FILE1 [DATA_FILE2 ...]
```

Running `java -jar StatAlign.jar` or `StatAlignParallel` with no arguments will output additional usage information and a list of available options.

2.2 Example Dataset

In this chapter we will investigate the phylogenetic relationship between cytoglobin [43, 44] and a set of nine other globin structures (Table 1). The functional role of cytoglobin is currently unknown, and there has been recent interest in determining its relationship to other globins [45–47].

For input to StatAlign, each PDB file should contain a single chain to be analyzed (*see* **Note 2**). To construct the example

Table 1
Protein structures used in example dataset, with heme coordination number and exogenous ligand shown

Structure	Protein	Organism	Coordination	Ligand
2oif	NsGb	*H. vulgare* (barley)	6	CN
1bin	Lhb	*G. max* (soybean)	5	Acetate
1lh1	Lhb	*L. luteus* (lupin bean)	5	Acetate
1urv	Cygb	*H. sapiens* (human)	6	None
2lhb	CycHb	*P. marinus* (lamprey)	5	CN
1myt	Mb	*T. albacares* (tuna)	5	None
2mm1	Mb	*H. sapiens* (human)	5	None
1psgA	α-Hb	*L. xanthurus* (spot croaker)	5	CO
2hhbA	α-Hb	*H. sapiens* (human)	5	None
2hhbB	β-Hb	*H. sapiens* (human)	5	None

dataset, the A-chain was extracted from the PDB file corresponding to each structure in Table 1; for *2hhbB*, the B-chain was used.

The PDB files corresponding to this dataset can be found in the `examples/10_globins` subdirectory distributed with StatAlign, along with a FASTA-formatted file `10_globins.fasta` containing only the primary sequences.

2.3 Analysis of StatAlign Output

We will make use of R in order to analyze the output files generated by StatAlign, and code is provided for each analysis step in this chapter. Unless otherwise specified, the example commands are to be run from within the directory where StatAlign is installed, with the path to this directory saved as a `STATALIGN_HOME` variable in the shell and R environments. Several R packages and some additional scripts will also be required; once installed, the requisite packages can be loaded using the code below:

```
packages = c('dplyr','coda','magrittr','ape','ggplot2',
'reshape2','data.table','gridExtra')
for (package in packages) require(package,character.only=TRUE)
```

3 Methods

3.1 Running StatAlign in Sequence-Only Mode

As a preliminary analysis, we will first analyze the globin dataset using the original sequence-only model in StatAlign; later we will go on to assess the effect of including structural information.

Table 2
Output files created by running StatAlign on the example dataset

File	Contents	Format
10_globins.fasta.ll	Sample number, sequence model log likelihood, total log likelihood	Tab-separated
10_globins.fasta.tree	Sampled trees	NEXUS
10_globins.fasta.coreModel.params	Sample number, parameters for the indel model (TKF92)	Tab-separated
10_globins.fasta.log	Specified via `-log` argument (by default includes sampled alignments, total log likelihood, and MCMC acceptance rates)	Modified FASTA
10_globins.fasta.length	Sample number, alignment length	Tab-separated

StatAlign can be invoked via the command below (*see* **Note 3**). Note that the backslashes below split up the command over multiple lines to avoid ambiguity, but this can also be run all on one line:

```
java -jar StatAlign.jar \
 -mcmc=1m,5m,200,0 \
 examples/10_globins/10_globins.fasta
```

The program options specify a burn-in period of one million iterations (during which the MCMC sampler is automatically tuned in order to improve convergence), followed by five million sampling iterations (during which the state is recorded every 200 moves), with zero pre-burn-in randomization period (setting this to a non-zero value is useful when assessing the sensitivity of the results to the initial state). On an Intel i7-4790K CPU (4.00GHz), this takes just over 2h to complete, producing the output files summarized in Table 2 (these will be located in the same directory as the input data file).

3.2 Analysis of MCMC Output

As discussed earlier, StatAlign uses an MCMC sampler to generate samples from the posterior distribution over alignments, trees, and model parameters. As with all MCMC analyses, the sampler will take time to converge to the posterior distribution, and it is important to assess whether convergence has been achieved before proceeding with downstream analysis.

Two basic summary statistics that can help to assess convergence are the total log likelihood and the alignment length. We can read in the output files and plot the posterior distribution of these quantities using the following commands:

```
base.name = paste0(STATALIGN_HOME,
                   "examples/10_globins/10_globins.fasta")
log.likelihood = paste0(base.name,".ll") %>%
                 fread %>% select(ll.all)
ali.length = paste0(base.name,".length") %>%
             fread %>% select(ali.length)
plot(mcmc(cbind(log.likelihood,ali.length)))
```

As shown in Fig. 1, although the mean values appear to have converged, both of these trace plots show significant autocorrelation. In the context of an MCMC sampler, this indicates what is referred to as "poor mixing," meaning that the sampler may not explore the posterior efficiently, requiring a long time to fully sample the relevant alignments, trees, and model parameters. We can quantify this more rigorously by examining the autocorrelation and effective sample size (ESS) using the `acf` and `effectiveSize` functions in R (*see* Fig. 2). The effective sample size of 109 for the alignment length indicates poor mixing in this case.

```
layout(t(1:2))
acf(log.likelihood)
acf(ali.length)
effectiveSize(ali.length)
## ali.length
## 108.9655
```

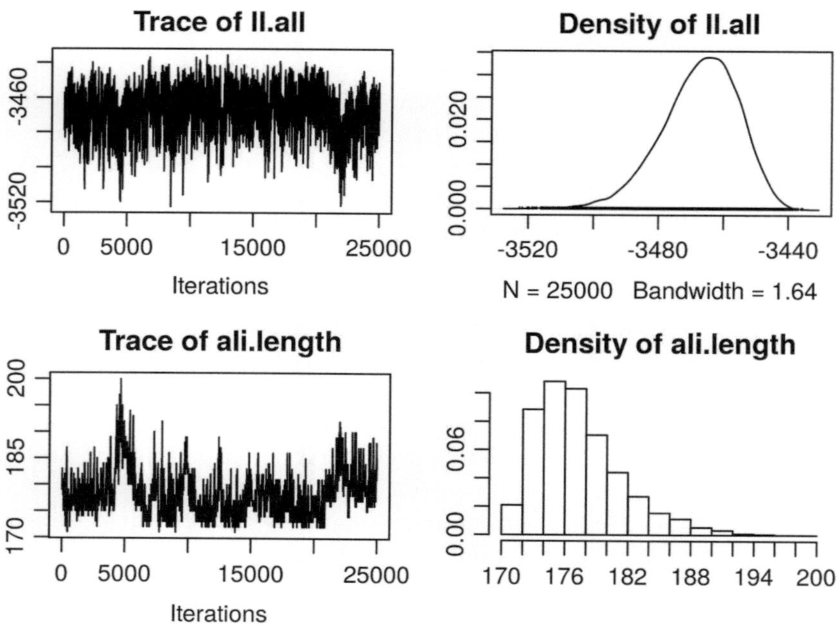

Fig. 1 Trace plots for the log likelihood and alignment length, illustrating some noticeable autocorrelation

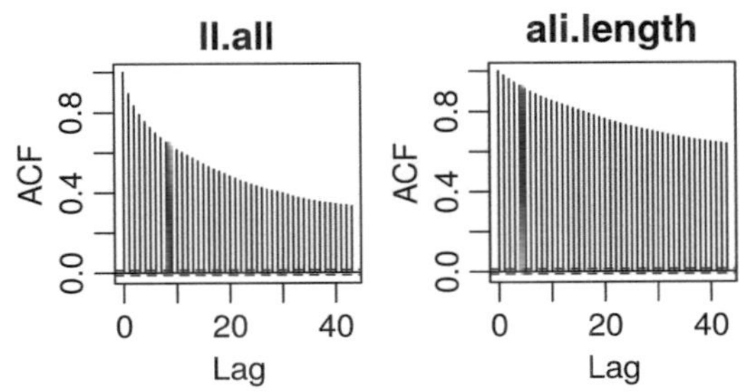

Fig. 2 Autocorrelation plots for the log likelihood and alignment length, indicating poor mixing

3.3 Using Parallel Tempering to Improve Mixing

Slow mixing between different regions of the posterior is a common issue when sampling high-dimensional hierarchical models involving discrete parameters such as sequence alignments and tree topologies. Although StatAlign employs a number of advanced proposal distributions to efficiently explore the parameter space, transitions between modes of similar posterior density may still require traversal of configurations that are highly unfavorable, making these transitions very infrequent.

One way to address this is to use a multiple-chain MCMC sampler where each chain has an associated heat parameter, t, used to flatten out the posterior surface in order to increase the frequency of transitions between configurations [48]. Under this scheme, the chain with $t = 1$ (the "cold chain") generates samples from the true posterior, and the heated chains (with $t > 1$) sample from flattened versions thereof, which can be traversed more easily. By swapping heats between chains, accepting/rejecting heat proposals according to the appropriate Metropolis-Hastings ratio, the stationary distribution of the cold chain is maintained.

Running several chains in parallel at different heats, parallel tempering can be implemented efficiently by swapping temperatures between specified chains according to a random sequence that is shared between all chains [49]. Empirically we have found linear spacings between adjacent inverse temperature values to be effective, with the default step size set to 0.01 in StatAlign v3.3 (this can be modified via the `-tempDiff` argument to StatAlign). As well as exchanging temperature parameters, StatAlign also swaps parameters that determine the variance of MCMC proposal distributions, ensuring that optimal acceptance ratios for individual moves are maintained.

3.4 Running the Parallel Version of StatAlign

Before running the parallel version on the test dataset, we will first copy the input files to a new directory, so that the new output does not overwrite the existing output files:

```
mkdir -p mpi_output
cp examples/10_globins/10_globins.fasta mpi_output
```

StatAlign can then be run in parallel via the StatAlignParallel wrapper script:

```
StatAlignParallel 8 \
 -mcmc=1m,5m,200,0,20 \
 -tempDiff=0.02 \
 mpi_output/10_globins.fasta
```

The additional (fifth) MCMC argument specifies that temperature exchanges should be proposed between parallel chains every 20 iterations. This ensures that the chains do not drift too far apart at different temperatures, increasing the acceptance rate of temperature swaps. Running StatAlign with these parameters takes approximately 5h on a machine with eight 2.3GHz Xeon processors.

3.5 Analyzing Parallel Output

When running in parallel mode, each chain generates a separate set of output files, indicated by `chainX` in the filename (*see* **Note 4**). The `.coreModel.params` files now contain the inverse temperature parameter (beta) as the second column. We will first extract this information in order to aggregate the samples based on the chain temperature:

```
n.chains = 8
chains = 0:(n.chains-1)
base.name = paste0(STATALIGN_HOME,
           "/mpi_output/10_globins.fasta")
# read in core model (TKF92) MCMC output file
coreModel.list = lapply(chains,function(x)
 paste0(base.name,".chain",x,".coreModel.params") %>% fread
)
coreModel = do.call(rbind,coreModel.list)
# extract the values of the beta (inverse temperature)
# parameter
beta.values = as.character(sort(unique(coreModel$beta)))
```

We can then read in the per-sample log likelihood and alignment length for each chain:

```
log.likelihood = lapply(chains, function(x)
 paste0(base.name,".chain",x,".ll") %>% fread
) %>% do.call(rbind,.)
ali.length = lapply(chains, function(x)
 paste0(base.name,".chain",x,".length") %>% fread
) %>% do.call(rbind,.)
```

These values can then be aggregated based on the beta (inverse temperature) parameter:

```
log.likelihood =
 Map(function(i)
   log.likelihood %>%
   filter(coreModel$beta==i) %>%
   arrange(sample) %>%
   select(ll.all),
 beta.values)
ali.length =
 Map(function(i)
   ali.length %>%
   filter(coreModel$beta==i) %>%
   arrange(sample) %>%
   select(ali.length),
 beta.values)
```

In order to assess the mixing between temperatures, we can create a table indicating the frequencies with which each chain samples the different temperature parameters. For a well-mixing parallel tempering scheme, each chain should be sampling multiple temperature parameters with high frequency. As shown in Table 3, for the example dataset, each chain samples a range of temperature values with high frequency, indicating very good mixing across temperatures:

```
beta.freqs = lapply(coreModel.list,with,beta) %>%
             lapply(.,table) %>%
             do.call(cbind,.)
```

Table 3
Frequency of inverse temperature parameters sampled in each MCMC chain

	1	2	3	4	5	6	7	8
0.86	3200	2686	3177	2748	3472	3578	3477	2662
0.88	3171	2982	3252	2901	3216	3350	3381	2747
0.90	3033	3102	3395	3021	3223	3141	3222	2863
0.92	3137	3319	3311	3280	3100	3111	2932	2810
0.94	3227	3161	3119	3209	3041	2973	3051	3219
0.96	3153	3239	3057	3218	2978	2962	2967	3426
0.98	3126	3153	2901	3294	3036	2979	2979	3532
1.00	2953	3358	2788	3329	2934	2906	2991	3741

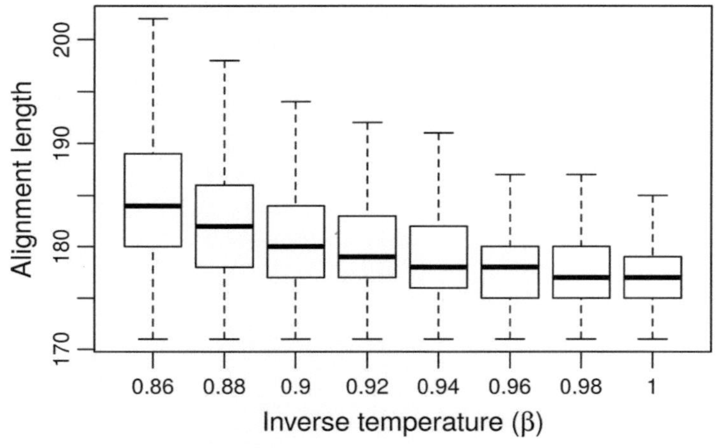

Fig. 3 Distribution of alignment lengths as a function of the temperature parameter

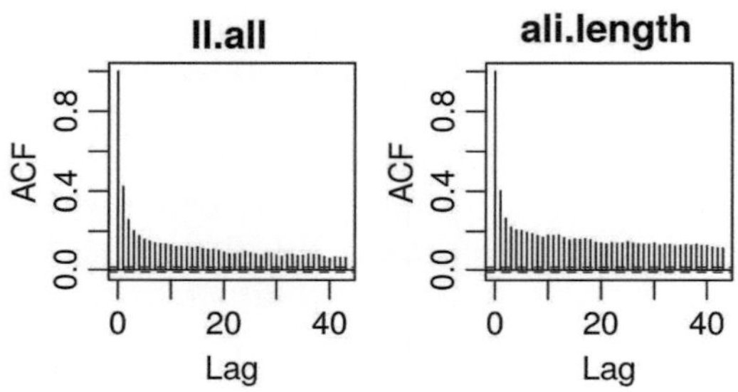

Fig. 4 Autocorrelation plots for the log likelihood and alignment length, sampled using parallel tempering

As shown in Fig. 3, at higher temperature (lower inverse temperature) values, the distribution of alignment lengths broadens, including much longer alignments, indicating that varying the temperature is allowing the MCMC sampler to explore a wider variety of configurations:

```
do.call(cbind,ali.length) %>%
 boxplot(outline = FALSE,
         names = beta.values,
         xlab = expression("Inverse temperature ("*beta*")"),
         ylab = "Alignment length",
         cex.axis = 0.8)
```

Examining the autocorrelation for the log likelihood and alignment length, we see that it has reduced significantly (Fig. 4), and the effective sample size has increased more than eightfold. This indicates that, as well as now moving between different regions of the space more efficiently, we are now obtaining more information about the posterior distribution than would be obtained from eight independent chains running with beta = 1:

```
layout(t(1:2))
acf(log.likelihood$`1`)
acf(ali.length$`1`)
effectiveSize(mcmc(ali.length$`1`))
## ali.length
## 970.4661
```

The effective sample size for the insertion-deletion model parameters is also very high, indicating good mixing across the parameter space:

```
tkf = coreModel %>% filter(beta==1) %>%
      select(-c(sample,beta))
effectiveSize(mcmc(tkf))

##         R     Lambda       Theta
## 2588.848  5630.171  12228.428
```

The R parameter governs the geometric distribution on indel lengths, with the expected fragment length equal to $1/R$; λ is the insertion rate per time unit (as defined by the substitution model, usually substitutions per site), μ is the deletion rate, and $\theta = \lambda/(\lambda + \mu) < 0.5$ is the insertion rate relative to the total indel rate [50].

3.6 Running with Different Random Seeds to Assess Convergence

In addition to basic checks on the trace plots and autocorrelation functions of individual parameters, a common approach for assessing convergence is to run multiple MCMC samplers with different random seeds or starting configurations. This can be accomplished by rerunning StatAlign using a different value for the `-seed` argument, storing the output into a separate directory for each run. If each of these samplers ends up sampling from the same posterior distribution, it is a good indication that they have converged. Agreement between the independent runs can be quantified more rigorously via the Gelman-Rubin potential scale reduction factor [51]. We will return to this when analyzing the output of StructAlign.

3.7 Consensus Trees

The `.tree` files generated by StatAlign contain samples from the posterior distribution over phylogenetic trees. To summarize this distribution, a majority consensus tree can be generated by using StatAlign's ConsensusTree plugin, which can be called from the command line on the MCMC output via the following command (*see* **Note 5**):

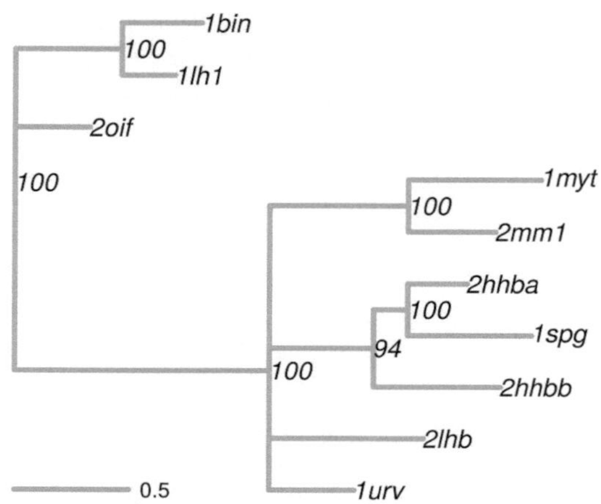

Fig. 5 Consensus tree for trees sampled under the sequence-only model

```
java -cp $STATALIGN_JAR \
 statalign.postprocess.plugins.contree.CTMain \
 $SEQ_DIR/$FASTA.chain{0..7}.tree \
 > $SEQ_DIR/$FASTA.ctree
```

This command will create a `.ctree` file containing a Newick-formatted consensus tree, which can then be examined in R using the `ape` package:

```
# read in the consensus tree created by StatAlign
consensus = read.tree(paste0(STATALIGN_HOME,
 "/mpi_output/10_globins.fasta.ctree"))
# collapse splits with less than 60% posterior support into
# polytomies
source(paste0(STATALIGN_HOME,"/scripts/apply.min.support.R"))
consensus %<>% apply.min.support(tol=60)

root(consensus,"2oif") %>%
 plot(show.node.label = TRUE,
      use.edge.length = TRUE,
      edge.color = "grey",
      cex = 1.2,
      show.tip.label = TRUE,
      edge.width = 4,
      no.margin = TRUE)
add.scale.bar(lwd=4,lcol="grey")
```

As shown in Fig. 5, the consensus tree contains a four-way polytomy, indicating high uncertainty regarding the relative placement of the hemoglobin, myoglobin, and cytoglobin/cyclostome clades.

3.8 Including Protein Structures

In order to better resolve the relationships between the different clades, we can utilize the StructAlign plugin for StatAlign, which models structural divergence using a continuous-time stochastic process on C-alpha coordinates, combined with a Markov model of sequence evolution [41]. The rate of structural divergence along each branch is modeled via a diffusivity parameter, σ. To account for non-evolutionary sources of structural variability (e.g., due to conformational flexibility, differences in experimental conditions, or technical noise), each residue has an intrinsic baseline variability parameterized based on the crystallographic B-factors (*see* **Note 6**).

StructAlign can read protein structure coordinates directly from PDB files and can be run on the example globin dataset using the following command (*see* **Note 7**):

```
pdb_files=(1bina.pdb 1lh1.pdb 1myt.pdb 1spga.pdb 1urv.pdb
2hhbA.pdb 2hhbB.pdb 2lhb.pdb 2mm1.pdb 2oif.pdb)
StatAlignParallel 8 \
 -plugin:structal[printTree,printRmsd] \
 -mcmc=1m,5m,200,0,20 \
 -seed=1 \
 -tempDiff=0.01 \
 "${pdb_files[@]}"
```

This run takes approximately 7h on a machine with eight 2.8GHz Xeon processors and generates the same output files as for the sequence-only model, plus several additional files pertaining to the structural component of the model (*see* Table 4). The `printTree` option causes the `.struct.tree` file to be generated, which is useful for examining how rates of structural drift vary over the tree. The `printRmsd` option generates additional files that can be used to examine the relationship between sequence and

Table 4
Additional output files generated by StructAlign

File	Contents	Format
*.struct.params	Structural model parameters	Tab-separated
*.struct.tree	Trees with branch lengths equal to structural diffusivity	NEXUS
*.msd	Pairwise mean-squared deviation between each structure, for every MCMC sample	Tab-separated
*.mle.fasta	Maximum likelihood alignment	FASTA
*.mle.rmsd	Per-site RMSD for MLE alignment	Single-column
*.mle.bfactors	Per-site average B-factors (weighted by $\sqrt{3\varepsilon}$) for MLE alignment	Single-column
*.mle.super.pdb	Maximum likelihood structural superposition	PDB

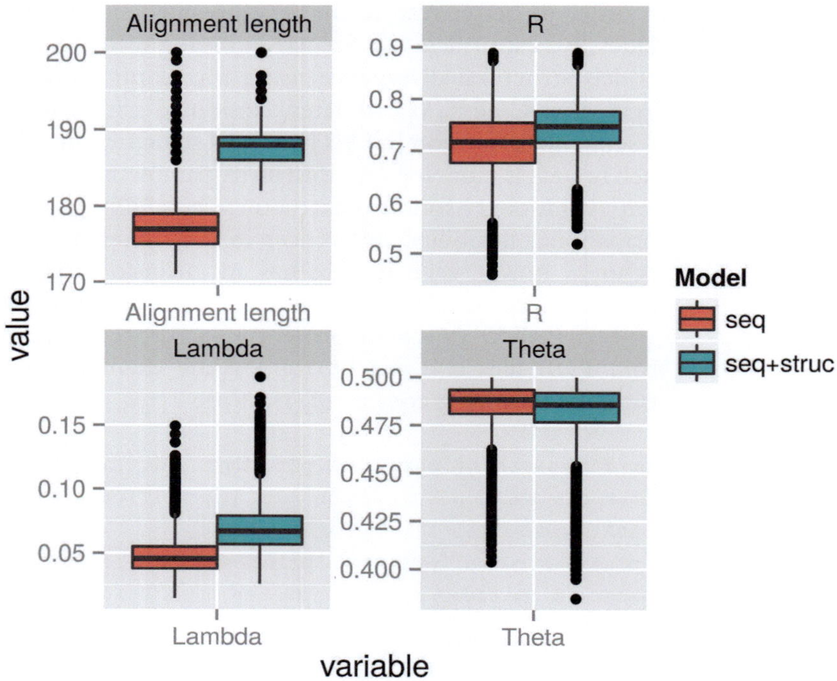

Fig. 6 Comparison of posterior distributions for alignment length and TKF92 model parameters under the sequence-only and sequence + structure models

structural divergence. We will return to these files in more detail later. The output files when running StructAlign will all be prefixed by the name of the first PDB file in the list passed at the command line, which in this case is "1bina.pdb".

As before, we can examine the effective sample size for the TKF92 parameters. The ESS values are now reduced, since we are simultaneously sampling a number of additional parameters for the structural model, although still remain very high, indicating good mixing within the TKF92 component of the model:

```
effectiveSize(mcmc(tkf.struc))
```

```
##        R    Lambda      Theta
## 1528.826  3241.599  11782.182
```

Comparing the posterior distributions for the model parameters, we can see that the structural model favors longer alignments, resulting in an increase in the estimated values for R and λ (*see* Fig. 6):

```
comparison =
 rbind(
  cbind('Alignment length'=ali.length.seq$'1'$ali.length,
        tkf.seq,
        Model=rep("seq",nrow(tkf.seq))
  ),
  cbind('Alignment length'=ali.length.struc$'1'$ali.length,
        tkf.struc,
        Model=rep("seq+struc",nrow(tkf.struc))
  )
 )
df.m = melt(comparison, id.var="Model")
ggplot(df.m, aes(x=variable, y=value)) +
 geom_boxplot(aes(fill=Model)) +
 facet_wrap( ~ variable, scales="free")
```

This same observation was noted by Herman et al. [41] and arises mainly from differences in the alignment at loop regions: while the sequence-only model tries to align the loops and encounters high uncertainty due to the higher sequence divergence in these regions, the structural model tends to favor indels when the structural divergence is significantly higher than would be expected due to baseline variability alone.

3.9 Effect of Structural Information on Sequence Alignments

We can visualize the change in distribution over alignments in more detail by computing a summary alignment annotated with associated posterior probabilities for each column. To do so, we will utilize the program WeaveAlign, which is distributed along with StatAlign. WeaveAlign takes as input a file or files containing multiple alignments for the same set of sequences and computes the summary alignment that maximizes the expected accuracy under a chosen scoring scheme [7].

When running StatAlign in parallel mode, the alignment samples generated by each chain must be combined into a single file before running WeaveAlign. This can be achieved using the `combine_logfiles.pl` script distributed with StatAlign, run as shown below (with the environment variables set to the appropriate values):

```
scripts/combine_logfiles.pl $SEQ_DIR/$FASTA.chain{0..7}.log
scripts/combine_logfiles.pl $STRUC_DIR/$PDB.chain{0..7}.log
```

WeaveAlign can then be run on the combined logfile; here we will generate an alignment that maximizes the expected total column score, i.e., the proportion of correct columns in the alignment [7], using the commands below:

```
java -jar WeaveAlign.jar -optgi -outgi \
 $SEQ_DIR/$FASTA.combined.log
java -jar WeaveAlign.jar -optgi -outgi \
 $STRUC_DIR/$PDB.combined.log
```

As well as a summary alignment file, WeaveAlign also generates a graphical representation of the alignment, annotated with posterior probabilities. As shown in Fig. 7, the addition of structural information into the model results in decreased uncertainty for the majority of columns, as indicated by the higher posterior probabilities (shown as a blue line above the alignment). In addition, there are several regions where the alignment changes significantly, which we will examine in further detail.

WeaveAlign can be used to generate figures for specific regions of the alignment, defined via column ranges or via subsequences of a particular sequence. As an example, the region of the alignment containing the first occurrence of the subsequence AWEVAYDE (Ala128-Glu135) in the structure *1bin* (corresponding to the center of the H-helix) can be displayed using the command below:

```
java -cp WeaveAlign.jar \
 alignshow.Show \
 -f imagefile.png \
 -r=1bin,AWEVAYDE \
 -t $SEQ_DIR/$FASTA.combined.log.scr \
 $SEQ_DIR/$FASTA.combined.log.fsa
```

Under the sequence-only model, the three plant globins are determined to have a length-6 indel in this region, whereas the structural model lines up all the sequences in this region, with very low uncertainty (*see* Fig. 8). Examining the corresponding structures in these regions (Fig. 9), we can see that the structural model aligns contiguous regions, whereas the sequence-only model infers an indel in the middle of the H-helix, which is a very unlikely scenario.

Other notable differences in the structurally informed alignment occur in the region corresponding to residues Val82-Ala87 in *1bin*, comprising the EF-loop region. Under the sequence-only model, the three plant globins align with each other in this region, with a length-3 indel relative to the other sequences, corresponding to Val83-Asp85 in *1bin*, Val85-Asp87 in *1lh1*, and Val93-Glu95 in *2oif*. The leghemoglobin structure *1bin* is also inferred to have a shortened loop region relative to the other plant globins, with a length-3 indel at the start of the F-helix, between Ala87 and Leu89.

This is very similar to the alignment reported by Hoy et al. [53], who proposed a length-4 indel between Ala87 and Leu89 in *1bin* relative to the *2oif* structure at the start of the F-helix, positing that this deletion, along with others, may have led to reduced conformational flexibility, causing the leghemoglobins to lose the

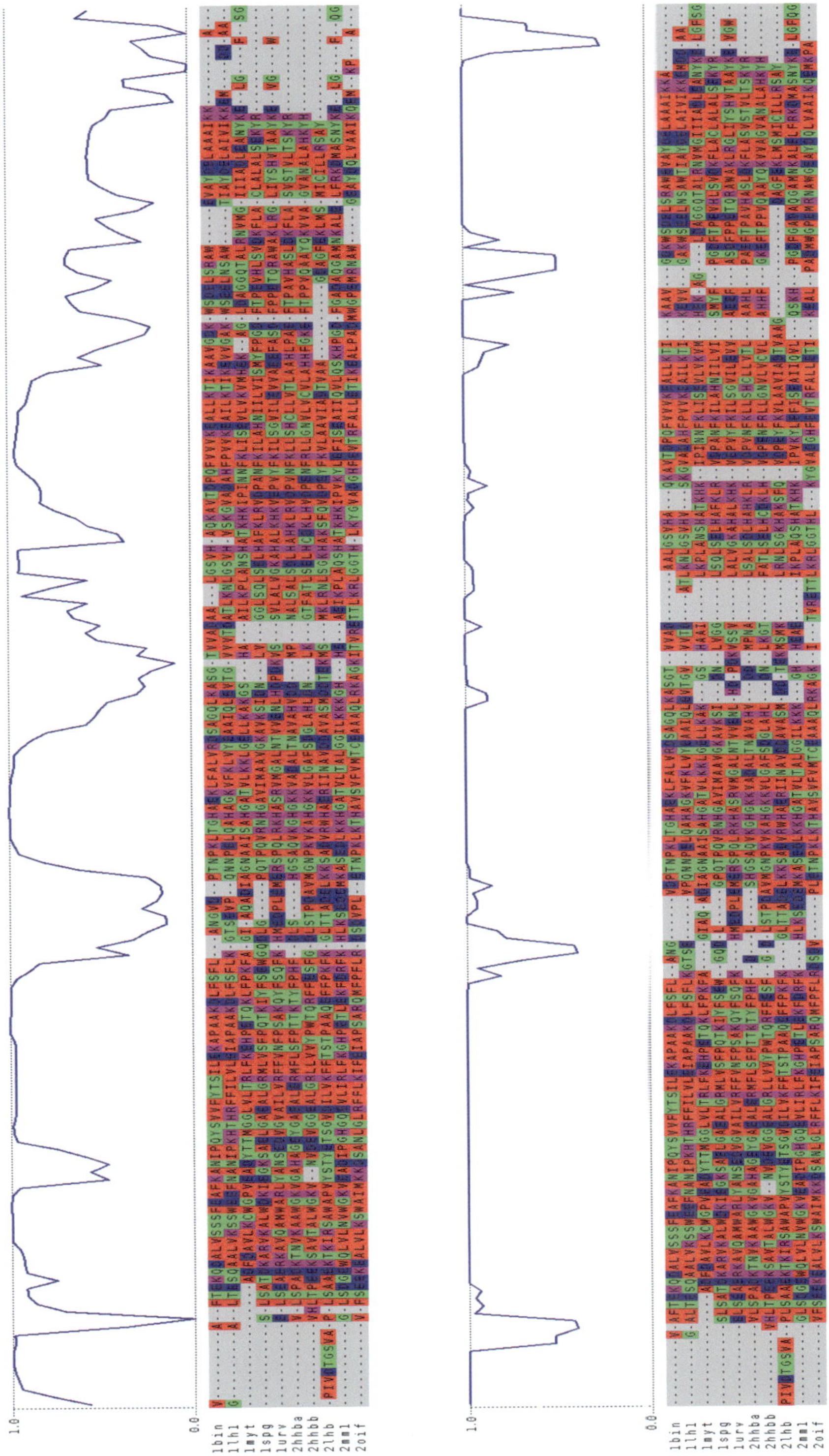

Fig. 7 Summary alignment under sequence-only model (left) and structural model (right), annotated with posterior probabilities for each column (blue lines)

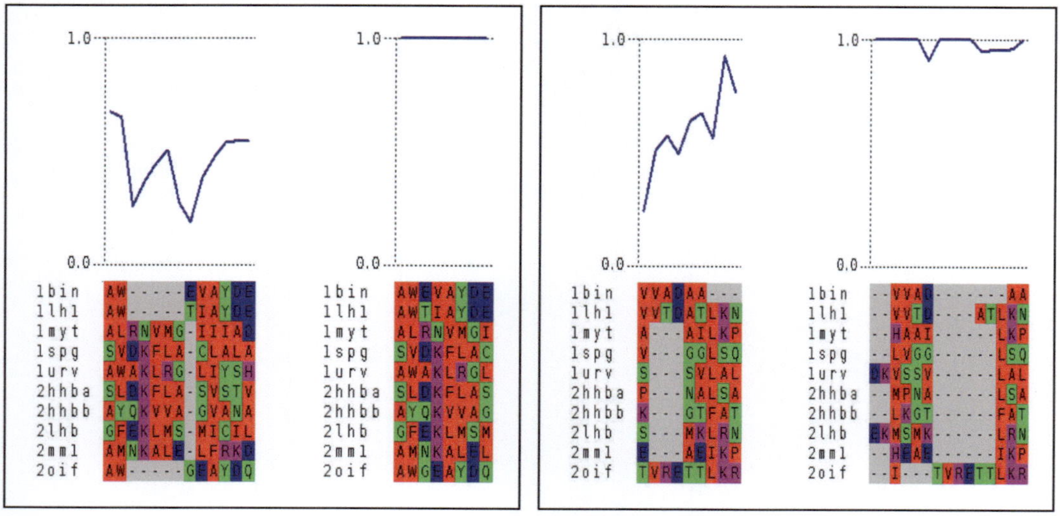

Fig. 8 Zoomed-in view of two regions of the summary alignment corresponding to Ala128-Glu135 and Val82-Ala87 in the sequence for *1bin* (left and right panels, respectively), aligned under the sequence-only model (left within each panel) and structural model (right within each panel)

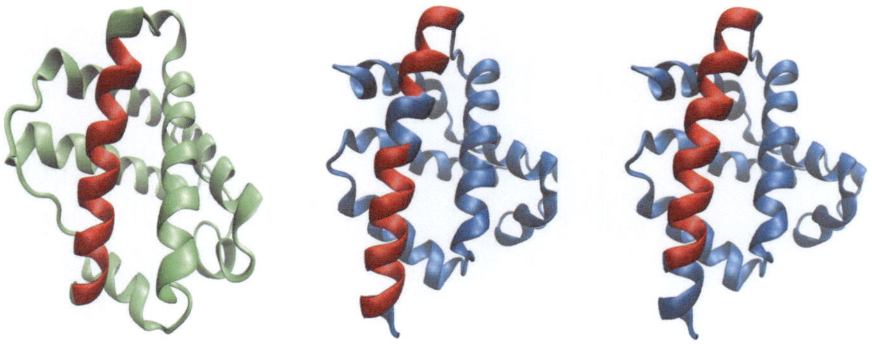

Fig. 9 Aligned regions of helix H for leghemoglobin structure *1bin* (green) and myoglobin structure *1myt* (blue), with the corresponding aligned regions shown in red using the sequence-only model (center), and structural model (right). Figures generated using VMD [52]

ability to adopt a hexacoordinate configuration with both distal and proximal histidines in coordination with the heme.

In contrast, the structural model favors aligning *1bin*:Val83-Asp85 and *1lh1*:Val85-Asp87 with the other eight globin sequences, treating the start of the F-helix in *2oif* as an indel, due to the very high structural deviation observed in the region between Thr92 and Thr97.

We can visualize the structural variability at this region via the maximum likelihood structural superposition generated by StructAlign, which is outputted into the `.mle.super.pdb` files

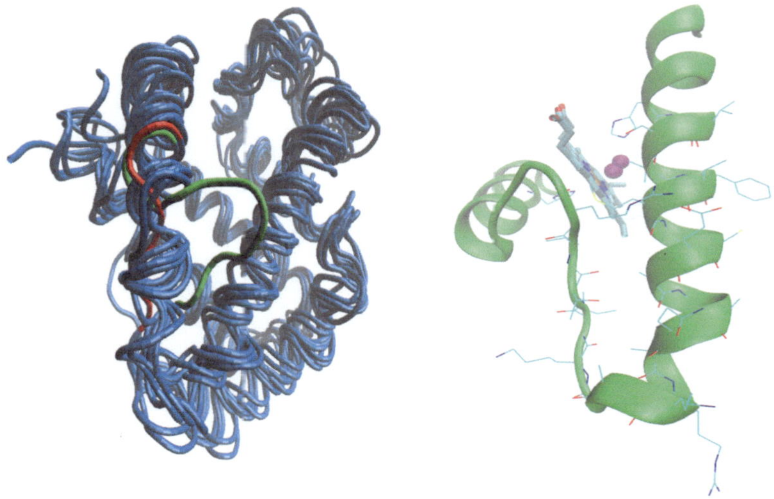

Fig. 10 Maximum likelihood superposition for the ten globin structures in the example dataset, oriented with the E-helix descending from top right to bottom left, and the F-helix ascending vertically (left), and a view of the EF-loop section of *2oif*, including heme and cyanide ligand (right). Highlighted in red and green on the left panel are sections of the structures *1bin* and *2oif*, illustrating the large deformation that occurs in the latter at the start of the F-helix, which may stabilize the ligand-bound conformation. Figures generated using VMD [52]

(*see* **Note 8**). As shown in Fig. 10, the structure *2oif* exhibits a large deformation at the start of the F-helix, with a number of side chains coming into contact with residues within the E-helix, for example, Arg94 with Glu80. As discussed by Hoy et al. [53], this deformation may help to stabilize the conformation in which the exogenous ligand displaces His70.

3.10 Posterior Distribution of Structural Model Parameters

As mentioned in the previous section, the large structural displacement in the EF-loop region of *2oif* relative to the other structures causes StructAlign to treat Thr92-Thr97 as an indel. To understand how the model makes this decision and how this affects parameter inference, we can examine the individual parameters of the structural model in more detail.

The structural parameters can be read in from the `.struct.params` output files in a manner similar to the TKF92 model parameters. Here we will illustrate reading in parameters from four independent runs conducted with different random number seeds, each executed in its own separate `'run_x'` subdirectory, in order to assess the consistency of the parameter estimates across runs:

```
# for each of four independent runs with different starting
# seeds
run = 1:4

# for each run, read in structural model parameters
struct.list =
 lapply(run,
  function(r) {
   base = paste0(STRUC_DIR,"/run_",r,"/",PDB)

   core =
    lapply(chains,function(x)
     fread(paste0(base,".chain",x,".coreModel.params"))
    ) %>%
    do.call(rbind,.)

   struct.params =
    lapply(chains,function(x)
     fread(paste0(base,".chain",x,".struct.params"))
    ) %>%
    do.call(rbind,.) %>%
    filter(core$beta==1) %>%
    select(c(tau,eps,s2_g,nu))

   return(struct.params)
  }
 )
```

Before analyzing the individual parameters, we can first assess the reliability of the inference by comparing the posterior distributions across runs. As shown in Fig. 11, the inferred posteriors are very similar for all four global structural parameters, indicating that the chains have converged:

```
comparison =
 lapply(run,
  function(r) cbind(struct.list[[r]],Seed=r)
 ) %>%
 do.call(rbind,.)
comparison$Seed = factor(comparison$Seed)
df.m = melt(comparison, id.var="Seed")
ggplot(df.m, aes(x=variable, y=value)) +
 geom_boxplot(aes(fill=Seed)) +
 facet_wrap( ~ variable, scales="free")
```

We can quantify this convergence more rigorously using the Gelman-Rubin potential scale reduction factor (PSRF), which should be close to unity for chains that have converged. As shown

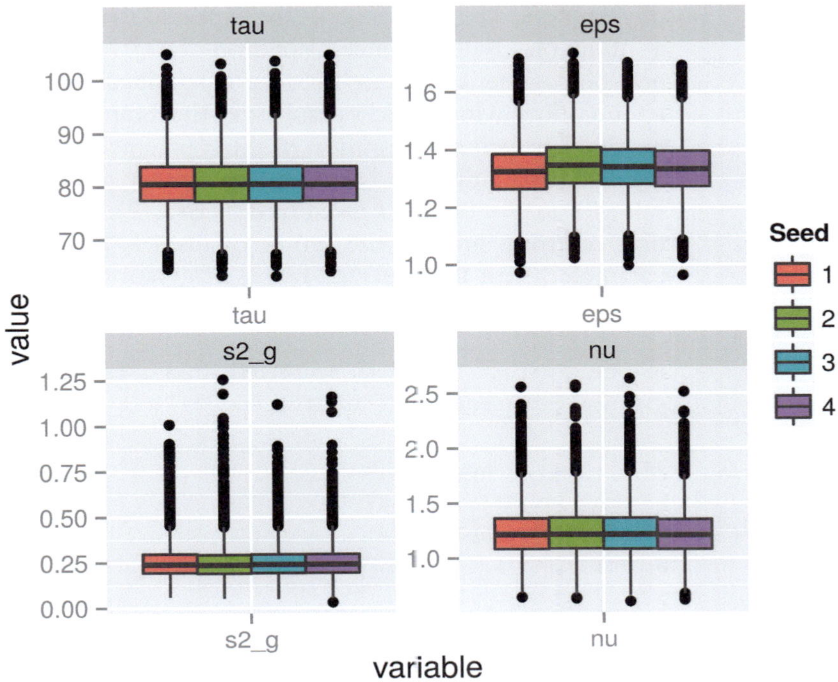

Fig. 11 Comparison of posterior distributions for structural model parameters with four different starting seeds

Table 5
Gelman-Rubin potential scale reduction factors for the structural model parameters

	Point est.	Upper C.I.
tau	1	1.001
eps	1.009	1.028
s2_g	1	1.001
nu	1.001	1.002

in Table 5, for each of the four structural parameters, the upper bound for the PSRF is very close to unity, indicating excellent convergence:

```
gelman = gelman.diag(mcmc.list(lapply(struct.list,mcmc)))$psrf
```

3.11 Interpreting the Structural Parameters

StructAlign models the structural variation at each residue as a combination of a baseline non-phylogenetic term (usually parameterized via crystallographic *B*-factors) and a branch-specific phylogenetic term that governs the expected structural deviation as a function of sequence variation [41]. The amount of non-phylogenetic baseline variance is specified via the parameter ε, and the phylogenetic variance via the branch-specific σ parameters. The prior distribution of the σ parameters is governed by a global σ_g parameter and a variance parameter, ν, with the expected value for σ_k^2 for branch k given by $\sigma_g^2\ exp(\nu/2)$. The prior distribution of atoms around the center of mass is governed by the τ parameter, with the expected radius of gyration equal to $\sqrt{3\tau}$. In our example, this can be computed from the posterior samples via

```
struct = struct.list[[1]] # use only run 1 for this analysis
HPDinterval(mcmc(sqrt(3*struct[,'tau'])))
```

```
##          lower    upper
## tau  14.66535  16.48737
## attr(,"Probability")
## [1] 0.95
```

The top end of this range includes the value of 16.4 ± 0.2 reported by Lobanov et al. [54] for all-alpha proteins of comparable size taken from SCOP.

The ε parameter acts as a multiplier on the baseline variance associated with each alignment column (estimated via squared normalized *B*-factors), with $B_i\sqrt{3\varepsilon}$ yielding the expected standard deviation for site i arising from non-phylogenetic sources of structural variability (including uncertainty in the structural superposition). We can examine the correlation between these predicted values and the observed per-site RMSD via three of the additional files generated via the `printRmsd` option to StructAlign, i.e., the `.mle.fasta` alignment file and the `.mle.rmsd` and `.mle.bfactors` files that contain the RMSD and *B*-factor-based predictions, respectively. When running in parallel mode, each chain again generates its own version of these files; we will select the chain with the highest likelihood MLE (in our example case, chain 7) for further analysis. WeaveAlign can again be used to generate an image with these annotations plotted above the alignment. As shown in Fig. 12, the correlation between predicted and observed values is very high, with higher structural variability occurring mostly in the areas of high alignment uncertainty in the loop regions (*see* **Note 9**):

```
java -cp WeaveAlign.jar \
 alignshow.Show \
 -t $STRUC_DIR/$PDB.chain7.mle.rmsd \
 -t $STRUC_DIR/$PDB.chain7.mle.bfactors \
 -c=RED -c=GREEN -png \
 $STRUC_DIR/$PDB.chain7.mle.fasta
```

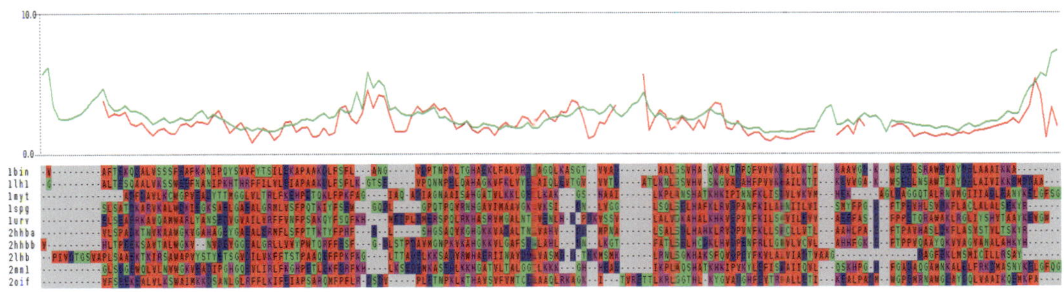

Fig. 12 Maximum likelihood alignment, annotated with predicted (green) and observed (red) per-site RMSD

The magnitude of σ has a more direct interpretation, since $\sigma^2 t$ is the expected mean-squared deviation along each of the three spatial coordinates occurring from phylogenetic structural drift over a time interval of length t.

Previous studies have tried to model the relationship between substitutions per site and RMSD via a global linear regression with a fixed constant of proportionality. Using this approach, Illergård et al. [29] derived a linear coefficient of 0.312 for the globin family.

In contrast, StructAlign models the MSD (squared RMSD) as increasing linearly with sequence divergence, as implied by a diffusion model, but permits the rate of structural divergence to vary in a piecewise linear fashion across the tree, allowing for more complex relationships between sequence and structure to be modeled. As we shall see in the following section, the resulting fit between observed and predicted pairwise RMSD is typically very good.

3.12 Effect of Structural Information on Consensus Trees

We can compute consensus trees on the StructAlign output as previously, now running the ConsensusTree plugin both on the `.tree` files and the `.struct.tree` files, in order to create consensus trees with branch weighted by branch length and structural diffusivity (σ^2), respectively:

```
java -cp StatAlign.jar \
 statalign.postprocess.plugins.contree.CTMain \
 $STRUC_DIR/$PDB.chain{0..7}.tree \
 > $STRUC_DIR/$PDB.ctree

java -cp StatAlign.jar \
 statalign.postprocess.plugins.contree.CTMain \
 $STRUC_DIR/$PDB.chain{0..7}.struct.tree \
 > $STRUC_DIR/$PDB.struct.ctree
```

Plotting the consensus tree as before, we can see that most of the uncertainty present in the tree generated by the sequence-only model has been resolved, with the cyclostome (*2lhb*) and cytoglobin (*1urv*) clade placed next to the plant globins, and the

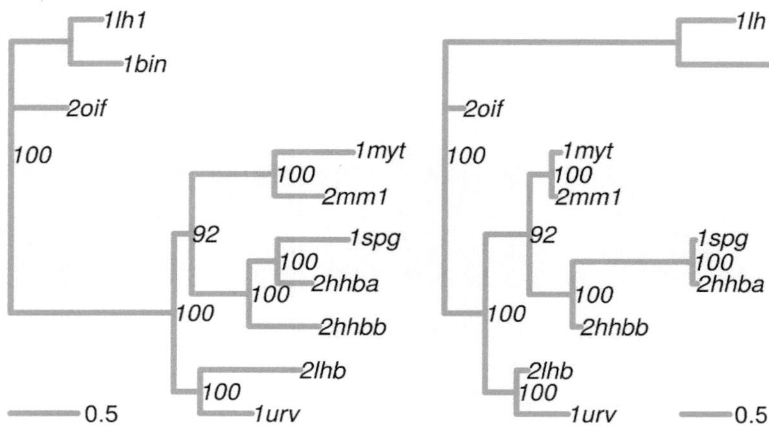

Fig. 13 Consensus tree for trees sampled under the sequence+structure model, with branches scaled according to branch length (left) and structural diffusivity (right)

myoglobins and hemoglobins clustering closer to each other (*see* Fig. 13). This matches the observations of Herman et al. [41] and Christensen et al. [55], who inferred the same relationship between these clades based on analysis of larger sets of sequences and structures.

Comparing the inferred branch lengths under the sequence and sequence+structure models, most are of very similar length under both models. We can examine this more closely by looking both at the branch lengths at the tips of the tree and the pairwise distances between leaves of the tree:

```
consensus.seq = read.tree(paste0(SEQ_DIR,"/10_globins.fasta.ctree"))
consensus.struc = read.tree(paste0(STRUC_DIR,"/1bina.pdb.ctree"))
n = length(consensus.struc$tip.label)
# map ordering of sequences in consensus.seq
# to that of consensus.struc
map = pmatch(consensus.struc$tip.label,consensus.seq$tip.label)

# extract edge lengths for branches leading to tips of tree
tip.edges.struc =
 consensus.struc$edge.length[consensus.struc$edge[,2]<=n]
tip.edges.seq =
 consensus.seq$edge.length[consensus.seq$edge[,2]<=n][map]

# combine into data.frame
df.tips = data.frame(name=consensus.struc$tip.label,
                     seq=tip.edges.seq,
                     struc=tip.edges.struc)

# find distances between leaves along edges of the tree
dist.seq = cophenetic(consensus.seq)
seqs = colnames(dist.seq)
```

```
# for consensus.struc we reorder rows and cols to match
dist.struc = cophenetic(consensus.struc)[seqs,seqs]

# extract off-diagonal elements and combine into data.frame
df.pw = data.frame(seq=dist.seq[lower.tri(dist.seq)],
                   struc=dist.struc[lower.tri(dist.struc)])
```

Plotting these quantities for the sequence-only versus sequence +structure models, most values are seen to be very similar (*see* Fig. 14), indicating that the structural model does not systematically distort the estimated branch lengths. The tip branches leading to *2lhb* and *2oif* are lengthened slightly in the sequence+structure model, reflecting the fact that the structural model infers additional indels in these sequences:

```
outliers = subset(df.tips,name %in% c('2oif','2lhb'))

# plot edge lengths for tip branches
p1 = ggplot(df.tips,aes(x=seq,y=struc)) +
     geom_point() +
     geom_point(data=outliers,shape=1,size=3) +
     xlab("Tip branch length (seq-only)") +
     ylab("Tip branch length (seq + struc)") +
     geom_abline(intercept=0,slope=1) +
     geom_text(data=outliers,aes(label=name),
               size=4,hjust=1.5,vjust=1)
# plot pairwise distances
p2 = ggplot(df.pw,aes(x=seq,y=struc)) +
      geom_point() +
      xlab("Tree distance (seq-only)") +
      ylab("Tree distance (seq + struc)") +
      geom_abline(intercept=0,slope=1)

grid.arrange(p1,p2,ncol=2)
```

3.13 Variation of Rate of Structural Evolution Across the Tree

As discussed above, although inclusion of the structural model significantly alters the distribution over alignments and trees, it does not significantly impact the inferred branch lengths. This is due to the fact that the model allows the rate of structural evolution to vary independently along each branch, such that the branch lengths do not need to adjust in order to reflect the structural deviation across the tree. This allows the well-calibrated sequence model to be the primary determinant of divergence times, while making use of the structural model to improve homology inference.

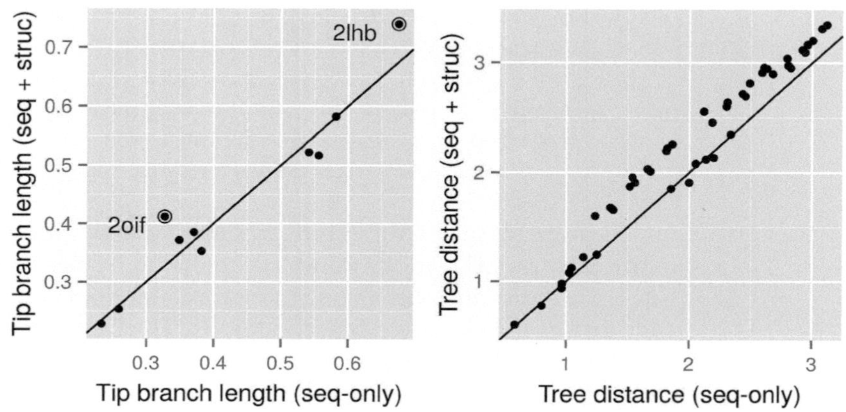

Fig. 14 *Left*: branch lengths for branches leading to tips of the tree. *Right*: Pairwise tree distances (in substitutions per site) between each pair of structures for the consensus tree computed under the sequence-only and sequence + structure models

As shown in Fig. 13, the rate of structural evolution varies significantly across the tree, with very low diffusivity parameters observed within the myoglobin, α-hemoglobin and β-hemoglobin clades, and larger rates inferred between clades. The low rates within clades may reflect the need to preserve structure in order to maintain function, whereas the increased rates between clades may represent duplication/divergence events, leading to decreased selective pressure on structure. A notable example of the latter occurs at the branch between the hexacoordinate nonsymbiotic plant globin, *2oif*, and the two pentacoordinate leghemoglobins, *1bin* and *1lh1* [56].

It is also notable that the structural diffusivity along the branch to *2oif* is very low, explaining why the model preferentially creates an indel at the start of the F-helix, where the structural deviation is much higher than would be expected given the level of sequence divergence.

3.14 Pairwise Structural Diffusion Distance Is Linearly Related to RMSD

As discussed in the introduction, permitting the rate of structural deviation to vary over the tree allows for more complex relationships between sequence and structural divergence. In order to examine the model fit, we can first use the MCMC output to estimate the expected RMSD between each pair of structures, averaged over alignments and structural superpositions. This can then be compared to the pairwise structural diffusion distances using the consensus tree, corresponding to the expected pairwise deviation under the StructAlign model.

We will focus here on one of the four runs, since all four appeared to be very similar. First we read in the core model parameters again, in order to locate the cold chain at each iteration:

```
base = paste0(STRUC_DIR,"/run_1/",PDB)
core = lapply(chains, function(x)
     fread(paste0(base,".chain",x,".coreModel.params"))
     ) %>% do.call(rbind,.)
```

Next, we will read in the pairwise mean-squared deviation for each MCMC sample, and compute the square root of the posterior mean, to yield an estimate of the RMSD between each pair of structures:

```
rmsd =
 lapply(chains,
  function(x) fread(paste0(base,".chain",x,".msd"))
 ) %>%
 do.call(rbind,.) %>%
 filter(core$beta==1) %>% colMeans %>% sqrt
```

The distance between each leaf on the tree can then be computed using the `cophenetic` function from `ape`:

```
d1 = cophenetic(consensus)
```

The structural diffusion distance under the StructAlign model can also be computed this way, weighting each branch by the product of the edge length and the structural diffusivity parameter, σ_k^2. The expected mean-squared deviation under the model is then given by three times this pairwise distance, which includes a contribution from *x*, *y*, and *z* coordinates:

```
# tree distance weighted by structural diffusivity
ct = consensus
ct$edge.length = ct$edge.length * consensus.struct$edge.length
# convert from MSD in 1D to RMSD in 3D
d2 = sqrt(3*cophenetic(ct))
```

We can now plot the two tree distances versus RMSD for each pair of proteins:

```
# combine into a data.frame for plotting
st = do.call(rbind,strsplit(names(rmsd),"_"))
df = data.frame(rmsd,diffusion=d2[st],dist=d1[st])

p1 = ggplot(df,aes(x=dist,y=rmsd)) +
     geom_point() +
     xlab("Substitutions per site") +
```

```
    ylab(expression(paste("Pairwise RMSD / ",ring(A)))) +
    # relationship inferred by Illergard et al. (2009)
    geom_abline(intercept=0.734,slope=0.312)

# non-phylogenetic variability (0.25 gives a good fit)
baseline.sd = 0.25

p2 = ggplot(df,aes(x=diffusion,y=rmsd)) +
    geom_point() +
    xlab("Weighted tree distance") +
    ylab(expression(paste("Pairwise RMSD / ",ring(A)))) +
    # relationship implied by StructAlign model
    geom_abline(intercept=baseline.sd,slope=1)

grid.arrange(p1,p2,ncol=2)
```

As shown in Fig. 15, the estimated pairwise RMSD from the structural model closely matches the empirically observed values and is a much better fit to the data than a linear fit to sequence distance in substitutions per site. This may explain why previous studies have encountered difficulties modeling the relationship between sequence divergence and structural variability using a single linear model [29].

3.15 Distinguishing Structural Drift from Conformational Change

In the case of *2oif*, much of the local structural deviation observed in the EF-loop region can be attributed to the effect of binding the exogenous cyanide ligand [53]. In contrast, a rice globin with very similar sequence was crystallized in the hexacoordinate form [57] and does not display this large deviation at the start of the F-helix.

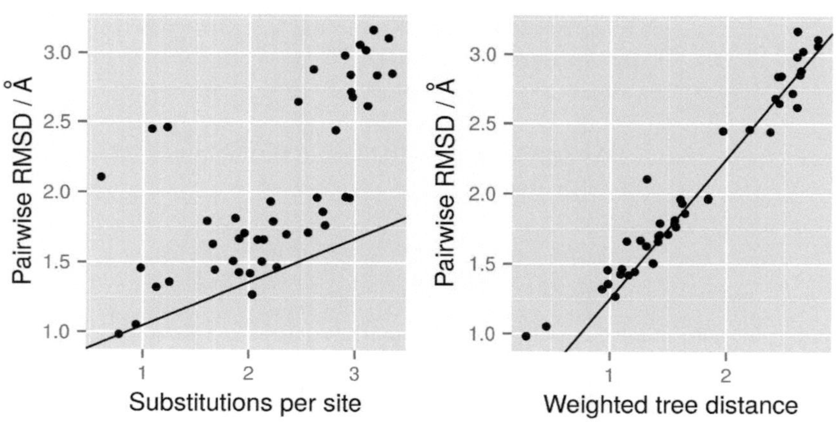

Fig. 15 Structural deviation as a function of evolutionary distance computed using the tree distance, before and after weighting by the branch-specific structural diffusivity parameters (left and right, respectively). The line on the left plot shows the linear relationship inferred by Illergård et al. [29] for the globins; the line on the right shows $y = x + 0.25$

Recent studies have highlighted how the existence of multiple functional conformations may constrain sequence evolution [58]; inclusion of this information when parameterizing the baseline variability modeled by StructAlign may help to further refine the estimates of model parameters derived under the joint sequence-structure model. In addition, modeling structural deviation via an angle-based model rather than a coordinate-based model may improve the ability to detect localized structural changes in cases where structures cannot easily be superposed due to large conformational changes [59, 60].

4 Notes

1. On Linux systems the JAR may need to first be made executable by running

   ```
   chmod a+x StatAlign.jar
   ```

2. There is currently no support for multiple chains; hence, any ATOM entries corresponding to other chains will be interpreted as alternative conformations for the first chain. If alternative conformations are present in the PDB file, only the first conformation will be used.
3. By default, the Dayhoff substitution model is used; this can be changed via the `-subst` command-line argument to StatAlign. The list of available substitution models can be obtained by running

   ```
   java -jar StatAlign.jar -list:subst
   ```

4. The default behavior is for all chains to output model parameters at every sample point, with tree and alignment output generated only when the temperature of the chain is equal to unity, to simplify downstream processing. In order to save disk space, the `-reportOnlyColdChain=true` argument can be used, such that all postprocessing output will also be restricted to the cold chain ($t = 1$).
5. The `-start=N` option to the ConsensusTree plugin can be used to specify that the first N trees should be ignored when computing the consensus. This is useful in cases where inspection of the log likelihood trace suggests that convergence has not occurred by the end of the burn-in period, but occurs some number of iterations thereafter.
6. For structures derived using NMR spectroscopy, a measure of baseline coordinate variability can be obtained from the S2 order parameters; this can be added as an additional column in

a `.coor` file for input to StructAlign (more details regarding these `.coor` files can be found in the StatAlign documentation).

7. Note that for the analysis under the structural model we have decreased the spacing between inverse temperature parameters to 0.01, since the addition of the structural likelihood increases the effect of changing the temperature parameter.
8. When running in parallel mode, each chain generates its own `.mle.super.pdb` file. Typically we are most interested in the superposition corresponding to the chain that samples the maximum likelihood configuration, which can be determined by inspecting the contents of the `.ll` files. The MLE superposition files are PDB-formatted files, with the aligned C-alpha coordinates of each structure corresponding to a separate chain. These aligned residues are numbered starting at 1; hence, an offset may need to be applied in order to highlight specific residues based on the original indexing.
9. The observed RMSD values for each column are computed based on the aligned structures at each site, using a single structural superposition, and constitute only a portion of the non-phylogenetic variance explained via ε. Hence the predicted baseline variability is typically larger than the empirical average pairwise RMSD values.

References

1. Godzik A (1996) The structural alignment between two proteins: is there a unique answer? Protein Sci 5:1325–1338
2. Sela I, Ashkenazy H, Katoh K, Pupko T (2015) GUIDANCE2: accurate detection of unreliable alignment regions accounting for the uncertainty of multiple parameters. Nucleic Acids Res 43:W7–W14
3. Morrison DA, Ellis JT (1997) Effects of nucleotide sequence alignment on phylogeny estimation: a case study of 18S rDNAs of apicomplexa. Mol Biol Evol 14:428–441
4. Ogden TH, Rosenberg MS (2006) Multiple sequence alignment accuracy and phylogenetic inference. Syst Biol 55:314–328
5. Wong KM, Suchard MA, Huelsenbeck JP (2008) Alignment uncertainty and genomic analysis. Science 319:473–476
6. Lunter G, Rocco A, Mimouni N, Heger A, Caldeira A, Hein J (2008) Uncertainty in homology inferences: assessing and improving genomic sequence alignment. Genome Res 18:298–309
7. Herman JL, Novák Á, Lyngsø R, Szabó A, Miklós I, Hein J (2015) Efficient representation of uncertainty in multiple sequence alignments using directed acyclic graphs. BMC Bioinformatics 16:108
8. Nelesen S, Liu K, Zhao D, Linder CR, Warnow T (2008) The effect of the guide tree on multiple sequence alignments and subsequent phylogenetic analyses. In: Proceedings of the 2008 Pacific Symposium on Biocomputing. World Scientific. p 25–36
9. Lunter G, Drummond AJ, Miklós I, Hein J (2005) Statistical alignment: recent progress, new applications, and challenges. In: Statistical Methods in Molecular Evolution. Statistics for Biology and Health. Springer, New York, NY
10. Redelings BD, Suchard MA (2005) Joint Bayesian estimation of alignment and phylogeny. Syst Biol 54:401–418
11. Westesson O, Lunter G, Paten B, Holmes I (2012) Accurate reconstruction of insertion-deletion histories by statistical phylogenetics. PLoS One 7:e34572
12. Holmes IH (2017) Historian: accurate reconstruction of ancestral sequences and evolutionary rates. Bioinformatics 33:1227–1229
13. Redelings BD (2014) Erasing errors due to alignment ambiguity when estimating positive selection. Mol Biol Evol 31:1979–1993

14. Satija R, Pachter L, Hein J (2008) Combining statistical alignment and phylogenetic footprinting to detect regulatory elements. Bioinformatics 24:1236–1242
15. Satija R, Novák Á, Miklós I, Lyngsø R, Hein J (2009) BigFoot: Bayesian alignment and phylogenetic footprinting with MCMC. BMC Evol Biol 9:217
16. Philippe H, Brinkmann H, Lavrov DV, Littlewood DTJ, Manuel M, Wörheide G, Baurain D (2011) Resolving difficult phylogenetic questions: why more sequences are not enough. PLoS Biol 9:e1000602
17. Kumar S, Filipski AJ, Battistuzzi FU, Kosakovsky Pond SL, Tamura K (2012) Statistics and truth in phylogenomics. Mol Biol Evol 29:457–472
18. Talavera G, Castresana J (2007) Improvement of phylogenies after removing divergent and ambiguously aligned blocks from protein sequence alignments. Syst Biol 56:564–577
19. Wu M, Chatterji S, Eisen JA (2012) Accounting for alignment uncertainty in phylogenomics. PLoS One 7:e30288
20. Gatesy J, DeSalle R, Wheeler W (1993) Alignment-ambiguous nucleotide sites and the exclusion of systematic data. Mol Phylogenet Evol 2:152–157
21. Lee MS (2001) Unalignable sequences and molecular evolution. Trends Ecol Evol 16:681–685
22. Löytynoja A, Goldman N (2008) Phylogeny-aware gap placement prevents errors in sequence alignment and evolutionary analysis. Science 320:1632–1635
23. Hasegawa H, Holm L (2009) Advances and pitfalls of protein structural alignment. Curr Opin Struct Biol 19:341–348
24. Johnson MS, Šali A, Blundell TL (1990) Phylogenetic relationships from three-dimensional protein structures. Methods Enzymol 183:670–690
25. Bujnicki JM (2000) Phylogeny of the restriction endonuclease-like superfamily inferred from comparison of protein structures. J Mol Evol 50:39–44
26. Lundin D, Poole AM, Sjöberg B-M, Högbom M (2012) Use of structural phylogenetic networks for classification of the ferritin-like superfamily. J Biol Chem 287:20565–20575
27. Chothia C, Lesk AM (1986) The relation between the divergence of sequence and structure in proteins. EMBO J 5:823
28. Panchenko AR, Wolf YI, Panchenko LA, Madej T (2005) Evolutionary plasticity of protein families: coupling between sequence and structure variation. Proteins 61:535–544
29. Illergård K, Ardell DH, Elofsson A (2009) Structure is three to ten times more conserved than sequence: a study of structural response in protein cores. Proteins 77:499–508
30. Echave J, Spielman SJ, Wilke CO (2016) Causes of evolutionary rate variation among protein sites. Nat Rev Genet 17:109–121
31. Worth CL, Gong S, Blundell TL (2009) Structural and functional constraints in the evolution of protein families. Nat Rev Mol Cell Biol 10:709–720
32. Gilson AI, Marshall-Christensen A, Choi J-M, Shakhnovich EI (2017) The role of evolutionary selection in the dynamics of protein structure evolution. Biophys J 112:1350–1365
33. Choi SC, Hobolth A, Robinson DM, Kishino H, Thorne JL (2007) Quantifying the impact of protein tertiary structure on molecular evolution. Mol Biol Evol 24:1769–1782
34. Kleinman CL, Rodrigue N, Lartillot N, Philippe H (2010) Statistical potentials for improved structurally constrained evolutionary models. Mol Biol Evol 27:1546–1560
35. Rodrigue N, Philippe H, Lartillot N (2006) Assessing site-interdependent phylogenetic models of sequence evolution. Mol Biol Evol 23:1762–1775
36. Sadowski M, Taylor W (2010) On the evolutionary origins of “fold space continuity”: a study of topological convergence and divergence in mixed alpha-beta domains. J Struct Biol 172:244–252
37. Rackovsky S (2015) Nonlinearities in protein space limit the utility of informatics in protein biophysics. Proteins 83:1923–1928
38. Sadreyev RI, Kim B-H, Grishin NV (2009) Discrete–continuous duality of protein structure space. Curr Opin Struct Biol 19:321–328
39. Holzgräfe C, Wallin S (2014) Smooth functional transition along a mutational pathway with an abrupt protein fold switch. Biophys J 107:1217–1225
40. Challis CJ, Schmidler SC (2012) A stochastic evolutionary model for protein structure alignment and phylogeny. Mol Biol Evol 29:3575–3587
41. Herman JL, Challis CJ, Novák Á, Hein J, Schmidler SC (2014) Simultaneous Bayesian estimation of alignment and phylogeny under a joint model of protein sequence and structure. Mol Biol Evol 31:2251–2266
42. Novák Á, Miklós I, Lyngsø R, Hein J (2008) StatAlign: an extendable software package for joint Bayesian estimation of alignments and evolutionary trees. Bioinformatics 24:2403–2404

43. Burmester T, Ebner B, Weich B, Hankeln T (2002) Cytoglobin: a novel globin type ubiquitously expressed invertebrate tissues. Mol Biol Evol 19:416–421
44. de Sanctis D, Dewilde S, Pesce A, Moens L, Ascenzi P, Hankeln T, Burmester T, Bolognesi M (2004) Crystal structure of cytoglobin: the fourth globin type discovered in man displays heme hexa-coordination. J Mol Biol 336:917–927
45. Hoffmann FG, Opazo JC, Storz JF (2010) Gene cooption and convergent evolution of oxygen transport hemoglobins in jawed and jawless vertebrates. Proc Natl Acad Sci U S A 107:14274–14279
46. Hoffmann FG, Opazo JC, Storz JF (2011) Differential loss and retention of cytoglobin, myoglobin, and globin-e during the radiation of vertebrates. Genome Biol Evol 3:588–600
47. Hoffmann FG, Opazo JC, Hoogewijs D, Hankeln T, Ebner B, Vinogradov SN, Bailly X, Storz JF (2012) Evolution of the globin gene family in deuterostomes: lineage-specific patterns of diversification and attrition. Mol Biol Evol 29:1735–1745
48. Geyer C (2011) Importance sampling, simulated tempering, and umbrella sampling. In: Brooks S, Gelman A, Jones G, Meng X (eds) Handbook of Markov Chain Monte Carlo. Chapman & Hall/CRC, Boca Raton, pp 295–311
49. Altekar G, Dwarkadas S, Huelsenbeck JP, Ronquist F (2004) Parallel Metropolis coupled Markov chain Monte Carlo for Bayesian phylogenetic inference. Bioinformatics 20:407–415
50. Thorne JL, Kishino H, Felsenstein J (1992) Inching toward reality: an improved likelihood model of sequence evolution. J Mol Evol 34:3–16
51. Gelman A, Rubin DB (1992) Inference from iterative simulation using multiple sequences. Stat Sci 7:457–472
52. Humphrey W, Dalke A, Schulten K (1996) VMD: visual molecular dynamics. J Mol Graph 14:33–38
53. Hoy JA, Robinson H, Trent JT, Kakar S, Smagghe BJ, Hargrove MS (2007) Plant hemoglobins: a molecular fossil record for the evolution of oxygen transport. J Mol Biol 371:168–179
54. Lobanov M, Bogatyreva N, Galzitskaia O (2008) Radius of gyration is indicator of compactness of protein structure. Mol Biol 42:701–706
55. Christensen AB, Herman JL, Elphick MR, Kober KM, Janies D, Linchangco G, Semmens DC, Bailly X, Vinogradov SN, Hoogewijs D (2015) Phylogeny of echinoderm hemoglobins. PLoS One 10:e0129668
56. Gupta KJ, Hebelstrup KH, Mur LA, Igamberdiev AU (2011) Plant hemoglobins: important players at the crossroads between oxygen and nitric oxide. FEBS Lett 585:3843–3849
57. Hargrove MS, Brucker EA, Stec B, Sarath G, Arredondo-Peter R, Klucas RV, Olson JS, Phillips GN (2000) Crystal structure of a nonsymbiotic plant hemoglobin. Structure 8:1005–1014
58. Sharir-Ivry A, Xia Y (2017) The impact of native state switching on protein sequence evolution. Mol Biol Evol 34:1378–1390
59. Maadooliat M, Zhou L, Najibi SM, Gao X, Huang JZ (2016) Collective estimation of multiple bivariate density functions with application to angular-sampling-based protein loop modeling. J Am Stat Assoc 111:43–56
60. Golden M, García-Portugués E, Sørensen M, Mardia KV, Hamelryck T, Hein J (2017) A generative angular model of protein structure evolution. Mol Biol Evol 34:2085–2100

Chapter 11

The Influence of Protein Stability on Sequence Evolution: Applications to Phylogenetic Inference

Ugo Bastolla and Miguel Arenas

Abstract

Phylogenetic inference from protein data is traditionally based on empirical substitution models of evolution that assume that protein sites evolve independently of each other and under the same substitution process. However, it is well known that the structural properties of a protein site in the native state affect its evolution, in particular the sequence entropy and the substitution rate. Starting from the seminal proposal by Halpern and Bruno, where structural properties are incorporated in the evolutionary model through site-specific amino acid frequencies, several models have been developed to tackle the influence of protein structure on sequence evolution. Here we describe stability-constrained substitution (SCS) models that explicitly consider the stability of the native state against both unfolded and misfolded states. One of them, the mean-field model, provides an independent sites approximation that can be readily incorporated in maximum likelihood methods of phylogenetic inference, including ancestral sequence reconstruction. Next, we describe its validation with simulated and real proteins and its limitations and advantages with respect to empirical models that lack site specificity. We finally provide guidelines and recommendations to analyze protein data accounting for stability constraints, including computer simulations and inferences of protein evolution based on maximum likelihood. Some practical examples are included to illustrate these procedures.

Key words Stability-constrained substitution models, Mean-field substitution model, Protein folding stability, Protein evolution, Ancestral protein reconstruction

1 Introduction

Mathematical models of protein evolution not only improve our understanding of the evolutionary process [1] but also have practical applications such as the design of therapies [2–4] and novel enzymatic properties [5–7]. Traditional substitution models of protein evolution are based on empirical amino acid substitution matrices such as JTT, WAG, or HIVb (*see* [8] for a review). However, these models assume that protein sites evolve independently of each other under the same substitution process, while it is well known that natural selection targets the structure and the stability of the native state of the protein, which is achieved through physical

Tobias Sikosek (ed.), *Computational Methods in Protein Evolution*, Methods in Molecular Biology, vol. 1851, https://doi.org/10.1007/978-1-4939-8736-8_11,

interactions between amino acids at different sites (for a review *see* [9–12]). This suggests that protein evolution models must explicitly represent the selective constraints on the structure and the stability of the native state.

A variety of models have been developed for this purpose. They belong to two main classes, depending on how selective constraints are implemented. The first group is that of stability-constrained protein evolution models (for a review *see* [13–15]), which put the focus on the stability of the native state. These models attempt to estimate the folding free energy ΔG of the native state of a mutated protein under the assumption that the mutation maintains the coarse-grained structure of the native state (typically, the contact matrix) and changes the interactions in the native and non-native states (although only few models consider non-native states). The fitness of the mutant is modeled as the fraction of correctly folded protein, $f = \frac{1}{1+e^{\Delta G/RT}}$. The second class corresponds to structurally constrained models of protein evolution, introduced by Julian Echave [16], which estimate the structural change due to the mutation through elastic network models (ENMs) [17] and model the fitness as a function of the structural distance between the mutated structure and a target structure. In this model the change in stability is neglected, also because ENMs do not allow for estimating it. The selective importance of the native structure, assumed by structurally constrained models, is justified by the fact that the precise native structure determines the functional dynamics of the protein, as ENM-based studies have shown [18]. Under realistic situations, mutations modify both the structure and the stability of the native state, and both of them are targeted by natural selection. However, for the moment, there are no models that try to estimate the effect of both changes on the protein fitness.

In this chapter we focus our attention on stability-constrained substitution models (SCS), which have been investigated in computer simulations since many years [19, 20]. However, these models are not yet well-established in current methods for phylogenetic inference, mainly because they require the computation of protein stability, which involves physical interactions and induces probabilistic dependencies between protein sites. In contrast, current software for phylogenetic inference computes the likelihood of the observed data under the assumption that protein sites evolve independently of each other. that likelihood computations are necessary for selecting the most supported substitution model, which is a crucial step for phylogenetic inference [21, 22]. Consequently, we believe that it is urgent not only to design novel models of protein evolution but also to implement these models in software tools useful for the community.

Unfortunately, substitution models with dependencies between sites increase enormously the complexity of likelihood computations, so that they can be managed only through sampling methods such as Monte Carlo [23, 24] that have inherent limitations in computer efficiency and may get trapped in local maxima. An alternative is to derive a model with independent sites that effectively enforces stability constraints, in the spirit of mean-field models from physics. Of course, the resulting model will be less realistic than a model with dependencies between sites but still may represent real data better than empirical substitution models that neglect stability constraints. One of these models is the mean-field model (MF) [25, 26].

In the present chapter we describe models of protein evolution with explicit stability constraints and models that effectively incorporate these constraints into site-independent substitution matrices, highlighting their implementation in phylogenetic frameworks. We also present some applications of these frameworks in the simulation and evolutionary analysis of diverse protein data. We finally provide guidelines and recommendations to use the presented frameworks.

2 Simulation of Protein Evolution with Stability Constraints

Models of protein evolution that explicitly represent selection on protein stability allow for describing the coevolution between protein sites [27, 28]. Here we describe two of these models, designed in our lab, that have been implemented in a phylogenetic framework to simulate the evolution of protein sequences.

2.1 Modeling Protein Evolution with Stability Constraints

The thermodynamic model adopted in the simulator of protein sequence evolution *ProteinEvolver* [29] estimates the stability of the native state not only against the unfolded state but also against compact, wrongly folded conformations (misfolded states) that are usually neglected in other models of protein stability. The characteristics of protein sequences that weaken the stability of frequently formed misfolded conformations are referred to as negative design, and its evolutionary importance is recognized through statistical analysis of protein sequences [29], and it was proposed to have important evolutionary consequences (for a review *see* [15, 30]).

The stability of the native state is estimated from the contact matrix representation of one native structure in the Protein Data Bank (PDB), $C_{ij} = 1$ if any two atoms in residues i and j are closer than 4.5Å and 0 otherwise:

$$G^{\text{nat}}(C^{\text{nat}}, A) = \sum_{ij} C_{ij}^{\text{nat}} U(A_i, A_j) \tag{1}$$

where A_i is the amino acid at site i (for instance leucine), C^{nat} is the native contact matrix, and $U(a, b)$ are the 210 contact interaction parameters derived in [31]. Contacts with $|i - j| < 4$ are not considered because they are formed in almost all alternative conformations. The conformational entropy of the native state is not considered because it is assumed to be almost the same as in other misfolded conformations. The folding free energy is computed as:

$$\Delta G = G^{\text{nat}}(C^{\text{nat}}, A) + kT \ln\left(e^{-G^{\text{misf}}/kT} + e^{-G^{\text{unf}}/kT}\right) \quad (2)$$

where $\frac{G^{\text{unf}}}{kT} = LS_{\text{unf}}$ is the free energy of the unfolded state, considered independent of the protein sequence of L residues, and G^{misf} is the free energy of the misfolded state, which depends on the sequence. It is expected that the free energy of unfolding is negligible with respect to misfolding for hydrophobic sequences and long proteins, since G^{misf} is expected to increase faster than L with the number of residues. G^{misf} is estimated from a statistical mechanical model of the misfolded state as [32]:

$$\begin{aligned} G^{\text{misf}}(A) = {} & \sum\nolimits_{ij} \langle C_{ij} \rangle U_{ij} - \frac{1}{2kT} \sum_{ijkl} \left(\langle C_{ij} C_{kl} \rangle - \langle C_{ij} \rangle \langle C_{kl} \rangle \right) U_{ij} U_{kl} - kTLS_c \\ & + \frac{1}{6(kT)^2} \sum_{ijklmn} \langle (C_{ij} - \langle C_{ij} \rangle)(C_{kl} - \langle C_{kl} \rangle)(C_{mn} - \langle C_{mn} \rangle) \rangle U_{ij} U_{kl} U_{mn} \end{aligned} \quad (3)$$

where $U_{ij} = U(A_i, A_j)$ and $\langle C_{ij} \rangle$ represents the frequency of contacts between residues at sequence distance $|i - j|$ in compact structures of L residues and $\langle C_{ij} C_{kl} \rangle$ represents contact correlations, which are precomputed from a representative subset of the PDB. The program *DeltaGREM* computes stabilities ΔG for sequence-structure pairs in the PDB and a list of user-supplied mutations or for multiple sequence alignments that include the PDB sequence. It is freely available from https://ub.cbm.uam.es/index.php.

Given the estimate of ΔG, two alternative models are used to compute the acceptance probability of a mutation.

1. In the *neutral* model, all sequences with $\Delta G < \Delta G_{\text{thr}}$ are considered viable and equally fit, and all other sequences are eliminated by negative selection. The threshold is chosen as 98% of the ΔG of the sequence in the PDB, so that this sequence would be selected and less stable sequences would be discarded.
2. In the *fitness* model, the fitness of the protein sequence is computed as the fraction of the folded protein:

$$f = \frac{1}{1 + e^{\Delta G/kT}} \quad (4)$$

that, for low temperature and large protein sequences, f tends to be a sigmoidal function, $f = 1$ if $\Delta G < 0$ and $f = 0$ otherwise. This binary fitness function enforces neutral evolution that is unable to distinguish between proteins with ΔG with the same sign. In general, if the starting sequence has fitness f^{wt} and the mutated sequence has fitness f^{mut}, the acceptance probability is computed as the fixation probability in a population of N individuals:

$$P_{\text{fix}}(f^{\text{mut}}, f^{\text{wt}}, N) = \frac{(f^{\text{wt}}/f^{\text{mut}}) - 1}{(f^{\text{wt}}/f^{\text{mut}})^N - 1} \quad (5)$$

As extensively discussed elsewhere [33, 34], the above fitness function establishes a formal analogy between molecular evolution and statistical physics, in the sense that an evolving population (in the limit of very low mutation rate, in which Eq. 5 is valid, under an unbiased mutation process) reaches a Boltzmann-like distribution in sequence space, $P(A_1 \ldots A_L) \propto e^{N \ln f(A)}$, in which $-\ln f(A)$ plays the role of energy (sequences with higher fitness are more frequently found) and $1/N$ plays the role of evolutionary temperature (small populations are more tolerant to slightly deleterious mutations and attain lower fitness). This analogy with statistical mechanics plays a key role in the development of the MF model.

2.2 Implementation in the Computer Simulator ProteinEvolver

We implemented these SCS models in the computer program *ProteinEvolver*. This framework simulates protein sequence evolution along phylogenetic trees. that computer simulations are very useful in population genetics and evolution for hypothesis testing, validation of analytical methods, model selection, and estimation of evolutionary parameters [35, 36].

ProteinEvolver implements the following steps. First, a phylogenetic tree is either specified by the user or is internally simulated under the coalescent model [37] extended with recombination (including recombination hotspots following Posada and Wiuf [38] and an adaptation of the intracodon recombination algorithm [39, 40] to simulate protein evolution with recombination (*see* Fig. 1)), demographics (population growth rate and demographic periods), longitudinal sampling, and user-specified populations structure with migration [41, 42]. Second, a protein sequence is assigned to the most recent common ancestor (MRCA), or grand MRCA (GMRCA) if recombination is simulated, and is evolved forward in time, from the root to the tip nodes, along the phylogeny (Fig. 1) [43]. The number of simulated substitution events depends on the branch lengths, and the kind of simulated substitutions depends on the applied substitution model of evolution. In addition to the SCS substitution models described above, *ProteinEvolver* implements a variety of empirical substitution models of protein evolution. *ProteinEvolver* is freely available from https://github.com/MiguelArenas/proteinevolver.

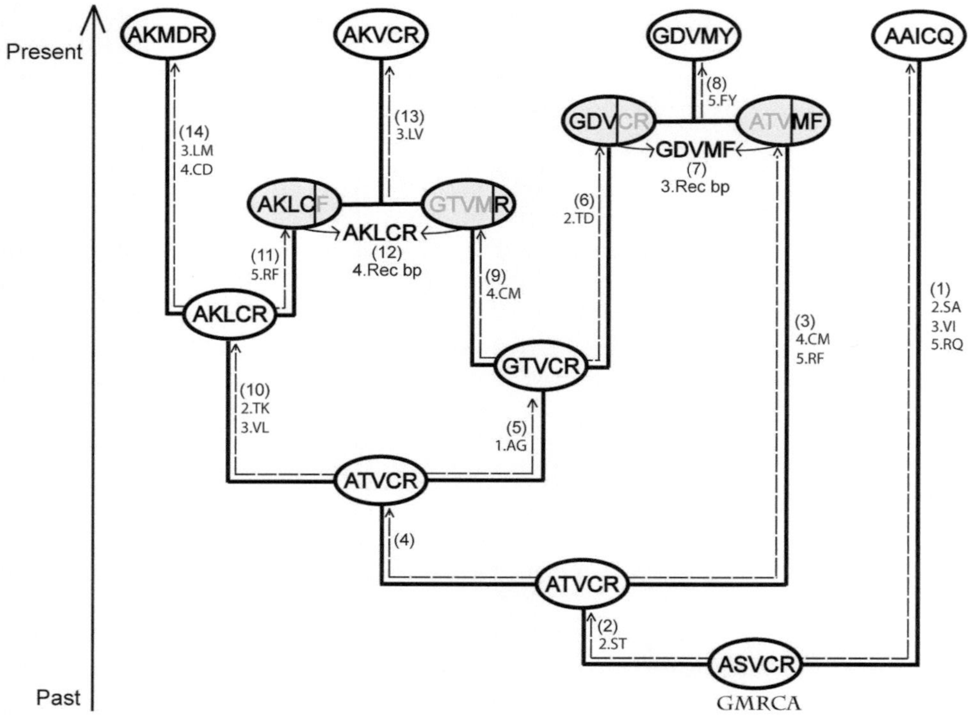

Fig. 1 Illustrative example of the recursive algorithm to simulate protein evolution along an ancestral recombination graph based on two recombination events. White and gray circles correspond to coalescence and recombination nodes, respectively. (1) The evolution starts from the GMRCA node; the protein is evolved along branches according to the SCS substitution model and the branch lengths. (3) The simulation reaches a recombinant node and because its parental recombinant node has not been assigned to a protein yet, the evolutionary process continues toward other direction (4). (6) The simulation reaches a parental recombinant node, and because its parental has already been assigned to a protein, (7) the simulation combines the two proteins according to the recombination breakpoint at position 3. (9) Another recombinant node is reached, and because its parental node has not been reached yet, a protein is assigned to this node and the simulation continues in the other direction (10). (11) The parental node is reached, and (12) the recombinant fragments are combined according to the recombination breakpoint at position 4. At the end of the process, a sequence was simulated for every internal and tip node

2.3 SCS Models Outperform Empirical Substitution Models in Terms of Distribution of Frequencies Among Sites and Maximum Likelihood

We tested whether the SCS models improve results obtained with traditional empirical substitution models analyzing ten protein families (phototactive yellow proteins, triosephosphate isomerases, rubredoxins, kinesins, phage lysozymes, ferredoxins, DNA ligases, heat shock proteins, oxysterol-binding proteins, and retroviral aspartyl proteases) [29]. For each protein family, we downloaded from the *Pfam* database a multiple sequence alignment (MSA), together with its associated phylogenetic tree and a representative protein structure deposited in the PDB. We also selected the best-fitting empirical amino acid substitution model with *ProtTest* [44].

We then performed 200 simulations of protein evolution along the *Pfam* tree under the best-fitting empirical amino acid substitution model, the neutral SCS model, and the fitness SCS model.

Using the Kullback-Leibler (KL) divergence [45], we found that the simulated amino acid distributions based on the SCS models were closer to the real distribution than the simulated amino acid distributions based on the empirical substitution model [29]. We also found that the neutral SCS model was robust in generating similar results under different thermodynamic features, while the fitness SCS model was more dependent on the thermodynamic parameters.

In addition, protein structures reconstructed with homology modeling [46] from simulations under the SCS models generated a better sequence-structure pair (protein folding stability closer to that from the real protein) than proteins simulated with empirical amino acid substitution models [29]. Altogether, substitution models that consider protein stability provide a better approximation of the real evolutionary process.

3 The Mean-Field Substitution Model Accounts for Stability Constraints While Adopting Independent Sites

3.1 The Mean-Field Model

As we discussed above, modeling selection on protein stability induces dependencies between protein sites that make the computation of the likelihood function extremely cumbersome. A possible shortcoming consists in developing a model with independent sites that effectively enforces protein stability, in the spirit of mean-field models in statistical physics. Specifically, we assume independence between sites, $P(A_1 \ldots A_L) = \Pi_i P_i(A_i)$, and we compute the site-specific amino acid distributions $P_i(A_i)$ obtained when the evolutionary process becomes stationary imposing two conditions:

1. The stationary distribution has minimum KL divergence with respect to the site-unspecific distribution $P^{\mathrm{mut}}(a)$ obtained under mutation alone, i.e., we minimize $\Sigma_i P_i(A_i)(\log(P_i(A_i)) - \log(P^{\mathrm{mut}}(A_i)))$.
2. This minimization is performed for a given average value of the fitness or, which is equivalent, for a given value of the average stability attained in the sequence ensemble, $\sum_{A_1 \ldots A_L} \Pi_i P_i(A_i) \Delta G(C^{\mathrm{nat}}, A_1 \ldots A_L) = X$, where ΔG is computed with Eqs.1–3 considering stability against both unfolding and misfolding.

These conditions are analogous to those that determine the Boltzmann distribution, which is the maximum entropy distribution (i.e., the distribution with minimum KL divergence with respect to the uniform distribution) for a given average value of the energy. The constrained minimization is performed through the global Lagrange multiplier Λ that represents natural selection, analogous to $1/T$ in a physical system. The main difference between a physical system and molecular evolution is that the reference distribution is not the uniform distribution, but it is the distribution induced by mutation alone. The final site-specific amino acid distributions are given by:

$$P_i(A_i) = \frac{P^{\text{mut}}(A_i)e^{\Lambda\phi_i(A_i)}}{\sum_a P^{\text{mut}}(a)e^{\Lambda\phi_i(a)}} \tag{6}$$

Given $P^{\text{mut}}(a)$ and Λ, the site-specific selective factors $\phi_i(a)$ are computed recursively without any new free parameter in a time that increases only quadratically with the number of residues L.

For a given $P^{\text{mut}}(a)$, the selective parameter Λ is computed by maximizing the likelihood of the sequence in the PDB with respect to the model, i.e., we maximize Σ_i log ($P^{\text{PDB}}{}_i(A_i, \Lambda$)). For the choice of the mutational distribution, three options are left to the user.

(a) The first option identifies each frequency $P^{\text{mut}}(a)$ with the frequency of the amino acid a in the PDB sequence or in a user-supplied MSA that contains the PDB sequence.

(b) The $P^{\text{mut}}(a)$ are obtained from a mutation model at the nucleotide level, with three parameters that represent the equilibrium nucleotide frequencies and a fourth one that represents the transition to transversion ratio. The stationary distributions of the 61 sense codons (excluding stop codons, which have fixation probability equal to zero) are computed, and the amino acid distribution is obtained summing over the codons of each amino acid. The four mutational parameters are optimized by maximizing the likelihood of the observed amino acids, i.e., $\Sigma_a n(a)$ log ($P^{\text{mut}}(a, \mu$)), where μ des the mutational parameters.

(c) $P^{\text{mut}}(a)$ is the mean between (a) and (b); (d) $P^{\text{mut}}(a)$ is obtained as in (b), but with mutational parameters that are input by the user. In particular, in this way we can fix the value of the nucleotide frequency and, consequently, the hydrophobicity of the protein sequence, which is correlated with the thymine (T) content since T at second codon position is almost only found in codons that code for hydrophobic amino acids.

Adopting the MF model, we verified the evolutionary importance of selection against misfolded conformations by comparing results obtained with the full model with results obtained with a reduced model in which stability against misfolding is not implemented, i.e., $\Delta G = G^{\text{nat}}(C^{\text{nat}}, A) + kTLS_{\text{unf}}$. We call this reduced model the native model. This model is inferior to the full model under several aspects: (i) if misfolding is not considered, the resulting sequences are on the average more hydrophobic than sequences in the PDB; (ii) in particular, exposed sites with few contacts are more hydrophobic than it is observed, indicating that it is negative design against misfolding that acts to limit the hydrophobicity of exposed sites; (iii) the likelihood of observed sequences is much higher with the full model than with the native model; (iv) the average folding free energy (taking into account both unfolded and misfolded states) is negative with the full model but positive with the reduced model, i.e., the sequences produced with the reduced model are not stable. These results confirm that it is important to impose stability against misfolding in SCS models.

3.2 The Wild-Type (WT) Model

Another possibility to develop a model with independent sites that implements stability constraints consists in computing the effect on stability and fitness of any possible mutation at site i starting from the wild-type sequence. We thus compute site-specific amino acid frequencies from Eq. 6 with $\phi_i(a) = \log f(A_1^{\mathrm{WT}} \ldots A_L^{\mathrm{WT}}, A_i = a)$, i.e., the wild-type sequence with the mutation $A_i^{\mathrm{WT}} \to a$. that the WT evolutionary model is only valid one mutation away from the sequence in the PDB, while the MF model is designed to remain valid after a long evolutionary divergence. The parameters $P^{\mathrm{mut}}(a)$ and Λ are determined as in the MF model.

3.3 The Substitution Process

To fully specify the substitution process, the site-specific amino acid frequencies $P_a^i = P_i(a)$ modeled with Eq. 6 must be complemented with site-specific exchangeability matrices E_{ab}^i, and the site-specific substitution rates that define the substitution process used to compute the likelihood function are computed as $Q_{ab}^i = E_{ab}^i P_b^i$. The exchangeability matrices that characterize the dynamics of the substitution process are assumed to be symmetric; thus, the detailed balance is satisfied, and P_a^i are the stationary distributions. The matrices E_{ab}^i are computed with the method of Halpern and Bruno as the product between a global exchangeability matrix E_{ab}^{mut} that represents the mutation process and a fixation probability analogous to Eq. 5, such that the site-specific frequency of amino acid a is the power of its site-specific fitness [47]. Specifically, if we write $P_a^i = P_a^{\mathrm{mut}} F_a^i$, where F_a^i are site-specific selective factors, the exchangeability matrices are given by:

$$E_{ab}^i = E_{ab}^{\mathrm{mut}} \frac{\log F_a^i - \log F_b^i}{F_a^i - F_b^i} \tag{7}$$

which is also a symmetric matrix that fulfills detailed balance. The substitution rates are maximal if the two amino acids have the same selective factors, in which case the fixation probability tends to

1, the selective factors are large, and the mutational exchangeability is large.

We allow three different models for the global exchangeability matrix. (a) E_{ab}^{mut} is equal to an empirical exchangeability matrix, WAG [48] or JTT [49]. We call it as *emp exchangeability model.* (b) The average flux between each pair of amino acids, $\frac{1}{L}\sum_i P_a^i E_{ab}^i P_b^i$ is equal to the flux of the empirical model, $E_{ab}^{\text{emp}} P_a^{\text{emp}} P_b^{\text{emp}}$ (*flux exchangeability model*). (c) E_{ab}^{mut} is computed from a mutation process at the nucleotide level, with parameters that are either optimized to fit the observed amino acid frequencies or imposed as input (*mut exchangeability model*).

3.4 Implementation in the Ancestral Sequence Reconstruction Framework ProtASR

The MF model was implemented in the computer simulator *ProtEvol* and in the ancestral sequence reconstruction (ASR) framework *ProtASR* [50].

ProtEvol computes global (whole protein) and local (site-specific) amino acid frequencies and exchangeability matrices that satisfy stability of the native state against both unfolding and misfolding. The program is freely available from https://ub.cbm.uam.es/index.php.

ProtASR is an evolutionary framework to infer ancestral protein sequences from a multiple sequence alignment (MSA) of proteins, a rooted phylogenetic tree, a protein structure representative of the proteins of the MSA, and a set of thermodynamic parameters. Internally, *ProtASR* runs *ProtEvol* to generate global and local amino acid frequencies and exchangeability matrices. Next, these frequencies and matrices are transferred to the well-established program *PAML* [51] where the ASR is performed under *joint* or *marginal* maximum likelihood (ML) approaches [52]. *ProtASR* is freely available from https://github.com/MiguelArenas/protasr.

3.5 Ancestral Proteins Reconstructed Under SCS Models Present More Realistic Folding Stabilities than Those Reconstructed Under Empirical Substitution Models

Ancestral sequence reconstruction (ASR) is a useful tool of evolutionary biology [53, 54] with a wide variety of applications such as HIV vaccine development [2, 3, 55] or reconstruction of proteins of extinct organisms [56, 57]. The accuracy of ASR methods is crucial to obtain realistic sequences and thus considering realistic substitution models is recommended.

In this example, we analyzed the performance of *ProtASR* in reconstructing ancestral proteins under MF and empirical models in terms of protein stability [50].

We analyzed a total of six protein families present in different bacterial species (D-ala-D-ala ligases, chaperone proteins dnaK, triosephosphate isomerases, tryptophan synthases α chain, thioredoxins I, and SH2 domain) whose folding stability had been previously studied [58]. After obtaining the corresponding MSA, a ML phylogenetic tree was inferred and rooted using a Eukaryotic protein as the out-group. Next, for each protein family, a total of 50 computer simulations were performed with *ProteinEvolver* by evolving a representative protein (for which there is a PDB structure) of the MSA along the corresponding phylogeny and under a SCS model (*see* Subheading 2). We also performed computer simulations with the SCS model described by Williams et al. [59].

The simulated MSA was later used to perform ASR with *ProtASR* under MF and empirical substitution models. In addition, *ProtASR* adopting the MF model was compared with *PhyloBayes* [60] adopting CAT models [61]. Next, the folding free energy of the inferred ancestral protein sequences was computed with the program *DeltaGREM* described above [32].

We found that ancestral sequences inferred with *ProtASR* under MF generated free energies significantly closer to those of the simulated sequences than ancestral protein sequences reconstructed with empirical models. We also found that the reconstructed sequences were more stable than the simulated sequences, a bias that was previously observed in [59]. However, ASR adopting the MF model reduced this bias with respect to empirical substitution models, which is an apparently counterintuitive result since the MF model enforces folding stability, whereas empirical models do not consider this condition [50].

3.6 Ancestral Prokaryotic Proteins Reconstructed Under SCS Models Present Different Energy Fluctuations over Time

In this example, we reconstructed the ancestral sequences of five extant prokaryotic protein families (D-ala-D-ala ligases, chaperone proteins dnaK, triosephosphate isomerases, tryptophan synthases α chain, and thioredoxins I) with *ProtASR* under the MF model. These protein families were selected because they were used in a previous study that evaluated the evolution of folding thermodynamic properties [58].

We found that the folding free energies varied broadly across evolution [50]. All protein families presented periods of increase, conservation, and decrease of free energies following a seascape model of protein evolution [62].

4 Guidelines, Recommendations, and Practical Examples for Using *ProteinEvolver* and *ProtASR*

In this section we present some guidelines and recommendations to simulate protein evolution with *ProteinEvolver* and to infer ancestral protein sequences with *ProtASR* under SCS models. A practical example for each framework is also described.

ProteinEvolver is a computer program written in C that runs from the command line. Its input is very simple with just a main input file that calls secondary input files.

4.1 Guidelines and Recommendations for Simulating Protein Evolution with ProteinEvolver

As for any computer simulator, the first step is to design the simulation study including the choice of the parameters to mimic the desired evolutionary scenario, the required number of simulations, and the output format. Second, *ProteinEvolver* includes detailed documentation and several examples, which we recommend to read in detail. Next, we describe the input and output information of this framework.

Since the simulation of molecular evolution is a stochastic process [43], the user has to indicate the number of computer simulations to be performed. The simulation of protein evolution is performed upon a phylogeny. This phylogeny can be user-specified or can be simulated with *ProteinEvolver* under the coalescent with recombination, demographics, longitudinal sampling, population structure, and migration (*see* Subheading 2.2). For the latter, the user has to specify the sample size (number of protein sequences of the simulated MSA), population size, and, optionally, other population genetics parameters (i.e., recombination rate, distributions for recombination hotspots, population growth rate, demographic periods, number of populations and migration rate, among others). Next, the user has to specify a substitution model of protein evolution, which could be empirical or stability-constrained. Concerning SCS models, the user has to indicate a protein structure, a representative set of alternative contact matrices (already included in the package), and some thermodynamic parameters (*see* Subheading 2.2). Proportion of invariable sites and additional rate heterogeneity among sites can be optionally specified. Finally, a sequence for the MRCA node can also be user-specified or, alternatively, internally computed by sampling from the amino acid frequencies.

Concerning the outputs, the program generates a MSA of proteins of the sample (and, optionally, of proteins of ancestral nodes) that can be written in formats such fasta, phylip, or nexus. Optionally, the program also outputs the simulated recombination breakpoints and folding energies of the simulated proteins.

Next, we describe a practical example to simulate data with *ProteinEvolver* under a site-dependent SCS model. We apply the second example (simulation of protein sequences under the neutral site-dependent SCS model) included in the program package.

1. Setting up the input files. First, we can explore the file *parameters*, which is the main input file. In this file, that text in brackets is ignored by the program. The specifications by default in this example indicate the simulation of two replicates. Since the setting *input tree/s file* is empty, the program will perform a coalescent simulation. The coalescent simulation considers a sample of 8 individuals (proteins) with length 255 amino acids. Effective population size is 1000 individuals, and its variation over time is considered with the specification of a population

growth rate. Longitudinal sampling is not specified in this example. A homogeneous recombination rate along the sequence is specified, also a substitution rate and an out-group with a fixed branch length of 0.1. The settings of the substitution model are specified in the file *Pop_evol.in*. There, a PDB file *1TRE.pdb*, its chain, a list of alternative contact matrices *structures.in*, several thermodynamic parameters (temperature and configurational entropies) that we recommend do not alter [29], the specification of the neutral SCS model and other minor information, are specified. Coming back to the settings file, the user can indicate the desired output information such as the format of the simulated MSA, coalescent trees and network, coalescent times, or recombination breakpoints.

2. Running the computer simulations. First, the program must be compiled. In the directory *src* of the package just type *make all*, some warnings without importance may appear. Next, the executable file *ProteinEvolver1.2.0* should be placed in the same directory of the input files and to run it one has to type ./ *ProteinEvolver1.2.0*. The program will automatically recognize the input file *parameters*, and the simulation may take a few seconds. Simulating a larger sample size, sequence length, population size, substitution rate, and/or recombination rate will increase the computer time.
3. Analyzing the results. A folder named *Results* will be created in the working directory and will include all the output data. For each replicate (#), the output file *sequences#* provides the simulated protein MSA, and *NetworkFile#* provides the simulated recombination network in branch list format [63]. The output file *breakpoints* presents a list of simulated recombination breakpoints, *times* presents a list of times of coalescent events, and *trees* presents the simulated coalescent tree/s in Newick format. A folder named *ProteinStability* presents the folding energy for each simulated protein at every ancestral and tip node.

4.2 Guidelines and Recommendations for Inferring Ancestral Protein Sequences with ProtASR

ProtASR is a computer program written in C and Perl that runs on the command line. The program includes detailed documentation and several examples, which we also recommend to read in detail. Its input is very simple with just a main input file that calls secondary input files. The input files are a MSA of protein sequences, a rooted phylogenetic tree for the MSA, a PDB protein that should be representative of the MSA, and a series of parameters to specify the desired substitution model. For beginners we recommend applying the parameters provided by default in the examples included in the package since those parameters have provided a good fitting with diverse real data [25, 29, 50].

Next, we describe a practical example to infer ancestral protein sequences with *ProtASR* under the MF model. We apply the example of rubredoxins that is included in the program package.

1. Setting up the input files. In the file *Settings*, the user has to specify the alignment file (in nexus format) that must include a rooted phylogenetic tree (in Newick format), a substitution model (in this example, it is MF), a PDB file and chain, and a variety of thermodynamic parameters that we recommend to use with values provided by default.
2. Running the inferences. First, the program must be compiled. In the directory *src* of the package just type *make all*, some warnings without importance may appear, and the compilation should take less than a minute. Next, all the material (files and folders) generated after the compilation should be placed in the directory of the input files (or vice versa), and there just type *perl ProtASR_main.pl Settings.txt* to run the ASR. The analysis will take several seconds. Datasets with a higher sample size and/or sequence length will increase the computer time.
3. Analyzing the results. A folder named RESULTS will be created in the working directory. There the output file *InferredAncestralProteins.txt* presents the inferred ancestral protein sequence for each node of the phylogenetic tree, and *LocalResultsLikelihood.txt* presents the estimated ML at local (site) and global (entire protein) levels (further information is included in the output directories *Global_ASR_ML* and *Local_ASR_ML*). The output directory *Meanfield* includes additional information generated by the MF model such as folding free energies, local and global amino acid frequencies, and exchangeability matrices.

5 Concluding Remarks

Protein evolution is a complex process where different evolutionary forces occur to generate new variants upon which selection operates (e.g., toward stable proteins). As a consequence, substitution models of evolution that incorporate structural properties of the native state, such as secondary structure and solvent accessibility, have produced a better fitting to real data than traditional empirical substitution models. A number of SCS models have been developed, but mainly due to their complexity, they have not been implemented yet in useful frameworks for evolutionary biologists.

In this chapter we described some SCS models and their evaluation and implementation in freely available frameworks to simulate protein evolution and to reconstruct ancestral proteins. We believe that the future of SCS models should of course continue developing realistic models but also implementing such models in useful frameworks for the evolutionary analysis, as we proposed in the different studies described in this chapter.

Acknowledgments

M.A. was supported by the grant "Ramón y Cajal" RYC-2015-18241 from the Spanish Government. U.B. is supported by the grant BIO2016-79043 from the Spanish Ministry of Economy.

References

1. Schmitt AO, Schuchhardt J, Ludwig A, Brockmann GA (2007) Protein evolution within and between species. J Theor Biol 249 (2):376–383. https://doi.org/10.1016/j.jtbi.2007.08.001
2. Gao F, Bhattacharya T, Gaschen B, Taylor J, Moore JP, Novitsky V, Yusim K, Lang D, Foley B, Beddows S, Alam M, Haynes B, Hahn BH, Korber B (2003) Consensus and ancestral state HIV vaccines. Science 299 (5612):1515–1518
3. Arenas M, Posada D (2010) Computational design of centralized HIV-1 genes. Curr HIV Res 8(8):613–621
4. Wilson C, Agafonov RV, Hoemberger M, Kutter S, Zorba A, Halpin J, Buosi V, Otten R, Waterman D, Theobald DL, Kern D (2015) Kinase dynamics. Using ancient protein kinases to unravel a modern cancer drug's mechanism. Science 347(6224):882–886. https://doi.org/10.1126/science.aaa1823
5. Perez-Jimenez R, Ingles-Prieto A, Zhao ZM, Sanchez-Romero I, Alegre-Cebollada J, Kosuri P, Garcia-Manyes S, Kappock TJ, Tanokura M, Holmgren A, Sanchez-Ruiz JM, Gaucher EA, Fernandez JM (2011) Single-molecule paleoenzymology probes the chemistry of resurrected enzymes. Nat Struct Mol Biol 18(5):592–596
6. Wijma HJ, Floor RJ, Janssen DB (2013) Structure- and sequence-analysis inspired engineering of proteins for enhanced thermostability. Curr Opin Struct Biol 23(4):588–594. https://doi.org/10.1016/j.sbi.2013.04.008
7. Cole MF, Gaucher EA (2011) Utilizing natural diversity to evolve protein function: applications towards thermostability. Curr Opin Chem Biol 15(3):399–406. https://doi.org/10.1016/j.cbpa.2011.03.005
8. Arenas M (2015) Trends in substitution models of molecular evolution. Front Genet 6:319. https://doi.org/10.3389/fgene.2015.00319
9. Liberles DA, Teichmann SA, Bahar I, Bastolla U, Bloom J, Bornberg-Bauer E, Colwell LJ, de Koning AP, Dokholyan NV, Echave J, Elofsson A, Gerloff DL, Goldstein RA, Grahnen JA, Holder MT, Lakner C, Lartillot N, Lovell SC, Naylor G, Perica T, Pollock DD, Pupko T, Regan L, Roger A, Rubinstein N, Shakhnovich E, Sjolander K, Sunyaev S, Teufel AI, Thorne JL, Thornton JW, Weinreich DM, Whelan S (2012) The interface of protein structure, protein biophysics, and molecular evolution. Protein Sci 21 (6):769–785
10. Bastolla U (2014) Detecting selection on protein stability through statistical mechanical models of folding and evolution. Biomol Ther 4:291–314
11. Wilke CO (2012) Bringing molecules back into molecular evolution. PLoS Comput Biol 8(6):e1002572
12. Sikosek T, Chan HS (2014) Biophysics of protein evolution and evolutionary protein biophysics. J R Soc Interface 11(100):20140419. https://doi.org/10.1098/rsif.2014.0419
13. Goldstein RA (2011) The evolution and evolutionary consequences of marginal thermostability in proteins. Proteins 79(5):1396–1407
14. Serohijos AW, Shakhnovich EI (2014) Merging molecular mechanism and evolution: theory and computation at the interface of biophysics and evolutionary population genetics. Curr Opin Struct Biol 26:84–91. https://doi.org/10.1016/j.sbi.2014.05.005
15. Bastolla U, Dehouck Y, Echave J (2017) What evolution tells us about protein physics, and protein physics tells us about evolution. Curr Opin Struct Biol 42:59–66. https://doi.org/10.1016/j.sbi.2016.10.020
16. Echave J (2008) Evolutionary divergence of protein structure: the linearly forced elastic network model. Chem Phys Lett 457

(4):413–416. https://doi.org/10.1016/j.cplett.2008.04.042

17. Tirion MM (1996) Large amplitude elastic motions in proteins from a single-parameter, atomic analysis. Phys Rev Lett 77 (9):1905–1908
18. Bahar I, Rader AJ (2005) Coarse-grained normal mode analysis in structural biology. Curr Opin Struct Biol 15(5):586–592. https://doi.org/10.1016/j.sbi.2005.08.007
19. Bornberg-Bauer E, Chan HS (1999) Modeling evolutionary landscapes: mutational stability, topology, and superfunnels in sequence space. Proc Natl Acad Sci U S A 96 (19):10689–10694
20. Bastolla U, Porto M, Eduardo Roman MH, Vendruscolo MH (2003) Connectivity of neutral networks, overdispersion, and structural conservation in protein evolution. J Mol Evol 56(3):243–254
21. Lemmon AR, Moriarty EC (2004) The importance of proper model assumption in bayesian phylogenetics. Syst Biol 53(2):265–277
22. Zhang J (1999) Performance of likelihood ratio tests of evolutionary hypotheses under inadequate substitution models. Mol Biol Evol 16(6):868–875
23. Bordner AJ, Mittelmann HD (2013) A new formulation of protein evolutionary models that account for structural constraints. Mol Biol Evol 31(3):736–749
24. Rodrigue N, Lartillot N, Bryant D, Philippe H (2005) Site interdependence attributed to tertiary structure in amino acid sequence evolution. Gene 347(2):207–217
25. Arenas M, Sanchez-Cobos A, Bastolla U (2015) Maximum likelihood phylogenetic inference with selection on protein folding stability. Mol Biol Evol 32(8):2195–2207. https://doi.org/10.1093/molbev/msv085
26. Bastolla U, Porto M, Roman HE, Vendruscolo M (2006) A protein evolution model with independent sites that reproduces site-specific amino acid distributions from the Protein Data Bank. BMC Evol Biol 6:43
27. Anishchenko I, Ovchinnikov S, Kamisetty H, Baker D (2017) Origins of coevolution between residues distant in protein 3D structures. Proc Natl Acad Sci U S A 114:9122–9127. https://doi.org/10.1073/pnas.1702664114
28. Wang ZO, Pollock DD (2005) Context dependence and coevolution among amino acid residues in proteins. Methods Enzymol 395:779–790. https://doi.org/10.1016/S0076-6879(05)95040-4
29. Arenas M, Dos Santos HG, Posada D, Bastolla U (2013) Protein evolution along phylogenetic histories under structurally constrained substitution models. Bioinformatics 29 (23):3020–3028
30. Echave J, Wilke CO (2017) Biophysical models of protein evolution: understanding the patterns of evolutionary sequence divergence. Annu Rev Biophys 46:85–103. https://doi.org/10.1146/annurev-biophys-070816-033819
31. Bastolla U, Farwer J, Knapp EW, Vendruscolo M (2001) How to guarantee optimal stability for most representative structures in the Protein Data Bank. Proteins 44(2):79–96
32. Minning J, Porto M, Bastolla U (2013) Detecting selection for negative design in proteins through an improved model of the misfolded state. Proteins 81(7):1102–1112. https://doi.org/10.1002/prot.24244
33. Sella G, Hirsh AE (2005) The application of statistical physics to evolutionary biology. Proc Natl Acad Sci U S A 102(27):9541–9546
34. Mustonen V, Lassig M (2005) Evolutionary population genetics of promoters: predicting binding sites and functional phylogenies. Proc Natl Acad Sci U S A 102(44):15936–15941. https://doi.org/10.1073/pnas.0505537102
35. Arenas M (2012) Simulation of molecular data under diverse evolutionary scenarios. PLoS Comput Biol 8(5):e1002495
36. Hoban S, Bertorelle G, Gaggiotti OE (2012) Computer simulations: tools for population and evolutionary genetics. Nat Rev Genet 13 (2):110–122
37. Kingman JFC (1982) The coalescent. Stoch Process Appl 13:235–248
38. Posada D, Wiuf C (2003) Simulating haplotype blocks in the human genome. Bioinformatics 19(2):289–290
39. Arenas M, Posada D (2010) Coalescent simulation of intracodon recombination. Genetics 184(2):429–437
40. Arenas M (2013) Computer programs and methodologies for the simulation of DNA sequence data with recombination. Front Genet 4:9
41. Arenas M, Posada D (2014) Simulation of genome-wide evolution under heterogeneous substitution models and complex multispecies coalescent histories. Mol Biol Evol 31 (5):1295–1301
42. Hudson RR (1998) Island models and the coalescent process. Mol Ecol 7(4):413–418
43. Yang Z (2006) Computational molecular evolution. Oxford University Press, Oxford

44. Abascal F, Zardoya R, Posada D (2005) ProtTest: selection of best-fit models of protein evolution. Bioinformatics 21(9):2104–2105
45. Kullback S, Leibler RA (1951) On information and sufficiency. Ann Math Stat 22(1):79–86
46. Marti-Renom MA, Stuart AC, Fiser A, Sanchez R, Melo F, Sali A (2000) Comparative protein structure modeling of genes and genomes. Annu Rev Biophys Biomol Struct 29:291–325
47. Halpern AL, Bruno WJ (1998) Evolutionary distances for protein-coding sequences: modeling site-specific residue frequencies. Mol Biol Evol 15(7):910–917
48. Whelan S, Goldman N (2001) A general empirical model of protein evolution derived from multiple protein families using a maximum-likelihood approach. Mol Biol Evol 18(5):691–699
49. Jones DT, Taylor WR, Thornton JM (1992) The rapid generation of mutation data matrices from protein sequences. Comput Appl Biosci 8 (3):275–282
50. Arenas M, Weber CC, Liberles DA, Bastolla U (2017) ProtASR: an evolutionary framework for ancestral protein reconstruction with selection on folding stability. Syst Biol 66:1054–1064. https://doi.org/10.1093/sysbio/syw121
51. Yang Z (2007) PAML 4: phylogenetic analysis by maximum likelihood. Mol Biol Evol 24 (8):1586–1591
52. Yang Z (1997) PAML: a program package for phylogenetic analysis by maximum likelihood. Comput Appl Biosci 13(5):555–556
53. Merkl R, Sterner R (2016) Ancestral protein reconstruction: techniques and applications. Biol Chem 397(1):1–21. https://doi.org/10.1515/hsz-2015-0158
54. Liberles DA (2007) Ancestral sequence reconstruction. Oxford University Press, Oxford
55. Kothe DL, Li Y, Decker JM, Bibollet-Ruche F, Zammit KP, Salazar MG, Chen Y, Weng Z, Weaver EA, Gao F, Haynes BF, Shaw GM, Korber BT, Hahn BH (2006) Ancestral and consensus envelope immunogens for HIV-1 subtype C. Virology 352(2):438–449
56. Gaucher EA, Govindarajan S, Ganesh OK (2008) Palaeotemperature trend for Precambrian life inferred from resurrected proteins. Nature 451(7179):704–707
57. Hobbs JK, Shepherd C, Saul DJ, Demetras NJ, Haaning S, Monk CR, Daniel RM, Arcus VL (2012) On the origin and evolution of thermophily: reconstruction of functional precambrian enzymes from ancestors of Bacillus. Mol Biol Evol 29(2):825–835. https://doi.org/10.1093/molbev/msr253
58. Bastolla U, Moya A, Viguera E, van Ham RC (2004) Genomic determinants of protein folding thermodynamics in prokaryotic organisms. J Mol Biol 343(5):1451–1466
59. Williams PD, Pollock DD, Blackburne BP, Goldstein RA (2006) Assessing the accuracy of ancestral protein reconstruction methods. PLoS Comput Biol 2(6):e69
60. Lartillot N, Lepage T, Blanquart S (2009) PhyloBayes 3: a Bayesian software package for phylogenetic reconstruction and molecular dating. Bioinformatics 25(17):2286–2288. https://doi.org/10.1093/bioinformatics/btp368
61. Lartillot N, Philippe H (2004) A Bayesian mixture model for across-site heterogeneities in the amino-acid replacement process. Mol Biol Evol 21(6):1095–1109
62. Mustonen V, Lassig M (2009) From fitness landscapes to seascapes: non-equilibrium dynamics of selection and adaptation. Trends Genet 25(3):111–119. https://doi.org/10.1016/j.tig.2009.01.002
63. Arenas M, Patricio M, Posada D, Valiente G (2010) Characterization of phylogenetic networks with NetTest. BMC Bioinformatics 11 (1):268

Chapter 12

Navigating Among Known Structures in Protein Space

Aya Narunsky, Nir Ben-Tal, and Rachel Kolodny

Abstract

Present-day protein space is the result of 3.7 billion years of evolution, constrained by the underlying physicochemical qualities of the proteins. It is difficult to differentiate between evolutionary traces and effects of physicochemical constraints. Nonetheless, as a rule of thumb, instances of structural reuse, or focusing on structural similarity, are likely attributable to physicochemical constraints, whereas sequence reuse, or focusing on sequence similarity, may be more indicative of evolutionary relationships. Both types of relationships have been studied and can provide meaningful insights to protein biophysics and evolution, which in turn can lead to better algorithms for protein search, annotation, and maybe even design.

In broad strokes, studies of protein space vary in the entities they represent, the similarity measure comparing these entities, and the representation used. The entities can be, for example, protein chains, domains, supra-domains, or smaller protein sub-parts denoted themes. The measures of similarity between the entities can be based on sequence, structure, function, or any combination of these. The representation can be global, encompassing the whole space, or local, focusing on a particular region surrounding protein(s) of interest. Global representations include lists of grouped proteins, protein networks, and maps. Networks are the abstraction that is derived most directly from the similarity data: each node is the protein entity (e.g., a domain), and edges connect similar domains. Selecting the entities, the similarity measure, and the abstraction are three intertwined decisions: the similarity measures allow us to identify the entities, and the selection of entities influences what is a meaningful similarity measure. Similarly, we seek entities that are related to each other in a way, for which a simple representation describes their relationships succinctly and accurately. This chapter will cover studies that rely on different entities, similarity measures, and a range of representations to better understand protein structure space. Scholars may use publicly available navigators offering a global representation, and in particular the hierarchical classifications SCOP, CATH, and ECOD, or a local representation, which encompass structural alignment algorithms. Alternatively, scholars can configure their own navigator using existing tools. To demonstrate this DIY (do it yourself) approach for navigating in protein space, we investigate substrate-binding proteins. By presenting sequence similarities among this large and diverse protein family as a network, we can infer that one member (pdb ID 4ntl; of yet unknown function) may bind methionine and suggest a putative binding mechanism.

Key words Protein space navigation, Structure space, Evolutionary relationships in protein space

Tobias Sikosek (ed.), *Computational Methods in Protein Evolution*, Methods in Molecular Biology, vol. 1851, https://doi.org/10.1007/978-1-4939-8736-8_12, © Springer Science+Business Media, LLC, part of Springer Nature 2019

1 Introduction

1.1 Protein Structure Space

Protein structure space is an abstract model which we use when we study large, representative, sets of protein structures and their interrelationships. Inspecting these large datasets allows us to better understand protein evolution and biophysics. While protein space is not real, the entities that populate it are: for example, these can be protein chains or domains; furthermore, their comparisons are meaningful. Thus, the first and essential step when studying protein structure space is to decide on the set of entities and the measure of similarity among them (coupled with a method to compute it). We can then calculate all-against-all comparisons of these entities to construct the initial dataset. Because the abstract model is derived from these comparisons, it is essential that this initial set is as accurate and comprehensive as possible. Navigating in protein structure space is in many ways navigating within this initial dataset, and we can do this either locally or globally.

1.2 Navigation Modes

Navigating "locally" or "globally" in protein structure space is a metaphor, which describes how we study the dataset. By "local," we mean that we identify small sets of comparisons, which we deem relevant. Given a query protein chain, or query protein domain, we think of the comparisons of that protein and its near structural neighbors (i.e., other proteins in the dataset that are similar to it) as covering its local region in structure space. Navigating locally is moving between (overlapping) local regions, akin to moving between landmarks when using a navigation app. By "global," we mean that we derived a model which integrates information from many (possibly all) comparisons and explore this model. Alternatively, we can think of this model as a data structure that organizes all entities based on the relationships between them. Navigating globally means that we either explore the properties of this data structure, akin to staring at a map, or move between proteins based on their location in the data structure.

1.3 The Potential of Studying Protein Structure Space

Studying protein structure space can help us better understand protein evolution and biophysics. It may also have a practical value: insights could be used in protein structure prediction, protein function prediction, and protein design. By way of motivation, we list a few examples; there are many more (e.g., those listed in [1, 2].) Evolution scholars have navigated protein space looking for clues in the remnants of evolutionary processes [3, 4]. For example, Choi et al. [5] derive the "multiple birth model" for proteins from maps, Dokholyan et al. [6] offered support for all proteins evolving from a few precursors, Alva et al. [7] studied the relationship between convergent and divergent evolution, Farias-Rico et al. traced the evolutionary relationships between ancient superfolds

[8], and Nepomnyachiy et al. [9] highlighted the complex nature of reuse patterns, which often overlap with each other. Studying protein structure space also revealed biophysical properties of proteins: examples include the work of Skolnick et al. [10], Nepomnyachiy et al. [11], and Mackenzie et al. [12]. Understanding the space of all structures can help in protein structure prediction and in better organizing the databases for structure search [13]. A global perspective also offered a hint to the relationship between protein structure and function, showing that there is a localized region of high function diversity [14]. Notice that one size does not fit all: different insights were gained from representations of protein space that varied in the sets of entities curated and in the way the entities were compared to each other.

2 Materials and Methods

2.1 The Entities

The entities are derived from the proteins of known structure in the Protein Data Bank (PDB) [15] and can be parts of proteins of different scales, depending on the question at hand. With minimal processing, these can be protein complexes or protein chains. One could also consider protein domains [16, 17] (or even supra-domains [18]), or meaningful sub-domain entities: protein fragments (e.g., [19, 20]), protein themes [9], protein interfaces [21], protein-peptide complexes [22], repetitive secondary structure elements (e.g., Smotifs [23]), or tertiary structural motifs (TERMS) [12]. Alternatively, the structures could possibly be predictions [24], or homology models [25]. Typically, one would use datasets that were curated by others (e.g., the domain sets in SCOP [26], CATH [27], or ECOD [28]). It is important to consider if the entities are mutually exclusive, or not. For example, domains are mutually exclusive because when partitioning chains to domains, each residue is associated with only a single domain; in contrast, themes cover multiple (nested) segments in a protein chain.

2.2 Relating the Entities

Comparing proteins can be based on their sequences, structures, or functions. The most straightforward measure is *sequence* similarity, which suggests shared evolutionary ancestor(s) [29]. Sequence alignment tools vary in sensitivity: less sensitive methods rely directly on the protein sequences (e.g., BLAST). More sensitive methods rely on an enriched version of the sequences: either sequence profiles (e.g., PSI-BLAST) or HMMs (e.g., HHSearch [30] or HHMER [31]); these are probabilistic models that include not only the protein sequence but also sequences of its close homologues [30, 31]. Using sensitive sequence aligners like HHSearch or HHMER reveals more distant evolutionary relationships. To avoid relating pairs of proteins that have diverged beyond what we would consider similar, scholars add an additional restriction that the structures of the aligned residues be similar [11, 32]; it is not impossible that structural changes emerged upon evolution though (and anyway, proteins often undergo conformational changes [33, 34]). Note that using profile or HMM-based sequence aligners requires calculating these profiles or HMMs; one can use pre-calculated ones (which influences the set of entities available). Alternatively, it is possible to compare the *structures* of the proteins. Structure similarity is often viewed as a method for relating proteins that were similar further back in evolutionary history, with sequences that diverged beyond the point where one can identify their common ancestry; for example, the SCOP "fold," CATH "Architecture," and ECOD "X" levels are based on structure similarity. This is akin to using a more powerful telescope to look back in time [35]. A concern when relying only on structure similarity to study protein evolution is that these proteins share structures because these structures are especially favorable from a biophysical perspective. In other words, that what we see is merely a consequence of the biophysical properties and constraints [36], perhaps due to convergent evolution. To compare structures, we use one of many structural alignment methods. In fact, structural alignment is a vast field with many intricacies, far beyond the scope of this chapter. For more details, *see* [37–40] and below in the section highlighting structural alignment servers.

The similarity measure (be it based on sequence or on structure) can be local or global.[1] In global similarity, the proteins are considered in their entirety. In contrast, in local similarity, we consider subsections, so that proteins can be identified as similar even if there is only a partial match. The disadvantage of using a global similarity measure is that to be meaningful, we must first segment our proteins to pieces, which are similar in their entirety (e.g., domains); this creates a chicken-and-egg situation, because we want to segment the proteins in a way that we can find globally meaningful similarities. The disadvantage of using a local similarity measure is that it leads to non-transitive relationships: protein A that is locally similar to protein B, protein B that is locally similar to protein C, and at the same time proteins A and C have nothing in common ([1] has an illustration of this). Non-transitive relationships are counterintuitive when we think of the notion of similarity and especially when we integrate all these relationships into a unified (global) model of protein space.

2.3 Addressing Redundancy

The PDB is redundant, and some proteins are far more abundant than others (e.g., due to research interests of the scholars studying these proteins) [41]. This suggests that when seeking a global

[1] Notice that the terms used here characterize the similarity measure, not the style of navigation in protein space, to use the same terms as in the Needleman–Wunsch and Smith–Waterman sequence alignment algorithms.

perspective, one should either rely on nonredundant datasets or alternatively remove, or cull, the redundancy on their own. Notice that we consider an entity redundant if the dataset includes another copy of that entity: i.e., one that is (globally) similar to it. Hence, both the definition of the entities and the measures of similarity influence this redundancy removal process. There are software packages, and servers, that implement algorithms for removing redundancy: two popular ones are CD-HIT [42] and PISCES [43].

2.4 Data Structures for Global Representation

For a global perspective, one must derive a data structure, or an abstract model, from the dataset of all proteins and their comparisons. Scholars used three types of models: (1) networks, (2) classifications, and (3) maps (for a review of these, *see* [2]). A network is the data structure closest to the raw data. To construct it, one only needs to list the meaningful similarities, and the network is a straightforward representation of the entities (as nodes) and the similarities (as edges connecting these nodes.) A classification groups the entities into nonoverlapping sets of proteins. It is assumed that proteins in the same set in the classification (i.e., with the same classification) are similar to each other, while those not in the same set are not (or less so). The classifications are hierarchical, and proteins are grouped with decreasing degrees of similarity. Hence, to construct a classification, one needs to weight the importance of the similarities identified among the protein entities: emphasizing the ones that are within a set and downplaying the ones between sets. Finally, in a map, each protein is represented by a point, and the points are positioned in two or three dimensions, so that the distance between them approximates the dissimilarity between the proteins they represent. The mapping is calculated by first converting the measures of similarity between the protein entities to an all-by-all dissimilarity matrix, followed by a multidimensional scaling (MDS) to project this matrix to a lower (two or three) dimension. Because the position of a protein is not indicative of its relationship to other proteins in a straightforward manner, maps were not used for local navigation. Rather, the insights were derived from a global perspective [5, 14, 35, 44, 45].

2.5 Publicly Available Navigators for Protein Structure Space

Defining a meaningful nonredundant set of entities, calculating the relationships between them, and collecting all this information to a centralized data structure require both ingenuity and computational resources. Even more so, as the database of all protein structures (the PDB) is constantly growing, the calculations need to be routinely updated. Consequently, many groups have set up web servers with data for navigating protein structure space; these navigators have datasets which were curated, compared, and organized—some at a single time point (but possibly with a more elaborate organization)—while others are maintained up-to-date.

The navigators enable users to move in protein structure space as if they are using a navigation app. Some of the navigators offer their users a global perspective of protein structure space as well.

2.6 Navigators with a Global Perspective

The most established resources for navigating protein structure space are the hierarchical classifications; the popular ones are SCOP from the Murzin lab, CATH from the Orengo lab, and ECOD from the Grishin Lab; another popular classification—Pfam [46]—is not discussed here because it is based on sequence rather than structure. For a recent and extensive review of the classifications, *see* [47]. The classifications organize the data in a hierarchy: a user can gain a perspective of the whole space by drilling down, starting at the top. For example, starting at the highest level of SCOP, we see that structure space has regions of all-alpha domains, all-beta domains, alpha+beta domains, and alpha/beta domains, where the two latter classes include both alpha and beta elements, separated or intertwined, respectively [48]. Alternatively, one can search for a specific protein and consider the classification of its domains and the list of all its related proteins—ones whose domains are classified similarly (at different levels of the hierarchy.) In short, the data structure that is used in the hierarchies is a collection of sets (or lists), organized as a tree; each entity is classified in several (nested) sets (depending on the height of the hierarchy). The similarity measure used is based on the sequences (at the lower levels of the hierarchy) and structures (at the higher levels of the hierarchy). The entities classified are domains: nonoverlapping subsections of the protein chains, which cover all chain residues (or, in other words, each PDB chain is segmented into one or more domains such that each residue is part of exactly one domain). There is much discussion, and controversy, on what is the correct definition of domains [49–51]; that there are several domains databases (rather than one) is a clear indication of this.

In practical terms, domains are the entities classified in SCOP, CATH, ECOD, or in servers curating domains like CDD [52]. More formally, there are several (not necessarily overlapping) definitions of a domain [16, 17, 53]: (1) a structurally distinct region (perhaps a compact unit) [54], (2) a segment that is identified as an evolutionary unit based on observations of reuse in protein space, (3) an independently folding unit, and (4) a section with assigned biochemical function. The domains in the hierarchical classifications are defined based on reuse. Unfortunately, these domains, which are classified in the different databases, are not the same ones (for comparisons, *see* [50, 51, 55, 56]); a recent study estimates that only 60% of CATH domains have a similar SCOP counterpart [53]. Nonetheless, the domains in the hierarchical classifications have similar lengths of approximately 100 residues; this is the average for the distributions of domain lengths in the

SCOP, CATH, and ECOD (*see* Fig. 8b in [28]). Indeed, splitting a protein chain into domains is challenging [49], leading to many algorithmic methods devoted to this task (e.g., [54, 57–59]), and a significant amount of human intervention in some of the classifications (rather than only relying on automatic domain assignment procedures). Regardless of how automatic the procedure for identifying the domain boundaries, a fundamental problem remains if the domains are defined based on reuse: the reuse patterns in protein space are not simply reuse of segments of an appropriate length (~100 residues). Rather, it is a complicated pattern of nested segments that are reused to different extents [9, 27]. Consequently, there is more than one way to reduce this complex pattern into domain definitions. Due to this very same complexity, once the domains are defined, there are many instances of common parts (segments) between domains that are not wholly similar and are thus classified differently (at different levels of the hierarchy) [11, 29, 60–62].

The classification hierarchies maintain an up-to-date dataset representing the complete and current PDB, with an intuitive user interface. In CATH and ECOD, one can drill down the tree to explore different members of the sets; CATH also has a sunburst visualization, which indicates the relative sizes of the classified sets. Since the last version (1.75 in 2009) of the classic SCOP, the classification diverged into two variants: SCOPe and SCOP2. SCOPe [63] is a continuously and (mostly) automatically updated extension of classic SCOP. In contrast, SCOP2 [64] changed the data structure: rather than the classic tree of sets, it uses a network; the network representation (sometimes called graphs) is implemented with a web tool based on the visualization software Graphviz [65]. In all classifications, the user can search for a specific protein chain or domain and explore the local context of that protein within the data structure (typically, within the hierarchy), allowing the user to see proteins of similar sequence (with the same classification at the lower levels) and of similar structure (with the same classification at higher levels.)

2.7 Publicly Available Navigators for Local Environments of Structure Space

Another way of navigating protein structure space is zooming into a local region, while ignoring the global view, and exploring, by moving between such local environments. Starting from the protein of interest, we think of its local environment as a list of its structural neighbors (sorted from near to distant ones); we can then move in space by selecting one of these neighbors to see its slightly shifted local environment (centered on this neighbor.) We think of this process as navigating in protein structure space, like a driver following a navigation app without seeing the full landscape. For this, all one must have is the list of neighbors for each protein in the dataset. The entities considered are typically both PDB chains and domains (either taken from the classifications or calculated with an automatic domain parser). Because the overall data structure is not considered, the structural alignment remains the most important computational component. Thus, such navigators were often set up by groups developing structural alignment methods. What transforms a structural alignment server into a useful navigator is speed: to navigate comfortably, the server must be fast. This is because when navigating, we search for structural neighbors repeatedly, each time starting at a different protein. Indeed, significant sophistication is needed to build servers that are up-to-date, fast, and comprehensive.

The differences between the structural alignment servers are largely due to the differences between the structural alignment methods. We list examples of structural alignment servers that allow users to locally navigate in protein structure space. The PDB website has precomputed structural alignments for a representative nonredundant dataset, calculated using the FATCAT aligner [66]. The European PDB website has PDBeFold [67], a structural alignment server based on the SSM aligner [68]. NCBI's server is called VAST+ and is based on the aligner VAST [69]. PhyreStorm [70] is a new server, which relies on TM-align [71] and offers a very comfortable navigation experience. Another new server is TopSearch (using the structural aligner TopMatch), which has the unique feature that it considers larger entities of protein oligomer [72].

2.8 DIY: Build-Your-Own Navigator

There are several reasons why scholars may want to customize their own navigator to explore protein structure space, or parts of it. First, the entities they wish to include may be specific to their problem: a set of proteins that is not covered in the public servers (perhaps a more redundant one), unpublished structures, or even predicted ones. Also, one may want to study subsections of proteins, which are different from chains or domains, for example, shorter themes [9] or loops [73]. Second, scholars may want to compare the entities themselves, as it gives them flexibility in the choice of a specific sequence or structure alignment program, full control over the parameters used, and the ability to enforce additional conditions when comparing proteins (e.g., a minimal alignment length). In some cases, even though there is a publicly available structural alignment server, it is not fast enough for navigating structure space; for these, one may prefer to pre-calculate all-against-all comparisons (e.g., using the parallel power of a computer cluster). We list just a few examples of comparison methods that were used in a similar context: HHSearch [30], Matt [74], CE [75], Mammoth [76], 3D-BLAST [77], FragBag [78], TM-align [71], SSM [68], GRASP [79], and STRUCTAL [80]. Third, the structural alignment servers do not offer a global perspective of structure space, only a local one, and one may be interested in this global perspective. Finally, scholars have different preferences when exploring structures in a molecular viewer, both in terms of the viewer they are using and its configuration.

If the navigator is based on a network data structure, it is easy to build your own navigator with the network visualization tool Cytoscape [81] and its molecular viewer configuration apps CyToStruct [82] or structureViz [83]. To represent a part of protein space as a network, one needs to define the list of nodes (entities to be compared and the edges that connect them (pairs of entities that are similar). This is very easy to do with Cytoscape: a (fantastic) open-source network analysis and visualization tool. Given the list of nodes and edges, Cytoscape visualizes the information as a well-laid two-dimensional network; one can configure this visualization easily and extensively. For example, the color of the nodes may depend on the structural class of the entities they represent, and the thickness of the edges may depend on the similarity of the entities they connect. This provides the global perspective. To gain a local perspective, one would like to use a molecular viewer to study the nodes or edges and the structures or structural alignments they represent.

Molecular viewers need to be configured: these are sophisticated software tools, with many alternative settings. By configuring the molecular viewer, one can display and highlight the relevant parts in the protein structure. Popular molecular viewers are PyMOL [84], UCSF Chimera [85], JMol [86], VMD [87], and recently NGL—a particularly fast web-based viewer [88]; for a review of these and more, *see* [89]. There are two methods of configuring molecular viewers: (1) manually, using the graphical user interface (GUI) and (2) by running a script in the language specific to that viewer. Configuring the viewer manually is easier for a novice but far more tedious; configuring it via scripts requires command of the scripting language but facilitates repeated visualizations dramatically. To link the entities in Cytoscape with a molecular viewer, one can install one of two Cytoscape apps: structureViz or CyToStruct. structureViz is tightly coupled with UCSF Chimera. In structureViz, node attributes can specify PDB names, so that the corresponding pdb file opens in UCSF Chimera; the molecular viewer can also be configured via its GUI. In contrast, CyToStruct is suited for users who configure the molecular viewers via scripts; it is very powerful in that it allows using any molecular viewer, and within that viewer configuring anything that can be specified via a script, or equivalently, computed with that software.

CyToStruct can run any molecular viewer (and any external program in general) from all nodes and edges (a menu opens when right-clicking on it), with scripts that are tailored to each node or edge. To configure CyToStruct, the user has to specify the external program, a template of script to be run, and a file with node- or edge-specific data for that template. CyToStruct then creates the runnable script by infusing the node- or edge-specific data into the template and runs the molecular viewer with a copy of this script. The source code of CyToStruct is publicly available (https://bitbucket.org/sergeyn/cytostruct/wiki/Home), along a series of demos that users can rely on as a starting point. The demos include visualization using the four popular molecular viewers (each with their own syntax), configuring the visualization of complete structures, protein interfaces, structurally aligning multiple structures, and selecting specific residues. CyToStruct can also be used within the web-based version of Cytoscape (Cytoscape.js), to provide an online visualization combining a network and a molecular viewer.

We present two examples for DIY navigators. The first is the navigator that Nepomnyachiy et al. customized for a global view of protein structure space [11]. The entities, or nodes in the network, are 9710 SCOP domains (70% nonredundant set). These domains were compared using the structural aligner SSM [68]; for sufficiently meaningful alignments, Nepomnyachiy et al. calculated measures of the similarity of the domains. Then, they define several networks, each characterized by its edges, which connect all domain pairs that were aligned with parameters better than some fixed thresholds: a minimal alignment length (55, 75 residues), maximal RMSD (2, 2.5, and 3 Å), and minimal percent sequence similarity (30, 40, and 50%). By coloring the nodes based on their SCOP class, all-alpha, all-beta, alpha/beta, and alpha+beta, they could see that protein structure space has a continuous region (the alpha/beta domains) and discrete regions [11]. The Cytoscape networks provide a global view, but navigating in specific regions of structure space is also interesting. Nepomnyachiy et al. link and configure the molecular viewer using CyToStruct [82] to see the domains and the alignments and package and distribute the data and configuration files (http://cs.haifa.ac.il/~trachel/domain_motif_networks/), allowing anyone to study protein structure space in this way.

2.9 Case Study

We present here a new example, where Cytoscape and CyToStruct are used to navigate protein space for function inference. The navigator helps because a careful examination of populated regions in the protein universe can help decipher unknown qualities of proteins found in these regions. Here, we demonstrate this using substrate-binding proteins (SBPs) [90]. SBPs are involved in transport of substrates into the cell, where their role is to recognize the substrate and relay it to its transmembrane transporter. Although they vary in size and share relatively low sequence similarity, they share a similar, highly conserved, fold. In general, their shape is a lung-like structure, formed of two structurally similar globular domains, connected by a hinge. The hinge facilitates alteration between substrate-free and substrate-bound conformations; substrate binding to a cavity between the two domains brings them closer to one another, into a bound, or "closed," conformation.

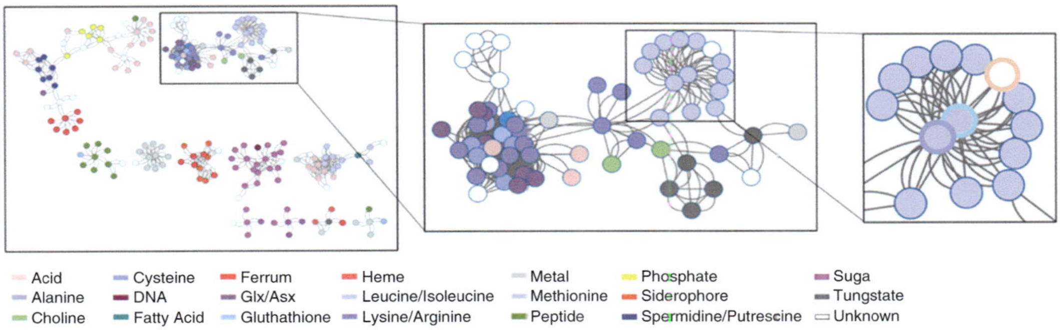

Fig. 1 Navigating protein structure space to study proteins with unknown function. *Left panel*: network of substrate-binding proteins. Each node represents a single PDB chain; two nodes are connected by an edge if they share some sequential and structural similarity. The nodes are colored according to the substrate; see color-code at the bottom. White nodes represent proteins of unknown function. Middle panel: zooming-in on the top-right cluster. This cluster is composed mostly of amino acid binding proteins. *Right panel*: zooming-in on one connected component. Violet nodes represent methionine binding proteins. 4ntl, represented here by a white node encircled in orange, has no bound substrate, and its function is unknown. It is connected to the two central nodes, 4qhq and 3tqw (encircled in blue and purple). The figure was created using Cytoscape [94]

A dataset of binding proteins was collected from the 70% NR PDB, by using the website text search. This dataset was extended by adding proteins that share at least 30% of their sequence, over a segment of at least 35 residues, with an RMSD lower than 3.5 Å, with the proteins in the initial dataset. Cytoscape generated the network (Fig. 1, left panel), where each node represents a protein in the dataset and two nodes are connected if the proteins are deemed related (more than 30% sequence similarity, over more than 35 residues, with less than 3.5 Å RMSD). With this particular choice, several clusters are formed, so that in general SBPs which bind similar substrates (as evident in their PDB structures) belong to the same cluster (Fig. 1, left panel). Thus, their binding preferences and modes of interaction with the substrate can be predicted by the cluster they are found in. For example, one cluster is formed by SBPs that bind amino acids (Fig. 1, middle panel). A connected component within this cluster contains SBPs that generally bind methionine (Fig. 1, right panel). The substrate of one of these SBPs (white, encircled in orange, pdb 4ntl) is unknown. However, in this case we can suggest a likely hypothesis is that it also binds methionine. The sequence identity between the query and its neighbors is less than 40%; thus this functional inference, which is in keeping with the conjecture listed in CDD [52], is not trivial [91].

Using CyToStruct [82] and our molecular viewer of choice, we can examine this hypothesis in detail. Reassuringly, comparison of this query protein with its first neighbors in protein space (Fig. 1, right panel, the two nodes at the center of the cluster, encircled in cyan and green) supports this inference, as they share high structural similarity to the query (Fig. 2a). As both neighbors (pdb 4qhq

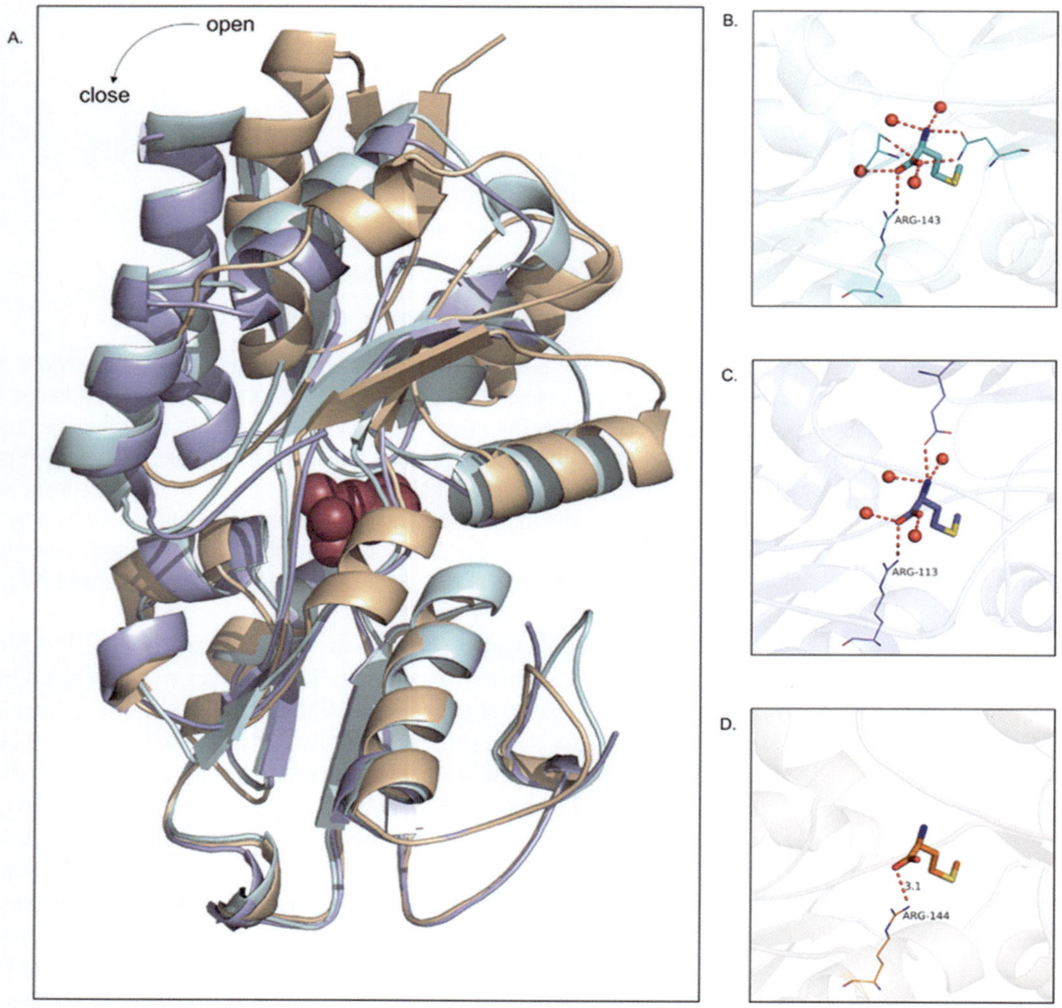

Fig. 2 Methionine binding in the SBPs 4ntl, 4qhq, and 3tqw. (**A**) Structural superposition of the 4ntl query (orange) with 4qhq and 3tqw (blue and purple), respectively. The superposition is over the C-terminal lobe to highlight the conformational change between the bound (close; 4qhq and 3tqw) and unbound (open; query) states of the SBPs. The bound methionine is shown in red spheres. (**B**) The methionine binding site in 4qhq. Methionine is presented using sticks model, and the polar residues of the binding site are depicted as wireframes. The hydrogen bonds that mediate methionine's interactions with these residues and with water molecules (red sphere) are marked as red dashed lines. The highly conserved Arg143 is also marked. (**C**) The methionine binding site in 3tqw. The highly conserved Arg113, equivalent of Arg143 in panel **B**, is marked. (**D**) Putative encounter complex between methionine and the query. Arg144 (depicted as wireframe) has the same location and rotameric state as its equivalents: Arg144 of 4qhq and Arg113 of 3tqw. The dashed line shows the putative hydrogen bond, which could form between the arginine and the methionine carbonyl group. The figure was created using the Pymol molecular viewer [84]

and pdb 3qwl) have a bound methionine in their PDB structure (Fig. 2b, c), a superposition of the structures can even be used to suggest a putative binding site (Fig. 2d). Evolutionary analysis, using ConSurf [92, 93], shows that the binding cavity is highly conserved, providing further support for the inferred function and binding mode. In particular, the three binding sites feature a highly conserved arginine residue (conservation grade of 9 on a 1–9 scale). Furthermore, in all three proteins, the arginine populates the exact same rotameric state, which allows it to form a hydrogen bond with the methionine substrate (Fig. 2b–d). In addition, water molecules that participate in the binding are also found in all the structures. However, not all the interactions that are found in the two bound states have equivalents in the query, and the structural superposition indicates that it is in an open conformation (Fig. 2a). It suggests that binding may follow the population shift theory, where methionine is initially recognized by the conserved arginine residue in the open conformation. This interaction may induce a shift of the protein to its closed conformation, where additional residues interact with methionine. Further investigation is needed to examine this suggestion.

3 Conclusions and Outlook

How did proteins emerge in evolution, and how do they evolve? Theoretically, a protein could emerge and evolve by linking one amino acid after another. Scholars believe that this approach is doomed, because the vast majority of polypeptide chains would not even fold. Thus, we presume that proteins emerged by mixing and matching short amino acid fragments (peptides) from the primordial soup, evolving by recombination, decoration, and mutation. Lupas et al. wrote an insightful review of this [29]. While most protein scientists would agree with this suggested scenario, the mechanics and details of the process which gave rise to proteins, and that govern their evolution, is still yet to be understood.

This leads to two observations: (1) We can look for clues to address these fundamental questions in current proteins by studying the reuse patterns in all proteins of known structure. (2) We can mine the evolutionary signal to identify common ancestry and improve methods of protein similarity search, function annotation, and design. For both of these, navigating in protein space can be very useful.

References

1. Kolodny R, Pereyaslavets L, Samson AO, Levitt M (2012) On the universe of protein folds. Annu Rev Biophys 42:559. https://doi.org/10.1146/annurev-biophys-083012-130432
2. Ben-Tal N, Kolodny R (2014) Representation of the protein universe using classifications, maps, and networks. Israel J Chem 54:1286
3. Zeldovich KB, Shakhnovich EI (2008) Understanding protein evolution: from protein

physics to Darwinian selection. Annu Rev Phys Chem 59:105–127

4. Trifonov EN, Berezovsky IN (2003) Evolutionary aspects of protein structure and folding. Curr Opin Struct Biol 13(1):110–114
5. Choi IG, Kim SH (2006) Evolution of protein structural classes and protein sequence families. Proc Natl Acad Sci U S A 103 (38):14056–14061. https://doi.org/10.1073/pnas.0606239103
6. Dokholyan NV, Shakhnovich B, Shakhnovich EI (2002) Expanding protein universe and its origin from the biological big bang. Proc Natl Acad Sci 99(22):14132–14136. https://doi.org/10.1073/pnas.202497999
7. Alva V, Remmert M, Biegert A, Lupas AN, Söding J (2010) A galaxy of folds. Protein Sci 19(1):124–130. https://doi.org/10.1002/pro.297
8. Farías-Rico JA, Schmidt S, Höcker B (2014) Evolutionary relationship of two ancient protein superfolds. Nat Chem Biol 10 (9):710–715. https://doi.org/10.1038/nchembio.1579 http://www.nature.com/nchembio/journal/v10/n9/abs/nchembio.1579.html#supplementary-information
9. Nepomnyachiy S, Ben-Tal N, Kolodny R (2017) Complex evolutionary footprints revealed in an analysis of reused protein segments of diverse lengths. Proc Natl Acad Sci U S A 114:11703
10. Skolnick J, Arakaki AK, Lee SY, Brylinski M (2009) The continuity of protein structure space is an intrinsic property of proteins. Proc Natl Acad Sci 106:15690. https://doi.org/10.1073/pnas.0907683106
11. Nepomnyachiy S, Ben-Tal N, Kolodny R (2014) Global view of the protein universe. Proc Natl Acad Sci 111:11691. https://doi.org/10.1073/pnas.1403395111
12. Mackenzie CO, Zhou J, Grigoryan G (2016) Tertiary alphabet for the observable protein structural universe. Proc Natl Acad Sci U S A 113(47):E7438–E7447
13. Kolodny R, Petrey D, Honig B (2006) Protein structure comparison: implications for the nature of 'fold space', and structure and function prediction. Curr Opin Struct Biol 16 (3):393–398
14. Osadchy M, Kolodny R (2011) Maps of protein structure space reveal a fundamental relationship between protein structure and function. Proc Natl Acad Sci 108 (30):12301–12306. https://doi.org/10.1073/pnas.1102727108
15. Berman HM, Westbrook J, Feng Z, Gilliland G, Bhat TN, Weissig H, Shindyalov IN, Bourne PE (2000) The Protein Data Bank. Nucleic Acids Res 28(1):235–242
16. Koehl P (2006) Protein structure classification. In: Reviews in Computational Chemistry. John Wiley & Sons, Inc., New York, pp 1–55. https://doi.org/10.1002/0471780367.ch1
17. Ponting CP, Russell RR (2002) The natural history of protein domains. Annu Rev Biophys Biomol Struct 31(1):45–71. https://doi.org/10.1146/annurev.biophys.31.082901.134314
18. Vogel C, Berzuini C, Bashton M, Gough J, Teichmann SA (2004) Supra-domains: evolutionary units larger than single protein domains. J Mol Biol 336(3):809–823. https://doi.org/10.1016/j.jmb.2003.12.026
19. Kolodny R, Koehl P, Guibas L, Levitt M (2002) Small libraries of protein fragments model native protein structures accurately. J Mol Biol 323(2):297–307
20. Vanhee P, Verschueren E, Baeten L, Stricher F, Serrano L, Rousseau F, Schymkowitz J (2011) BriX: a database of protein building blocks for structural analysis, modeling and design. Nucleic Acids Res 39(Suppl 1):D435–D442
21. Davis FP, Sali A (2005) PIBASE: a comprehensive database of structurally defined protein interfaces. Bioinformatics 21(9):1901–1907
22. Vanhee P, Reumers J, Stricher F, Baeten L, Serrano L, Schymkowitz J, Rousseau F (2009) PepX: a structural database of non-redundant protein–peptide complexes. Nucleic Acids Res 38(Suppl 1):D545–D551
23. Fernandez-Fuentes N, Dybas JM, Fiser A (2010) Structural characteristics of novel protein folds. PLoS Comput Biol 6(4):e1000750
24. Ovchinnikov S, Park H, Varghese N, Huang P-S, Pavlopoulos GA, Kim DE, Kamisetty H, Kyrpides NC, Baker D (2017) Protein structure determination using metagenome sequence data. Science 355(6322):294–298
25. Pieper U, Eswar N, Davis FP, Braberg H, Madhusudhan MS, Rossi A, Marti-Renom M, Karchin R, Webb BM, Eramian D (2006) MODBASE: a database of annotated comparative protein structure models and associated resources. Nucleic Acids Res 34(Suppl 1): D291–D295
26. Lo Conte L, Ailey B, Hubbard TJP, Brenner SE, Murzin AG, Chothia C (2000) SCOP: a structural classification of proteins database. Nucleic Acids Res 28(1):257–259
27. Orengo C, Michie A, Jones S, Jones D, Swindells M, Thornton J (1997) CATH-a hierarchic classification of protein domain structures. Structure 5(8):1093–1108

28. Cheng H, Schaeffer RD, Liao Y, Kinch LN, Pei J, Shi S, Kim B-H, Grishin NV (2014) ECOD: an evolutionary classification of protein domains. PLoS Comput Biol 10(12): e1003926. https://doi.org/10.1371/journal.pcbi.1003926
29. Lupas AN, Ponting CP, Russell RB (2001) On the evolution of protein folds: are similar motifs in different protein folds the result of convergence, insertion, or relics of an ancient peptide world? J Struct Biol 134 (2–3):191–203
30. Soding J (2005) Protein homology detection by HMM-HMM comparison. Bioinformatics 21(7):951–960
31. Eddy SR (2009) A new generation of homology search tools based on probabilistic inference. Genome Inform 1:205–211
32. Alva V, Söding J, Lupas AN (2016) A vocabulary of ancient peptides at the origin of folded proteins. elife 4:e09410
33. Kosloff M, Kolodny R (2008) Sequence-similar, structure-dissimilar protein pairs in the PDB. Proteins 71(2):891–902
34. Narunsky A, Nepomnyachiy S, Ashkenazy H, Kolodny R, Ben-Tal N (2015) ConTemplate suggests possible alternative conformations for a query protein of known structure. Structure 23(11):2162–2170
35. Holm L, Sander C (1996) Mapping the protein universe. Science 273(5275):595–603
36. Skolnick J, Gao M, Zhou H (2014) On the role of physics and evolution in dictating protein structure and function. Israel J Chem 54 (8–9):1176–1188
37. Hasegawa H, Holm L (2009) Advances and pitfalls of protein structural alignment. Curr Opin Struct Biol 19(3):341–348
38. Kolodny R, Koehl P, Levitt M (2005) Comprehensive evaluation of protein structure alignment methods: scoring by geometric measures. J Mol Biol 346(4):1173–1188
39. Kolodny R, Linial N (2004) Approximate protein structural alignment in polynomial time. Proc Natl Acad Sci U S A 101 (33):12201–12206
40. Carugo O (2007) Recent progress in measuring structural similarity between proteins. Curr Protein Pept Sci 8(3):241
41. Yanover C, Vanetik N, Levitt M, Kolodny R, Keasar C (2014) Redundancy-weighting for better inference of protein structural features. Bioinformatics 30(16):2295–2301
42. Li W, Godzik A (2006) Cd-hit: a fast program for clustering and comparing large sets of protein or nucleotide sequences. Bioinformatics 22(13):1658–1659
43. Wang G, Dunbrack RL (2003) PISCES: a protein sequence culling server. Bioinformatics 19 (12):1589–1591. https://doi.org/10.1093/bioinformatics/btg224
44. Choi I-G, Kim S-H (2007) Global extent of horizontal gene transfer. Proc Natl Acad Sci 104(11):4489–4494. https://doi.org/10.1073/pnas.0611557104
45. Orengo CA, Flores TP, Taylor WR, Thornton JM (1993) Identification and classification of protein fold families. Protein Eng 6 (5):485–500. https://doi.org/10.1093/protein/6.5.485
46. Finn RD, Bateman A, Clements J, Coggill P, Eberhardt RY, Eddy SR (2014) Pfam: the protein families database. Nucleic Acids Res 42: D222. https://doi.org/10.1093/nar/gkt1223
47. Pearl FMG, Sillitoe I, Orengo CA (2015) Protein structure classification. In: eLS. John Wiley & Sons, Ltd., New York. https://doi.org/10.1002/9780470015902.a0003033.pub3
48. Levitt M, Chothia C (1976) Structural patterns in globular proteins. Nature 261 (5561):552–558
49. Holland TA, Veretnik S, Shindyalov IN, Bourne PE (2006) Partitioning protein structures into domains: why is it so difficult? J Mol Biol 361(3):562–590
50. Hadley C, Jones DT (1999) A systematic comparison of protein structure classifications: SCOP, CATH and FSSP. Structure 7 (9):1099–1112
51. Day R, Beck DAC, Armen RS, Daggett V (2003) A consensus view of fold space: combining SCOP, CATH, and the Dali Domain Dictionary. Protein Sci 12(10):2150–2160. https://doi.org/10.1110/ps.0306803
52. Marchler-Bauer A, Lu S, Anderson JB, Chitsaz F, Derbyshire MK, DeWeese-Scott C, Fong JH, Geer LY, Geer RC, Gonzales NR (2010) CDD: a conserved domain database for the functional annotation of proteins. Nucleic Acids Res 39(Suppl 1):D225–D229
53. Kelley LA, Sternberg MJ (2015) Partial protein domains: evolutionary insights and bioinformatics challenges. Genome Biol 16(1):1–3. https://doi.org/10.1186/s13059-015-0663-8
54. Veretnik S, Gu J, Wodak S (2009) Identifying structural domains in proteins. In: Gu G, Bourne P (eds) Structural bioinformatics, 2nd edn. Wiley-Blackwell, Hoboken, NJ, pp 485–513
55. Schaeffer RD, Jonsson AL, Simms AM, Daggett V (2011) Generation of a consensus protein domain dictionary. Bioinformatics 27

(1):46–54. https://doi.org/10.1093/bioinformatics/btq625

56. Csaba G, Birzele F, Zimmer R (2009) Systematic comparison of SCOP and CATH: a new gold standard for protein structure analysis. BMC Struct Biol 9(1):23

57. Redfern OC, Harrison A, Dallman T, Pearl FM, Orengo CA (2007) CATHEDRAL: a fast and effective algorithm to predict folds and domain boundaries from multidomain protein structures. PLoS Comput Biol 3(11): e232. https://doi.org/10.1371/journal.pcbi.0030232

58. Zhou H, Xue B, Zhou Y (2007) DDOMAIN: dividing structures into domains using a normalized domain–domain interaction profile. Protein Sci 16(5):947–955. https://doi.org/10.1110/ps.062597307

59. Alexandrov N, Shindyalov I (2003) PDP: protein domain parser. Bioinformatics 19 (3):429–430. https://doi.org/10.1093/bioinformatics/btg006

60. Krishna SS, Grishin NV (2005) Structural drift: a possible path to protein fold change. Bioinformatics 21(8):1308–1310

61. Pascual-García A, Abia D, Ortiz ÁR, Bastolla U (2009) Cross-over between discrete and continuous protein structure space: insights into automatic classification and networks of protein structures. PLoS Comput Biol 5(3): e1000331. https://doi.org/10.1371/journal.pcbi.1000331

62. Edwards H, Deane CM (2015) Structural bridges through fold space. PLoS Comput Biol 11(9):e1004466

63. Fox NK, Brenner SE, Chandonia J-M (2014) SCOPe: structural classification of proteins—extended, integrating SCOP and ASTRAL data and classification of new structures. Nucleic Acids Res 42(D1):D304–D309. https://doi.org/10.1093/nar/gkt1240

64. Andreeva A, Howorth D, Chothia C, Kulesha E, Murzin AG (2013) SCOP2 prototype: a new approach to protein structure mining. Nucleic Acids Res 42:D310. https://doi.org/10.1093/nar/gkt1242

65. Ellson J, Gansner E, Koutsofios L, North SC, Woodhull G (2001) Graphviz—open source graph drawing tools. In: International symposium on graph drawing. Springer, Heidelberg, pp 483–484

66. Prlić A, Bliven S, Rose PW, Bluhm WF, Bizon C, Godzik A, Bourne PE (2010) Pre-calculated protein structure alignments at the RCSB PDB website. Bioinformatics 26 (23):2983–2985. https://doi.org/10.1093/bioinformatics/btq572

67. Krissinel E, Henrick K (2003) Protein structure comparison in 3D based on secondary structure matching (SSM) followed by C-alpha alignment, scored by a new structural similarity function. Proceedings of the 5th International Conference on Molecular Structural Biology, Vienna, vol. 88

68. Krissinel E, Henrick K (2004) Secondary-structure matching (SSM), a new tool for fast protein structure alignment in three dimensions. Acta Crystallogr D 60(Pt 12 Pt 1):2256–2268

69. Madej T, Lanczycki CJ, Zhang D, Thiessen PA, Geer RC, Marchler-Bauer A (2014) MMDB and VAST+: tracking structural similarities between macromolecular complexes. Nucleic Acids Res D42:D297. https://doi.org/10.1093/nar/gkt1208

70. Mezulis S, Sternberg MJE, Kelley LA (2016) PhyreStorm: a web server for fast structural searches against the PDB. J Mol Biol 428 (4):702–708. https://doi.org/10.1016/j.jmb.2015.10.017

71. Zhang Y, Skolnick J (2005) TM-align: a protein structure alignment algorithm based on the TM-score. Nucleic Acids Res 33 (7):2302–2309. https://doi.org/10.1093/nar/gki524

72. Wiederstein M, Gruber M, Frank K, Melo F, Sippl Manfred J (2014) Structure-based characterization of multiprotein complexes. Structure 22(7):1063–1070. https://doi.org/10.1016/j.str.2014.05.005

73. Berezovsky IN, Guarnera E, Zheng Z (2017) Basic units of protein structure, folding, and function. Prog Biophys Mol Biol 128:85–99. https://doi.org/10.1016/j.pbiomolbio.2016.09.009

74. Menke M, Berger B, Cowen L (2008) Matt: local flexibility aids protein multiple structure alignment. PLoS Comput Biol 4(1):e10

75. Shindyalov I, Bourne P (1998) Protein structure alignment by incremental combinatorial extension (CE) of the optimal path. Protein Eng 11(9):739–747

76. Ortiz A, Strauss C, Olmea O (2002) MAMMOTH (matching molecular models obtained from theory): an automated method for model comparison. Protein Sci 11(11):2606–2621

77. Tung CH, Huang JW, Yang JM (2007) Kappa-alpha plot derived structural alphabet and BLOSUM-like substitution matrix for rapid search of protein structure database. Genome Biol 8(3):R31

78. Budowski-Tal I, Nov Y, Kolodny R (2010) FragBag, an accurate representation of protein structure, retrieves structural neighbors from

the entire PDB quickly and accurately. Proc Natl Acad Sci U S A 107(8):3481–3486. https://doi.org/10.1073/pnas.0914097107

79. Petrey D, Xiang Z, Tang CL, Xie L, Gimpelev M, Mitros T, Soto CS, Goldsmith-Fischman S, Kernytsky A, Schlessinger A, Koh IY, Alexov E, Honig B (2003) Using multiple structure alignments, fast model building, and energetic analysis in fold recognition and homology modeling. Proteins 53(Suppl 6):430–435. https://doi.org/10.1002/prot.10550

80. Subbiah S, Laurents DV, Levitt M (1993) Structural similarity of DNA-binding domains of bacteriophage repressors and the globin core. Curr Biol 3(3):141–148

81. Saito R, Smoot ME, Ono K, Ruscheinski J, Wang P-L, Lotia S, Pico AR, Bader GD, Ideker T (2012) A travel guide to Cytoscape plugins. Nat Methods 9(11):1069–1076

82. Nepomnyachiy S, Ben-Tal N, Kolodny R (2015) CyToStruct: augmenting the network visualization of cytoscape with the power of molecular viewers. Structure 23(5):941–948

83. Morris JH, Huang CC, Babbitt PC, Ferrin TE (2007) structureViz: linking Cytoscape and UCSF chimera. Bioinformatics 23 (17):2345–2347. https://doi.org/10.1093/bioinformatics/btm329

84. Schrodinger, LLC (2010) The PyMOL molecular graphics system, Version 1.3r1. Schrodinger, LLC, New York

85. Pettersen EF, Goddard TD, Huang CC, Couch GS, Greenblatt DM, Meng EC, Ferrin TE (2004) UCSF chimera—a visualization system for exploratory research and analysis. J Comput Chem 25(13):1605–1612

86. Jmol: an open-source java viewer for chemical structure in 3D. http://www.jmol.org/

87. Humphrey W, Dalke A, Schulten K (1996) VMD: visual molecular dynamics. J Mol Graph 14(1):33–38

88. Rose AS, Hildebrand PW (2015) NGL viewer: a web application for molecular visualization. Nucleic Acids Res 43(Web Server issue): W576–W579. https://doi.org/10.1093/nar/gkv402

89. O'Donoghue SI, Goodsell DS, Frangakis AS, Jossinet F, Laskowski RA, Nilges M, Saibil HR, Schafferhans A, Wade RC, Westhof E (2010) Visualization of macromolecular structures. Nat Methods 7:S42–S55

90. Berntsson RP-A, Smits SH, Schmitt L, Slotboom D-J, Poolman B (2010) A structural classification of substrate-binding proteins. FEBS Lett 584(12):2606–2617

91. Radivojac P, Clark WT, Oron TR, Schnoes AM, Wittkop T, Sokolov A, Graim K, Funk C, Verspoor K, Ben-Hur A (2013) A large-scale evaluation of computational protein function prediction. Nat Methods 10 (3):221–227

92. Glaser F, Pupko T, Paz I, Bell RE, Bechor-Shental D, Martz E, Ben-Tal N (2003) ConSurf: identification of functional regions in proteins by surface-mapping of phylogenetic information. Bioinformatics 19(1):163–164

93. Ashkenazy H, Abadi S, Martz E, Chay O, Mayrose I, Pupko T, Ben-Tal N (2016) ConSurf 2016: an improved methodology to estimate and visualize evolutionary conservation in macromolecules. Nucleic Acids Res 44(W1): W344–W350

94. Shannon P, Markiel A, Ozier O, Baliga NS, Wang JT, Ramage D, Amin N, Schwikowski B, Ideker T (2003) Cytoscape: a software environment for integrated models of biomolecular interaction networks. Genome Res 13(11):2498–2504. https://doi.org/10.1101/gr.1239303

Chapter 13

A Graph-Based Approach for Detecting Sequence Homology in Highly Diverged Repeat Protein Families

Jonathan N. Wells and Joseph A. Marsh

Abstract

Reconstructing evolutionary relationships in repeat proteins is notoriously difficult due to the high degree of sequence divergence that typically occurs between duplicated repeats. This is complicated further by the fact that proteins with a large number of similar repeats are more likely to produce significant local sequence alignments than proteins with fewer copies of the repeat motif. Furthermore, biologically correct sequence alignments are sometimes impossible to achieve in cases where insertion or translocation events disrupt the order of repeats in one of the sequences being aligned. Combined, these attributes make traditional phylogenetic methods for studying protein families unreliable for repeat proteins, due to the dependence of such methods on accurate sequence alignment.

We present here a practical solution to this problem, making use of graph clustering combined with the open-source software package HH-suite, which enables highly sensitive detection of sequence relationships. Carrying out multiple rounds of homology searches via alignment of profile hidden Markov models, large sets of related proteins are generated. By representing the relationships between proteins in these sets as graphs, subsequent clustering with the Markov cluster algorithm enables robust detection of repeat protein subfamilies.

Key words Repeat proteins, Sequence homology, Graph clustering, Profile-HMM alignment, Protein families, Evolution

1 Introduction

Proteins comprising tandem structural motif repeats are ubiquitous and can be separated into five broad classes [1, 2]; these range from low-complexity domains with repeat units less than ten residues in length to large "beads on a string" type proteins such as titin (Fig. 1). Due to their sequence diversity and functional importance, the two most important classes of repeat proteins are those that form open, solenoid structures, such as leucine-rich repeat (LRR) proteins [3], and those that form closed, toroidal structures, such as the WD40 domain [4].

While common in all domains of life, repeat proteins are particularly prevalent in eukaryotes, and many families are ancient and

Tobias Sikosek (ed.), *Computational Methods in Protein Evolution*, Methods in Molecular Biology, vol. 1851, https://doi.org/10.1007/978-1-4939-8736-8_13,

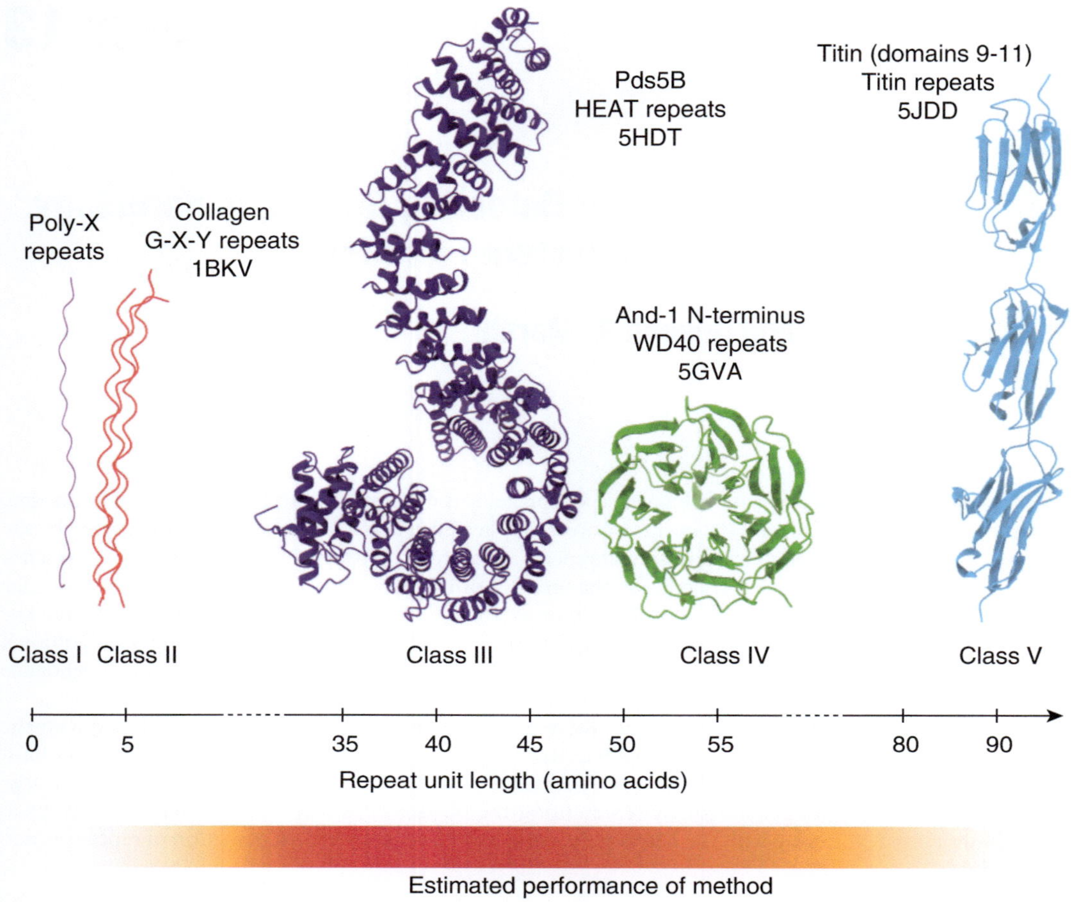

Fig. 1 Repeat protein classes. Repeat proteins can be roughly categorized according to the length of the repeated sequence motif. The very simplest repeats are simply long tracts of a single amino acid, most commonly alanine or glutamine. These are often associated with disease, most famously the poly-Q tracts in Huntington's disease, but are nonetheless prevalent and have recently been shown to play an important role in facilitating rapid protein divergence in eukaryotes [28]. The most diverse classes, both structurally and functionally, are III and IV, which include families of solenoid proteins such as the HEAT repeat containing Hawk family [22], of which Pds5B is a member, and ubiquitous domains such as the WD40 beta-propeller [4]. At the other extreme are proteins such as titin, which comprises hundreds of repeated domains joined by short linker regions. The method described here is likely to perform best on those proteins in classes III and IV

highly conserved [5, 6]. However, the precise sequence of the repeating unit tends to be of secondary importance to its structure. Indeed, the sequence similarity between many homologous repeat proteins approaches that of random sequences, even while their structures remain highly similar [7, 8]. Additional repeats tend to be gained via internal duplications [9], and provided the overall structure of the protein is not disrupted, the insertion/deletion rate can be quite rapid, particularly in the case of proteins where the repeats are structurally independent of each other [10].

An observation with important implications for this method is that, despite considerable evidence for the high degree of conservation of repeat proteins across eukaryotes, within individual proteins, the repeats themselves are often highly divergent. Attractive work from the Koonin lab has shown that this is because recent duplications are generally subject to a period of relaxed purifying selection, leading to rapid sequence divergence prior to fixation [11].

However, this presents serious difficulties when attempting to reconstruct the evolutionary history of these proteins. Accurate phylogenetic trees are critically dependent on good sequence alignments, and in the case of repeat proteins, these are often impossible to produce, particularly when multiple internal insertion or deletion events have occurred. Numerous attempts have been made to develop methods that can be used to improve the quality of repeat proteins' multiple sequence alignments, and thus phylogenetic trees [12–16]. However, an independent benchmark of several such programs has shown them to be highly dependent on the type of repeat unit under investigation and concluded that characteristics of detected repeats (unit length, number of repeats, etc.) varied more with the algorithm being used than with the underlying data [17].

In this chapter, we describe a conceptually simple method that can be used to reveal within-species evolutionary relationships between highly divergent repeat proteins when phylogenetic analysis is not possible. This method makes extensive use of the open-source software HH-suite [18, 19], available from the Söding research group or GitHub (https://github.com/soedinglab/hh-suite), and also the Markov cluster (MCL) algorithm [20], which has been in widespread use in computational biology, including for detection of protein families [21]. Though computationally intensive, the method that follows can be applied straightforwardly by a user familiar with UNIX-based command line applications and with basic programming experience.

Briefly, multiple rounds of homology searches in HH-suite are used to generate a large set of putatively related repeat proteins from those of initial interest. This larger set of proteins is represented as a graph, in which edges between proteins are weighted according to the rank of the best alignment between them, relative to all other homologous proteins detected. By clustering this graph using the MCL algorithm, different subfamilies of repeat proteins are revealed. These clusters are robust to parameter changes, and their statistical significance can be assessed with permutation tests of the underlying graph.

The principal problem motivating the development of this method lies in the fact that proteins sharing multiple copies of the same repeat motif may produce alignments that erroneously suggest a close evolutionary relationship where none exists. As an example, one can imagine a scenario in which a highly diverged repeat protein produces high-scoring alignments with a larger protein containing many well-conserved copies of the canonical motif, simply because the larger protein provides many opportunities for such alignments. We reasoned that the likelihood of obtaining false positives in this manner could be reduced by performing reciprocal searches, using both proteins in any given pair as queries. Unlike the pair in the scenario just described, pairs directly related to each other would be expected to produce mutually high-ranking alignments, regardless of which was used as the query.

By generating a network out of many such reciprocal searches, families of related repeat proteins can then be obtained by clustering. The underlying assumption behind this idea being that, on average, closely related proteins will produce more high-ranking alignments within their immediate family than they will with proteins outside the family, even if individual alignments are sometimes misleading.

We have previously used this approach to demonstrate the common ancestry of members of the Hawk family (HEAT repeat proteins associated with kleisins)—a set of widely conserved proteins that arose in the last common ancestor of eukaryotes [22] and which play an essential role in regulating the activity of condensin and cohesin. The protocol as presented here is the same as that used to describe the Hawk family, but is directly applicable to other families of repeat proteins with qualitatively similar structures, for example, leucine-rich repeat proteins (Fig. 2). For proteins from different repeat classes, such as beta-propellers, adjustments to certain method parameters will likely be needed, as discussed in the Notes section.

2 Methods

2.1 Preliminary Setup and Database Building

HH-suite, specifically the command-line programs hhblits and hhsearch, works by aligning profile hidden Markov models (profile HMMs) using the Viterbi algorithm [23, 24]. Each profile HMM is essentially a condensed version of a multiple sequence alignment, unique to the protein of interest. Since the quantity of information in profile HMMs is much greater than the corresponding raw amino acid sequences, homologous relationships between proteins can be detected with much higher sensitivity than methods such as BLAST or PSI-BLAST [25, 26].

Before building the network, a search database must first be built for the species of interest. This is by some margin the most time-consuming step in this method, but fortunately only needs doing once, after which it can be saved for reuse. To begin with, the UniProt database itself must be clustered and converted into a set of profile HMMs; pre-compiled versions can be downloaded directly from the Söding lab (http://wwwuser.gwdg.de/

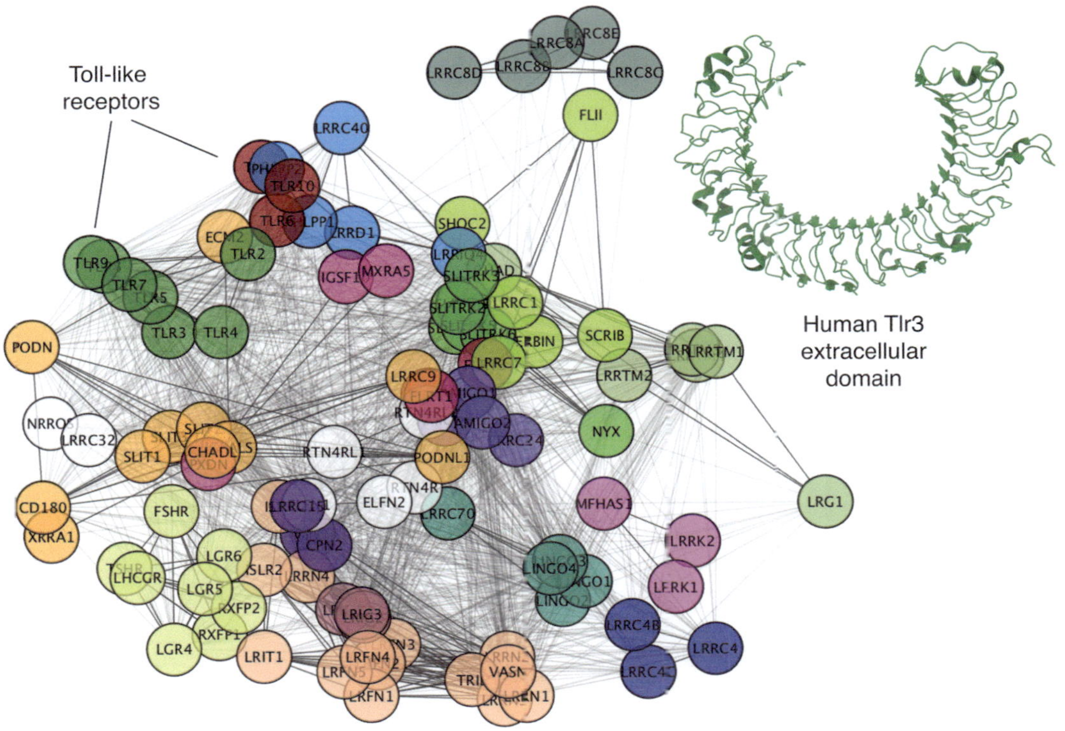

Fig. 2 Example network showing leucine-rich repeat subfamilies. Starting with leucine-rich repeat protein 1 (LRR1, not shown) to initiate searches, a network showing different subgroups within the large and diverse LRR family was generated, using the protocol that follows, with an MCL inflation parameter of 3.0 for the final clustering. Several known families are recapitulated here, most notably the Toll-like receptors, highlighted top-left. The networks generated are very dense, with the majority of proteins being connected to tens to hundreds of others. To aid visualization, only those with a reported true positive probability of 1.0 are shown. Darker edges represent higher mutual ranks

~compbiol/data/hhsuite/), and using these is strongly recommended.

The species-specific database is then built using the clustered UniProt database. As the practical details of this step are beyond the scope of this method description, readers are instead referred to the excellent documentation on the matter provided with the HH-suite package. For smaller proteomes consisting of fewer than 10,000 sequences, this takes about a day on a powerful desktop or laptop, but for larger proteomes—such as those of mammals—it is simpler to build the database in batches on a computer cluster.

2.2 Generating Homology Networks

Once the required species databases have been built, we are ready to begin carrying out the sequence searches needed to generate our network. This can be initiated using either multiple candidate sequences or just a single representative of the family of interest (*see* **Note 1** for additional information on selecting seed sequences). For each sequence, generate a profile HMM using hhblits searched against the clustered UniProt database, with the output format set to either hhm or a3m if the multiple sequence alignment is required. Multiple iterations can be carried out for increased sensitivity, but the default value of two iterations is almost certainly sufficient at this stage.

Next, using the multi-species profile HMM that this produces, search the species-specific database that you generated previously using hhsearch—this is equivalent to using hhblits with a single iteration, with the exception of some additional steps in the latter. The species-specific results from this step should be saved in the default hhresults format, but the profile HMMs or sequence alignments are not required. Within this file, there will be up to 500 proteins that have been identified as sharing significant sequence similarity (assuming default hhsearch settings); the amino acid sequence of each of these proteins must then be downloaded, a step most easily achieved programmatically, for obvious reasons. Having downloaded these sequences, the first round of homologue searches is complete.

Each of these new sequences is then subjected to the same procedure as before, i.e., searched against the clustered UniProt database to generate a profile HMM and then searched against the species-specific database. However, once the results files from this second round of searches have been acquired, it is no longer necessary to download the new protein sequences that have appeared. At this point, all the proteins that will be represented in the final network are present, along with many more which will be discarded.

The set of results files must now be processed and converted into graph format. Each protein, whether used as a query sequence or returned as a hit, can be represented as a node in the graph. Similarly, the significant alignments between each query protein and the hits can be represented as edges from the former to the latter, weighted by the rank of the alignment. Importantly, many edges will be redundant, corresponding to alignments between different sections of the repetitive sequence; in these cases, simply use the highest-ranking alignment between the protein pair and discard the rest.

The resulting graph will be directed, and likely quite large. In order to reduce the size, we can filter out edges and nodes that do not meet certain significance criteria. The choice of these thresholds is partly case-dependent (*see* **Note 2**), but for demonstration purposes, discard all edges with an expect value of greater than 0.01 (thus controlling the false-positive rate) and a true positive probability less than 0.15. These values are given with each alignment in the results files for query proteins.

Since some proteins will have many significant homologues and others very few, it is necessary to normalize the ranks (and therefore the edge weights) of each alignment. Do so according to the following formula:

$$f(r) = \frac{r_{\max} - r_{\min}}{99r + r_{\max} - 100r_{\min}}, 1 \le r \le 500$$

where r is the rank of the alignment/edge in question and $r_{\max}$ and $r_{\min}$ are the maximum and minimum ranks in the result file describing the alignment. This ensures that the normalized edge weights lie between 1.0 and 0.01, with the former being the best possible rank and the latter the worst (*see* **Note 3** for alternative edge weighting possibilities).

To simplify the subsequent clustering step, we now need to make the network undirected. At this point, each pair of proteins will have two edges between them or a single edge; the latter can occur either because an alignment did not meet the significance thresholds imposed on it, or much more commonly because one of the proteins involved only appeared in the second round of searches and was thus never queried. Since these proteins are only of secondary importance to the family of interest, if a node in the network has a degree of less than two, then discard it. For each pair of remaining nodes, collapse the edges between them into a single, undirected edge using the geometric mean. Since the geometric mean is always lower than the arithmetic mean, this avoids giving too much weight to high-ranking, but low significance alignments from proteins with few detectable homologues.

2.3 Clustering Networks and Assessing Significance

At this stage, the network is undirected and has been trimmed down considerably from the size that would otherwise be produced. Clustering is carried out with the MCL algorithm, which can be downloaded as a standalone program from Stijn van Dongen's personal website https://micans.org/mcl/index.html, or as implemented in other programs such as Cytoscape [27]. A sensible inflation parameter should be used (roughly speaking, this is a measure of the granularity of the resulting clusters): $I = 2.5$ is a good starting point but may need to be changed depending on the properties of your network (*see* **Note 4**). The workflow leading to this point is summarized graphically in Fig. 3.

Once you have obtained clusters containing your proteins of interest, their statistical significance can be assessed with permutation tests of the underlying network. Specifically, the probability of obtaining a specific cluster by chance can be calculated by randomizing the ranks of the underlying alignments in each result file, regenerating the network and clustering it. This is then repeated as many times as is computationally practical—on a powerful desktop computer (tested on ×8 Intel® Core™ i7-4790 K CPUs @ 4.00 GHz), a network containing

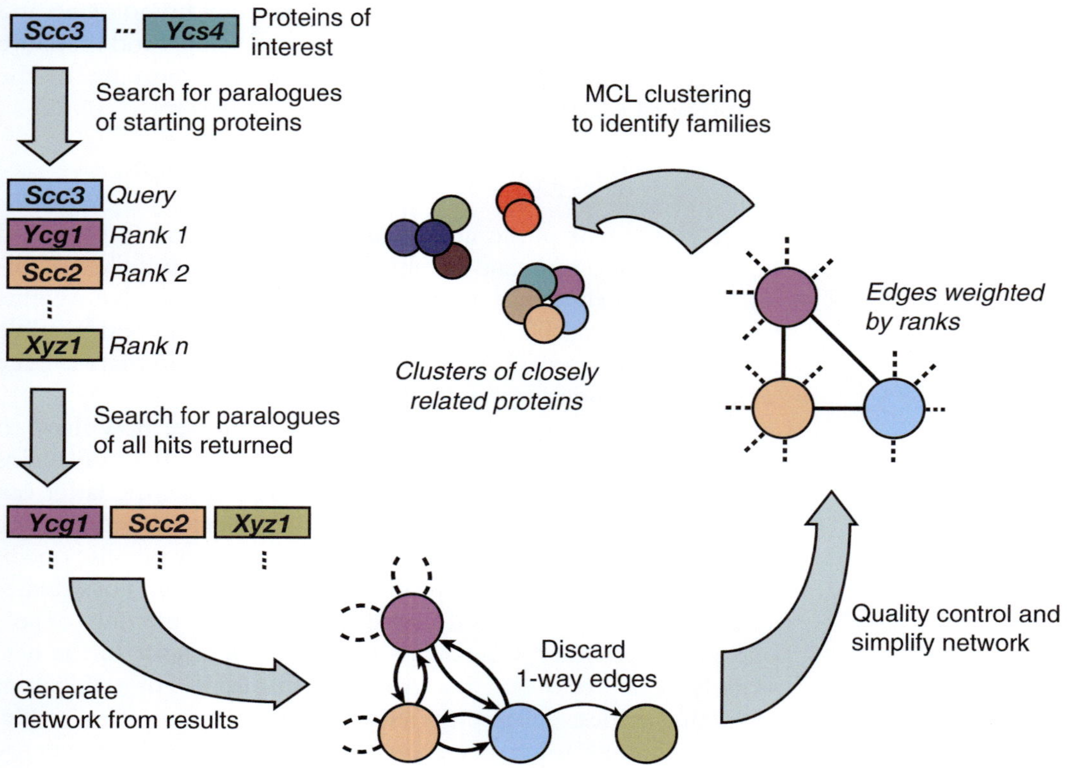

Fig. 3 Summary of network construction. First choose a small number of proteins to initiate the searches. Use hhblits and the clustered UniProt database to generate profile HMMs for each sequence, and then use this with hhsearch to search the species database. Carry out a second round of searches using the results from the first round as queries in the second. After choosing appropriate thresholds for alignment significance, build a network using the alignment ranks as edge weights. Simplify this network by removing nodes with degree = 1 and collapse edge pairs using the geometric mean. Cluster this network using the MCL algorithm to reveal subfamilies within the larger repeat motif family. Figure adapted from Wells et al. [22]

~200 nodes can be regenerated and clustered on the order of 10^5 times overnight, but for larger networks, access to a computer cluster may be necessary.

3 Discussion

Although qualitatively different in terms of their output, there are some interesting parallels between phylogenetic trees and the clustered graphs produced here. In an ideal graph, different clusters are equivalent to monophyletic clades within a tree-based representation of the true evolutionary history. In the same fashion, varying the inflation parameter is analogous to producing sub-trees by cutting internal branches at specific time points.

However, less welcome similarities are also likely to exist, and though we have not investigated the process, it seems probable that the formation of some clusters in the graph may involve a process akin to long-branch attraction. Specifically, if large groups of highly divergent proteins are present, none of which exhibit distinct similarities to each other, then these may cluster together as a result of being densely connected, yet evenly weighted across the group. Concerted evolution leading to homogenization of repeats has also been observed in repeat proteins [11], and this too is likely to confound analyses in cases where it occurs. Given this, independent validation of clusters should be sought wherever possible, for example, by looking for enrichment of GO terms or shared structural similarities.

In summary, the method described here is a robust, semiquantitative alternative to traditional tree-based descriptions of phylogeny and is particularly powerful for repeat protein families for which it is not possible to generate accurate multiple sequence alignments. Although originally designed to demonstrate the shared ancestry of a single group of proteins, due to fact that many distantly related proteins are acquired during construction of the network, it can also be used more broadly as a tool for generating new hypotheses.

4 Notes

There are several points to make about the method described above. As mentioned in the introduction, different classes of repeat proteins will likely require different parameters and should be approached on a case-by-case basis. The following notes offer some additional guidance on how different parameter choices affect the resulting networks:

1. Beginning with selection of seed proteins for the first round of sequence searches, it may be necessary to remove extraneous domains if present. While the method will work for both repetitive and non-repetitive proteins, if the core domains of interest are non-repetitive, then traditional phylogenetic methods are likely to be more appropriate. Otherwise, manually removing such domains will avoid the possibility of subsequent searches being led off track by high-scoring alignments in non-repetitive regions.
2. When choosing alignment significance thresholds for inclusion in the network (e.g., expect value, true positive probability, alignment length), as long as the values are reasonable, then their main effect will be on the size of the network. More stringent thresholds will decrease the overall size of the network, at the expense of distant family members in each cluster, whereas relaxed thresholds will increase the computational requirements in carrying out permutation tests for each cluster.

3. Attributes other than the relative rank alignments can also be used to weight edges for clustering, for example, true positive probability. However, since we are more interested in relative relationships, using the rank is the most robust metric. To understand this, it helps to imagine two hypothetical protein families whose average rates of divergence differ and consider how a relative metric such as rank would behave versus absolute metrics.

4. An important strength of MCL is the fact that in most cases the only parameter that needs to be explicitly set by the user is the inflation parameter, which affects the granularity of the resulting clustered graph. As a rule of thumb, values in the region of 1.2–6.0 are reasonable, with 2.0 being the default [20]; larger values will tend to produce smaller clusters and vice versa. To adhere to best scientific practice, once this parameter has been set, it should not be changed, and to avoid researcher bias, the value should ideally be settled on before revealing node labels. This is not unrealistic given prior knowledge about the average size of species gene families. For example, an inflation parameter that gave clusters with a median size of two in a human network would almost certainly be inappropriate, since most repeat protein families are considerably larger.

References

1. Kajava AV (2001) Review: proteins with repeated sequence—structural prediction and modeling. J Struct Biol 134:132–144. https://doi.org/10.1006/jsbi.2000.4328
2. Kajava AV (2012) Tandem repeats in proteins: from sequence to structure. J Struct Biol 179:279–288. https://doi.org/10.1016/j.jsb.2011.08.009
3. Kobe B, Deisenhofer J (1994) The leucine-rich repeat: a versatile binding motif. Trends Biochem Sci 19:415–421
4. Neer EJ, Schmidt CJ, Nambudripad R, Smith TF (1994) The ancient regulatory-protein family of WD-repeat proteins. Nature 371:297–300. https://doi.org/10.1038/371297a0
5. Marcotte EM, Pellegrini M, Yeates TO, Eisenberg D (1999) A census of protein repeats. J Mol Biol 293:151–160. https://doi.org/10.1006/jmbi.1999.3136
6. Schaper E, Gascuel O, Anisimova M (2014) Deep conservation of human protein tandem repeats within the eukaryotes. Mol Biol Evol 31:1132–1148. https://doi.org/10.1093/molbev/msu062
7. Andrade MA, Petosa C, O'Donoghue SI et al (2001) Comparison of ARM and HEAT protein repeats. J Mol Biol 309:1–18. https://doi.org/10.1006/jmbi.2001.4624
8. Sutherland TD, Campbell PM, Weisman S et al (2006) A highly divergent gene cluster in honey bees encodes a novel silk family. Genome Res 16:1414–1421. https://doi.org/10.1101/gr.5052606
9. Björklund ÅK, Ekman D, Elofsson A (2006) Expansion of protein domain repeats. PLoS Comput Biol 2:0959–0970. https://doi.org/10.1371/journal.pcbi.0020114
10. Schüler A, Bornberg-Bauer E (2016) Evolution of protein domain repeats in Metazoa. Mol Biol Evol 33:3170
11. Persi E, Wolf YI, Koonin EV (2016) Positive and strongly relaxed purifying selection drive the evolution of repeats in proteins. Nat Commun 7:13570. https://doi.org/10.1038/ncomms13570
12. Szklarczyk R, Heringa J (2004) Tracking repeats using significance and transitivity. Bioinformatics 20(Suppl 1):i311–i317. https://doi.org/10.1093/bioinformatics/bth911

13. Söding J, Remmert M, Biegert A, Lupas AN (2006) HHsenser: exhaustive transitive profile search using HMM-HMM comparison. Nucleic Acids Res 34:374–378. https://doi.org/10.1093/nar/gkl195
14. Newman AM, Cooper JB (2007) XSTREAM: a practical algorithm for identification and architecture modeling of tandem repeats in protein sequences. BMC Bioinformatics 8:382. https://doi.org/10.1186/1471-2105-8-382
15. Vo A, Nguyen N, Huang H (2010) Solenoid and non-solenoid protein recognition using stationary wavelet packet transform. Bioinformatics 26:i467–i473. https://doi.org/10.1093/bioinformatics/btq371
16. Szalkowski AM, Anisimova M (2013) Graph-based modeling of tandem repeats improves global multiple sequence alignment. Nucleic Acids Res 41:e162–e162. https://doi.org/10.1093/nar/gkt628
17. Schaper E, Kajava AV, Hauser A, Anisimova M (2012) Repeat or not repeat?--Statistical validation of tandem repeat prediction in genomic sequences. Nucleic Acids Res 40:10005–10017. https://doi.org/10.1093/nar/gks726
18. Soding J, Söding J (2005) Protein homology detection by HMM-HMM comparison. Bioinformatics 21:951–960. https://doi.org/10.1093/bioinformatics/bti125
19. Remmert M, Biegert A, Hauser A, Söding J (2011) HHblits: lightning-fast iterative protein sequence searching by HMM-HMM alignment. Nat Methods 9:173–175. https://doi.org/10.1038/nmeth.1818
20. Van Dongen S (2000) A cluster algorithm for graphs. Rep Inf Syst 10:1–40
21. Enright AJ, Van Dongen S, Ouzounis CA (2002) An efficient algorithm for large-scale detection of protein families. Nucleic Acids Res 30:1575–1584
22. Wells JN, Gligoris TG, Nasmyth KA, Marsh JA (2017) Evolution of condensin and cohesin complexes driven by replacement of kite by hawk proteins. Curr Biol 27:R17–R18. https://doi.org/10.1016/j.cub.2016.11.050
23. Eddy SR (1998) Profile hidden Markov models. Bioinformatics 14:755–763
24. Viterbi A (1967) Error bounds for convolutional codes and an asymptotically optimum decoding algorithm. IEEE Trans Inf Theory 13:260–269. https://doi.org/10.1109/TIT.1967.1054010
25. Altschul SF, Gish W, Miller W et al (1990) Basic local alignment search tool. J Mol Biol 215:403–410. https://doi.org/10.1016/S0022-2836(05)80360-2
26. Altschul SF, Madden TL, Schäffer AA et al (1997) Gapped BLAST and PSI-BLAST: a new generation of protein database search programs. Nucleic Acids Res 25:3389–3402
27. Cline MS, Smoot M, Cerami E et al (2007) Integration of biological networks and gene expression data using Cytoscape. Nat Protoc 2:2366–2382. https://doi.org/10.1038/nprot.2007.324
28. Chavali S, Chavali PL, Chalancon G et al (2017) Constraints and consequences of the emergence of amino acid repeats in eukaryotic proteins. Nat Struct Mol Biol 24:765–777. https://doi.org/10.1038/nsmb.3441

Chapter 14

Exploring Enzyme Evolution from Changes in Sequence, Structure, and Function

Jonathan D. Tyzack, Nicholas Furnham, Ian Sillitoe, Christine M. Orengo, and Janet M. Thornton

Abstract

The goal of our research is to increase our understanding of how biology works at the molecular level, with a particular focus on how enzymes evolve their functions through adaptations to generate new specificities and mechanisms. FunTree (Sillitoe and Furnham, Nucleic Acids Res 44:D317–D323, 2016) is a resource that brings together sequence, structure, phylogenetic, and chemical and mechanistic information for 2340 CATH superfamilies (Sillitoe et al., Nucleic Acids Res 43:D376–D381, 2015) (which all contain at least one enzyme) to allow evolution to be investigated within a structurally defined superfamily.

We will give an overview of FunTree's use of sequence and structural alignments to cluster proteins within a superfamily into structurally similar groups (SSGs) and generate phylogenetic trees augmented by ancestral character estimations (ACE). This core information is supplemented with new measures of functional similarity (Rahman et al., Nat Methods 11:171–174, 2014) to compare enzyme reactions based on overall bond changes, reaction centers (the local environment atoms involved in the reaction), and the structural similarities of the metabolites involved in the reaction. These trees are also decorated with taxonomic and Enzyme Commission (EC) code and GO annotations, forming the basis of a comprehensive web interface that can be found at http://www.funtree.info. In this chapter, we will discuss the various analyses and supporting computational tools in more detail, describing the steps required to extract information.

Key words FunTree, Enzyme evolution, CATH, EC-Blast, Phylogenetic tree

1 Introduction

FunTree [1] is a resource for exploring the evolution of protein function through relationships in sequence, structure, phylogeny, and function. It catalogues 2340 CATH superfamilies with over 400,000 representative sequences (selected to cover taxonomic lineage and function), over 70,000 structural domains, and 2358 EC (Enzyme Commission) codes.

FunTree can be used to place structures and sequences in the context of their structural and functional evolution, allowing the investigation of how novel enzyme functions have evolved within a

Tobias Sikosek (ed.), *Computational Methods in Protein Evolution*, Methods in Molecular Biology, vol. 1851, https://doi.org/10.1007/978-1-4939-8736-8_14,

structurally similar group (SSG). This can also be helpful to identify new but currently unobserved reactions and substrates for known enzymes, as well as possible reactions for enzyme sequences/structures of unknown function.

Often CATH [2] superfamilies can be structurally highly diverse, hindering the confident atomic superimposition of all structures in the superfamily. A key step in the generation of FunTree is the sub-clustering of each superfamily into distinct SSGs, where all interstructure SIMAX scores are less than 9 angstroms (where SIMAX is the RMSD between two domains multiplied by the number of residues in the larger domain divided by the number of aligned residues).

The core output of FunTree is a phylogenetic tree for each SSG (discussed in Subheading 2.2.1), calculated from structure-guided sequence alignments using a novel agglomerative clustering technique. The resulting alignments are provided to TreeBest [4], along with a species tree derived from species relationships in the NCBI Taxonomic definitions, to generate a maximum likelihood-based phylogenetic tree.

The phylogenetic trees are decorated with information such as EC code, GO annotations, and multidomain architecture (MDA) and augmented with various ancillary analyses describing the diversity in areas such as enzyme chemistry and taxa distribution. These will all be described in more detail in the Methods section; however it is important to note that some annotations such as GO and EC are assigned to entire gene products rather than the individual structural domains included in the SSGs. Most functions can be ascribed to a single domain, but many are a product of domain combinations or multiple gene products. Thus, as FunTree is a domain-centric resource, some annotations might be relevant at the protein rather than domain level. The FunTree pipeline describing the various steps in collecting, processing, and presenting the data is shown in Fig. 1.

The trees generated in FunTree can become difficult to navigate due to their size and mixed media content. To facilitate easy navigation, a web interface has been constructed using the JavaScript D3 libraries [5] to provide intuitive and user-friendly functionality (e.g., using the mouse wheel to zoom and dragging images to pan) and interaction with the trees (e.g., collapsing and expanding nodes by clicking).

2 Methods

FunTree [1] can be browsed by CATH [2] superfamily or searched by CATH superfamily, UniProtKB accession, and EC code or by entering a text string for a fuzzy search. Overview statistics and

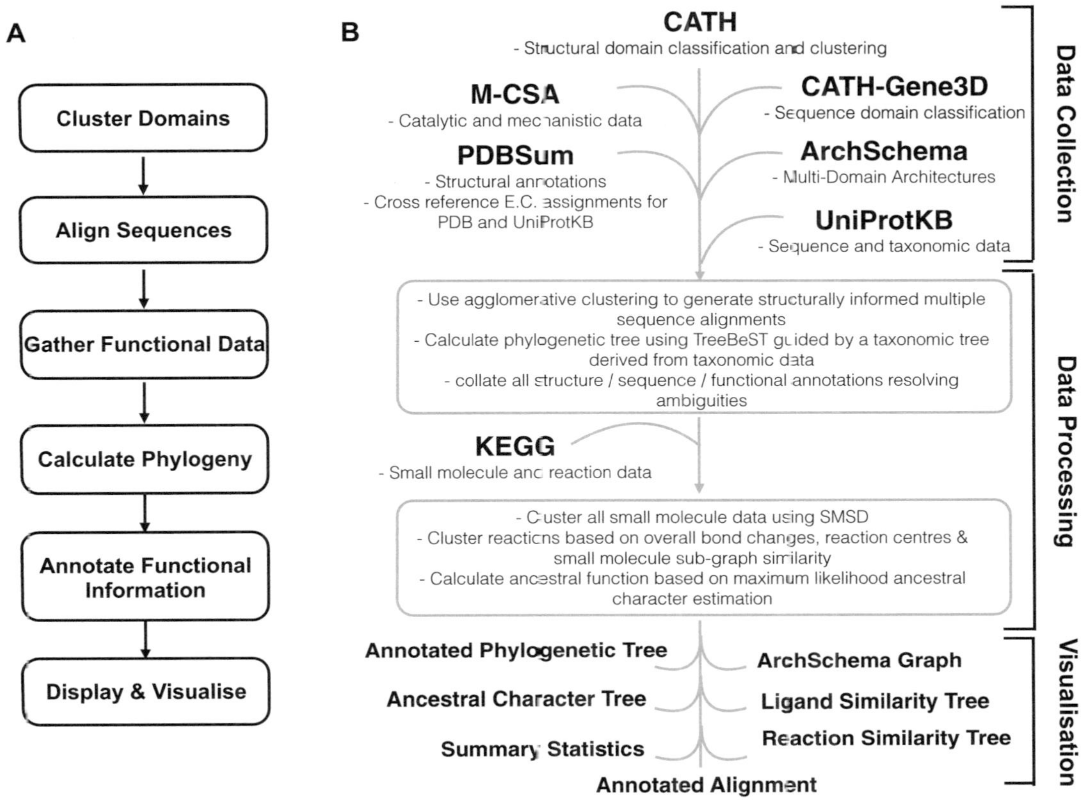

Fig. 1 The FunTree pipeline. (**a**) An overview of the workflow for collecting and processing sequence, structure, and functional information in FunTree. (**b**) A detailed schematic representation of the various steps in data collection, processing, and visualization in FunTree

high-level results are produced at the CATH superfamily level, with the phylogenetic trees and lower level results produced at the structurally similar group (SSG) level. These are discussed in more detail in the remainder of this section.

2.1 CATH Superfamily Results Gateway

This page is the gateway for results at the CATH superfamily level for the selected domain (Fig. 2), where each thumbnail provides a link to a detailed analysis of the selected results. The SSGs within the superfamily are shown in Clusters with a link to lower level results for that SSG.

2.1.1 Domain Architectures

This page shows an interactive force directed graph generated by ArchSchema [6] of the multidomain architectures (MDAs) associated with the current search, with the current domain shown at the center connected to increasingly more complicated architectures.

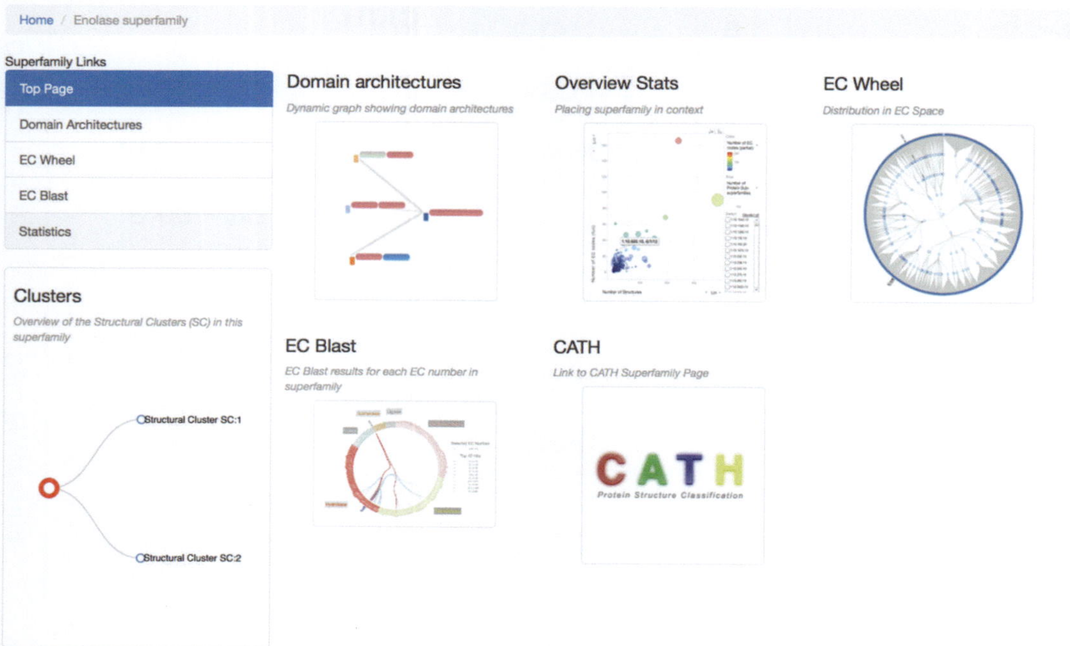

Fig. 2 Superfamily gateway. CATH superfamily results for CATH 3.20.20.120 Enolase. Each thumbnail provides a link to a detailed analysis of the selected results. The SSGs within the superfamily are shown in Clusters with a link to lower level results for that SSG

1. The colored graph nodes represent the different MDAs and can be dragged to reorganize the graph. Hovering over the graph nodes shows the following information for that MDA:
   (a) Number of sequences
   (b) Number of structures
   (c) List of EC codes (annotated by UniProtKB [7])
   (d) List of structures
2. The colored domain bars show the domain composition, where hovering over the bar reveals the domain name and clicking opens the webpage for that CATH superfamily.

2.1.2 *Overview Stats*

This page contains a dynamic, interactive plot allowing various properties of CATH superfamilies to be plotted on two axes (Fig. 3). The different properties that can be plotted on either a linear or log scale and also used to scale and color the data points are as follows:

1. Alphabetical order (x-axis only)
2. Average conservation score for each position in the alignment (ScoreCons) [8] for SSGs

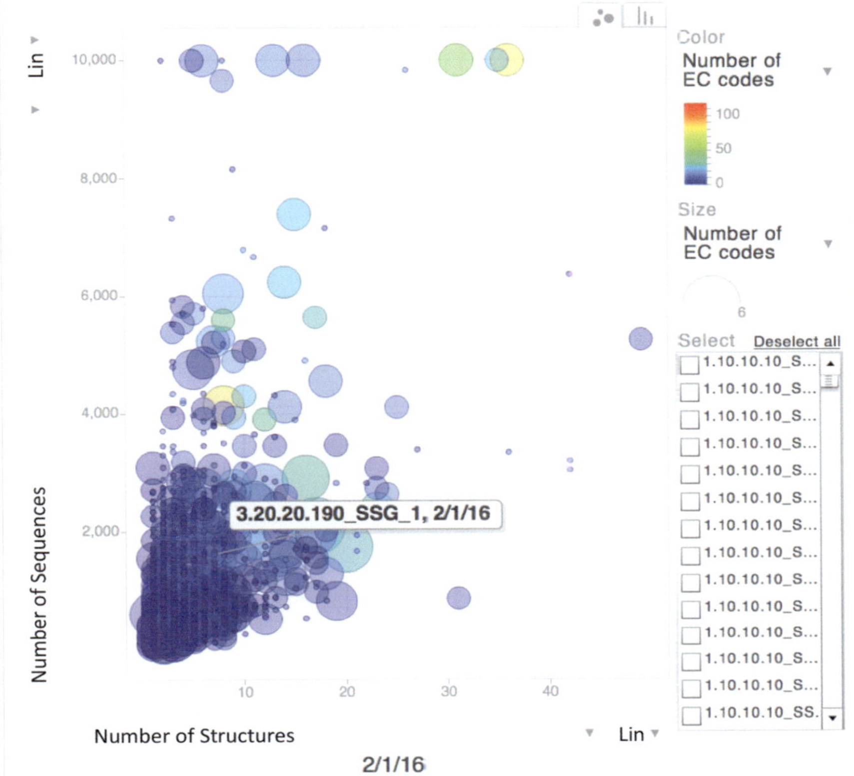

Fig. 3 CATH superfamily Overview Stats. The plot shows the number of sequences on the y-axis against the number of structures on the x-axis with color representing the number of EC codes and size representing the number of partial EC codes

3. Number of multidomain architectures (MDAs)
4. Number of full Enzyme Commission (EC) codes
5. Number of partial EC codes
6. Number of sequences
7. Number of structures
8. Number of structurally similar groups (SSGs)

2.1.3 EC Wheel

This page shows the EC hierarchy as an unrooted tree with EC codes within the superfamily labelled outside the wheel.

Nodes/leaves for class, sub-class, sub-subclass, and numerical identifier are highlighted for the enzymes found in the superfamily.

2.1.4 EC-Blast

This page shows the EC classification rendered as a circular rooted tree.

1. Leaves represent EC code and are colored by primary EC class. EC codes that are found in the superfamily are pushed out of the circle and are colored blue.

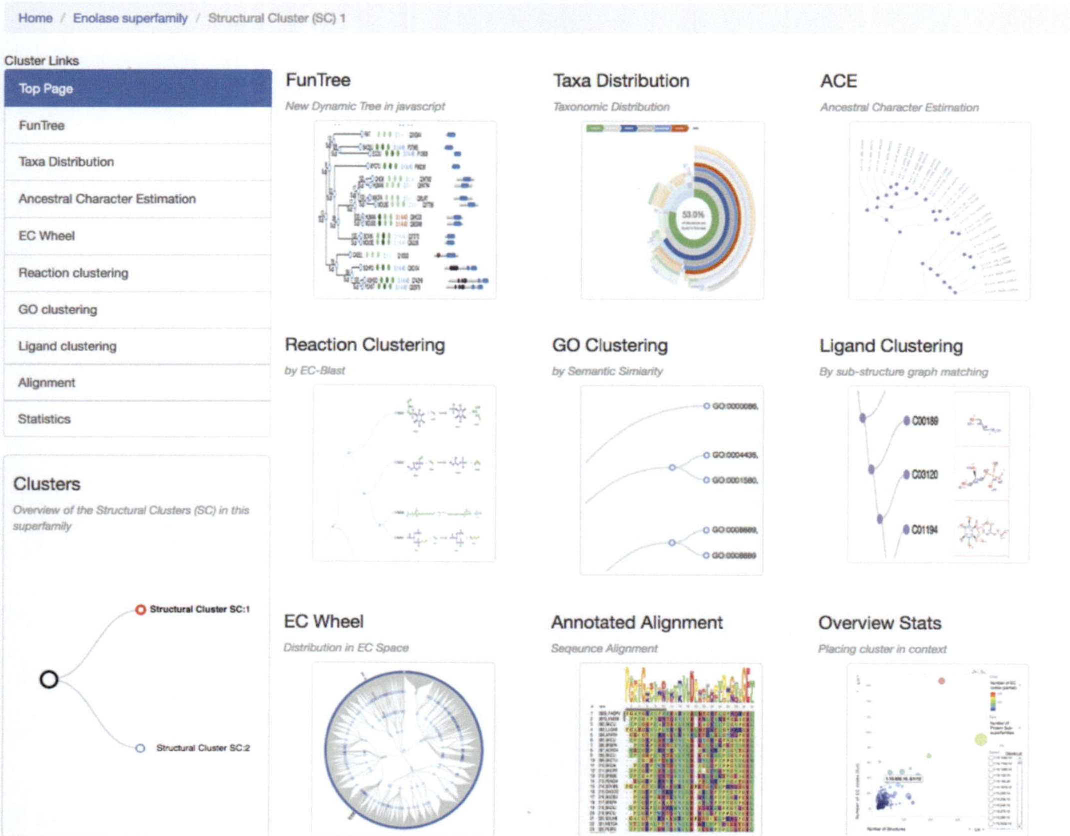

Fig. 4 SSG (structurally similar group) Gateway: SSG results for CATH 3.20.20.120 SSG1. Each thumbnail provides a link to a detailed analysis of the selected results. The SSGs within the superfamily are shown in Clusters with a link to lower level results for that SSG

2. Links between EC codes found in the superfamily and their 10 most similar functions as calculated by EC-Blast [3] are highlighted in blue, tracing the path through the tree between them.
3. Hovering over an EC code highlights in red connections to the top 10 most similar reactions, which are also listed on the right of the page.

2.1.5 CATH

This is a link to the CATH page [2] for that superfamily containing further information on structure and function.

2.2 Structurally Similar Group (SSG) Results Gateway

This page is the gateway for results at the SSG level for the selected domain (Fig. 4). Each thumbnail provides a link to a detailed analysis of the selected results.

2.2.1 FunTree: Rooted Phylogenetic Tree

This page contains a rooted phylogenetic tree for the SSG selected, with annotations and links embedded in the nodes and leaves (Fig. 5).

1. Navigation is implemented using the mouse wheel to zoom, dragging the image to pan, clicking on a node to collapse/expand that node, clicking on text for links to data sources, and hovering over text/images for more information.
2. At each node to the tree, a confidence score can be found. This is the confidence bootstrap score provided by TreeBest for bifurcation at the node. Please note that as these trees are automatically generated, some of the bifurcations might have low confidence scores and should be considered with caution.
3. The annotations at the end of each leaf are as follows:
   (a) The first number/text section is the node name (internal to FunTree) made up of a reference and the taxonomic code.

Fig. 5 Rooted phylogenetic tree for SSG1 in CATH 3.20.20.120 Enolase. Each node contains a score that measures the confidence in the bifurcation. Each leaf contains labels for reaction similarity represented as green circles, EC code/function (where available), UniProtKB sequence, representative PDB domain (where available), and a domain bar representing the multidomain architecture (MDA). *See* Subheading 2.1 for further details

(b) If the leaf represents an enzyme, the next three circles show the similarity between reactions in the EC code on a bond change, reaction center, and sub-structure basis, respectively. Coloring is based on the degree of similarity as calculated by EC-Blast [3].

(c) Primary EC code, containing a hyperlink to the IntEnz database.

(d) UniProtKB identifier, containing a link to the UniProtKB record.

(e) If the sequence represents a known structural superfamily, then the PDB (linked to PDBe entry [9]) and CATH domain (linked to the CATH superfamily page) are shown.

(f) The MDA of the protein at each leaf is depicted showing the domains as uniquely colored bars along a line, the position and length of which are proportional to the total sequence. Hovering over each bar shows the CATH superfamily, and clicking navigates to the CATH superfamily page.

2.2.2 Taxa Distribution

The taxa distribution shows the distribution of taxonomic classes within the SSG tree.

1. Hovering over the band reveals the taxonomic lineage (shown top left) as well as the percentage of sequences in the tree that belong to that group.

2.2.3 Ancestral Character Estimation (ACE) Tree

This is a circular representation of the phylogenetic tree based on SSG alignments showing likelihoods of functions at ancestral nodes.

1. Hovering over a node shows the EC code/function with the maximum likelihood for that node.
2. Hovering over leaves of the tree shows the contribution to function annotation from each internal node in the lineage.

2.2.4 Reaction Clustering

This page shows a tree representing the similarities between reactions based on bond changes calculated by EC-Blast [3], where the clustering is made using the PVClust [10] methods as implemented in *R* (Fig. 6).

1. The tree can be zoomed using the mouse wheel or moved/panned by dragging the image.
2. Each leaf shows a schematic of the reaction with color coding highlighting the atoms that are involved in the reaction.

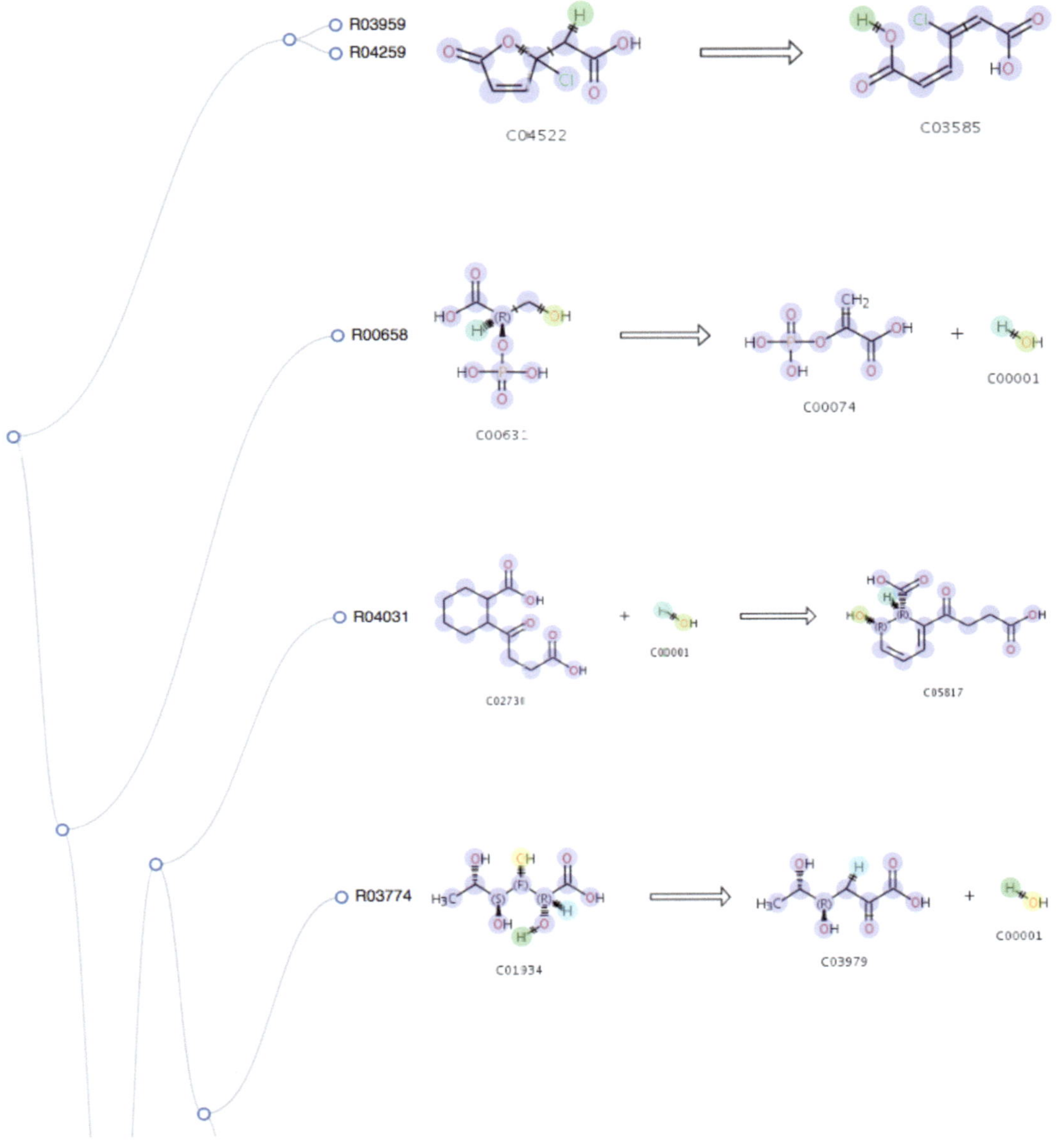

Fig. 6 Reaction Clustering for SSG1 in CATH 3.2.2.120 Enolase. A tree representing the similarities between reactions based on bond changes calculated by EC-Blast, where the clustering is made using the PVClust methods as implemented in R

2.2.5 GO Clustering

This page shows a tree representing the similarities between GO annotations using a semantic similarity score.

1. The tree can be zoomed using the mouse wheel or moved/panned by dragging the image.

2.2.6 Ligand Clustering

This page shows a similarity tree of all the small molecules found in all the reactions in the SSG. The similarities are calculated using SMSD [11], and the clustering is made using the PVClust methods as implemented in *R*.

1. The tree can be zoomed using the mouse wheel or moved/panned by dragging the image.
2. By hovering over the leaves of the tree, the reaction is displayed, and the other ligands in the reaction are highlighted.

2.2.7 EC Wheel

The functionality is as described in Subheading 2.1.3 but for data at the SSG level.

2.2.8 Annotated Alignment

This page shows the multiple sequence alignment generated with the BioJS [12] module (Fig. 7) that was used to build the phylogenetic tree. The sequences in the alignment are annotated by secondary structure where available and catalytic residues as catalogued in the M-CSA [13] (bright red if from the curated M-CSA, light red if from the predicted M-CSA).

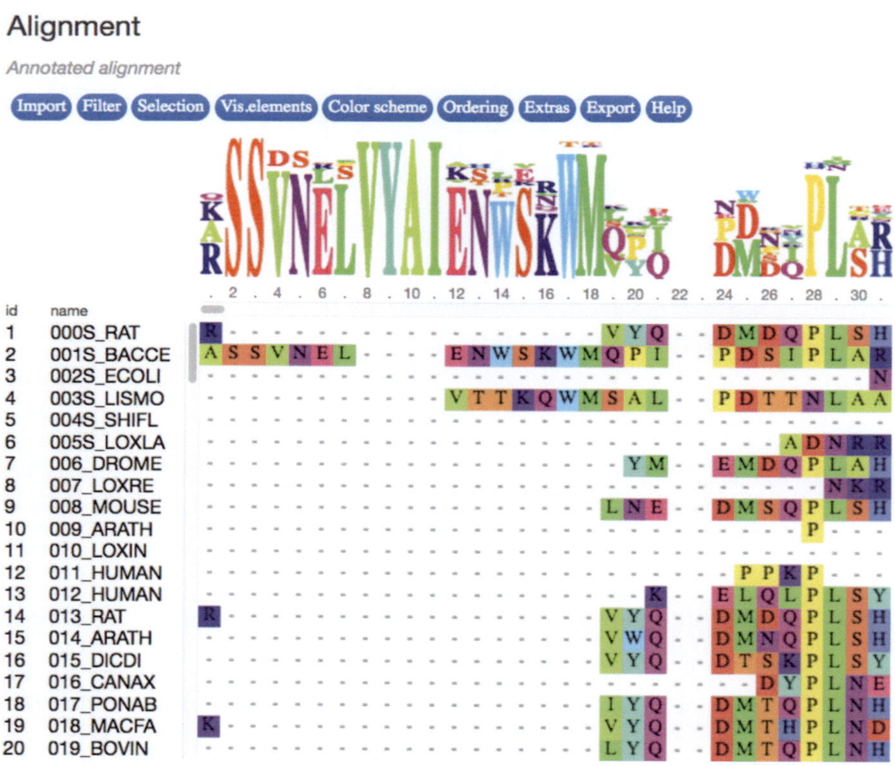

Fig. 7 Annotated alignment for phosphatidylinositol (PI) phosphodiesterase

1. The alignments can be scrolled vertically (to show more sequences) and horizontally (to show different parts of the sequence).
2. The sequences can be selected, ordered, and filtered by the various data fields including sequence identity.
3. Other formatting options include editing the color scheme and hiding/showing visual elements such as labels and headers.
4. There is also functionality to import and export data for external analysis.

2.2.9 Overview Stats

The functionality is as described in Subheading 2.1.2 but for data at the SSG level.

2.3 Examples of the Application of FunTree

As FunTree [1] holds data across many domain superfamilies, it is possible to use FunTree to make large-scale general observations about how enzymes have evolved their function [14]. These observations can be made at the domain and residue level, exploring how function is modulated via the addition/removal of domains within a multi-domain architecture or adaptations of the catalytic/binding pocket. This allows analyses to be prepared comparing the number and types of evolutionary steps observed within domain superfamilies [15].

Furthermore, detailed analysis within a single superfamily or for a specific enzyme can be undertaken. An example of this is the evolution of functionality within the phosphatidylinositol-phosphodiesterase superfamily (CATH 3.20.20.190), which is summarized briefly here but discussed more comprehensively in reference [16]. This superfamily shows relatively high structural conservation, presenting just one structurally similar group, but the phylogenetic tree generated within FunTree reveals three clades (*see* Fig. 8). Clades C1 and C3 show hydrolase activity (EC: 3.1.4) using a metal cofactor, whereas Clade 2 exhibits a transition to lyase activity (EC: 4.6.1). The structure-informed sequence alignment reveals that none of the three metal-chelating residues are conserved in Clade 2, so that a metal is no longer bound, resulting in the cyclic intermediate leaving the active site prior to hydrolysis and giving the change from hydrolase to lyase functionality. The mechanistic changes that give rise to this change in functionality can be explored further using the Mechanism and Catalytic Site Atlas (M-CSA [13], formerly called MACiE [17] and CSA [18]).

FunTree is an important resource providing a comprehensive analysis of the evolution of enzyme functionality within structurally similar subdivisions of CATH superfamilies. Not only will this improve our understanding of the link between enzyme structure and function but, coupled with FunTree's various supporting analyses such as structural alignments and measures of molecular

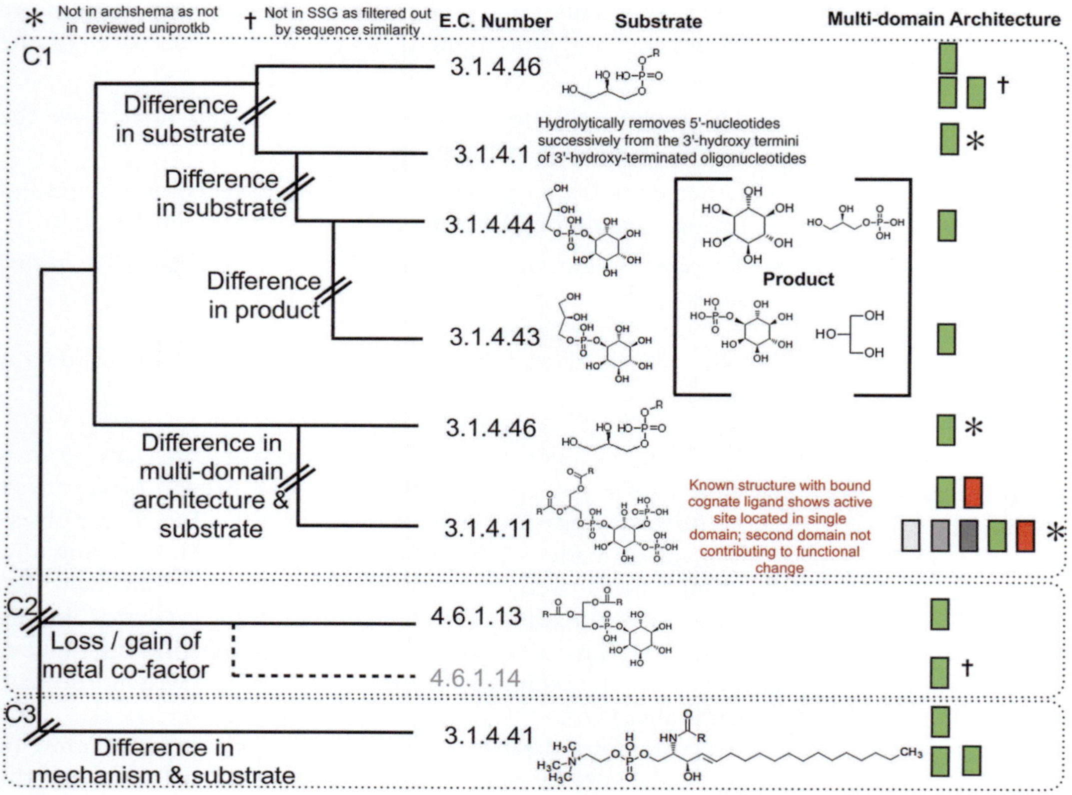

Fig. 8 Summary of phylogenetic, functional, metabolite, and multidomain architectures for the phosphatidylinositol-phosphodiesterase superfamily (3.20.20.190) [16]. This shows a diagrammatic representation of the FunTree phylogenetic tree with associated functional data and multidomain architectures. Domain 3.20.20.190 performs all molecular functionality and is represented in green in the multidomain architecture analysis. Three major clades (C1–C3) are highlighted. Within the first group, a number of functional sub-groups can be observed, with differences in function defined by changes in substrate or product formed

similarity, offers potential to inform de novo enzyme design, annotate sequences/structures of unknown function, and propose novel indications for drugs.

References

1. Sillitoe I, Furnham N (2016) FunTree: advances in a resource for exploring and contextualising protein function evolution. Nucleic Acids Res 44:D317–D323. https://doi.org/10.1093/nar/gkv1274
2. Sillitoe I, Lewis TE, Cuff A et al (2015) CATH: comprehensive structural and functional annotations for genome sequences. Nucleic Acids Res 43:D376–D381. https://doi.org/10.1093/nar/gku947
3. Rahman SA, Cuesta SM, Furnham N et al (2014) EC-BLAST: a tool to automatically search and compare enzyme reactions. Nat Methods 11:171–174. https://doi.org/10.1038/nmeth.2803
4. Ruan J, Li H, Chen Z et al (2007) TreeFam: 2008 update. Nucleic Acids Res 36:D735–D740. https://doi.org/10.1093/nar/gkm1005
5. Bostock M (2017) https://d3js.org

6. Tamuri AU, Laskowski RA (2010) ArchSchema: a tool for interactive graphing of related Pfam domain architectures. Bioinformatics 26:1260–1261. https://doi.org/10.1093/bioinformatics/btq119
7. Uniprot Consortium (2009) The universal protein resource (UniProt) 2009. Nucleic Acids Res 37:D169–D174. https://doi.org/10.1093/nar/gkn664
8. Valdar WSJ (2002) Scoring residue conservation. Proteins Struct Funct Genet 48:227–241. https://doi.org/10.1002/prot.10146
9. Gutmanas A, Alhroub Y, Battle GM et al (2014) PDBe: Protein Data Bank in Europe. Nucleic Acids Res 42:D285–D291. https://doi.org/10.1093/nar/gkt1180
10. Suzuki R, Shimodaira H (2006) Pvclust: an R package for assessing the uncertainty in hierarchical clustering. Bioinformatics 22:1540–1542. https://doi.org/10.1093/bioinformatics/btl117
11. Rahman S, Bashton M, Holliday GL et al (2009) Small Molecule Subgraph Detector (SMSD) toolkit. J Cheminform 1:12. https://doi.org/10.1186/1758-2946-1-12
12. Yachdav G, Goldberg T, Wilzbach S et al (2015) Anatomy of BioJS, an open source community for the life sciences. elife 4:e07009. https://doi.org/10.7554/eLife.07009
13. Ribeiro AJM, Holliday GL, Furnham N et al (2018) Mechanism and Catalytic Site Atlas (M-CSA): a database of enzyme reaction mechanisms and active sites. Nucleic Acids Res 46(D1):D618–D623
14. Furnham N, Dawson NL, Rahman SA et al (2016) Large-scale analysis exploring evolution of catalytic machineries and mechanisms in enzyme superfamilies. J Mol Biol 428:253–267. https://doi.org/10.1016/j.jmb.2015.11.010
15. Tyzack JD, Furnham N, Sillitoe I et al (2017) Understanding enzyme function evolution from a computational perspective. Curr Opin Struct Biol 47:131–139. https://doi.org/10.1016/j.sbi.2017.08.003
16. Furnham N, Sillitoe I, Holliday GL et al (2012) Exploring the evolution of novel enzyme functions within structurally defined protein superfamilies. PLoS Comput Biol 8:e1002403. https://doi.org/10.1371/journal.pcbi.1002403
17. Holliday GL, Bartlett GJ, Almonacid DE et al (2005) MACiE: a database of enzyme reaction mechanisms. Bioinformatics 21:4315–4316. https://doi.org/10.1093/bioinformatics/bti693
18. Furnham N, Holliday GL, de Beer TAP et al (2014) The catalytic site atlas 2.0: cataloging catalytic sites and residues identified in enzymes. Nucleic Acids Res 42:D485–D489. https://doi.org/10.1093/nar/gkt1243

Chapter 15

Identification of Protein Homologs and Domain Boundaries by Iterative Sequence Alignment

Dustin Schaeffer and Nick V. Grishin

Abstract

Evolutionary domains are protein regions with observable sequence similarity to other known domains. Here we describe how to use common sequence and profile alignment algorithms (i.e., BLAST, HHsearch) to delineate putative domains in novel protein sequences, given a reference library of protein domains. In this case, we use our database of evolutionary domains (ECOD) as a reference, but other domain sequence libraries could be used (e.g., SCOP, CATH). We describe our domain partition algorithm along with specific notes on how to avoid domain indexing errors when working with multiple data sources and software algorithms with differing outputs.

Key words Protein domains, Homologs, Sequence alignment

1 Introduction

Protein domains are regions sharing a common origin (and sometimes function) that may be observed shuffled among homologous proteins by recombination, deterioration, fusion, and other evolutionary events [1, 2]. Clear identification of protein domains aids in the identification of novel homologs and can give insight to protein function [3, 4]. The detection of homologous proteins and the delineation of their domain boundaries can be complicated by multidomain proteins in the search space [5]. Multidomain proteins in the domain search space can lead to nonhomologous domains being classified as homologous, and potential novel insertions or deletions may be missed. Discontinuous domains caused by insertion may also complicate partition and assignment of boundaries [6]. By careful selection of a reference database of domains, combined with use of publically available sequence alignment software, the detection of similarity can be used to determine domain boundaries in a set of input proteins [7]. Here we demonstrate how the Basic Local Alignment Search Tool (BLAST) can be combined with profile-profile detection (HHsearch) to determine

Tobias Sikosek (ed.), *Computational Methods in Protein Evolution*, Methods in Molecular Biology, vol. 1851,
https://doi.org/10.1007/978-1-4939-8736-8_15,

domain boundaries in an input protein [8–10]. Additionally, we show how a database of proteins with known domain architectures can be used to efficiently partition multidomain proteins. For our domain database, we use our Evolutionary Classification Of protein Domains (ECOD), but this method could be adapted to work with other common domain databases [11].

2 Materials

2.1 Software

1. BLAST+ version 2.2.25+ or greater (ftp://ftp.ncbi.nlm.nih.gov/blast/executables/blast+/LATEST/).
2. HHsuite v2.0.15 or greater (https://github.com/soedinglab/hh-suite).
3. PSIPRED (http://bioinfadmin.cs.ucl.ac.uk/downloads/psipred/).

2.2 Databases

1. ECOD domain description flat file (http://prodata.swmed.edu/ecod/distributions/ecod.latest.domains.txt).
2. Local PDB with PDBml or mmCIF no-atom headers (ftp://ftp.wwpdb.org/pub/pdb/data/structures/divided).
3. Non-redundant sequence set for generation of reference profiles (http://wwwuser.gwdg.de/~compbiol/data/hhsuite/databases/hhsuite_dbs/).

3 Methods

3.1 Preparation of Domain and Protein Sequence Databases

Given a set of domain sequence ranges, generate a set of domain sequences in FASTA formats. In this case we will translate modified or unnatural residues to unknown residues, although it is possible in some cases to identify parent amino acids and translate accordingly (*see* **Note 1**). If prepared sequence databases are used, this step can be skipped. This protocol assumes basic scripting knowledge (e.g., Python, Perl, or Ruby) and the ability to parse and write structured data formats (e.g., XML, mmCIF) [12]. We will use a sample workspace as illustrated in Fig. 1; directory structures can clearly be adapted for individual computing needs and infrastructures. The overall workflow of this domain partition is illustrated in Fig. 2. The workflow presented here assumes the sequence inputs are sourced from PDB structure depositions.

1. Download the PDBml no-atom headers and place into workspace (*see* **Note 2**):
   (a) */data/pdb*
2. Download the ECOD domain description file and place into workspace:
   (b) */work/domain_search*

External sequence and structure databases

```
/data/pdb/
  ./data/structures/divided/XML-noatom/*
  ./data/structures/divided/mmCIF/*
/data/sequences/
  ./uniref/UniRef30.fasta
```

Reference database sequences and definitions

```
/work/domain_search/
  ./ecod.latest.domains.txt
  ./domain_ref_seq.fasta
  ./protein_ref_seq.fasta
  ./F40_domain_ref_seq.hhm
```

Software

```
/programs/blast/
  ./bin/makeblastdb
  ./bin/blastp
/programs/hhsuite/
  ./bin/hhblits
  ./bin/hhsearch
/programs/psipred/
```

Individual protein search results

```
/work/domain_sequences/
  ./PDB_WEEK/
    ./1foo_A/
      ./1foo_A.fa
      ./1foo_A.hhm
      ./1foo_A.domain_blast.xml
      ./1foo_A.protein_blast.xml
      ./1foo_A.hh_result
      ./1foo_A.hh_result.xml
      ./1foo_A.summary.xml
      ./1foo_A.domains.xml
```

Fig. 1 A sample workspace for domain partition. We delineate directories for storage of external databases (top left), the reference domain database against which we search (top right), necessary downloadable software programs (bottom left), and the contents of the search directory for a chain A found in the PDB deposition 5XCT

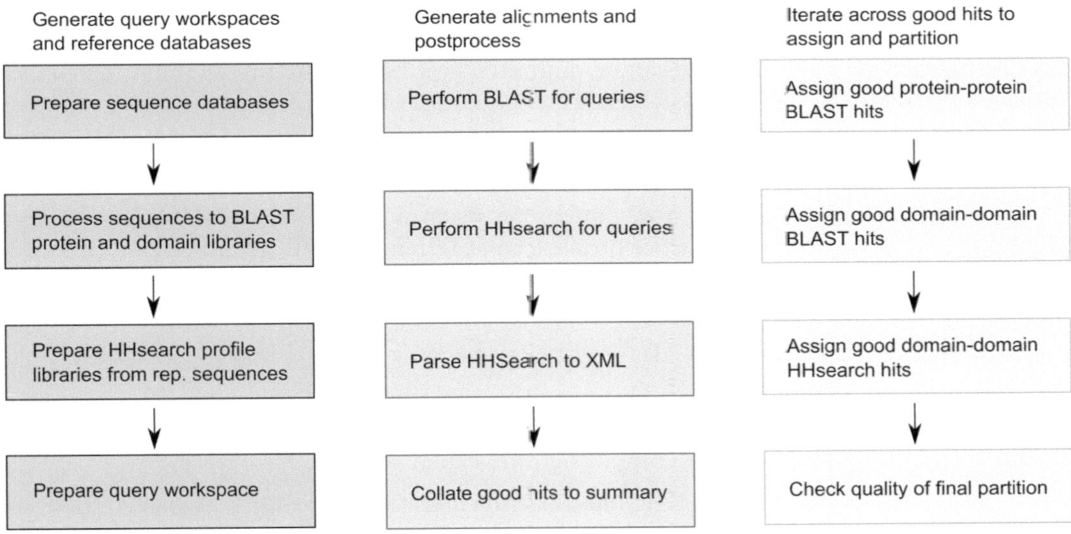

Fig. 2 Workflow for domain partition by iterative sequence alignment. Briefly, the workflow can be split into three large components. The search databases and the query workspace are prepared based on your domain definitions and external protein database (left). Alignments are generated for the query proteins against the reference databases, and the subsequent alignments are post-processed into structured data files containing only well-covered hits (middle). Well-covered hits are used to iteratively assign and partition domain boundaries using protein-protein sequence hits with the highest precedence and domain-domain profile hits with the lowest (right)

3. Using domain ranges (PDB seqid indices) and the PDBml no-atom XML files, prepare a set of domain FASTA sequences and place into a single text file (*see* **Note 3**):

   (c) */work/domain_search/domain_ref_seq.fasta*

4. For each PDB protein sequence containing domains in the ECOD file, using the structured residues with small (<20 residue) unstructured gaps included, prepare a set of protein FASTA sequences and place into a single flat file (*see* **Note 4**):

   (d) */work/domain_search/protein_ref_seq.fasta*.

3.2 Preparation of BLAST Query Sequence Databases

1. Generate a BLAST query database given an input FASTA library of domain sequences:

   (a) */programs/blast/bin/makeblastdb -in /work/domain_search/domain_ref_seq.fasta -out domain_ref_seq -title domain_ref_seq*

2. Generate a BLAST query database given an input FASTA library of protein sequences:

   (b) */programs/blast/bin/makeblastdb -in /work/domain_search/protein_ref_seq.fasta -out protein_ref_seq -title protein_ref_seq*

3.3 Preparation of HHsearch Reference Profile Database

This step can be omitted if you have chosen to use a pre-generated profile database. Select a subset of your original sequence database that is more sparsely populated. We will use the ECOD F40 representatives in our example (*see* **Note 4**).

1. For each reference domain sequence, generate a reference sequence profile using HHblits queried against a non-redundant protein sequence database (e.g., UniRef30, nr, RefSeq, etc.):

   (a) Use PSIPRED secondary structure prediction to aid with HHsearch alignments.

   (b) HHblits can be allocated to use multiple CPUs using the –cpu switch; select a value that is appropriate for your local computing infrastructure.

   (c) We find that three iterations are sufficient (−n 3) to locate close sequence homologs.

   (d) */programs/hhsuite/bin/hhblits -i /data/domain_data/e1mppA2/ef1ooA1.fasta -d /data/hhsuite/lib/UniRef30.fa -ohhm /data/domain_data/e1mppA2/e1mppA2.hhm -n 3 -cpu 8 -addss -psipred /programs/psipred -psipred_data /data/psipred*

2. Concatenate the set of domain profiles (HHM files) into a single file:

   (a) *cat /data/domain_data/*/*.hhm > /work/domain_search/domain_ref_seq_40.hhm*

3.4 Prepare Query Workspace

1. Download or prepare the set of query sequences. We will demonstrate with a recent week of PDB depositions (20170929) with a single protein (5xct_B). If a set of query sequences is highly redundant, it is appropriate to cluster the set using CD-HIT or blastclust to reduce the size of the search set [13].
2. Create a subdirectory for each query sequence:

   (a) */work/domain_sequences/20170929/5xct_B*
3. Distribute a FASTA file for each query sequence into each subdirectory:

   (b) */work/domain_sequences/20170929/5xct_B/5xct_B.fa*
4. Create a sequence profile for each query sequence using HHblits:

   (c) */work/domain_sequences/20170929/5xct_B/5xct_B.hhm*

3.5 Performing BLAST and HHsearch Queries

1. For each query FASTA sequence, perform a BLAST search against both reference sequence BLAST databases (protein and domain) using the XML output format (−outfmt 5):

   (a) */programs/blast/bin/blastp -query /work/domain_sequences/20170929/5xct_B/5xct_B.fa -db /work/domain_search/domain_ref_seq.fa -outfmt 5 -num_alignments 5000 -evalue 0.002 > /work/domain_sequences/20170929/5xct_B/5xct_B. protein_blast.xml*

   (b) */programs/blast/bin/blastp -query /work/domain_sequences/20170929/5xct_B/5xct_B.fa -db /work/domain_search/protein_ref_seq.fa -outfmt 5 -num_alignments 5000 -evalue 0.002 > /work/domain_sequences/20170929/5xct_B/5xct_B. domain_blast.xml*
2. For each query HMM, perform an HHsearch against the reference sequence profile database:

   (a) */programs/hhsuite/bin/hhsearch -i /work/domain_sequences/20170929/5xct_B/5xct_B.fa -db /work/domain_search/domain_ref_seq_40.hhm -o /work/domain_search/20170929/5xct_B/5xct_B.hh_result -cpu 8*

3.6 Collate HHsearch Output to Structured Data Format

To better work with HHsearch results, it is convenient to parse them to a structured data format, so that inconsistencies and errors in batch jobs can be identified early in the process. It is possible but not recommended to work directly from the standard HHsearch result format in interpretation of results.

1. Locate the completed HHsearch outputs from the query sequence workspace.
2. Record the hit number, HH probability of homology (%), and HH *E*-value from the HH summary result block.
3. Using the original domain reference ranges, index aligned positions in hit alignments to original domain residues indices (*see* **Note 3**).
4. Convert alignments into ranges of aligned residue indices from both the query sequence and the reference sequence.
5. Calculate the residue coverage of the reference alignment over the reference domain sequence.
6. For hits with more than 70% of the reference domain aligned to the query, deposit the query aligned range, the reference aligned range, the HH probability of homology, the HH *E*-value, and the coverage of the template sequence into a structured data format in the query workspace directory.

3.7 Collate Protein BLAST Hits

In the previous step, we chose an XML format for BLAST output. Data locations for BLAST results are presented as an XPath statement.

1. For each protein BLAST query result, record the following:
   (a) Database used (**//BlastOutput/BlastOutput-db**)
   (b) Query submitted (**//BlastOutput/BlastOutput-query_def**)
   (c) Query length (**//BlastOutput/BlastOutput-query_len**)
2. Iterate over the protein BLAST hits and their high-scoring segment pairs (HSPs) and determine whether the hit is of sufficient quality for further consideration.
3. For each hit, record the hit number (**Hit/hit_num**), the hit length (**Hit/hit_len**), and the hit definition (**Hit/hit_def**). For protein queries conducted against a protein reference database containing a set of PDB chains, the hit definition is a four-character PDB identifier and a chain identifier of up to four characters.
4. For each high-scoring segment pair (**Hit-hsps/Hsp**), record the hsp *E*-value (**Hsp/Hsp-evalue**) and generate the aligned range for the query sequence (**Hsp-query_from .. Hssp-query_to**) as well as the aligned range for the reference sequence (**Hsp-hit_from .. Hsp-hit_to**).

5. Considering HSPs from lowest to highest *E*-value, if any of the query aligned range overlaps more than five residues with a previous HSP, the reference aligned range overlaps more than ten residues with a previous HS, or if the HSP exceeds an *E*-value threshold of 5e−3, discard it. Otherwise, aggregate the retained HSP query and reference aligned ranges for the hit.
6. If neither the respective aggregate query aligned range nor reference aligned range differs from the total query length or total reference length by more than 50 residues, retain the protein-protein hit and the query and reference aligned ranges for further consideration.

3.8 Collate Domain BLAST Hits

Collation of domain BLAST hits is similar to that of protein BLAST, with some small modifications to tighten constraints on hsp overlaps arising from discontinuous domains.

1. As in protein BLAST, record the database used, the query submitted, and the query length.
2. For each domain hit, record the reference domain identifier (**Hit/hit_def**), the hit length (**Hit/hit_len**), and the hit number (**Hit/hit_num**).
3. For each HSP in a hit, allow no more than five residues overlap between query aligned residues and no more than ten residues overlap between reference aligned residues. HSPs must have an *E*-value lower than 2e−3 to be accepted. The total coverage of aligned reference residues over the reference sequence must be 70% or greater for the hit to be accepted.
4. Collect the protein BLAST, domain BLAST, and domain HHsearch results into a single structured data format, where each method contains a list of hits with the respective query aligned range, reference aligned range, reference aligned coverage, and quality score (*E*-value for BLAST, HH probability of homology for HHsearch) associated with each hit.

3.9 Domain Partition by Iteration Over Alignments

Given a set of well-scoring hits to protein sequences, domain sequences, and domain profiles, we are prepared to partition the query sequence into domains.

1. For a query protein, process alignments in the following order: protein sequence hits, domain sequence hits, and domain profile hits. If either less than 5% of the query sequence is unassigned or less than ten residues remain unassigned, the partition is complete.
2. For each protein sequence alignment, if at least 90% of the query aligned residues have not been assigned and less than 5% of the query aligned residues have not been assigned to a previous putative domain, then define domains based on this protein sequence alignment.

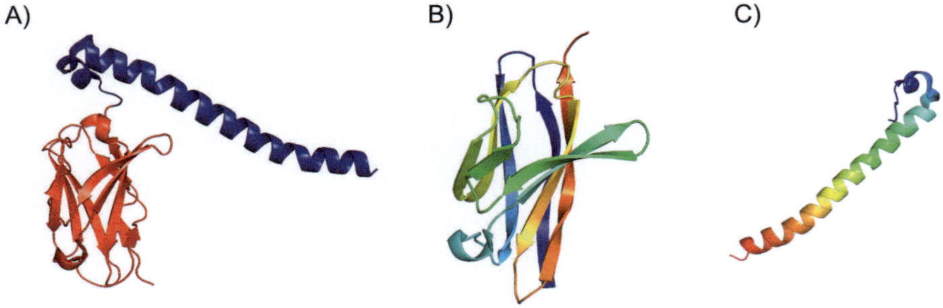

Fig. 3 An example domain partition using multiple aligner. A domain partition using iterative sequence alignment of a fusion structure of Fv and MST1 coiled coil (5xct_B). This novel domain architecture (**a**) was partitioned into its components by hits against a Fv Ig beta-sandwich (**b**) domain(1mfa_L) and a coiled-coil domain (**c**) from MST1 kinase (2jo8_B)

3. To define domains from a protein-protein alignment, consider the sequence ranges from the domains which comprise the reference protein. Generate a 1:1 mapping between aligned positions, omitting those positions in the reference protein which align to a gap. Using this mapping, generate domain ranges for the query protein sequence based on the domains of the reference proteins. Discard any domains which are shorter than the global gap tolerance (20 residues).
4. For each domain sequence alignment, if at least 90% of the query aligned residues have not been assigned and less than 5% of the query aligned residues have not been assigned to a previous putative domain, then define a domain based on this domain sequence alignment.
5. To define domains from a domain-domain alignment, simply assign the query aligned residues to the reference domain as a putative domain. Retain the family identification of the reference domain; this is indicative of the homologous link between the putative domain and the reference domain.
6. For each domain profile alignment, if at least 90% of the query aligned residues have not been assigned, less than 20 of the query aligned residues overlap with a previous putative domain, the HH probability of homology is greater than 90%, and if the alignment itself is longer than 20 residues, then use the aligned residues to define a putative domain.
7. For each putative domain, close small gaps (length < 20 residues) in the putative aligned domains internal to the defined range.
8. Check partition completeness by examining total coverage of the query sequence (<20 residues is considered good) or the length of the longest uncovered segment. *See* Fig. 3 for a sample domain partition result.

4 Notes

1. Depositions in the Protein Databank often contain modified or unnatural amino acids. In some cases, such as selenomethionine or alkylated lysines, modifications reflect a change from a natural amino acid to more aid with purification or structural determination. Where a parent compound has been recorded for a modified amino acid, it is possible to revert an amino acid to its natural form. Individual modified amino acids are recorded in the *PDBx:pdbx_struct_mod_residue* datablock in the PDB. Residues in this datablock are indexed by both PDB seqid (*PDBx:label_seq_id*) and the more commonly used PDB residue number (*PDBx:auth_seq_id*). The parent compound of the modified residue is recorded in the *PDBx:parent_comp_id* record.
2. A common problem using protein structures is that many resources still use the 80-column PDB legacy format. Both the mmCIF and PDBml formats offer richer data in a more easily parsed format [12]. Additionally, many recently released EM structures are simply too large to be described by the legacy format. There are advantages and disadvantages to choosing either. Although mmCIF is far more compact (due to a less verbose loop data structure), there are less off-the-shelf parsing packages available for commonly used scripting and programming languages. XML parsing tools for PDBml files are commonly available in many languages, but XML is extremely verbose, and care must be taken to avoid ruinous performance traps when using these files on the large scale. One easy option is to use the "no-atom" PDBml files for those analyses using only structure metadata (such as sequence).
3. Users of the PDB 80-column legacy format are accustomed to sequence data in PDB depositions being organized by PDB chain identifier and PDB residue number. PDB residue number has many properties which make it unsuitable for use as a domain index: it is not necessarily sequential, may or may not incorporate insertion codes, and in rare cases is not unique within a chain. Internally, both mmCIF and PDBml representations of structures use a sequence index (seqid) to represent the position of a residue within a polymer, an entity index (entity_id) to indicate a specific chemical compound with a deposition, and an asymmetric index (asym_id) to represent a specific instance of a chemical compound in 3D space within the asymmetric unit. The PDBx:pdbx_poly_seq_scheme datablock provides a single location wherein each of these concepts is collectively represented.
4. When we refer to structured residue, we refer to those residues that are structurally resolved (i.e., present in ATOM records).

In addition, we remove small gaps (<20 residues) that are unresolved in sequence. We choose this length because, in general, domains of shorter than 20 residues are not observed. This structured residue model is thus a subset of the residues sequence described. In the subsequent workflow, it is important to accurately resolve positions in subsequent alignments to this structured residue model, in order to avoid indexing errors.

Acknowledgments

This work was supported in part by the National Institutes of Health (GM094575 to NVG) and the Welch Foundation (I-1505 to NVG).

References

1. Soding J, Lupas AN (2003) More than the sum of their parts: on the evolution of proteins from peptides. BioEssays 25(9):837–846
2. Leipe DD, Aravind L, Grishin NV, Koonin EV (2000) The bacterial replicative helicase DnaB evolved from a RecA duplication. Genome Res 10(1):5–16
3. Tyzack JD, Furnham N, Sillitoe I, Orengo CM, Thornton JM (2017) Understanding enzyme function evolution from a computational perspective. Curr Opin Struct Biol 47 (Suppl C):131–139. https://doi.org/10.1016/j.sbi.2017.08.003
4. Cheng H, Schaeffer RD, Liao Y, Kinch LN, Pei J, Shi S, Kim BH, Grishin NV (2014) ECOD: an evolutionary classification of protein domains. PLoS Comput Biol 10(12): e1003926. https://doi.org/10.1371/journal.pcbi.1003926
5. Song N, Sedgewick RD, Durand D (2007) Domain architecture comparison for multidomain homology identification. J Comput Biol 14(4):496–516. https://doi.org/10.1089/cmb.2007.A009
6. Holland TA, Veretnik S, Shindyalov IN, Bourne PE (2006) Partitioning protein structures into domains: why is it so difficult? J Mol Biol 361(3):562–590. https://doi.org/10.1016/j.jmb.2006.05.060
7. Andreeva A, Howorth D, Chandonia JM, Brenner SE, Hubbard TJ, Chothia C, Murzin AG (2008) Data growth and its impact on the SCOP database: new developments. Nucleic Acids Res 36(Database issue):D419–D425
8. Altschul SF, Madden TL, Schaffer AA, Zhang J, Zhang Z, Miller W, Lipman DJ (1997) Gapped BLAST and PSI-BLAST: a new generation of protein database search programs. Nucleic Acids Res 25(17):3389–3402
9. Soding J (2005) Protein homology detection by HMM-HMM comparison. Bioinformatics 21(7):951–960. https://doi.org/10.1093/bioinformatics/bti125
10. Remmert M, Biegert A, Hauser A, Soding J (2011) HHblits: lightning-fast iterative protein sequence searching by HMM-HMM alignment. Nat Methods 9:173. https://doi.org/10.1038/nmeth.1818
11. Cheng H, Liao Y, Schaeffer RD, Grishin NV (2015) Manual classification strategies in the ECOD database. Proteins 83(7):1238–1251. https://doi.org/10.1002/prot.24818
12. Westbrook J, Ito N, Nakamura H, Henrick K, Berman HM (2005) PDBML: the representation of archival macromolecular structure data in XML. Bioinformatics 21(7):988–992. https://doi.org/10.1093/bioinformatics/bti082
13. Fu L, Niu B, Zhu Z, Wu S, Li W (2012) CD-HIT: accelerated for clustering the next-generation sequencing data. Bioinformatics 28 (23):3150–3152. https://doi.org/10.1093/bioinformatics/bts565

Chapter 16

A Roadmap to Domain Based Proteomics

Carsten Kemena and Erich Bornberg-Bauer

Abstract

Protein domains are reusable segments of proteins and play an important role in protein evolution. By combining the elements from a relatively small set of domains into unique arrangements, a large number of distinct proteins can be generated. Since domains often have specific functions, changes in their arrangement usually affect the overall protein function. Furthermore, domains are well amenable to computational representations, e.g., by Hidden Markov Models (HMMs), and these HMMs are widely represented in various databases. Therefore, domains can be efficiently used for proteomic analyses. Here, we describe how domains are annotated using different domain databases and then how to assess the annotation quality of proteomes. We next show how functional annotations of domains in large-scale data such as whole genomes or transcriptomes can be used to analyze molecular differences between species. Furthermore, we describe methods to analyze the changes in domain content of proteins which significantly helps to characterize and reconstruct the modular evolution of proteins. Altogether, domain-based methods offer a computationally highly effective approach to analyze large amounts of proteomic data in an evolutionary setting.

Key words Protein domain, Molecular evolution

1 Introduction

Domains are modular building blocks of proteins. Protein domains have a conserved sequence, often describe a specific structure or function and, since they occur in different proteins and frequently in changing combination with other domains, they represent independent units of evolution [1, 2]. Furthermore, the alternative combinations of domains in a single sequence can generate new proteins with varying functions [1, 3]. Therefore, a relatively small number of domains can generate a much higher number of distinct proteins [4] and allow for a fast adaptation to changing conditions and the generation of new functions without the need to completely generate a new protein from scratch. Additionally, domains are much more conserved when compared to the surrounding linker regions and are therefore traceable even in sequences that are only very distantly related where, e.g., BLAST is not able to find matches anymore [5].

Tobias Sikosek (ed.), *Computational Methods in Protein Evolution*, Methods in Molecular Biology, vol. 1851, https://doi.org/10.1007/978-1-4939-8736-8_16,

Changes on the genome/gene level alter domain arrangements during evolution and these changes can be tracked, e.g., using DomRates to determine the evolutionary origin of a domain arrangement. DomRates is based on an algorithm previously described by Moore et al. [6]. The most common way to generate new protein domain arrangements is by fusion of two existing domain arrangements or the fission of a single domain arrangement into two separate ones. Another common way is to lose a domain at either end of the protein [7, 8].

Many domains have a known function. Gene Ontology [9] assigns formalized terms to them and thereby allows to assign possible functions to unknown sequences by identifying the domains in a sequence. Changes in the function of a protein are often reflected in the domains the protein contains. Therefore, studying the changes of domain arrangements and thereby the function of a protein is of biological importance.

Domains are identified by the maintainers of databases as Pfam and Superfamily either based on structural or sequence similarity. These databases usually store the domains in forms of HMMs, i.e., profile-based methods which have a high sensitivity *and* selectivity as they are trained on a data set of known family members. However, some databases (e.g., PROSITE [10]) use string patterns (regular expressions) to store a domain. The latter approach is much faster but only allows to find very conserved motifs and can only give a binary classification.

The databases can then be used to identify domain instances in a sequence that is scanned for the occurrences of a domain and thereby, for example, gain functional insights as described above.

This approach of course only allows to identify known domains. Other approaches like the analysis of hydrophobic clusters in alignments allow to identify new, formerly uncharacterized domains [11]. These approaches detect sequence fragments which are compatible with some basic structural elements, such as α-helices and β-sheets and assume that the resulting spatial proximity of hydrophobic residues (e.g., L, I, W) will indicate if a relatively compact and stable conformation exists. This method, therefore, does not require any prior homology information but does also not have any reliability measure such as a p-value.

Altogether domains and domain arrangements are therefore ideal candidates to be studied when one is interested in proteome evolution. In the next section we cover the steps needed to do domain-based proteomics, starting with the data preparation and then go on to different methods to do domain-based proteomics.

2 Materials

All the tools used in the method section are freely available. Below is a list of used programs with a short description of their purpose.

2.1 Databases

- **Gene Ontology** Gene Ontology (GO) [9] is an effort to provide a vocabulary to represent biological functions. Website: http://geneontology.org.
- **InterPro** InterPro [12] is a meta-domain database. It contains domains from 14 databases and groups identical domains from different databases into the same InterPro ID. Website: http://www.ebi.ac.uk/interpro/.
- **Pfam** Pfam [13] is a database of domains, with about 16,700 domains (versions 31). The domains are based on sequence conservation and are clustered into clans based on similarity of either sequence or structure. For each domain family an e-value threshold is defined to separate random hits from real domain instance occurrences. Website: http://pfam.xfam.org/.

Beside databases, several programs are needed to analyze the data. An overview can be found in Table 1.

Table 1
List of software programs needed

Program	Description
BioBundle	A small collection of programs we use to prepare the data. Website: https://github.com/CarstenK/BioBundle
DAMA	`DAMA`[14] annotates sequences with Pfam domains. The results are based on an existing (e.g., HMMER) annotation that is then improved by using different filter criteria. Website: http://www.lcqb.upmc.fr/DAMA/
DOGMA	`DOGMA` can assess the quality of proteomes and transcriptomes based on the occurrences of domains. Website: http://domainworld.uni-muenster.de/programs/dogma/
DomRates	Program to trace evolutionary changes in domain arrangements. Website: http://domainworld.uni-muenster.de/programs/domrates/
gffread	`gffread` is part of the `cufflinks` package[15] and is used to extract protein sequences from a genome based on GFF file. Website: http://cole-trapnell-lab.github.io/cufflinks/install/
HMMER	`HMMER` is a program suite containing programs to construct sequence HMMs of, e.g., domains. These HMMs can then be used in searches for further matches in other sequences. Website: http://hmmer.org/
InterProScan	Program to annotate proteins with domains contained in the InterPro domain database. Website: http://www.ebi.ac.uk/interpro/download.html
PfamScan	The database Pfam [13] provides a software to annotate sequences with Pfam domains. The software as well as the domain database are needed to annotate sequences. It uses the HMMER program suite to find domain matches and then uses the Pfam e-value thresholds to filter out overlaps and spurious hits. Website: ftp://ftp.ebi.ac.uk/pub/databases/Pfam/Tools/
RADIANT	A fast domain annotation program. Used here together with DOGMA for a fast quality assessment. Website: http://domainworld.uni-muenster.de/programs/radiant/

3 Methods

Here, we describe how to use different programs in a domain-based proteomics analysis. An overview of the different steps that will be performed is shown in Fig. 1. Among the many potential mishaps, we list those that we experience in our research as the most frequent ones and present solutions for them as well.

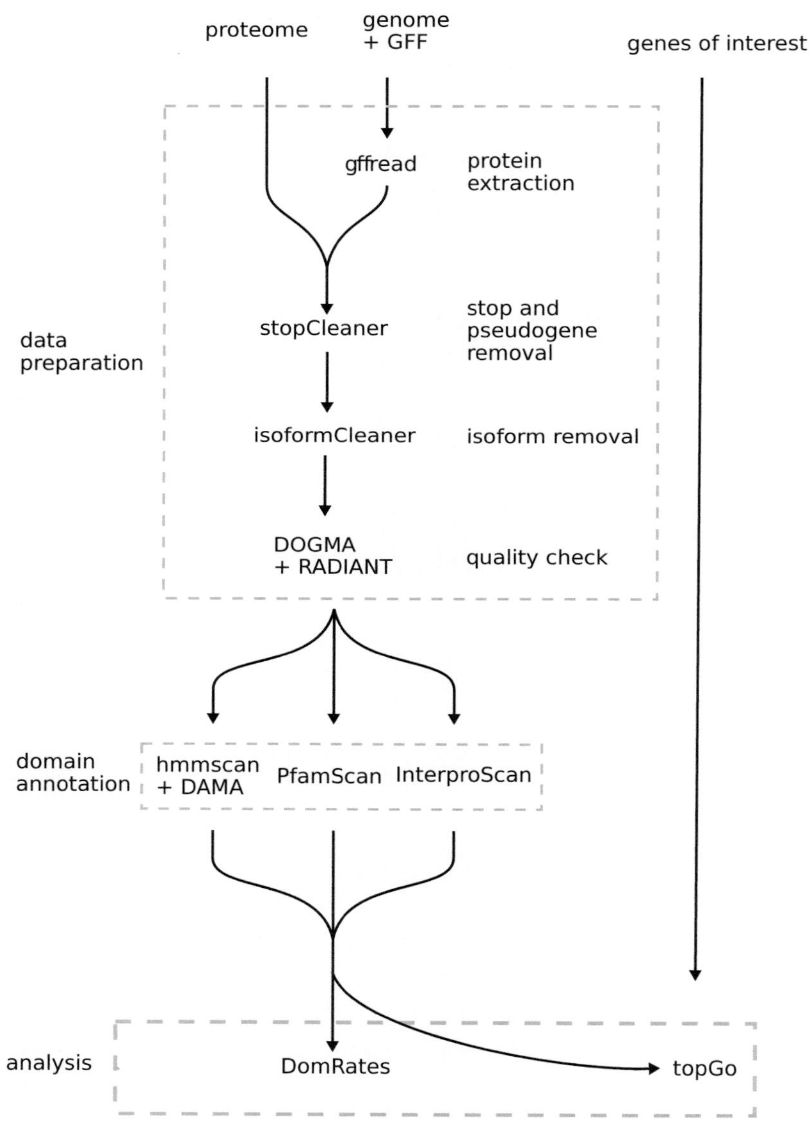

Fig. 1 Workflow of a domain-based proteome analysis. The steps "data preparation," "domain annotation," and the analysis itself are covered

3.1 Preparing a Data set for Domain Annotation and Subsequent Domain-Based Analyses

Proteomes for the species to be analyzed can be found in publicly available databases, e.g., on general portals (e.g., NCBI [16] or Ensembl [17]) or on more specialized websites for certain species groups (e.g., Hymenoptera genomes [18]) or single species. The simplest way to obtain a proteome set is to download the proteome directly but sometimes only a genome and a GFF file are available. In this case `gffread` can extract the mRNA from the genome and translate them into proteins.

It is important to make sure that the gene annotation version fits the genome version. If this is not the case the protein extraction might fail or, in the worst case, might extract incorrect proteins due to shifts in protein coordinates. Even if the versions match problems might occur. A possible error can be that two identical gene annotations exist (with same ID) or the same ID has been used twice for different genes. In these cases the gene annotation needs to be fixed manually either by removing the gene annotation (first case) or change the gene ID (second case).

In other cases, e.g., if scaffolds in the genome file are missing, the providers of the GFF/genome need to be contacted to ask for correction. Sometimes the GFF/genome files contain a prefix to the scaffold names in either the GFF or the genome sequence file but not in both. The solution is simply to remove the prefix (or add it to the other file).

On the first run of `gffread` on a genome file it creates an index file (< genome file >.fai) that contains the names and positions of the scaffolds for faster access. This file is not regenerated automatically when the genome file is changed. It is therefore important to delete the index file after manually having changed the genome as otherwise `gffread` will not recalculate the index.

```
# data: Harpegnathos saltator http://
    hymenopteragenome.org/harpegnathos/?q=genome_
    consortium_datasets

# removing prefix "gnl|Hsal_3.3|" in the scaffold
    names in the genome file
sed -i "s/gnl|Hsal_3.3|//' Hsal_3.3_scaffolds.fa

# extract CDS and translate them into proteins
gffread -y hsal_pep.fa -g Hsal_3.3_scaffolds.fa hsal
    _OGSv3.3.gff3
```

The terminating stop codon in coding sequences (CDS) is signified by a stop sign (".") at the end of the protein. The "." at the end is a minor technical issue which most programs will easily cope with, though not all. Additionally, sometimes one or several "." occur in the middle of a protein sequence either on purpose because the GFF contains pseudogenes or because the gene annotation is erroneous. These should be removed as, for example, a domain could be found after a recently obtained stop codon.

However, if the proteome was downloaded as protein sequences, the stop codons at the end might have been removed already but not the ones in the middle which might be masked by a different character (e.g., "U") as they usually cause less problems for programs. These genes should still be removed to avoid wrong domain annotations as mentioned above.

Additionally, it is important to realize that some read through genes (e.g., selenoproteins [19]) do contain a stop codon in the sequence which can be either replaced with an amino acid or is simply being ignored during translation. The cases cannot be handled automatically and need attention by the user. We use the `stopCleaner` program of the BioBundle package to remove the stop signs at the end as well as potentially problematic genes.

```
# remove "." at the end and genes with a "." in the
   middle
stopCleaner -i hsal_pep.fa -r -o hsal_clean_pep.fa
```

A quality check will be performed using DOGMA [20]. It searches for a set of conserved domain arrangements in the proteome or transcriptome of interest and computes a quality score based on the percentage of found arrangements. For the quality check as well as for the enrichment and DomRates analysis, that will be performed at the end of this chapter (*see* Fig. 1), each gene should be present with a single isoform only. Otherwise, a bias will be introduced. Since there is no generally accepted way to mark different isoforms, one first needs to determine how isoforms are marked. This has to be done manually. A common way is to mark the end of the protein IDs (e.g., -PA, -PB, . . ., or.t1,.t2, . . .). However, not always it is as simple as gene names might not contain the isoform information. In this case the GFF is needed, then the Parent field can be used to identify isoforms because isoforms of the same protein should have different IDs but identical parent IDs. To help with this procedure the `isoformCleaner` program can be used. It provides different options to keep only the longest isoform version of a protein. Here, we make use of the possibility to define a character (option "-s") which separates the gene name from the isoform identifier, e.g., the sequence ID "HSAL20401-RB" will be split into gene ID "HSAL20401" and isoform "RB".

```
# keep only longest isoform
isoformCleaner -s "-" -i hsal_clean_pep.fa -o hsal_
    clean_longestIso_pep.fa
```

A general problem is that the quality of available data can vary a lot. It is therefore important to check the quality of the selected proteomes. BUSCO [21] and DOGMA [20] are both programs to analyze proteome and transcriptome quality. Here, we describe only the usage of DOGMA, as it is based on domains and therefore perfectly suits our purpose, however, it can be used just as well for general quality checks even if no domain analyses are planned to be done. Both programs give very similar results.

DOGMA compares the domains found in the proteome to analyze with a predefined core set. For best results it is recommended to use the core set that is the smallest group containing the species to analyze. In the case of the ant used in this example the insect group is the best fit.

```
# running DOGMA with automatic annotation with
    RADIANT using the insect core set.
dogma.py proteome -i hsal_clean_longestIso_pep.fa  -
    d <path to database>/pfam31 -r insects -o hsal_
    dogma_insect.txt
```

The resulting quality score 94.55 is very good. A more specific core set should always give a quality score lower or equal to one with a less specific core set as more domain arrangements are checked. Therefore, the better the core set fits the more accurate the quality estimation will be.

In general one should try to have only proteomes which have a quality value of at least 75 (ideally higher) as the lower the number the less comparable are the proteomes.

3.1.1 Annotating Sequences with Domains

The first step to annotating the sequences with domains is to decide which database to use as many different ones exist. They differ in the number of domains they contain, and in the way they define them (e.g., more structural or sequence based). Here, for demonstration purposes, we use the Pfam and the InterPro database and apply it to the prepared file from the previous section. It contains 17,146 sequences that will be annotated with domains such that we can perform a domain-based functional enrichment or rearrangement analysis in the next step.

```
# annotation of sequences with Pfam domain
pfam_scan.pl -fasta hsal_clean_longestIso_pep.fa -
    dir <path to Pfam folder>  -outfile hsal.dom
```

This is the standard approach to annotate sequences with Pfam domains. However, there are two ways to further increase the domain coverage of a proteome. One simple approach would be to increase the e-value (i.e., lower the threshold) to allow for more domain hits. However, this increased sensitivity would increase as well the number of false positive hits. A better approach is to incorporate co-occurrence information as, for example, done in CODD [22] or DAMA [14]. Domains are often preferentially co-occurring together with specific other domains. This information can be used during the annotation of sequences. For example, in the case of two domains in a sequence one with an e-value threshold above the default one below. By default only one domain would be annotated. But if co-occurrence information suggest that both domains often occur together one can recover the one with the too high e-value.

In the next step it is shown how to use DAMA to improve domain coverage:

```
# begin with annotating domains with hmmscan
hmmscan --domtblout hsal_hmmscan.txt --cpu 4 <path
    to Pfam folder>/Pfam-A.hmm hsal_clean_longestIso
    _pep.fa  >/dev/null

# converting the hmmscan output into DAMA usable
    format
perl convertHmmscanOutput.pl -hmmscanFile hsal_
    hmmscan.txt -outputFile hsal_dama_input.txt

# run DAMA to filter domain list
DAMA -domainsHitFile hsal_dama_input.txt -
    knownArchFile DAMA/database/pfam31/pfam.
    knownArch.txt -outputFile hsal_dama.txt -
    domainsInfoFile DAMA/database/pfam31/pfam.
    domains -overlappingDomainFile DAMA/database/
    pfam31/pfam.overlapping
```

The second approach is to use InterPro, a meta database which combines several different domain databases into a single one. This will result in a higher coverage as well because different databases contain different domains although, in general, a large overlap exists between the models characterizing one domain. However, the results will often contain overlaps of domains from different databases making further post-processing necessary for many analyses (e.g., analysis of arrangement changes).

```
interproscan.sh -i hsal_clean_longestIso_pep.fa -f
    tsv -o hsal_interPro.tsv -goterms
```

3.2 DomRates: Analyzing Domain Arrangement Changes Along a Phylogenetic Tree

The analysis of domain arrangement changes can provide insights into the kind of events that were important for a new species. We traceback domain arrangement changes using the DomRates program. Based on a domain annotation and a given phylogeny it is able to reconstruct the events that lead to the extant species in the data set.

We will analyze a small set of hymenopterans. For each species we prepare the proteome as described above and put the final domain annotation together in one folder. The set additionally includes an outgroup (*Drosophila melanogaster*) to reconstruct the ancient state at the root of the hymenopteran branch. Furthermore, a phylogenetic tree in Newick format is needed. The labels in the tree correspond to the domain file names without a fixed suffix. For later visualization we will produce a statistics file which will be used in a subsequent step. With only six species the tree is very short. We therefore use the "-l" option to adjust the legend.

```
# running DomRates to reconstruct the history of
   rearrangement events
domRates -t speciesTree.nwk -a annotation/ -g
   Drosophila_melanogaster -e _iso.fa.dom -o domOut
   .txt -s statistics.txt

# visualize the changes
visualization_domrates_tree.py -t speciesTree.nwk -s
    statistics.txt  -o hymenopteraChanges -l
```

The output of the visualization script is a tree with pie charts showing the amount and proportion of rearrangement events which have happened along the branches (Fig. 2).

3.3 Functional Enrichment Analysis Based on Domain Annotations

It is often of great interest if gene sets (e.g., genes under positive selection) have a common function as this can help to find biological explanations. Domains, as known functional units, combined with a defined biological vocabulary (e.g., Gene Ontology) can be used together to characterize genes in respect of the molecular or biological processes they are involved in or the cellular component they are active in.

The Gene Ontology (GO) consortium provides mappings of GO terms to domains of different databases. Here, we will use the pfam2go mapping that assigns GO terms to numerous Pfam domains. The combination of the domain assignments is then used to identify the function of a protein and perform analyses of enrichment of certain terms in a set of genes. The R package topGO [23] provides several algorithms and statistical tests that can be used for the enrichment analysis. It compares the GO terms

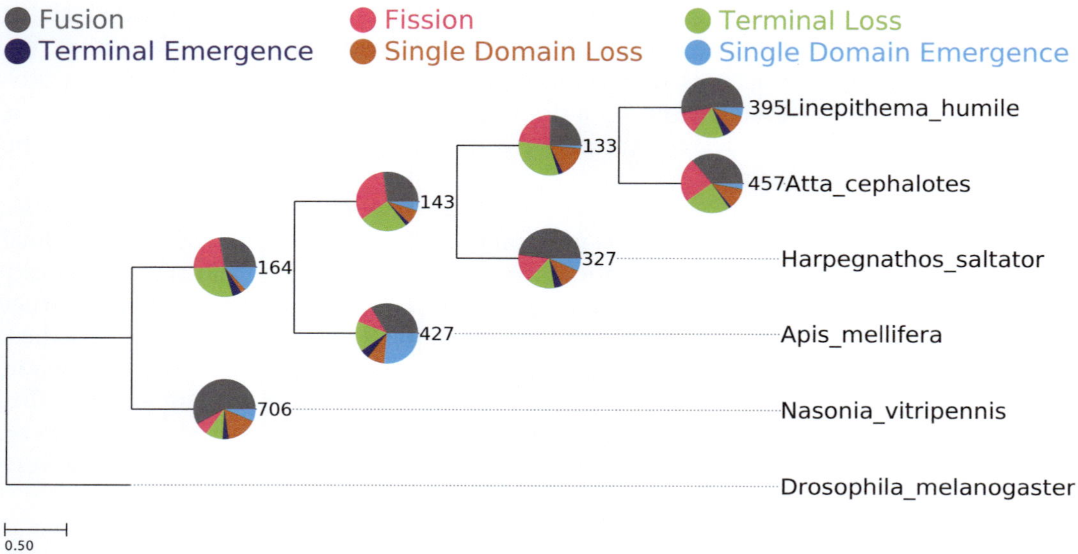

Fig. 2 Domain arrangement changes along a selected set of hymenopterans

of genes of interest with the larger "GO universe" to identify overrepresented terms.

The first step is to prepare the GO universe. For this, we can use a simple Python script which merges a domain annotation file with the pfam2go file into the needed format. The universe file is a simple two-column text file. The first column contains the gene names, the second one a comma separated list of GO-IDs associated with the gene. The file containing the genes of interest is a simple text file with one gene ID per line, for example, all the genes which are differentially expressed or have another common feature that is of interest.

```
# merge domain annotation file with pfam2go mapping
    using domain2topGo.py (BioBundle)
./domain2topGo.py -d hsal_dama.txt -g pfam2go -o
    universe.txt
```

We can now install (if necessary) the topGO package directly from R. Subsequently, we load the package using the library command.

```
# installing topGO
source("https://bioconductor.org/biocLite.R")
biocLite("topGO")

# load the topGO package
library("topGO")
```

The next step is to load the data from the files and prepare it for the following GO term enrichment analysis.

```
# load the GO universe
GOmap <- readMappings(file = "universe.txt")

# make list of gene names in the universe
genes_in_universe <- names(GOmap)

# loading and preparing of the genes of interest
subset <- scan(file='genesOfInterest.txt', what=
    character())
genes_of_interest <- factor(as.integer(genes_in_
    universe %in% subset))
names(genes_of_interest) <- genes_in_universe
```

An enrichment test can be performed for three different categories: molecular function (MF), biological process (BP), and cellular component (CC). The category analyzed in the analysis can be changed by simply changing the abbreviation used in the ontology parameter.

```
# performing the enrichment analysis for biological
    processes
GOdata <- new("topGOdata", description="GO-Analysis"
    , ontology="BP", allGenes=genes_of_interest,
    annot = annFUN.gene2GO, gene2GO = GOmap)

# run a significance test
resultParentchild <- runTest(GOdata, algorithm="
    parentchild", statistic="fisher")

# filter results for GO terms with a p-value <= 0.05
summary <- summary(attributes(resultParentchild)$
    score <= 0.05)
numsignif <- as.integer(summary[[3]])

if(numsignif > 0)
{
  results <- GenTable(GOdata, fisher =
    resultParentchild, topNodes = numsignif)
} else {
  results <- NULL
}

# write results to file in simple CSV format
write.csv(results, file='results.csv')
```

Table 2
Top 10 enriched GO terms based on the "parentChild" algorithm with a fisher test

	GO.ID	Term	Annotated	Significant	Expected	Fisher
1	GO:0032196	Transposition	104	18	4.30	9.5e−12
2	GO:0006259	DNA metabolic process	245	26	10.14	1.2e−10
3	GO:0044710	Single-organism metabolic process	689	59	28.52	1.3e−08
4	GO:0008152	Metabolic process	2302	120	95.27	4.5e−06
5	GO:0006508	Proteolysis	279	23	11.55	1.8e−05
6	GO:0034641	Cellular nitrogen compound metabolic pro...	952	41	39.40	5.3e−05
7	GO:0006313	Transposition, DNA-mediated	104	18	4.30	0.0013
8	GO:0006725	Cellular aromatic compound metabolic pro...	796	34	32.94	0.0029
9	GO:0046483	Heterocycle metabolic process	797	34	32.99	0.0030
10	GO:0006310	DNA recombination	128	20	5.30	0.0061

The "results.csv" file will now contain a list of all GO terms that are enriched in the gene set of interest and have a p-value ≤ 0.05. An example output is shown in Table 2.

In this chapter, we gave a basic overview why the analysis of domains is important. Additionally, we described the basic methods to prepare and analyze data within a protein domain context. Domains allow a fast evolutionary analysis of large data sets and by using GO term assignments allow to perform functional analyses as well. However, it is important to remember that not all domains have a known function and that not all proteins contain a domain which might influence the analysis. Additionally, the database used might have a species bias (e.g., contain an over-proportional amount of domains of eukaryotes) which will influence coverage and functional depth of analyses based on such annotations.

Acknowledgements

We would like to thank Mark Harrison and Ulrike Brandt for helpful suggestions.

References

1. Vogel C, Bashton M, Kerrison ND, Chothia C, Teichmann SA (2004) Structure, function and evolution of multidomain proteins. Curr Opin Struct Biol 14(2):208–216
2. Moore AD, Asa KB, Ekman D, Bornberg-Bauer E, Elofsson A (2008) Arrangements in the modular evolution of proteins. Trends Biochem Sci 33(9):444–451

3. Lees JG, Dawson NL, Sillitoe I, Orengo CA (2016) Functional innovation from changes in protein domains and their combinations. Curr Opin Struct Biol 38:44–52
4. Levitt M (2009) Nature of the protein universe. Proc Natl Acad Sci USA 106 (27):11079–11084
5. Remmert M, Biegert A, Hauser A, Soding J (2011) HHblits: lightning-fast iterative protein sequence searching by HMM-HMM alignment. Nat Methods 9(2):173–175
6. Moore AD, Grath S, Schüler A, Huylmans AK, Bornberg-Bauer E (2013) Quantification and functional analysis of modular protein evolution in a dense phylogenetic tree. Biochim Biophys Acta Proteins Proteomics 1834 (5):898–907
7. Moore AD, Bornberg-Bauer E (2012) The dynamics and evolutionary potential of domain loss and emergence. Mol Biol Evol 29 (2):787–796
8. Kersting AR, Bornberg-Bauer E, Moore AD, Grath S (2012) Dynamics and adaptive benefits of protein domain emergence and arrangements during plant genome evolution. Genome Biol Evol 4(3):316–329
9. Ashburner M, Ball CA, Blake JA, Botstein D, Butler H, Cherry JM, Davis AP, Dolinski K, Dwight SS, Eppig JT, Harris MA, Hill DP, Issel-Tarver L, Kasarskis A, Lewis S, Matese JC, Richardson JE, Ringwald M, Rubin GM, Sherlock G (2000) Gene ontology: tool for the unification of biology. The Gene Ontology Consortium. Nat Genet 25 (1):25–29
10. Sigrist CJA, Castro E, de Cerutti L, Cuche BA, Hulo N, Bridge A, Lydie B, Xenarios I (2013) New and continuing developments at PROSITE. Nucleic Acids Res 41(Database-Issue):344–347
11. Bitard-Feildel T, Heberlein M, Bornberg-Bauer E, Callebaut I (2015) Detection of orphan domains in *Drosophila* using "hydrophobic cluster analysis". Biochimie 119:244–253
12. Finn RD, Attwood TK, Babbitt PC, Bateman A, Bork P, Bridge AJ, Chang HY, Dosztanyi Z, El-Gebali S, Fraser M, Gough J, Haft D, Holliday GL, Huang H, Huang X, Letunic I, Lopez R, Lu S, Marchler-Bauer A, Mi H, Mistry J, Natale DA, Necci M, Nuka G, Orengo CA, Park Y, Pesseat S, Piovesan D, Potter SC, Rawlings ND, Redaschi N, Richardson L, Rivoire C, Sangrador-Vegas A, Sigrist C, Sillitoe I, Smithers B, Squizzato S, Sutton G, Thanki N, Thomas PD, Tosatto SC, Wu CH, Xenarios I, Yeh LS, Young SY, Mitchell AL (2017) InterPro in 2017–beyond protein family and domain annotations. Nucleic Acids Res 45(D1):D190–D199
13. Finn RD, Coggill P, Eberhardt RY, Eddy SR, Mistry J, Mitchell AL, Potter SC, Punta M, Qureshi M, Sangrador-Vegas A, Salazar GA, Tate J, Bateman A (2016) The Pfam protein families database: towards a more sustainable future. Nucleic Acids Res 44(D1): D279–D285
14. Bernardes JS, Vieira FR, Zaverucha G, Carbone A (2016) A multi-objective optimization approach accurately resolves protein domain architectures. Bioinformatics 32 (3):345–353
15. Trapnell C, Williams BA, Pertea G, Mortazavi A, Kwan G, van Baren MJ, Salzberg SL, Wold BJ, Pachter L (2010) Transcript assembly and quantification by RNA-Seq reveals unannotated transcripts and isoform switching during cell differentiation. Nat Biotechnol 28 (5):511–515
16. NCBI Resource Coordinators (2017) Database Resources of the National Center for Biotechnology Information. Nucleic Acids Res 45 (D1):D12–D17
17. Yates A, Akanni W, Amode MR, Barrell D, Billis K, Carvalho-Silva D, Cummins C, Clapham P, Fitzgerald S, Gil L, Giron CG, Gordon L, Hourlier T, Hunt SE, Janacek SH, Johnson N, Juettemann T, Keenan S, Lavidas I, Martin FJ, Maurel T, McLaren W, Murphy DN, Nag R, Nuhn M, Parker A, Patricio M, Pignatelli M, Rahtz M, Riat HS, Sheppard D, Taylor K, Thormann A, Vullo A, Wilder SP, Zadissa A, Birney E, Harrow J, Muffato M, Perry E, Ruffier M, Spudich G, Trevanion SJ, Cunningham F, Aken BL, Zerbino DR, Flicek P (2016) Ensembl 2016. Nucleic Acids Res 44 (D1):D710–D716
18. Elsik CG, Tayal A, Diesh CM, Unni DR, Emery ML, Nguyen HN, Hagen DE (2016) Hymenoptera Genome Database: integrating genome annotations in HymenopteraMine. Nucleic Acids Res 44(D1):793–800
19. Labunskyy VM, Hatfield DL, Gladyshev VN (2014) Selenoproteins: molecular pathways and physiological roles. Physiol Rev 94 (3):739–777
20. Dohmen E, Kremer LPM, Bornberg-Bauer E, Kemena C. (2016) DOGMA: domain-based transcriptome and proteome quality assessment. Bioinformatics 32(17):2577–2581

21. Simão FA, Waterhouse RM, Ioannidis P, Kriventseva EV, Zdobnov EM (2015) BUSCO: assessing genome assembly and annotation completeness with single-copy orthologs. Bioinformatics 31(19):3210–3212
22. Terrapon N, Gascuel O, Marechal E, Breehelin L (2009) Detection of new protein domains using co-occurrence: application to *Plasmodium falciparum*. Bioinformatics 25 (23):3077–3083
23. Alexa A, Rahnenführer J (2016) topGO: enrichment analysis for gene ontology. R package version 2.26.0

Chapter 17

Modeling of Protein Tertiary and Quaternary Structures Based on Evolutionary Information

Gabriel Studer, Gerardo Tauriello, Stefan Bienert, Andrew Mark Waterhouse, Martino Bertoni, Lorenza Bordoli, Torsten Schwede, and Rosalba Lepore

Abstract

Proteins are subject to evolutionary forces that shape their three-dimensional structure to meet specific functional demands. The knowledge of the structure of a protein is therefore instrumental to gain information about the molecular basis of its function. However, experimental structure determination is inherently time consuming and expensive, making it impossible to follow the explosion of sequence data deriving from genome-scale projects. As a consequence, computational structural modeling techniques have received much attention and established themselves as a valuable complement to experimental structural biology efforts. Among these, comparative modeling remains the method of choice to model the three-dimensional structure of a protein when homology to a protein of known structure can be detected.

The general strategy consists of using experimentally determined structures of proteins as templates for the generation of three-dimensional models of related family members (targets) of which the structure is unknown. This chapter provides a description of the individual steps needed to obtain a comparative model using SWISS-MODEL, one of the most widely used automated servers for protein structure homology modeling.

Key words Homology modeling, Oligomeric proteins, Quaternary structure, Protein structure prediction, Model quality assessment, Model quality estimates, SWISS-MODEL

1 Introduction

Homology modeling, or comparative protein structure modeling, is a technique to generate a three-dimensional model of a protein from its amino acid sequence (target) using the structures of related proteins as reference (templates) [1, 2]. The applicability and success of the approach mainly depend upon two factors: the extent of sequence similarity between the target protein and the template, and the extent of structural divergence during evolution [3]. Although confident results are expected in case of close

Tobias Sikosek (ed.), *Computational Methods in Protein Evolution*, Methods in Molecular Biology, vol. 1851, https://doi.org/10.1007/978-1-4939-8736-8_17,

homologs, each step of the modeling process should be carefully considered as it can affect the intended applications of the model.

In this context, the availability of automated servers with user-friendly web interfaces also allows nonspecialists to generate reliable 3D models without the need to install complex software and databases and by providing easy access to results, their visualization, and interpretation. SWISS-MODEL pioneered the field of automated modeling servers 25 years ago. Since then, it has been continuously improved and its functionality greatly extended following both methodological and conceptual advances [4–7].

From a technical point of view, comparative modeling consists of the following main steps:

1. The amino acid sequence of the target protein is compared to the sequences of homologous proteins of known structure in order to identify reliable templates.
2. The amino acid sequences of the target and the selected templates are aligned.
3. Three-dimensional models of the target are built based on the alignments.
4. The global and local quality of the resulting models is evaluated.

So far, steps 1 and 2 gathered a solid consensus of being the most critical ones [8] especially when only remote homologs with known structure are available. Accordingly, many efforts have been devoted over the past years to the development and improvement of dedicated methods. Position-specific sequence profiles (PSSM) and profile-based hidden Markov models (HMM) have represented a breakthrough in this context, being able to enhance template identification and alignment even in case of remote homology.

There is no risk of overstating the importance of a curated and annotated template library at this stage of the process. The SWISS-MODEL Template Library (SMTL) is a database of experimental structures derived from the Protein Data Bank (PDB) [9], which are thoroughly processed, annotated, and organized to support efficient query of high-quality template structure data [5]. The SMTL is searched with two sequence search methods: BLAST [10] and HHblits [11]. While the first delivers fast and accurate templates of closely related sequences, the second adds sensitivity in case of distant homology, taking advantage of HMM-HMM alignments and incorporating secondary structure information predicted by PSIPRED [12].

Once a list of potential templates is obtained, the next step is to select the best possible among the available ones. To this aim, any available information about the target and the template should be taken into account. In other words, sequence similarity is not the only factor to consider, especially if different templates with similar sequences exist. The experimental quality of the structure is an important parameter, as well as the environmental conditions in which it has been determined. Hence it can be beneficial for the planned application to select a template structure which is in complex with a given ligand or substrate or in a certain conformational state. If the target and the template share the same function, their active sites or functional residues are expected to be conserved and aligned [13]. However, this is not necessarily what a sequence alignment algorithm produces, i.e., the alignment that maximizes a certain similarity score. On the other hand, even small mistakes in the alignment can translate into prominent errors in the final model. Therefore, a careful inspection of the alignment is generally recommended since it provides a valid a priori estimate of the expected accuracy of the model and, when possible given the information known about the specific protein family, allows avoiding modeling errors by manually correcting alignment problems. SWISS-MODEL facilitates addressing these two aspects. Based on the analysis of different target-template alignment properties such as sequence identity, sequence similarity, and secondary structure agreement, SWISS-MODEL computes a Global Model Quality Estimation score (GMQE), indicating the expected quality of the model resulting from the given alignment [5]. Comprehensive protein annotations and display functionalities are also provided to aid the identification of problems in the alignment; for example template secondary structure information is shown to facilitate identification of errors due to incorrect placing of insertion/deletions in conserved regions.

The aspects described so far are all key elements to identify a reliable template to produce single-chain protein models. But it is well known that the function of a protein is often the result of its interaction with other proteins, forming either homo- or hetero-oligomeric complexes. Therefore, the oligomeric state of the target protein must be considered in order to generate biologically meaningful models. However, protein oligomeric states are difficult to characterize experimentally and, during evolution, show only limited conservation: i.e., the number of subunits and their binding modes in different complexes can vary substantially [14]. As a consequence, modeling a protein in its correct quaternary structure is still far from being a routine procedure. In SWISS-MODEL, we recently introduced a new strategy to infer the stoichiometry and the overall structure of oligomers by homology [15]. The method exploits a novel description of protein-protein interface conservation as a function of evolutionary distance and an efficient distance measure to structurally compare homologous multimeric protein complexes. Interface conservation scores, structural clustering, and classical interface descriptors are combined in a supervised machine learning algorithm to provide a quaternary structure quality estimate (QSQE). This information, together with annotation of the template quaternary structure, is provided in SWISS-MODEL to improve the selection of homologous protein templates for the subsequent modeling steps.

Immunoglobulins represent an important class of hetero-oligomeric proteins, for which classical comparative modeling approaches give only suboptimal results, especially for the variable loop regions responsible for epitope recognition. However, specific protocols exist that enable the modeling of this class of molecules with high accuracy based on the canonical structures of antibody loops [16]. In line with this, the pre-screening of the target sequence in SWISS-MODEL has been extended to identify whether a target sequence represents an immunoglobulin and, if a matching sequence signal is detected, sends data to the Prediction of Immunoglobulin Structure server PIGSPro [17] where the modeling job can proceed. Alternatively, modeling can also be performed using the standard SWISS-MODEL workflow. However, this is only recommended if the template structure is closely related; that is, there are no insertions/deletions occurring in the variable loop regions.

Given a target-template alignment, the model coordinates are built based on the assumption that aligned residues between the target and the template are structurally equivalent. In practice, this requires the transfer of information from the template and the modeling of parts where such information is missing, i.e., insertions and deletions in the alignment. SWISS-MODEL relies on the OpenStructure software framework [18] and the ProMod3 modeling engine (manuscript in preparation) to generate the atomic coordinates for all residues of the target protein that are in the range covered by the target-template alignment. This is achieved by fully automatically performing the following steps: (1) building an initial model, (2) loop modeling, (3) modeling of side chains, (4) energy minimization, and (5) model quality estimation. Each step is summarized below (*see* also Fig. 1).

1.1 Building an Initial Model

In this step, structural information from template residues is transferred to corresponding target residues as defined by the target-template alignment. Several algorithms have been developed to accomplish this task, based on different approaches which are reviewed elsewhere [19]. In SWISS-MODEL this is done by transferring the atomic coordinates from the corresponding template residues in Cartesian space. ProMod3 aims at inferring as many atom positions as possible from template structures, depending on the conservation of the corresponding residues between target and template. This usually results in an incomplete model with missing side-chain coordinates and gaps originating from amino acid insertions/deletions.

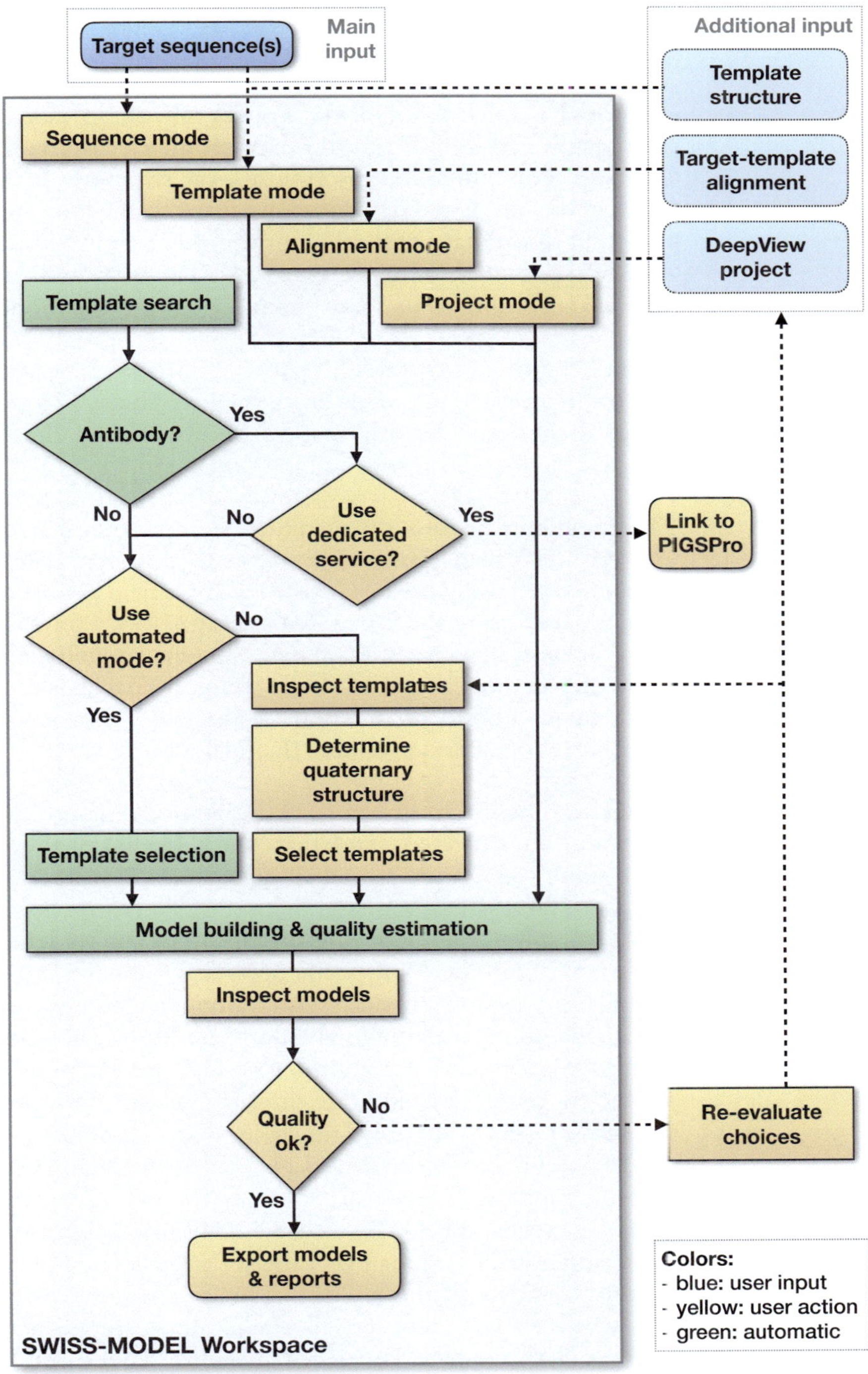

Fig. 1 Flowchart of the comparative protein structure modeling pipeline implemented in SWISS-MODEL

1.2 Loop Modeling

With the possible exception of antibody loops, as discussed later in this chapter, modeling protein loops is a challenging task and often a major source of modeling errors. Loop modeling methods can be categorized into two main groups: ab initio and database approaches [19–22]. ProMod3 uses geometric criteria to query a database containing high-resolution X-ray structures for matching loop candidates. Candidate loops are fitted to the loop stems using the cyclic coordinate descent algorithm [23] and scored based on statistical potentials of mean force [24]. The best candidate is then selected according to its score and inserted into the model.

1.3 Modeling of Side Chains

To model non-conserved side chains, ProMod3 extracts side-chain conformations from the Dunbrack rotamer library [25] and determines their optimal conformation by minimizing the SCWRL4 energy function [26] using a graph-based approach [27].

1.4 Energy Minimization

The modeling process can produce stereochemical irregularities and clashes, which ProMod3 resolves in this step. The model is parameterized using the CHARMM27 force field [28] and energies are evaluated using the OpenMM package [29]. ProMod3 applies short steepest descent and conjugate gradient minimization iteratively on the model until all stereochemical problems are resolved or an upper bound of iterations is reached. This concludes the modeling process and produces the final model.

1.5 Model Quality Estimation

Quality estimation tools aim to quantify modeling errors and give estimates on expected model accuracy both on a global and per-residue scale. From a modeling perspective, such estimates are useful to select the best model in a set of alternatives or detecting local errors. But, even more importantly, they aim to determine the usefulness of a model for a specific application at hand [30, 31]. Various tools assessing physical plausibility are routinely applied on models based on experimental data [32]. However, while stereochemistry is a necessary condition for a high-quality model, it is not a sufficient criterion to indicate similarity of a theoretical model to the native target structure. Knowledge-based approaches with statistical potentials of mean force [24] constitute a valid complement for estimating the expected accuracy of a theoretical model. SWISS-MODEL relies on QMEAN [33, 34] to assign global and per-residue quality estimates. QMEAN linearly combines four statistical potentials of mean force. Two of them evaluate pairwise distances, the first between all chemically distinguishable heavy atoms and the second between Cβ atoms. Two more potentials evaluate backbone torsion angles and packing of the model.

The accuracy of models generated by SWISS-MODEL is continuously assessed by the CAMEO project [35] based on weekly blind prediction of proteins from the upcoming PDB release.

1.6 Concluding Remarks

The availability of reliable and robust fully automated workflows for protein structure modeling has made homology modeling the method of choice to reliably generate three-dimensional models for proteins when experimental structures are not available. Easy-to-use interactive web servers and reliable model quality estimation tools allow also nonspecialists to successfully use protein models in structure-based applications in biomedical research.

2 Materials

1. A computer with access to the Internet and a web browser.
2. The amino acid sequence, either in FASTA or as a plain text, or the UniProtKB identifier of the target to be modeled.
3. The DeepView desktop application, available at https://spdbv.vital-it.ch/ (optional).
4. A DeepView project (optional).
5. The coordinates in PDB format of the structure to be used as a template (optional).
6. A target-template alignment, either in FASTA or Clustal format (optional).
7. An e-mail address (optional).

3 Methods

Figure 1 illustrates the workflow of the modeling pipeline implemented in SWISS-MODEL. A detailed description of the individual steps is provided in the following sections.

3.1 Access the SWISS-MODEL Workspace

The SWISS-MODEL website is available at https://swissmodel.expasy.org. From the homepage and any other page on the website, the user can create a password-protected user account, which is associated with an e-mail address. It is also possible to use SWISS-MODEL anonymously, i.e., without registration. However, in this case it is necessary to bookmark individual project URLs in order to access the results once the session has been closed. By default, projects are stored for 2 weeks on the server with an option to extend the project lifetime.

3.2 Start a New Modeling Project

Depending on the type of information at hand, there are different modes to start a new modeling project. If only the sequence of the target protein to be modeled is available, a first step is to search for templates, as described in Subheading (3.2.1.) Sequence mode. Alternatively, if the model should be based on a specific template structure, three different modes can be used according to template-related information available: (3.2.2.) Template mode, (3.2.3.) Alignment mode, and (3.2.4.) Project mode.

3.2.1 Sequence Mode: Starting from the Sequence of the Target Protein

1. Insert the amino acid sequence of the target protein into the main input box of the homepage. The sequence can be provided either as a plain text or in FASTA format. Alternatively, the UniProtKB identifier can be used.
2. Press the "Validate" button or the return key. A sequence validation step is performed to check for nonstandard amino acid codes and to reformat the input sequence. If the target UniProtKB identifier is provided as input, the protein sequence is automatically retrieved and validated. After validation, a non-editable wrapped view of the target sequence is displayed.
3. If the target protein is heteromeric, i.e., it consists of different protein chains as subunits, it is possible to enter an additional amino acid sequence by clicking the "Add Hetero Target" button. Repeat this step until all subunit sequences have been entered.
4. The next step is to identify reliable templates to be used for modeling. Two options are available to perform this task:
   (a) Manual template selection: This option allows the user to inspect the template search results before selecting one or more template structures for modeling, taking into account information such as quality of the experimental structure, oligomeric state, bound ligands, or crystallization conditions. To use this option, click the "Search for Templates" button and proceed to **step 3.3**.
   (b) Automatic template selection: Using this option, when the template search is complete, templates are ranked according to the expected quality of the resulting models and a number of templates are selected automatically. This option is especially useful for well-characterized protein families where target-template sequence similarity is expected to be sufficiently high to automatically generate unambiguous alignments and high-quality models. Note that also in this option, the full template search results can be inspected to select additional template structures in case the automated modeling results are not satisfactory. To use this option, click the "Build Model" button and proceed to **step 3.4** to access the modeling results.

   In both cases, as soon as the template search starts, an automatic scanning of the target sequence is performed to verify whether any immunoglobulin variable domain is present in the input. If this is the case, the user is provided with a link to the

Prediction of Immunoglobulin Structure server PIGSPro [17]. The link redirects to the PIGSPro server home page where the input form is pre-filled with the detected antibody variable domains and the modeling process can proceed using a protocol developed specifically for immunoglobulins (*see* **Note 1**).

3.2.2 Template Mode: Using a Specific Three-Dimensional Structure as Template

1. Click the "User Template" button.
2. Submit the target sequence as described in "Sequence mode."
3. Upload the template coordinates in PDB format.
4. Click the "Build Model" button and proceed to **step 3.4**.

3.2.3 Alignment Mode: Using a Specific Template from the SWISS-MODEL Template Library (SMTL) with a User-Defined Target-Template Alignment

1. Click the "Target-Template Alignment" button.
2. Provide the target-template alignment, either by copy/paste of the alignment or by uploading a file. The alignment must be in FASTA or Clustal format. Make sure that the provided template sequence and ID correspond to SEQRES and ID of the SMTL entry, respectively. Note that the content of the SMTL can be browsed manually (Menu: Modeling—Template Library).
3. Click the "Build Model" button and proceed to **step 3.4**.

3.2.4 Project Mode: Using a Specific Three-Dimensional Structure as Template by Manually Adjusting the Target-Template Alignment in the DeepView Desktop Application

The desktop application DeepView (available for Windows and Mac OS) [1] allows for visualization of one or more template structures, and manual editing of the target-template alignment. Projects generated in DeepView, or obtained in **step 3.4**, can be submitted for modeling after manual manipulation of the target-template alignment.

1. Click the "DeepView Project" button.
2. Upload the DeepView project file.
3. Click the "Build Model" button and proceed to **step 3.4**.

3.3 Template Identification

The Template Results page provides an overview of the available templates as well as interactive views and selection tools.

1. From the views below, select one or more template structures for modeling.
   (a) *Templates*. The main table displays the list of top 50 templates, ranked according to the expected quality of the resulting model. The complete list of templates is accessible by links at the bottom of the Template Results page. Features such as coverage, model quality estimates (GMQE and QSQE), oligomeric state, and bound ligands are shown in a condensed tabular form. Each of the table rows can be expanded to display additional information

and the target-template alignments. If the desired template could not be found or the templates do not cover the full target sequence, please refer to **Notes 2–4**. Note that selecting a template with a given bound ligand of interest does not guarantee that the ligand will be present in the final model (*see* **Note 5**).

(b) *Quaternary Structure.* The results of the quaternary structure analysis are provided for oligomeric targets. Templates are clustered and displayed in a decision tree according to their quaternary structure features: oligomeric state, stoichiometry, topology, and interface similarity. Each leaf of the tree corresponds to a template, with a bar indicating sequence identity and coverage to the target. In each cluster, templates are ranked according to their QSQE score. According to the selected clustering level, a protein-protein interaction (PPI) fingerprint plot informs about the conservation of template interfaces as a function of the evolutionary distance within the protein family. Please refer to **Notes 6** and **7** if no template with the expected oligomeric state is found.

(c) *Sequence Similarity.* A chart displays how templates relate to each other in the sequence similarity space. Each template is shown as a circle on an interactive plot, which allows selecting individual or groups of templates for further inspection and template superposition.

(d) *Alignment of Selected Templates.* Target-template alignments are shown with different coloring options for highlighting various sequence and structure properties along the alignment (i.e., secondary structure, hydrophobicity, solvent accessibility). Careful inspection of the alignment is recommended since this permits to identify, and possibly prevent, modeling errors beforehand (*see* **Note 8**).

2. Click the "Build Model" button to run the modeling using the selected templates and proceed to **step 3.4**.

3.4 Accessing Modeling Results

After completion of the modeling process, a detailed report of the modeling project is generated and can be accessed from the workspace. Model coordinates can be downloaded either formatted as PDB or DeepView project files.

The generated model(s) can be inspected in the model results page using the embedded structural viewer. The target-template alignment used for modeling is also shown and linked to the structure visualization such that hovering the mouse over the alignment highlights the corresponding residue in the viewer and vice versa (Fig. 2).

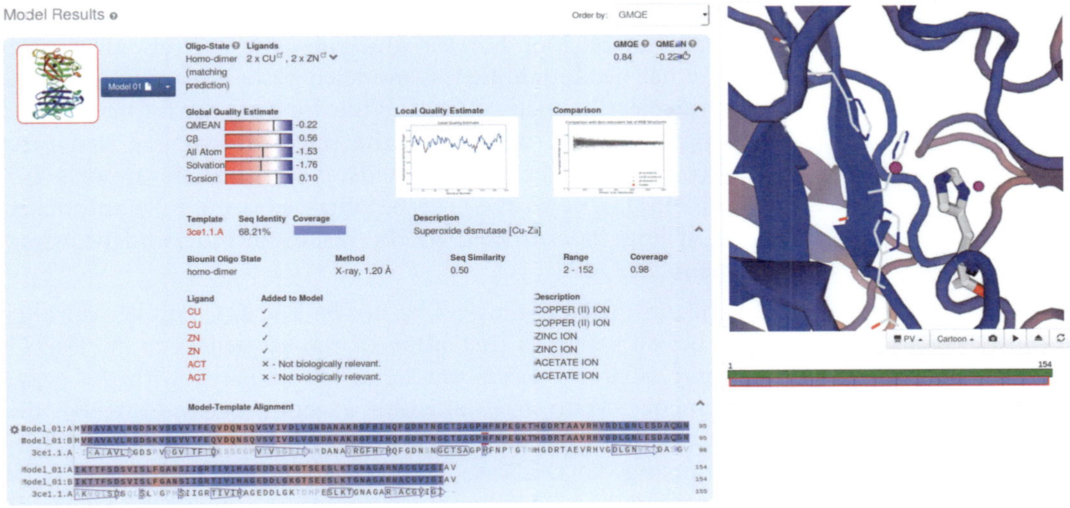

Fig. 2 Modeling results for the superoxide dismutase [Cu-Zn] protein from *S. pombe* (SOD1, UniProtKB AC: P28758) generated in automated mode in SWISS-MODEL. *Sp*SOD1 is predicted as homo-2-mer including 1 Zn ion and 1 Cu ion per subunit as cofactors based on the experimental structure of the deep-sea yeast *Cryptococcus liquefaciens* homologue (SMTL: 3ce1.1 A; [36]) as template

3.5 Model Quality Estimation

By default, models and alignments are colored based on the per-residue quality estimates from the QMEAN scoring function [37]. The color gradient ranges from red to blue, indicating low to high estimated per-residue quality. The same information is also available for every model in the form of a Local Quality plot, as well as in the B-factor column of the downloadable PDB file. The Global Quality plot gives an estimate of the overall model quality, based on four individual terms: Cβ, all atom, solvation, and torsion. The QMEAN score is also compared to what one would expect from experimentally determined protein structures of similar size using a Z-score scheme (hence 0.0 would be the optimal score). This is illustrated in the Comparison plot (Fig. 2).

4 Notes

1. While for most proteins classical homology modeling works well, for some protein families the results are not satisfying. For example, dedicated modeling methods for immunoglobulins exploit protein family-specific sequence and structure properties and are therefore more effective. In line with this, if an immunoglobulin sequence is present in the input, SWISS-MODEL provides a link to the Prediction of ImmunoGlobulin Structure server PIGSPro [17]. The PIGSPro core strategy consists of modeling the conserved region of the antibody by homology while the prediction of five of the six hypervariable

loops is based on the canonical structure model [38–40]. The hypervariable loop H3, for which no complete canonical structure has been identified, is modeled based on a different template selection method [41]. While for both the framework and loops the user can modify the selected template structures according to their own needs, very accurate models (C-α-RMSD close to 1 Å) can be obtained using fully automated modeling, as estimated by the results of independent assessments [17, 42].

2. The most frequent reason for not finding any template during a template search is that none of the sequences in the SMTL library shares a significant sequence similarity with the target sequence. With a few possible exceptions (*see* **Note 3**), this means that no suitable template is available for comparative modeling. In this case, de novo modeling methods may be considered as an alternative. However, one should keep in mind that the accuracy of such techniques is considerably lower than that achievable by comparative modeling [43, 44].

3. Stringent quality filters are applied to protein structures from PDB in order to be included in the SMTL. For example, structures only providing Cα coordinates are considered low quality and are excluded. Therefore, it can happen that a specific PDB entry is not listed in a template search despite sharing significant sequence similarity with the target. The availability of specific template structures can be checked by directly querying the SMTL (Menu: Modeling—Template Library).

   In case a model should be generated based on template coordinates not present in SMTL, e.g., for newly solved structures not yet deposited to PDB, it is possible to create a model based on such templates by either the "Template mode" or the "Project mode" options (details in Subheading 3.2).

4. Increasingly, multiple alternative experimental template structures are available for target proteins of interest, often covering different regions of the amino acid sequence, representing different conformational states, or harboring different ligands. Using this information from heterogeneous templates efficiently is an area of active research in the field of comparative modeling [45]. Structural information from multiple templates can, in some cases, be complementary and the modeling procedure potentially benefits from the added information from alternative templates [46]. The main improvement by multi-template modeling is generally due to an increased coverage of the target, i.e., increased size of the produced model. Such functionality is work in progress in SWISS-MODEL and currently not supported. Alternative solutions are available to this

purpose, both as stand-alone software and web servers [47, 48].

5. Biologically relevant ligands and cofactors are modeled based on a homology transfer approach from the templates identified in the SMTL. This approach is rather restrictive; that is, it requires a high conservation of ligand-binding site residues between target and template both in terms of sequence and structure. This implies that a given ligand which is present in the template will not necessarily be present in the final model. Docking strategies can be considered as a valuable solution in these cases. For this purpose, SWISS-MODEL provides a direct link to send models to SwissDock [49, 50] from the model results page.
6. Currently, the number of heteromeric complexes characterized experimentally at atomic resolution is limited. Consequently, there is a high chance that no suitable templates are identified for a heteromeric protein complex. In this case, a possible solution is to perform a template search for each subunit of the complex and model them separately as monomers. Resulting models can then be used as input for external docking software. The performance of current methods for protein docking is assessed during the Critical Assessment of PRediction of Interactions (CAPRI) [51].
7. In case a specific oligomeric state is expected (e.g., a homotetramer) but not found during the template search, the user can build a monomeric model and employ external software to enforce a specific stoichiometry [52] or symmetry [53] using the monomer as input. If further experimental information is available, i.e., interactions between subunits of the complex, a hybrid modeling approach can be suitable [54–56].
8. There are several different possible sources of modeling errors, which are typically detected by the provided local quality estimation tools. Some minor issues such as local stereochemical violations due to small structural distortions or incorrect modeling of side chains can be resolved by energy minimization. In other cases, further inspection of both the alignment and the model is needed. For alignments of templates with low sequence similarity, incorrect positioning of insertions and deletions may give rise to model errors; manual correction of the alignment may be a solution in some of these cases. For modeling very long insertions, library-based approaches have limitations and one has to recur to de novo modeling approaches such as Rosetta [57] and I-TASSER [58]. However, from a practical point of view, modeling long insertions is typically less reliable, and it is important to critically evaluate whether these efforts are necessary and sufficient for the

intended application of the model. Finally, deviations of the model from the native structure may be due to structural divergence between target and template during evolution. The structural variability of the protein family at hand typically provides a good estimate for the errors expected in the models [59]. One of the main limitations of comparative modeling today is the intrinsic dependence on information from template structures; that is, it is not able to "predict" structural divergence where the target deviates from the templates. Approaches for refining the model coordinates based on molecular modeling simulations have made significant progress in recent years for the task of refining models closer to the native structure, however, at a very high computational cost [60].

References

1. Guex N, Peitsch MC, Schwede T (2009) Automated comparative protein structure modeling with SWISS-MODEL and Swiss-PdbViewer: a historical perspective. Electrophoresis 30 Suppl 1:S162–S173
2. Sali A, Blundell TL (1993) Comparative protein modelling by satisfaction of spatial restraints. J Mol Biol 234:779–815
3. Chothia C, Lesk AM (1986) The relation between the divergence of sequence and structure in proteins. EMBO J 5:823–826
4. Arnold K, Bordoli L, Kopp J et al (2006) The SWISS-MODEL workspace: a web-based environment for protein structure homology modelling. Bioinformatics 22:195–201
5. Biasini M, Bienert S, Waterhouse A et al (2014) SWISS-MODEL: modelling protein tertiary and quaternary structure using evolutionary information. Nucleic Acids Res 42: W252–W258
6. Kiefer F, Arnold K, Kunzli M et al (2009) The SWISS-MODEL repository and associated resources. Nucleic Acids Res 37:D387–D392
7. Waterhouse A, Bertoni M, Bienert S et al (2018) SWISS-MODEL: homology modelling of protein structures and complexes. Nucleic Acids Research Res 46(W1):W296–W303
8. Kryshtafovych A, Venclovas C, Fidelis K et al (2005) Progress over the first decade of CASP experiments. Proteins 61(Suppl 7):225–236
9. Berman H, Henrick K, Nakamura H et al (2007) The worldwide protein data Bank (wwPDB): ensuring a single, uniform archive of PDB data. Nucleic Acids Res 35: D301–D303
10. Altschul SF, Madden TL, Schaffer AA et al (1997) Gapped BLAST and PSI-BLAST: a new generation of protein database search programs. Nucleic Acids Res 25:3389–3402
11. Remmert M, Biegert A, Hauser A et al (2011) HHblits: lightning-fast iterative protein sequence searching by HMM-HMM alignment. Nat Methods 9:173–175
12. Jones DT (1999) Protein secondary structure prediction based on position-specific scoring matrices. J Mol Biol 292:195–202
13. Sillitoe I, Cuff AL, Dessailly BH et al (2013) New functional families (FunFams) in CATH to improve the mapping of conserved functional sites to 3D structures. Nucleic Acids Res 41:D490–D498
14. Aloy P, Ceulemans H, Stark A et al (2003) The relationship between sequence and interaction divergence in proteins. J Mol Biol 332:989–998
15. Bertoni M, Kiefer F, Biasini M et al (2017) Modeling protein quaternary structure of homo- and hetero-oligomers beyond binary interactions by homology. Sci Rep 7:10480
16. Marcatili P, Olimpieri PP, Chailyan A et al (2014) Antibody modeling using the prediction of immunoglobulin structure (PIGS) web server [corrected]. Nat Protoc 9:2771–2783
17. Lepore R, Olimpieri PP, Messih MA et al (2017) PIGSPro: prediction of immunoGlobulin structures v2. Nucleic Acids Res 45:W17
18. Biasini M, Schmidt T, Bienert S et al (2013) OpenStructure: an integrated software framework for computational structural biology. Acta Crystallogr D Biol Crystallogr 69:701–709

19. Fiser A (2010) Template-based protein structure modeling. Methods Mol Biol 673:73–94
20. Choi Y, Deane CM (2010) FREAD revisited: accurate loop structure prediction using a database search algorithm. Proteins 78:1431–1440
21. Liang S, Zhang C, Zhou Y (2014) LEAP: highly accurate prediction of protein loop conformations by integrating coarse-grained sampling and optimized energy scores with all-atom refinement of backbone and side chains. J Comput Chem 35:335–341
22. Messih MA, Lepore R, Tramontano A (2015) LoopIng: a template-based tool for predicting the structure of protein loops. Bioinformatics 31:3767–3772
23. Canutescu AA, Dunbrack RL Jr (2003) Cyclic coordinate descent: a robotics algorithm for protein loop closure. Protein science: a publication of the protein. Society 12:963–972
24. Sippl MJ (1990) Calculation of conformational ensembles from potentials of mean force. An approach to the knowledge-based prediction of local structures in globular proteins. J Mol Biol 213:859–883
25. Shapovalov MV, Dunbrack RL Jr (2011) A smoothed backbone-dependent rotamer library for proteins derived from adaptive kernel density estimates and regressions. Structure 19:844–858
26. Krivov GG, Shapovalov MV, Dunbrack RL Jr (2009) Improved prediction of protein side-chain conformations with SCWRL4. Proteins 77:778–795
27. Xu J (2005) Rapid protein side-chain packing via tree decomposition. In: Miyano S, Mesirov J, Kasif S, Istrail S, Pevzner PA, Waterman M (eds) Research in computational molecular biology: 9th Annual International Conference, RECOMB 2005, Cambridge, MA, USA, May 14–18, 2005. Proceedings. Springer Berlin, Heidelberg, pp 423–439
28. Mackerell AD Jr, Feig M, Brooks CL 3rd (2004) Extending the treatment of backbone energetics in protein force fields: limitations of gas-phase quantum mechanics in reproducing protein conformational distributions in molecular dynamics simulations. J Comput Chem 25:1400–1415
29. Eastman P, Swails J, Chodera JD et al (2017) OpenMM 7: rapid development of high performance algorithms for molecular dynamics. PLoS Comput Biol 13:e1005659
30. Baker D, Sali A (2001) Protein structure prediction and structural genomics. Science 294:93–96
31. Schwede T, Sali A, Honig B et al (2009) Outcome of a workshop on applications of protein models in biomedical research. Structure 17:151–159
32. Read RJ, Adams PD, Arendall WB 3rd et al (2011) A new generation of crystallographic validation tools for the protein data bank. Structure 19:1395–1412
33. Benkert P, Biasini M, Schwede T (2011) Toward the estimation of the absolute quality of individual protein structure models. Bioinformatics 27:343–350
34. Benkert P, Kunzli M, Schwede T (2009) QMEAN server for protein model quality estimation. Nucleic Acids Res 37:W510–W514
35. Haas J, Roth S, Arnold K et al (2013) The protein model portal--a comprehensive resource for protein structure and model information. Database 2013:bat031
36. Teh AH, Kanamasa S, Kajiwara S et al (2008) Structure of cu/Zn superoxide dismutase from the heavy-metal-tolerant yeast Cryptococcus liquefaciens strain N6. Biochem Biophys Res Commun 374:475–478
37. Benkert P, Tosatto SC, Schomburg D (2008) QMEAN: a comprehensive scoring function for model quality assessment. Proteins 71:261–277
38. Chothia C, Lesk AM (1987) Canonical structures for the hypervariable regions of immunoglobulins. J Mol Biol 196:901–917
39. Morea V, Tramontano A, Rustici M et al (1998) Conformations of the third hypervariable region in the VH domain of immunoglobulins. J Mol Biol 275:269–294
40. Tramontano A, Chothia C, Lesk AM (1990) Framework residue 71 is a major determinant of the position and conformation of the second hypervariable region in the VH domains of immunoglobulins. J Mol Biol 215:175–182
41. Messih MA, Lepore R, Marcatili P et al (2014) Improving the accuracy of the structure prediction of the third hypervariable loop of the heavy chains of antibodies. Bioinformatics 30:2733–2740
42. Almagro JC, Teplyakov A, Luo J et al (2014) Second antibody modeling assessment (AMA-II). Proteins 82:1553–1562
43. Moult J (2005) A decade of CASP: progress, bottlenecks and prognosis in protein structure prediction. Curr Opin Struct Biol 15:285–289
44. Tai CH, Bai H, Taylor TJ et al (2014) Assessment of template-free modeling in CASP10 and ROLL. Proteins 82(Suppl 2):57–83
45. Meier A, Soding J (2015) Automatic prediction of protein 3D structures by probabilistic multi-template homology modeling. PLoS Comput Biol 11:e1004343

46. Larsson P, Wallner B, Lindahl E et al (2008) Using multiple templates to improve quality of homology models in automated homology modeling. Protein Sci 17:990–1002
47. Cheng J (2008) A multi-template combination algorithm for protein comparative modeling. BMC Struct Biol 8:18
48. Webb B, Sali A (2014) Comparative protein structure modeling using MODELLER. Curr Protoc Bioinformatics 47:5.6.1–5.6.32
49. Grosdidier A, Zoete V, Michielin O (2011) Fast docking using the CHARMM force field with EADock DSS. J Comput Chem 32:2149–2159
50. Grosdidier A, Zoete V, Michielin O (2011) SwissDock, a protein-small molecule docking web service based on EADock DSS. Nucleic Acids Res 39:W270–W277
51. Lensink MF, Velankar S, Wodak SJ (2017) Modeling protein-protein and protein-peptide complexes: CAPRI 6th edition. Proteins 85:359–377
52. Esquivel-Rodriguez J, Filos-Gonzalez V, Li B et al (2014) Pairwise and multimeric protein-protein docking using the LZerD program suite. Methods Mol Biol 1137:209–234
53. Pierce B, Tong W, Weng Z (2005) M-ZDOCK: a grid-based approach for Cn symmetric multimer docking. Bioinformatics 21:1472–1478
54. De Vries SJ, Van Dijk M, Bonvin AM (2010) The HADDOCK web server for data-driven biomolecular docking. Nat Protoc 5:883–897
55. Leaver-Fay A, Tyka M, Lewis SM et al (2011) ROSETTA3: an object-oriented software suite for the simulation and design of macromolecules. Methods Enzymol 487:545–574
56. Russel D, Lasker K, Webb B et al (2012) Putting the pieces together: integrative modeling platform software for structure determination of macromolecular assemblies. PLoS Biol 10: e1001244
57. Simons KT, Kooperberg C, Huang E et al (1997) Assembly of protein tertiary structures from fragments with similar local sequences using simulated annealing and Bayesian scoring functions. J Mol Biol 268:209–225
58. Yang J, Yan R, Roy A et al (2015) The I-TASSER suite: protein structure and function prediction. Nat Methods 12:7–8
59. Maghrabi AHA, Mcguffin LJ (2017) ModFOLD6: an accurate web server for the global and local quality estimation of 3D protein models. Nucleic Acids Res 45(W1): W416–W421
60. Heo L, Feig M (2018) What makes it difficult to refine protein models further via molecular dynamics simulations? Proteins 86(Suppl 1):177–188

Chapter 18

Interface-Based Structural Prediction of Novel Host-Pathogen Interactions

Emine Guven-Maiorov, Chung-Jung Tsai, Buyong Ma, and Ruth Nussinov

Abstract

About 20% of the cancer incidences worldwide have been estimated to be associated with infections. However, the molecular mechanisms of exactly how they contribute to host tumorigenesis are still unknown. To evade host defense, pathogens hijack host proteins at different levels: sequence, structure, motif, and binding surface, i.e., interface. Interface similarity allows pathogen proteins to compete with host counterparts to bind to a target protein, rewire physiological signaling, and result in persistent infections, as well as cancer. Identification of host-pathogen interactions (HPIs)—along with their structural details at atomic resolution—may provide mechanistic insight into pathogen-driven cancers and innovate therapeutic intervention. HPI data including structural details is scarce and large-scale experimental detection is challenging. Therefore, there is an urgent and mounting need for efficient and robust computational approaches to predict HPIs and their complex (bound) structures. In this chapter, we review the first and currently only interface-based computational approach to identify novel HPIs. The concept of interface mimicry promises to identify more HPIs than complete sequence or structural similarity. We illustrate this concept with a case study on *Kaposi's sarcoma herpesvirus (KSHV)* to elucidate how it subverts host immunity and helps contribute to malignant transformation of the host cells.

Key words Host-pathogen interaction prediction, Protein–protein interaction, Structural network, Superorganism network, Molecular mimicry, Interface mimicry

1 Introduction

1.1 Molecular Mimicry

Signaling pathways shape and convey the cell's responses to stimuli from its environment; however, pathogens can circumvent this response by "repurposing" host signaling. Pathogens can interact with the host through proteins, metabolites, small molecules, and nucleic acids [1]. Direct protein-protein interactions are the most common interaction type (*see* **Note 1**). By interfering with key pathways pathogens can reshape physiological signaling, subverting the immune system, altering the cytoskeletal organization [2, 3], modifying membrane and vesicular trafficking [2, 4, 5], boosting pathogen entry into the host cell, changing the cell cycle regulation

Tobias Sikosek (ed.), *Computational Methods in Protein Evolution*, Methods in Molecular Biology, vol. 1851, https://doi.org/10.1007/978-1-4939-8736-8_18, © Springer Science+Business Media, LLC, part of Springer Nature 2019

[6, 7], and modulating apoptosis [8]. All host-pathogen interactions (HPIs) aim to ensure pathogen survival within the host.

Pathogens evolved several strategies to cross-talk with their hosts. One powerful way is molecular mimicry, which has been extensively reviewed in our recent study [9]. There are four different levels of molecular (protein) mimicry: hijacking (1) both sequence and structure of a protein or a domain, (2) only structure without sequence homology, (3) sequence of a short motif—motif mimicry, and (4) structure of a binding surface without sequence similarity—interface mimicry. Global sequence and structural similarity is much rarer than interface similarity both within and across species. Thus, utilizing interface mimicry allows pathogens to target more host proteins. The concept of interface mimicry, proposed over two decades ago, suggested that proteins with different global structures can interact in similar ways, via similar binding surfaces [10–12]. Interfaces are frequently "reused" by distinct proteins [13], suggesting that these recurring architectures are favorable scaffolds [12].

Interface mimicry is often observed within (intraspecies/endogenous) [13–15] and across species (interspecies/host-pathogen/exogenous) [16, 17]. Similarity in endogenous and exogenous protein-protein interfaces permits pathogenic proteins to compete with their host counterparts [17], rewire host signaling, and cause infections, as well as cancer. Identification of the HPIs and the rewired host-pathogen superorganism protein interaction network, together with structural details, should provide critical insights into pathogenic virulence strategies underlying infections and pathogen-driven cancers, and hence help innovative therapeutics [18].

To date, the HPI networks show that different pathogens often target the same host pathway, and certain host pathways are attacked at several nodes to guarantee alteration of host cell signaling [19]. Although there are several available host-pathogen metaorganism interaction networks [19–29], there have been few attempts to integrate these HPI networks with the human 3D structural protein-protein interactions (PPIs) [17]. Traditional node-and-edge representation of the PPI networks simplifies the "big picture." They depict which proteins interact, but not how. Structural networks allow a higher resolution with mechanistic insights, showing which residues are involved in the interaction and thus which binary interactions can co-occur or are mutually exclusive [16, 30] (*see* **Note 2**). The power of structural networks in displaying the details of endogenous signaling pathways was demonstrated earlier [30–33]. They are also vital to comprehend the mechanisms exerted by pathogens to avert and subvert host cell signaling and circumvent immune response [18]. Structures exhibit which endogenous PPIs are ablated by the HPIs, whether the virulence factors in different strains of the same pathogenic species have distinct HPIs, and possible outcomes of mutations on either the host or the pathogenic proteins.

The challenging large-scale experimental characterization of HPIs [34, 35], coupled with the scarcity of experimentally confirmed HPI data, especially structural details, escalates the demand for efficient and robust computational approaches to predict HPIs along with their complex (bound) structures. In this chapter, we first review available computational approaches to predict HPIs and present the only interface-based computational approach available to identify novel HPIs and their complex structures. Then, we illustrate the usefulness of our approach with a case study on *Kaposi's sarcoma herpesvirus (KSHV)*.

1.2 Review of Available Computational Tools to Identify HPIs

Several HPI databases have been developed for experimentally identified HPIs, including PHISTO [36], HPIDB [37], Proteopathogen [38], PATRIC [39], PHI-base [40], PHIDIAS [41], HoPaCI-DB [42], VirHostNet [43], ViRBase [44], VirusMentha [45], and HCVpro [46]. These databases comprise only a limited number of pathogens. Given that at least hundreds of different species can infect the host, thousands of HPIs are still unknown. Enriching of the host-pathogen interactome and construction of comprehensive HPI networks will still mostly rely on computational models in the near future [47]. Numerous studies computationally identified large-scale HPIs and built HPI networks for viruses and bacteria [20, 24, 48–56].

Although prediction of human PPIs is a well-established area, modeling of interspecies interactions is comparably new. Still, several attempts focused on computational approaches to identify HPIs [34], most of which rely on sequence homology [49, 52, 54, 57–63]. Homology-based approaches are successful only if the sequence similarity is high, but not all virulence factors have homologs in human. For instance, a secreted protein of *H. pylori*, VacA, does not have sequence similarity with any other known viral, bacterial, or eukaryotic proteins [64], but it alters signaling through several host pathways [65]. Thus, sequence-based methods cannot detect VacA's HPIs, highlighting the importance of considering the 3D structures of proteins in predicting HPIs. There are also sequence-based comparative methods that consider structure [48, 55, 56, 61, 62, 66–70]; interologs (interacting homologs/conserved interactions) [71, 72]; and transcriptome data [73]. Available structure-based techniques often depend on global structural similarity rather than interface mimicry [55, 69]. One method combines interface data with sequence homology and gene expression, but the predicted interacting host and pathogenic proteins should satisfy a minimum of 80% sequence identity over at least 50% of template host PPI complexes [66]. To the best of our knowledge, none of the current approaches utilizes solely interface structures to model HPIs, except our recently developed interface-based method [74].

It has been suggested that the existing interface structures in PDB are diverse enough to cover majority of the endogenous PPIs [75–78] and hence success of template-based approaches to model endogenous PPIs is high [15] and expected to increase even more with the increase in the number of resolved PPI 3D structures [79] and advances in computational biology. Exogenous interactions are underrepresented in the PDB: there are not many exogenous interfaces. Since exogenous interfaces hijack endogenous ones, available endogenous and exogenous interfaces may represent most the structural host-pathogen interface space (*see* **Note 2**).

2 Methods

2.1 Modeling HPIs

Here, we review the first and to date only computational approach that utilizes solely interface mimicry to predict putative HPIs and their 3D structures as complexes [74]. Local structural resemblance is sufficient; there is no need for sequence similarity. This approach reveals not only targets of pathogenic proteins and how they interact, but also the host endogenous PPIs which may be disrupted by these potential HPIs. Figure 1 displays our workflow. Generally, the interacting protein partners are known from docking studies and the main purpose is to discern how they interact structurally. Therefore, inputs of the docking algorithms are structures of the two monomeric target proteins to be docked to each other. However, when dealing with HPIs, the main aim is to identify the interacting partners, as well as how they interact. Normally, the pathogenic proteins (one of the targets in a docking study) are known but not their partners in the host (second target). Hence, before performing docking, we need to identify those potential host interactors.

To accomplish this, we generate all known human interfaces—including endogenous and exogenous—in the PiFace interface database, as described in [14]. Each interface has two chains (partners/sides). There are 26,236 human interfaces in our template set. Then, we structurally align these interfaces with the pathogenic proteins by MultiProt [80]. The structural alignment thresholds for the number of matching interface residues and the hot spots follow the PRISM algorithm [81–84]. If the pathogenic protein is aligned with one side of the human interface, it may interact with the complementary side. Thus, the pathogenic protein can compete with the first side of the interface—with which it is structurally aligned—to bind to the second side, thereby abrogating the endogenous binary interaction in the template PPI (Fig. 1). Structural complementarity does not necessarily guarantee chemical complementarity and favorable interaction energy. For instance, 8 *KSHV*

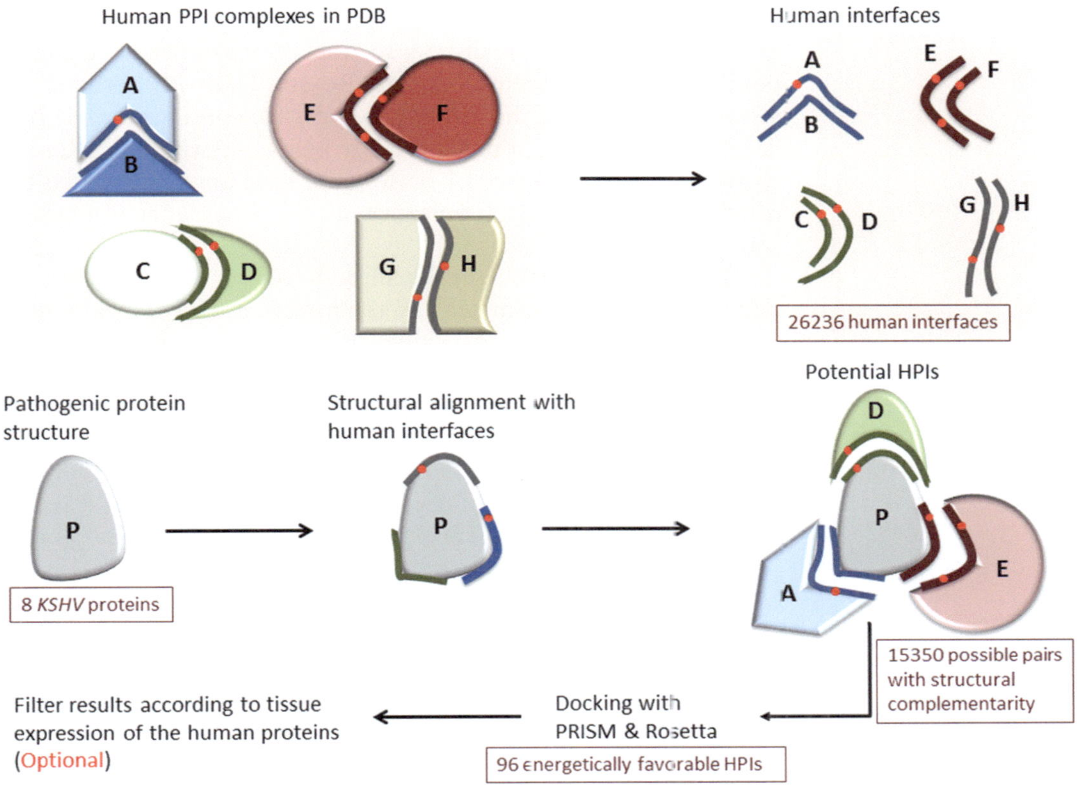

Fig. 1 Workflow of our interface-based HPI modeling approach. In the first step, we extract human interfaces from the PDB. Then, we obtain the structures of the pathogenic proteins from the PDB. Before docking, we need to identify the potential HPI pairs since docking programs require two target proteins. To do that, we structurally align the pathogenic proteins with the human interfaces in our template set. If the pathogenic protein is aligned with the B-side of the interface, it can interact with the complementary A-side. After determining potential HPI pairs, we perform docking of these pairs with PRISM [81–84] and Rosetta (local refinement) [85–87] to select the energetically favorable ones. We further assess the likelihood that the HPI models take place in the cell based on the percent match of the interface residues with the template interface and probability of the template interface being a real biological interface. In the final optional step, we filter our energetically favorable HPI results according to tissue expression of the human proteins by checking whether the interactors of the pathogenic proteins are expressed in the same tissue where the pathogen resides

proteins aligned with 15,350 interfaces, but only 96 of them are energetically favorable. So, after detection of the potential partners in humans with structural complementarity, we check whether these potential HPI pairs have favorable interaction energy. To do that we perform docking with two programs: PRISM [81–84] and Rosetta (local refinement) [85–87]. We take HPIs as energetically favorable only if their Rosetta interface scores (I_sc) are below −5 and total energy scores are below zero. We also calculate Rosetta I_sc for the endogenous template PPIs and compare them with those of modeled HPIs to determine whether the pathogenic protein will outcompete the endogenous partner to bind to a target host protein with a higher affinity. For some template PPIs, Rosetta gives extremely low unrealistic I_sc, due to intermolecular disulfide bonds. To correct this, we calculate Rosetta I_sc with both including and disregarding the disulfide bonds. We consider the HPIs as favorable interactions if they have I_sc below −5 with both Rosetta scorings. Note that Rosetta I_sc does not have units nor reflects the real binding free energy. It only gives an idea whether an interaction is favorable or not.

To further evaluate the likelihoods of our HPI models, we calculate the "percent match" of the interfaces by taking the ratio of the number of interface residues that are aligned with the pathogenic protein to the number of interface residues in the endogenous template PPI. Each template interface is assigned with a weight based on the size of the endogenous template interface such that larger interfaces have higher weights. If the template interfaces have less than 30 residues ($n < 30$), the weight is 0.5; if $30 < n < 50$, weight is 1; if $50 < n < 80$ weight is 1.5; and if $n > 80$ (very large interface), the weight is 2. Score1 given in Table 1 is the product of the interface percent match and the corresponding interface weight.

We employ the EPPIC (Evolutionary Protein-Protein Interface Classifier) [88], to evaluate whether the template interfaces are real biological interfaces or crystal artifacts. The EPPIC server gives the probability of a particular interface to be biological. Score2 in Table 1 is the product of Score1 and the probability of being a biological interface. The higher the Score2, the more confidence we have that a particular HPI model would take place in the cell, as they are better mimics of real biological endogenous interfaces (*see* **Note 3**).

Finally, with an optional step, the results can be filtered according to tissue expression, checking whether the host partners of the pathogenic proteins are expressed in the same tissue where the pathogen resides. We take the tissue expression data from the Human Protein Atlas, which includes 19,709 human proteins, mapping to 7106 human PDBs [89, 90]. If the pathogen is a bacterial species, it resides in only certain tissues. For instance, *Helicobacter pylori* is mainly in the stomach and gastrointestinal tract, making it reasonable to focus on human proteins that are expressed in these tissues. However, if the pathogen is a virus, it can infect several different—if not all—tissues. Therefore, filtering according to tissue expression is an optional step depending on the pathogen type (*see* **Note 4**).

2.2 Constructing the Structural Superorganism Network

As we have the complex (bound) structures of the predicted HPIs, it is possible to construct the structural interspecies interaction network. Our template set serves as the human endogenous binary interactions. 26,236 interfaces map to 3366 distinct human PPIs. The predicted HPIs serve as exogenous interactions. So, all

Table 1
HPIs for *KSHV* proteins

KSHV protein	*KSHV* protein PDB	Human protein	Human protein PDB	Template interface	I_sc of HPI	I_sc of PPI	# of residues aligned	# of residues in template interface	% Match	Weight	Sc1	Probability of template interface being a biological interface	Sc2
K4	2fhtA	CCL4	2x6lB	2x6lBD	−8.86	−9.53	29	35	82.9	1	82.9	0.9	74.6
K4	2fhtA	CXCR4	2k03D	2k03CD	−6.76	−11.26	25	55	45.5	1.5	68.2	0.48	32.7
K6	1zxtA	CCL5	1u4lB	1u4lAB	−8.90	−11.30	26	34	76.5	1	76.5	0.77	58.9
vCyclin	1g3nC	CDK4	3g33A	3g33AD	−6.07	−8.24	39	51	76.5	1.5	114.7	0.98	112.4
vCyclin	1g3nC	CDK2	1w98A	1w98AB	−5.41	−13.69	48	94	51.1	2	102.1	1	102.1
vIL6	1i1rB	IL12B	3duhB	3duhBD	−6.83	−13.47	26	50	52.0	1.5	78.0	0.9	70.2
vIL6	1i1rB	INAR1	3se4A	3se4AB	−6.08	−11.52	26	53	49.1	1.5	73.6	0.91	67.0
vIRF1	4hlxA	UBP21	3i3tG	3i3tGH	−5.88	−16.50	18	51	35.3	1.5	52.9	0.97	51.4
vFLIP	3cl3A	TNR6	3ezqI	3ezqIJ	−5.57	−11.04	19	54	35.2	1.5	52.8	0.11	5.8
vBCL2	1k3kA	ITA2B	2vdkA	2vdkAB	−5.19	−13.52	20	63	31.7	1.5	47.6	1	47.6

I_sc refers to Rosetta interface score, where we ignored disulfide bonds. If the I_sc of modeled HPI is lower than I_sc of template PPI, it means that the pathogenic protein may have higher affinity to target protein than the endogenous partner of the target pairwise interactions in the structural network will have structures as complexes. The topological features of the resulting superorganism network can be calculated by the NetworkAnalyzer [91] application in Cytoscape [92]. Functional annotation of pathogenic targets in the host can be performed by DAVID [93, 94].

To compare the pathogen of interest with other bacteria/viruses, we can also build the structural interspecies network for all known HPIs in PDB. There are 299 HPIs in PDB between human and different bacterial, yeast, and viral species.

2.3 Case Study

Our interface-based HPI modeling method was successfully applied to *H. pylori* before and can be applied to any commensal or pathogenic microorganism. As a case study to illustrate the utility of the concept, here we applied it to *KSHV*, infection of which is associated with a blood/lymph vessel cancer—Kaposi's sarcoma—and lymphoma [95]. We modeled its HPIs and constructed its structural superorganism network. We analyzed eight *KSHV* proteins, vCyclin, vFLIP, vBCL2, vIL6, vIRF1, vIRF2, and viral chemokines (K4 and K6). We found 96 putative HPIs. All our HPI models have 3D structures as complexes (*see* **Note 5**). Table 1 shows some examples from these 96 HPIs and Table 2 displays the human PPIs that are potentially disrupted by these HPIs.

Our HPI candidates may elucidate the roles of *KSHV* in modulation of host signaling and contribution to malignant transformation. For instance, we found that *KSHV* chemokines and cytokines, like K4, K6, and vIL6, target many human chemokine and cytokine receptors (Fig. 2). Signaling through the cytokine and chemokine receptors is critical for T-cell recruitment to the infected

Table 2
Potentially disrupted endogenous host PPIs due to predicted *KSHV* HPIs

KSHV protein	Human PPI disrupted by *KSHV* protein	PDB for the human PPI disrupted
K4	CCL4-CCL4	2x6lBD
K4	CXCR4-SDF1	2k03CD
K6	CCL5-CCL5	1u4lAB
vCyclin	CDK4-CCND3	3g33AD
vCyclin	CDK2-CCNE1	1w98AB
vIL6	IL12B-IL23A	3duhBD
vIL6	INAR1-IFNW1	3se4AB
vIRF1	UBP21-RL40	3i3tGH
vFLIP	TNR6-FADD	3ezqIJ
vBCL2	ITA2B-ITB3	2vdkAB

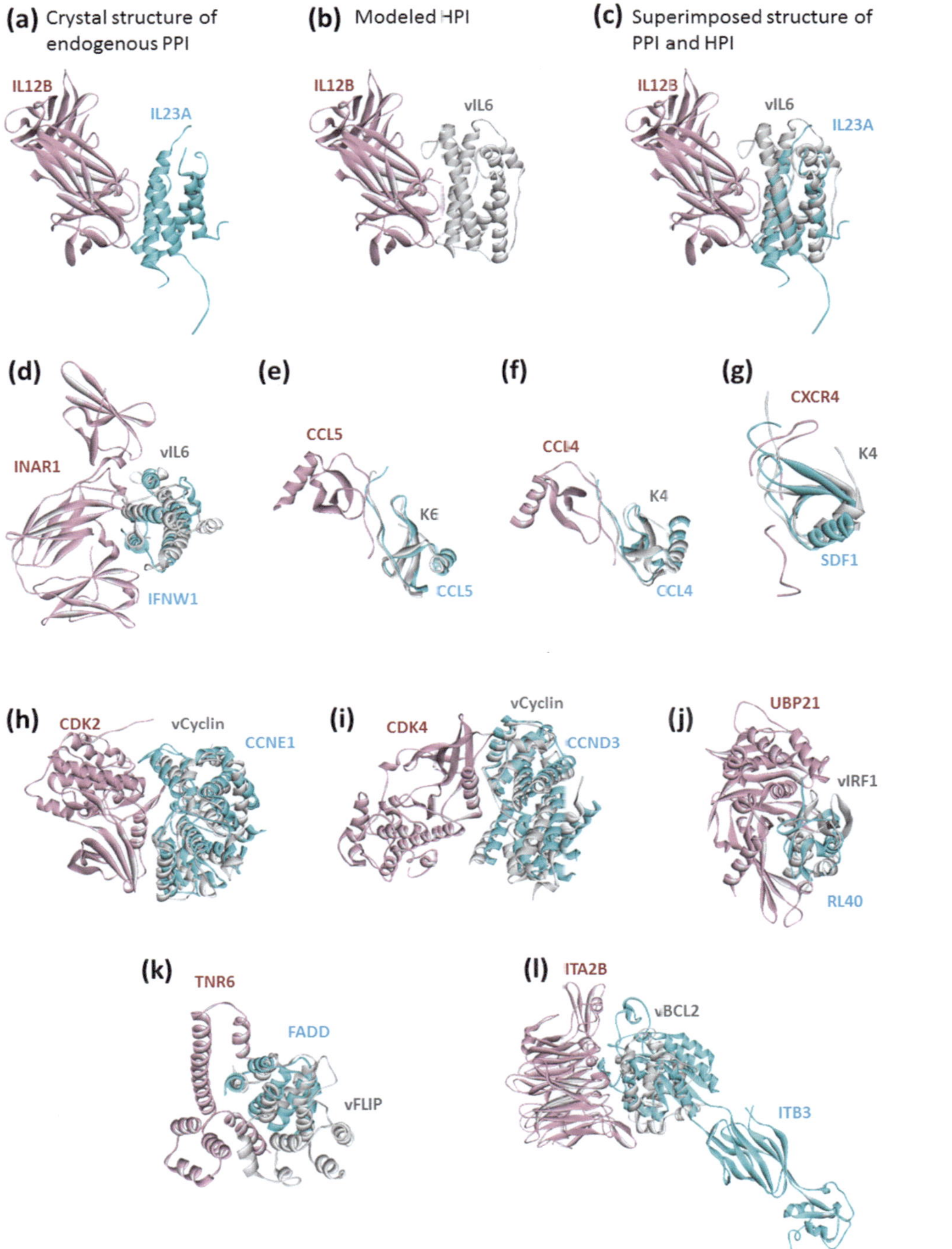

Fig. 2 *KSHV* proteins mimic the human protein-protein interfaces, blocking human PPIs. (**a**) Endogenous human PPI between IL12B and IL23A. (**b**) Our HPI model between vIL6 and IL12B. (**c**) Superimposed view of PPI and HPI shows that vIL6 almost perfectly mimics the interface on IL23A to bind to IL12B. (**d**) through (**l**) also show the superimposed structures of endogenous human PPIs and modeled HPIs. Human proteins are shown in cyan and pink; and *KSHV* proteins are shown in gray. Gray proteins bind to pink proteins by hijacking the interface on cyan proteins (only the interface similarity is enough, no need for global structural similarity). Thus, they may block the pink-cyan protein interactions

host tissue to eradicate the pathogens and for regulation of their activation and differentiation [96]. Blockage of these pathways by the *KSHV* proteins may underlie the molecular mechanisms of evading the immune system and persistence of infection. We also found that vCyclin interferes with several CDKs (Fig. 2), thereby disrupting normal host cell cycle regulation, which may contribute to aberrant proliferation in malignant transformation.

In addition to mimicked endogenous interfaces, hijacked exogenous interfaces can also be identified through our approach. A given pathogenic protein can mimic both human and pathogenic proteins from other species. For instance, we found that *KSHV* vCyclin mimics other viral vCyclin proteins to target human CDKs (Fig. 3).

We also constructed the structural superorganism network between human and *KSHV* (Fig. 4). The endogenous human PPIs are template PPIs and the exogenous virus-human interactions are HPI models. There are 3366 human PPIs and 96 HPIs in this network. Our results indicate that *KSHV* proteins can potentially target the highly connected part of the network and hub proteins, like CDK2 in the human PPI network. Hub proteins are critical to many cellular functions, establishing pathway cross talk. It is an ingenious pathogen strategy, since by attacking only a single

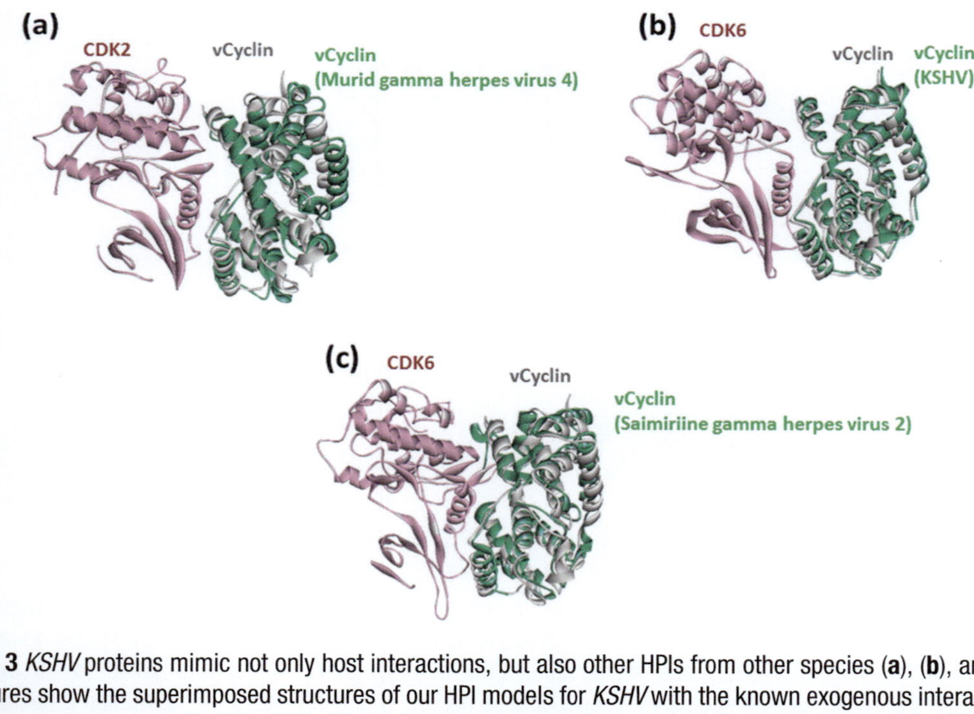

Fig. 3 *KSHV* proteins mimic not only host interactions, but also other HPIs from other species (**a**), (**b**), and (**c**). Figures show the superimposed structures of our HPI models for *KSHV* with the known exogenous interactions with proteins from other species. Pink proteins are from human, greens are proteins from other pathogens, and gray proteins are *KSHV* proteins. Gray proteins bind to pink proteins by hijacking the interfaces on green proteins

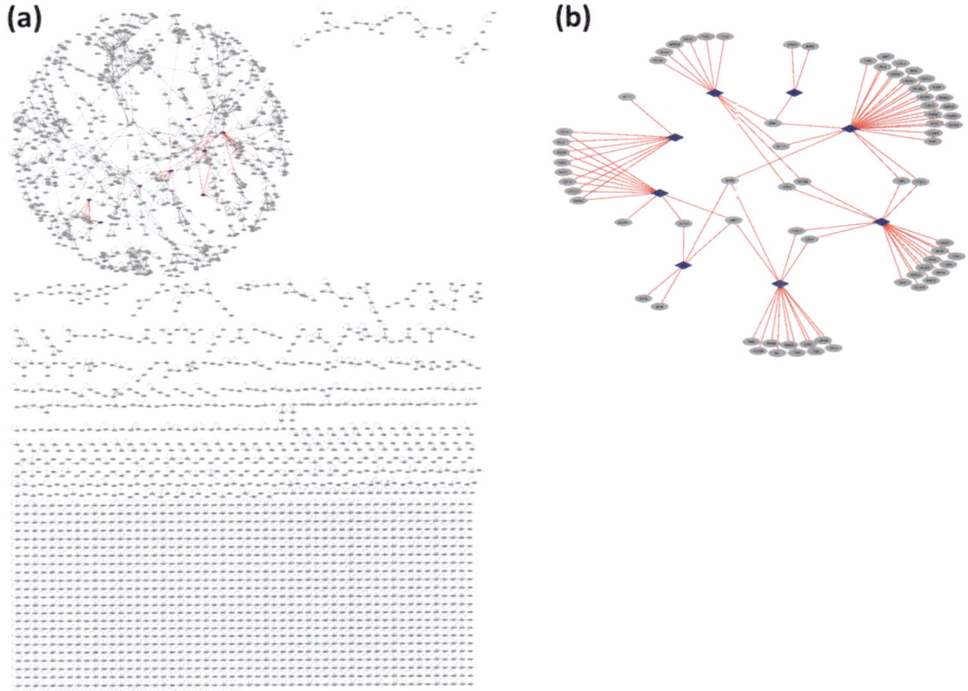

Fig. 4 Structural superorganism network for *KSHV* and human, where all binary interactions have structures as complexes. Endogenous human interactions (black edges) are obtained from crystal structures in PDB (template interface set), where human proteins are shown as gray circular nodes. Exogenous interactions (red edges) are our HPI models for 8 *KSHV* proteins, where viral proteins are shown as blue diamond nodes. (**a**) *KSHV* proteins target the highly connected part of the human PPI network. (**b**) Structural HPI network without the endogenous template interactions. Most targets of individual *KSHV* proteins are distinct, but some are shared across different *KSHV* proteins

protein they can interfere with several pathways. Functional annotation of the *KSHV*-targeted human proteins is enriched in 17 KEGG pathways (Table 3). Among the highly enriched, there are cytokine and chemokine signaling, and viral carcinogenesis pathways.

3 Concluding Remarks

Insight into mechanisms of infectious diseases and pathogen-driven cancers at the molecular level is limited. Identification of novel HPIs and their atomistic details may illuminate how virulence factors modulate host signaling, and stimulate innovative therapeutics. Large-scale detection of HPIs will rely on computational techniques in the near future due to current limitations of experimental methodologies. Most computational approaches rely on sequence homology which constrains the application of these tools to

Table 3
Functional enrichment of *KSHV*-targeted human proteins by DAVID [93, 94]

KEGG pathways	Number of genes enriched	%	*P* value	*KSHV*-targeted human proteins
Cytokine-cytokine receptor interaction	10	13.9	7.20E−05	CCL3, CCL2, CCL13, TNR6, CCL4, ACVR1, CCL5, CXCR4, INAR1, IL12B
Chemokine signaling pathway	9	12.5	9.80E−05	RHOA, CCL3, CCL2, CCL13, CCL4, CCL5, JAK2, CCL14, CXCR4
Herpes simplex infection	8	11.1	5.60E−04	CCL2, C1QBP, TNR6, CDK2, CCL5, JAK2, INAR1, IL12B
Measles	7	9.7	6.10E−04	TNR6, CCND3, CDK4, CDK2, JAK2, INAR1, IL12B
p53 signaling pathway	5	6.9	1.90E−03	TNR6, CCND3, CASP9, CDK4, CDK2
Influenza A	7	9.7	2.40E−03	CCL2, TNR6, CASP9, CCL5, JAK2, INAR1, IL12B
Pathways in cancer	10	13.9	3.50E−03	RHOA, ITA2B, FGFR2, TNR6, CASP9, CDK4, CDK2, CXCR4, ARHGB, BMP2
Hepatitis B	6	8.3	5.70E−03	CCNA2, TNR6, CASP9, CDK4, CDK2, INAR1
Chagas disease (American trypanosomiasis)	5	6.9	9.20E−03	CCL3, CCL2, TNR6, CCL5, IL12B
Toll-like receptor signaling pathway	5	6.9	9.80E−03	CCL3, CCL4, CCL5, INAR1, IL12B
PI3K-Akt signaling pathway	8	11.1	1.90E−02	ITA2B, FGFR2, CCND3, CASP9, CDK4, CDK2, JAK2, INAR1
African trypanosomiasis	3	4.2	2.80E−02	TNR6, HBA, IL12B
Small-cell lung cancer	4	5.6	3.00E−02	ITA2B, CASP9, CDK4, CDK2
Glutathione metabolism	3	4.2	6.20E−02	GSTA4, GSTP1, GSTM2
Cell cycle	4	5.6	7.60E−02	CCNA2, CCND3, CDK4, CDK2
Viral carcinogenesis	5	6.9	8.00E−02	CCNA2, RHOA, CCND3, CDK4, CDK2
Signaling pathways regulating pluripotency of stem cells	4	5.6	1.00E−01	FGFR2, ACVR1, JAK2, BMP2

pathogenic proteins that have no sequence homologs in human. Interface architectures are conserved within and across species regardless of the entire sequence and the structure of the proteins.

Here we reviewed the first and only available interface-based method to uncover novel HPIs and their complex 3D structures. This approach predicts not only the HPIs, but also the potentially disrupted endogenous human PPIs. It can be applied to any microbial organisms, including commensals and pathogens.

4 Notes

1. Our approach is based on the reasonable assumption that pathogenic proteins may alter host signaling. However, interactions through metabolites and small molecules also have roles in modulation of the host responses. Moreover, interaction of a particular pathogen with other microbial species in the microbiota and different combination of bacterial species also affects the overall response.
2. Some of the limitations are as follows: coverage of endogenous human PPIs is low; available endogenous protein structures are biased toward permanent, not transient, interactions; disordered proteins are underrepresented in the PDB; and most pathogenic proteins lack crystal structures.
3. Both experimental and computational methods have false positives with varying rates. Although HPIs predicted here may have false positives, we cannot calculate the exact false-positive rate due to limited experimental HPI data. We tried to minimize the error rates by calculating the percent match of the HPI models with the corresponding template PPI and incorporating the probability of template interfaces being real biological interfaces. Predicted models should be tested by experiments. Computational screening of big data can provide possible leads to experiments guiding functional characterization while avoiding testing millions of possible binary combinations of host and pathogenic proteins.
4. In addition to filtering by tissue expression, HPI models can also be filtered by subcellular localization of the host proteins. For instance, if the pathogenic protein is found in the cytoplasm of the host cell, then it cannot interact with the host nuclear proteins. Since the large-scale subcellular localization data for all proteins are not available, it is a choice of the researchers to do so.
5. Proteins often assemble into multi-protein complexes. Modeling only pairwise interactions between host and pathogenic proteins may not be sufficient.

Acknowledgments

This project has been funded in whole or in part with federal funds from the National Cancer Institute, National Institutes of Health, under contract number HHSN261200800001E. The content of this publication does not necessarily reflect the views or policies of the Department of Health and Human Services, nor does mention of trade names, commercial products, or organizations imply endorsement by the US Government. This research was supported (in part) by the Intramural Research Program of the NIH, National Cancer Institute, Center for Cancer Research. This study utilized the high-performance computational capabilities of the Biowulf PC/Linux cluster at the National Institutes of Health (NIH), Bethesda, MD (http://biowulf.nih.gov).

References

1. Durmus S, Cakir T, Ozgur A, Guthke R (2015) A review on computational systems biology of pathogen-host interactions. Front Microbiol 6:235. https://doi.org/10.3389/fmicb.2015.00235
2. Stebbins CE, Galan JE (2001) Structural mimicry in bacterial virulence. Nature 412 (6848):701–705. https://doi.org/10.1038/35089000
3. Sal-Man N, Biemans-Oldehinkel E, Finlay BB (2009) Structural microengineers: pathogenic Escherichia coli redesigns the actin cytoskeleton in host cells. Structure 17(1):15–19. https://doi.org/10.1016/j.str.2008.12.001
4. Kahn RA, Fu H, Roy CR (2002) Cellular hijacking: a common strategy for microbial infection. Trends Biochem Sci 27 (6):308–314. https://doi.org/10.1016/S0968-0004(02)02108-4
5. Finlay BB, McFadden G (2006) Anti-immunology: evasion of the host immune system by bacterial and viral pathogens. Cell 124 (4):767–782. https://doi.org/10.1016/j.cell.2006.01.034
6. Moody CA, Laimins LA (2010) Human papillomavirus oncoproteins: pathways to transformation. Nat Rev Cancer 10(8):550–560. https://doi.org/10.1038/nrc2886
7. Filippova M, Song H, Connolly JL, Dermody TS, Duerksen-Hughes PJ (2002) The human papillomavirus 16 E6 protein binds to tumor necrosis factor (TNF) R1 and protects cells from TNF-induced apoptosis. J Biol Chem 277(24):21730–21739. https://doi.org/10.1074/jbc.M200113200
8. Shirin H, Sordillo EM, Kolevska TK, Hibshoosh H, Kawabata Y, Oh SH, Kuebler JF, Delohery T, Weghorst CM, Weinstein IB, Moss SF (2000) Chronic helicobacter pylori infection induces an apoptosis-resistant phenotype associated with decreased expression of p27(kip1). Infect Immun 68(9):5321–5328
9. Guven-Maiorov E, Tsai CJ, Nussinov R (2016) Pathogen mimicry of host protein-protein interfaces modulates immunity. Semin Cell Dev Biol 58:136–145. https://doi.org/10.1016/j.semcdb.2016.06.004
10. Tsai CJ, Lin SL, Wolfson HJ, Nussinov R (1996) A dataset of protein-protein interfaces generated with a sequence-order-independent comparison technique. J Mol Biol 260 (4):604–620. https://doi.org/10.1006/jmbi.1996.0424
11. Tsai CJ, Lin SL, Wolfson HJ, Nussinov R (1996) Protein-protein interfaces: architectures and interactions in protein-protein interfaces and in protein cores. Their similarities and differences. Crit Rev Biochem Mol Biol 31 (2):127–152. https://doi.org/10.3109/10409239609106582
12. Keskin O, Nussinov R (2005) Favorable scaffolds: proteins with different sequence, structure and function may associate in similar ways. Protein Eng Des Sel 18(1):11–24. https://doi.org/10.1093/protein/gzh095
13. Keskin O, Nussinov R (2007) Similar binding sites and different partners: implications to shared proteins in cellular pathways. Structure 15(3):341–354. https://doi.org/10.1016/j.str.2007.01.007
14. Cukuroglu E, Gursoy A, Nussinov R, Keskin O (2014) Non-redundant unique interface structures as templates for modeling protein

interactions. PLoS One 9(1):e86738. https://doi.org/10.1371/journal.pone.0086738

15. Muratcioglu S, Guven-Maiorov E, Keskin O, Gursoy A (2015) Advances in template-based protein docking by utilizing interfaces towards completing structural interactome. Curr Opin Struct Biol 35:87–92. https://doi.org/10.1016/j.sbi.2015.10.001
16. Franzosa EA, Garamszegi S, Xia Y (2012) Toward a three-dimensional view of protein networks between species. Front Microbiol 3:428. https://doi.org/10.3389/fmicb.2012.00428
17. Franzosa EA, Xia Y (2011) Structural principles within the human-virus protein-protein interaction network. Proc Natl Acad Sci U S A 108(26):10538–10543. https://doi.org/10.1073/pnas.1101440108
18. Guven-Maiorov E, Tsai CJ, Nussinov R (2017) Structural host-microbiota interaction networks. PLoS Comput Biol 13(10):e1005579. https://doi.org/10.1371/journal.pcbi.1005579
19. Bhavsar AP, Guttman JA, Finlay BB (2007) Manipulation of host-cell pathways by bacterial pathogens. Nature 449(7164):827–834. https://doi.org/10.1038/nature06247
20. Uetz P, Dong YA, Zeretzke C, Atzler C, Baiker A, Berger B, Rajagopala SV, Roupelieva M, Rose D, Fossum E, Haas J (2006) Herpesviral protein networks and their interaction with the human proteome. Science 311(5758):239–242. https://doi.org/10.1126/science.1116804
21. von Schwedler UK, Stuchell M, Muller B, Ward DM, Chung HY, Morita E, Wang HE, Davis T, He GP, Cimbora DM, Scott A, Krausslich HG, Kaplan J, Morham SG, Sundquist WI (2003) The protein network of HIV budding. Cell 114(6):701–713
22. Calderwood MA, Venkatesan K, Xing L, Chase MR, Vazquez A, Holthaus AM, Ewence AE, Li N, Hirozane-Kishikawa T, Hill DE, Vidal M, Kieff E, Johannsen E (2007) Epstein-Barr virus and virus human protein interaction maps. Proc Natl Acad Sci U S A 104(18):7606–7611. https://doi.org/10.1073/pnas.0702332104
23. de Chassey B, Navratil V, Tafforeau L, Hiet MS, Aublin-Gex A, Agaugue S, Meiffren G, Pradezynski F, Faria BF, Chantier T, Le Breton M, Pellet J, Davoust N, Mangeot PE, Chaboud A, Penin F, Jacob Y, Vidalain PO, Vidal M, Andre P, Rabourdin-Combe C, Lotteau V (2008) Hepatitis C virus infection protein network. Mol Syst Biol 4:230. https://doi.org/10.1038/msb.2008.66
24. Shapira SD, Gat-Viks I, Shum BO, Dricot A, de Grace MM, Wu L, Gupta PB, Hao T, Silver SJ, Root DE, Hill DE, Regev A, Hacohen N (2009) A physical and regulatory map of host-influenza interactions reveals pathways in H1N1 infection. Cell 139(7):1255–1267. https://doi.org/10.1016/j.cell.2009.12.018
25. Zhang L, Villa NY, Rahman MM, Smallwood S, Shattuck D, Neff C, Dufford M, Lanchbury JS, Labaer J, McFadden G (2009) Analysis of vaccinia virus-host protein-protein interactions: validations of yeast two-hybrid screenings. J Proteome Res 8(9):4311–4318. https://doi.org/10.1021/pr900491n
26. Khadka S, Vangeloff AD, Zhang C, Siddavatam P, Heaton NS, Wang L, Sengupta R, Sahasrabudhe S, Randall G, Gribskov M, Kuhn RJ, Perera R, LaCount DJ (2011) A physical interaction network of dengue virus and human proteins. Mol Cell Proteomics 10(12):M111.012187. https://doi.org/10.1074/mcp.M111.012187
27. Jager S, Cimermancic P, Gulbahce N, Johnson JR, McGovern KE, Clarke SC, Shales M, Mercenne G, Pache L, Li K, Hernandez H, Jang GM, Roth SL, Akiva E, Marlett J, Stephens M, D'Orso I, Fernandes J, Fahey M, Mahon C, O'Donoghue AJ, Todorovic A, Morris JH, Maltby DA, Alber T, Cagney G, Bushman FD, Young JA, Chanda SK, Sundquist WI, Kortemme T, Hernandez RD, Craik CS, Burlingame A, Sali A, Frankel AD, Krogan NJ (2011) Global landscape of HIV-human protein complexes. Nature 481(7381):365–370. https://doi.org/10.1038/nature10719
28. Pichlmair A, Kandasamy K, Alvisi G, Mulhern O, Sacco R, Habjan M, Binder M, Stefanovic A, Eberle CA, Goncalves A, Burckstummer T, Muller AC, Fauster A, Holze C, Lindsten K, Goodbourn S, Kochs G, Weber F, Bartenschlager R, Bowie AG, Bennett KL, Colinge J, Superti-Furga G (2012) Viral immune modulators perturb the human molecular network by common and unique strategies. Nature 487(7408):486–490. https://doi.org/10.1038/nature11289
29. Rozenblatt-Rosen O, Deo RC, Padi M, Adelmant G, Calderwood MA, Rolland T, Grace M, Dricot A, Askenazi M, Tavares M, Pevzner SJ, Abderazzaq F, Byrdsong D, Carvunis AR, Chen AA, Cheng J, Correll M, Duarte M, Fan C, Feltkamp MC, Ficarro SB, Franchi R, Garg BK, Gulbahce N, Hao T, Holthaus AM, James R, Korkhin A, Litovchick L, Mar JC, Pak TR, Rabello S,

Rubio R, Shen Y, Singh S, Spangle JM, Tasan M, Wanamaker S, Webber JT, Roecklein-Canfield J, Johannsen E, Barabasi AL, Beroukhim R, Kieff E, Cusick ME, Hill DE, Munger K, Marto JA, Quackenbush J, Roth FP, DeCaprio JA, Vidal M (2012) Interpreting cancer genomes using systematic host network perturbations by tumour virus proteins. Nature 487(7408):491–495. https://doi.org/10.1038/nature11288

30. Guven Maiorov E, Keskin O, Gursoy A, Nussinov R (2013) The structural network of inflammation and cancer: merits and challenges. Semin Cancer Biol 23(4):243–251. https://doi.org/10.1016/j.semcancer.2013.05.003
31. Guven-Maiorov E, Keskin O, Gursoy A, VanWaes C, Chen Z, Tsai CJ, Nussinov R (2015) The architecture of the TIR domain signalosome in the toll-like Receptor-4 signaling pathway. Sci Rep 5:13128. https://doi.org/10.1038/srep13128
32. Guven-Maiorov E, Keskin O, Gursoy A, Nussinov R (2015) A structural view of negative regulation of the toll-like receptor-mediated inflammatory pathway. Biophys J 109(6):1214–1226. https://doi.org/10.1016/j.bpj.2015.06.048
33. Acuner-Ozbabacan ES, Engin BH, Guven-Maiorov E, Kuzu G, Muratcioglu S, Baspinar A, Chen Z, Van Waes C, Gursoy A, Keskin O, Nussinov R (2014) The structural network of Interleukin-10 and its implications in inflammation and cancer. BMC Genomics 15(Suppl 4):S2. https://doi.org/10.1186/1471-2164-15-S4-S2
34. Nourani E, Khunjush F, Durmus S (2015) Computational approaches for prediction of pathogen-host protein-protein interactions. Front Microbiol 6:94. https://doi.org/10.3389/fmicb.2015.00094
35. Brito AF, Pinney JW (2017) Protein-protein interactions in virus-host systems. Front Microbiol 8:1557. https://doi.org/10.3389/fmicb.2017.01557
36. Durmus Tekir S, Cakir T, Ardic E, Sayilirbas AS, Konuk G, Konuk M, Sariyer H, Ugurlu A, Karadeniz I, Ozgur A, Sevilgen FE, Ulgen KO (2013) PHISTO: pathogen-host interaction search tool. Bioinformatics 29(10):1357–1358. https://doi.org/10.1093/bioinformatics/btt137
37. Kumar R, Nanduri B (2010) HPIDB--a unified resource for host-pathogen interactions. BMC Bioinformatics 11(Suppl 6):S16. https://doi.org/10.1186/1471-2105-11-S6-S16
38. Vialas V, Nogales-Cadenas R, Nombela C, Pascual-Montano A, Gil C (2009) Proteopathogen, a protein database for studying Candida albicans--host interaction. Proteomics 9(20):4664–4668. https://doi.org/10.1002/pmic.200900023
39. Wattam AR, Abraham D, Dalay O, Disz TL, Driscoll T, Gabbard JL, Gillespie JJ, Gough R, Hix D, Kenyon R, Machi D, Mao C, Nordberg EK, Olson R, Overbeek R, Pusch GD, Shukla M, Schulman J, Stevens RL, Sullivan DE, Vonstein V, Warren A, Will R, Wilson MJ, Yoo HS, Zhang C, Zhang Y, Sobral BW (2014) PATRIC, the bacterial bioinformatics database and analysis resource. Nucleic Acids Res 42(Database issue):D581–D591. https://doi.org/10.1093/nar/gkt1099
40. Urban M, Pant R, Raghunath A, Irvine AG, Pedro H, Hammond-Kosack KE (2015) The Pathogen-Host Interactions database (PHI-base): additions and future developments. Nucleic Acids Res 43(Database issue):D645–D655. https://doi.org/10.1093/nar/gku1165
41. Xiang Z, Tian Y, He Y (2007) PHIDIAS: a pathogen-host interaction data integration and analysis system. Genome Biol 8(7):R150. https://doi.org/10.1186/gb-2007-8-7-r150
42. Bleves S, Dunger I, Walter MC, Frangoulidis D, Kastenmuller G, Voulhoux R, Ruepp A (2014) HoPaCI-DB: host-Pseudomonas and Coxiella interaction database. Nucleic Acids Res 42(Database issue):D671–D676. https://doi.org/10.1093/nar/gkt925
43. Guirimand T, Delmotte S, Navratil V (2015) VirHostNet 2.0: surfing on the web of virus/host molecular interactions data. Nucleic Acids Res 43(Database issue):D583–D587. https://doi.org/10.1093/nar/gku1121
44. Li Y, Wang C, Miao Z, Bi X, Wu D, Jin N, Wang L, Wu H, Qian K, Li C, Zhang T, Zhang C, Yi Y, Lai H, Hu Y, Cheng L, Leung KS, Li X, Zhang F, Li K, Li X, Wang D (2015) ViRBase: a resource for virus-host ncRNA-associated interactions. Nucleic Acids Res 43(Database issue):D578–D582. https://doi.org/10.1093/nar/gku903
45. Calderone A, Licata L, Cesareni G (2015) VirusMentha: a new resource for virus-host protein interactions. Nucleic Acids Res 43(Database issue):D588–D592. https://doi.org/10.1093/nar/gku830
46. Kwofie SK, Schaefer U, Sundararajan VS, Bajic VB, Christoffels A (2011) HCVpro: hepatitis C virus protein interaction database. Infect Genet Evol 11(8):1971–1977. https://doi.org/10.1016/j.meegid.2011.09.001
47. Arnold R, Boonen K, Sun MG, Kim PM (2012) Computational analysis of

interactomes: current and future perspectives for bioinformatics approaches to model the host-pathogen interaction space. Methods 57 (4):508–518. https://doi.org/10.1016/j.ymeth.2012.06.011

48. Doolittle JM, Gomez SM (2011) Mapping protein interactions between dengue virus and its human and insect hosts. PLoS Negl Trop Dis 5(2):e954. https://doi.org/10.1371/journal.pntd.0000954
49. Tyagi N, Krishnadev O, Srinivasan N (2009) Prediction of protein-protein interactions between Helicobacter pylori and a human host. Mol BioSyst 5(12):1630–1635. https://doi.org/10.1039/b906543c
50. Xu Q, Xiang EW, Yang Q (2011) Transferring network topological knowledge for predicting protein-protein interactions. Proteomics 11 (19):3818–3825. https://doi.org/10.1002/pmic.201100146
51. Remmele CW, Luther CH, Balkenhol J, Dandekar T, Muller T, Dittrich MT (2015) Integrated inference and evaluation of host-fungi interaction networks. Front Microbiol 6:764. https://doi.org/10.3389/fmicb.2015.00764
52. Evans P, Dampier W, Ungar L, Tozeren A (2009) Prediction of HIV-1 virus-host protein interactions using virus and host sequence motifs. BMC Med Genet 2:27. https://doi.org/10.1186/1755-8794-2-27
53. Zhang M, Su S, Bhatnagar RK, Hassett DJ, Lu LJ (2012) Prediction and analysis of the protein interactome in Pseudomonas aeruginosa to enable network-based drug target selection. PLoS One 7(7):e41202. https://doi.org/10.1371/journal.pone.0041202
54. Huo T, Liu W, Guo Y, Yang C, Lin J, Rao Z (2015) Prediction of host – pathogen protein interactions between Mycobacterium tuberculosis and Homo sapiens using sequence motifs. BMC Bioinformatics 16:100. https://doi.org/10.1186/s12859-015-0535-y
55. Doolittle JM, Gomez SM (2010) Structural similarity-based predictions of protein interactions between HIV-1 and Homo sapiens. Virol J 7:82. https://doi.org/10.1186/1743-422X-7-82
56. de Chassey B, Meyniel-Schicklin L, Aublin-Gex A, Navratil V, Chantier T, Andre P, Lotteau V (2013) Structure homology and interaction redundancy for discovering virus-host protein interactions. EMBO Rep 14 (10):938–944. https://doi.org/10.1038/embor.2013.130
57. Petrenko P, Doxey AC (2015) mimicMe: a web server for prediction and analysis of host-like proteins in microbial pathogens. Bioinformatics 31(4):590–592. https://doi.org/10.1093/bioinformatics/btu681
58. Krishnadev O, Srinivasan N (2011) Prediction of protein-protein interactions between human host and a pathogen and its application to three pathogenic bacteria. Int J Biol Macromol 48 (4):613–619. https://doi.org/10.1016/j.ijbiomac.2011.01.030
59. Dyer MD, Murali TM, Sobral BW (2007) Computational prediction of host-pathogen protein-protein interactions. Bioinformatics 23(13):i159–i166. https://doi.org/10.1093/bioinformatics/btm208
60. Doxey AC, McConkey BJ (2013) Prediction of molecular mimicry candidates in human pathogenic bacteria. Virulence 4(6):453–466. https://doi.org/10.4161/viru.25180
61. Mahajan G, Mande SC (2017) Using structural knowledge in the protein data bank to inform the search for potential host-microbe protein interactions in sequence space: application to Mycobacterium tuberculosis. BMC Bioinformatics 18(1):201. https://doi.org/10.1186/s12859-017-1550-y
62. Mariano R, Wuchty S (2017) Structure-based prediction of host-pathogen protein interactions. Curr Opin Struct Biol 44:119–124. https://doi.org/10.1016/j.sbi.2017.02.007
63. Becerra A, Bucheli VA, Moreno PA (2017) Prediction of virus-host protein-protein interactions mediated by short linear motifs. BMC Bioinformatics 18(1):163. https://doi.org/10.1186/s12859-017-1570-7
64. Jones KR, Whitmire JM, Merrell DS (2010) A tale of two toxins helicobacter pylori CagA and VacA modulate host pathways that impact disease. Front Microbiol 1:115. https://doi.org/10.3389/fmicb.2010.00115
65. Manente L, Perna A, Buommino E, Altucci L, Lucariello A, Citro G, Baldi A, Iaquinto G, Tufano MA, De Luca A (2008) The Helicobacter pylori's protein VacA has direct effects on the regulation of cell cycle and apoptosis in gastric epithelial cells. J Cell Physiol 214 (3):582–587. https://doi.org/10.1002/jcp.21242
66. Davis FP, Barkan DT, Eswar N, McKerrow JH, Sali A (2007) Host pathogen protein interactions predicted by comparative modeling. Protein Sci 16(12):2585–2596. https://doi.org/10.1110/ps.073228407
67. Drayman N, Glick Y, Ben-nun-shaul O, Zer H, Zlotnick A, Gerber D, Schueler-Furman O, Oppenheim A (2013) Pathogens use structural mimicry of native host ligands as a mechanism for host receptor engagement. Cell Host

Microbe 14(1):63–73. https://doi.org/10.1016/j.chom.2013.05.005

68. Aloy P, Bottcher B, Ceulemans H, Leutwein C, Mellwig C, Fischer S, Gavin AC, Bork P, Superti-Furga G, Serrano L, Russell RB (2004) Structure-based assembly of protein complexes in yeast. Science 303 (5666):2026–2029. https://doi.org/10.1126/science.1092645

69. Rajasekharan S, Rana J, Gulati S, Sharma SK, Gupta V, Gupta S (2013) Predicting the host protein interactors of Chandipura virus using a structural similarity-based approach. Pathog Dis 69(1):29–35. https://doi.org/10.1111/2049-632X.12064

70. Zhang A, He L, Wang Y (2017) Prediction of GCRV virus-host protein interactome based on structural motif-domain interactions. BMC Bioinformatics 18(1):145. https://doi.org/10.1186/s12859-017-1500-8

71. Lee SA, Chan CH, Tsai CH, Lai JM, Wang FS, Kao CY, Huang CY (2008) Ortholog-based protein-protein interaction prediction and its application to inter-species interactions. BMC Bioinformatics 9(Suppl 12):S11. https://doi.org/10.1186/1471-2105-9-S12-S11

72. Krishnadev O, Srinivasan N (2008) A data integration approach to predict host-pathogen protein-protein interactions: application to recognize protein interactions between human and a malarial parasite. In Silico Biol 8 (3–4):235–250

73. Schulze S, Henkel SG, Driesch D, Guthke R, Linde J (2015) Computational prediction of molecular pathogen-host interactions based on dual transcriptome data. Front Microbiol 6:65. https://doi.org/10.3389/fmicb.2015.00065

74. Guven-Maiorov E, Tsai CJ, Ma B, Nussinov R (2017) Prediction of host-pathogen interactions for helicobacter pylori by interface mimicry and implications to gastric cancer. J Mol Biol 429(24):3925–3941. https://doi.org/10.1016/j.jmb.2017.10.023

75. Zhang QC, Petrey D, Deng L, Qiang L, Shi Y, Thu CA, Bisikirska B, Lefebvre C, Accili D, Hunter T, Maniatis T, Califano A, Honig B (2012) Structure-based prediction of protein-protein interactions on a genome-wide scale. Nature 490(7421):556–560. https://doi.org/10.1038/nature11503

76. Zhang QC, Petrey D, Norel R, Honig BH (2010) Protein interface conservation across structure space. Proc Natl Acad Sci U S A 107 (24):10896–10901. https://doi.org/10.1073/pnas.1005894107

77. Gao M, Skolnick J (2010) Structural space of protein-protein interfaces is degenerate, close to complete, and highly connected. Proc Natl Acad Sci U S A 107(52):22517–22522. https://doi.org/10.1073/pnas.1012820107

78. Kundrotas PJ, Zhu Z, Janin J, Vakser IA (2012) Templates are available to model nearly all complexes of structurally characterized proteins. Proc Natl Acad Sci U S A 109 (24):9438–9441. https://doi.org/10.1073/pnas.1200678109

79. Franzosa EA, Xia Y (2012) Structural models for host-pathogen protein-protein interactions: assessing coverage and bias. Pac Symp Biocomput:287–298

80. Shatsky M, Nussinov R, Wolfson HJ (2004) A method for simultaneous alignment of multiple protein structures. Proteins 56(1):143–156. https://doi.org/10.1002/prot.10628

81. Tuncbag N, Gursoy A, Nussinov R, Keskin O (2011) Predicting protein-protein interactions on a proteome scale by matching evolutionary and structural similarities at interfaces using PRISM. Nat Protoc 6(9):1341–1354. https://doi.org/10.1038/nprot.2011.367

82. Keskin O, Nussinov R, Gursoy A (2008) PRISM: protein-protein interaction prediction by structural matching. Methods Mol Biol 484:505–521. https://doi.org/10.1007/978-1-59745-398-1_30

83. Baspinar A, Cukuroglu E, Nussinov R, Keskin O, Gursoy A (2014) PRISM: a web server and repository for prediction of protein-protein interactions and modeling their 3D complexes. Nucleic Acids Res 42 (Web Server issue):W285–W289. https://doi.org/10.1093/nar/gku397

84. Ogmen U, Keskin O, Aytuna AS, Nussinov R, Gursoy A (2005) PRISM: protein interactions by structural matching. Nucleic Acids Res 33 (Web Server):W331–W336. https://doi.org/10.1093/nar/gki585

85. Gray JJ, Moughon S, Wang C, Schueler-Furman O, Kuhlman B, Rohl CA, Baker D (2003) Protein-protein docking with simultaneous optimization of rigid-body displacement and side-chain conformations. J Mol Biol 331 (1):281–299

86. Wang C, Schueler-Furman O, Baker D (2005) Improved side-chain modeling for protein-protein docking. Protein Sci 14 (5):1328–1339. https://doi.org/10.1110/ps.041222905

87. Wang C, Bradley P, Baker D (2007) Protein-protein docking with backbone flexibility. J Mol Biol 373(2):503–519. https://doi.org/10.1016/j.jmb.2007.07.050

88. Duarte JM, Srebniak A, Scharer MA, Capitani G (2012) Protein interface classification by evolutionary analysis. BMC Bioinformatics 13:334. https://doi.org/10.1186/1471-2105-13-334

89. Uhlen M, Bjorling E, Agaton C, Szigyarto CA, Amini B, Andersen E, Andersson AC, Angelidou P, Asplund A, Asplund C, Berglund L, Bergstrom K, Brumer H, Cerjan D, Ekstrom M, Elobeid A, Eriksson C, Fagerberg L, Falk R, Fall J, Forsberg M, Bjorklund MG, Gumbel K, Halimi A, Hallin I, Hamsten C, Hansson M, Hedhammar M, Hercules G, Kampf C, Larsson K, Lindskog M, Lodewyckx W, Lund J, Lundeberg J, Magnusson K, Malm E, Nilsson P, Odling J, Oksvold P, Olsson I, Oster E, Ottosson J, Paavilainen L, Persson A, Rimini R, Rockberg J, Runeson M, Sivertsson A, Skollermo A, Steen J, Stenvall M, Sterky F, Stromberg S, Sundberg M, Tegel H, Tourle S, Wahlund E, Walden A, Wan J, Wernerus H, Westberg J, Wester K, Wrethagen U, Xu LL, Hober S, Ponten F (2005) A human protein atlas for normal and cancer tissues based on antibody proteomics. Mol Cell Proteomics 4(12):1920–1932. https://doi.org/10.1074/mcp.M500279-MCP200

90. Uhlen M, Fagerberg L, Hallstrom BM, Lindskog C, Oksvold P, Mardinoglu A, Sivertsson A, Kampf C, Sjostedt E, Asplund A, Olsson I, Edlund K, Lundberg E, Navani S, Szigyarto CA, Odeberg J, Djureinovic D, Takanen JO, Hober S, Alm T, Edqvist PH, Berling H, Tegel H, Mulder J, Rockberg J, Nilsson P, Schwenk JM, Hamsten M, von Feilitzen K, Forsberg M, Persson L, Johansson F, Zwahlen M, von Heijne G, Nielsen J, Ponten F (2015) Proteomics. Tissue-based map of the human proteome. Science 347(6220):1260419. https://doi.org/10.1126/science.1260419

91. Yang H, Ke Y, Wang J, Tan Y, Myeni SK, Li D, Shi Q, Yan Y, Chen H, Guo Z, Yuan Y, Yang X, Yang R, Du Z (2011) Insight into bacterial virulence mechanisms against host immune response via the Yersinia pestis-human protein-protein interaction network. Infect Immun 79(11):4413–4424. https://doi.org/10.1128/IAI.05622-11

92. Shannon P, Markiel A, Ozier O, Baliga NS, Wang JT, Ramage D, Amin N, Schwikowski B, Ideker T (2003) Cytoscape: a software environment for integrated models of biomolecular interaction networks. Genome Res 13(11):2498–2504. https://doi.org/10.1101/gr.1239303

93. Huang d W, Sherman BT, Lempicki RA (2009) Bioinformatics enrichment tools: paths toward the comprehensive functional analysis of large gene lists. Nucleic Acids Res 37(1):1–13. https://doi.org/10.1093/nar/gkn923

94. Huang d W, Sherman BT, Lempicki RA (2009) Systematic and integrative analysis of large gene lists using DAVID bioinformatics resources. Nat Protoc 4(1):44–57. https://doi.org/10.1038/nprot.2008.211

95. Dissinger NJ, Damania B (2016) Recent advances in understanding Kaposi's sarcoma-associated herpesvirus. F1000Res 5:F1000. https://doi.org/10.12688/f1000research.7612.1

96. Luther SA, Cyster JG (2001) Chemokines as regulators of T cell differentiation. Nat Immunol 2(2):102–107. https://doi.org/10.1038/84205

Chapter 19

Predicting Functions of Disordered Proteins with MoRFpred

Christopher J. Oldfield, Vladimir N. Uversky, and Lukasz Kurgan

Abstract

Intrinsically disordered proteins and regions are involved in a wide range of cellular functions, and they often facilitate protein-protein interactions. Molecular recognition features (MoRFs) are segments of intrinsically disordered regions that bind to partner proteins, where binding is concomitant with a transition to a structured conformation. MoRFs facilitate translation, transport, signaling, and regulatory processes and are found across all domains of life. A popular computational tool, MoRFpred, accurately predicts MoRFs in protein sequences. MoRFpred is implemented as a user-friendly web server that is freely available at http://biomine.cs.vcu.edu/servers/MoRFpred/. We describe this predictor, explain how to run the web server, and show how to interpret the results it generates. We also demonstrate the utility of this web server based on two case studies, focusing on the relevance of evolutionary conservation of MoRF regions.

Key words Intrinsic disorder, Prediction, Molecular recognition features, MoRFs, Protein-protein interactions, MoRFpred

1 Introduction

Intrinsically disordered proteins (IDPs) and protein regions (IDRs) are incompetent in forming stable three-dimensional structure, yet perform varied and vital biological functions [1–4]. The lack of the prerequisite of a stable structure for function creates several challenges in the study of IDPs and IDRs, both experimental and computational [3]. The crux of these challenges on the computational side is the lack of conservation in many IDRs relative to structured proteins. Without the need to maintain rigid structures many IDRs diverge drastically, even in closely related species [5]. Lack of conservation confounds established methods for function annotation that rely on sequence similarity to transfer functional annotations. Lack of conservation is not universal in IDRs; many IDRs may be conserved, or more commonly conserved in portions of their sequences [5].

Tobias Sikosek (ed.), *Computational Methods in Protein Evolution*, Methods in Molecular Biology, vol. 1851, https://doi.org/10.1007/978-1-4939-8736-8_19,

One mechanism of IDP function is short functional elements within IDRs. The evolutionary origin of these functional elements is seemingly idiosyncratic, where some examples have been found to be evolutionarily conserved [6–8], and others have been proposed to be emergent sequence features [9, 10]. A common function of these short functional elements is binding to molecular partners, often other proteins [11, 12]. These types of features are likely common across many biological processes [13], such as cell cycle regulation, modulation of cellular structure, and apoptosis.

One model of these functional elements is known as molecular recognition features (MoRFs) [14]. It models functional elements within IDRs as short regions of increased structural propensity within longer regions of intrinsic disorder [13]. Examples of these types of functional regions can readily be inferred from protein structures and sequence properties [11]. Several predictors of MoRFs have been developed [13–17]. Initial predictors relied on direct interpretation of the MoRF model, by scanning for patterns in prediction of intrinsic disorder and employing a second level of prediction over patterns of interest [13, 15]. The most recent MoRF predictors, including MoRFpred, relax the strict reliance on disorder prediction patterns while still directly considering disorder predictions [17]. Several other methods of MoRF prediction have been independently developed [13–16, 18–21]. In addition to MoRFs, several other related models of functional elements with IDRs have been proposed. Eukaryotic linear motifs (ELMs) model these elements as short sequence motifs which can be predicted by pattern matching and filtering spurious matches [22]. Though they are very different models, MoRF and ELM predictions are frequently coincident [23]. Further, several generalized models of binding regions within IDRs have been developed [24–27]. Relative to other methods, MoRFpred was developed on a well-defined dataset with short functional elements that bind to other proteins within larger regions of intrinsic disorder. Like all methods of this type, the specificity is difficult to assess exactly, but this predictor features a good estimated sensitivity [17]. MoRFpred is useful for gaining insight into the function of novel IDPs.

MoRFpred is available as a user-friendly web server at http://biomine.cs.vcu.edu/servers/MoRFpred/. This server has been extensively used by the community since it was released in early 2012. Usage data collected with the Google Analytics platform reveals that MoRFpred was utilized close to 9000 times by over 2700 unique users from 711 cities and 71 countries. The article that introduces this computational tool was already cited 175 times (source: Google Scholar on June 29, 2018).

2 Materials and Methods

2.1 Datasets

For training of MoRFpred, a set of MoRFs was constructed beginning with known binding regions from Protein Data Bank (PDB) [28]. Bound peptides from PDB were carefully filtered for clear binding to a longer protein chain and mapped back to their source proteins. This procedure resulted in a dataset of 842 MoRFs. To avoid training and testing on similar proteins, these MoRFs were grouped into 427 clusters and divided into testing and training sets. This gave training and testing sets with 421 and 419 MoRFs, respectively, with no protein more than 30% identical between the two sets (*see* **Note 1**).

A set of negative examples that do not contain MoRFs with near certainty were constructed from protein chains that have been completely structurally characterized by X-ray crystallography at a high resolution. The chance of intrinsic disorder in the negative set was minimized by only selecting monomeric proteins without large cofactors that contained no missing residues due to lack of electron density. Further, any protein with a significant amount of predicted intrinsic disorder, >30% of residues, was discarded. Filtering for proteins with less than 30% identity resulted in a set of 28 proteins.

2.2 Architecture

MoRFpred is a support vector machine (SVM) over a rich feature space merged with a sequence similarity-based prediction (Fig. 1). Features considered for the linear kernel SVM predictor included five disorder prediction methods [29–32], relative solvent accessible surface prediction [33], B-factor prediction [34], PSI-BLAST-generated position-specific scoring matrices (PSSMs), and amino acid propensity scales from AAindex [35]. Two broad sets of features were used from each of these methods: (1) per residue over a window of 25 residues and (2) values aggregated over a window. Aggregation methods included taking the difference over a window of 25 residues and a smaller window, which captures the features found to be useful for previous MoRF predictors. For example, previous MoRF predictors relied on elevated predicted disorder surrounding a predicted MoRF, but depressed values for the MoRF region itself. Indeed, the corresponding difference-based aggregation was found to be one of the strongest MoRF features.

Feature selection for the SVM predictor was based on a best-first iterative addition of ranked features. Features were ranked based on a combination of biserial correlations [36] and single-feature predictive performance, where poorly correlated or performing features were removed from consideration. Iterative addition of features was based on a modified fivefold cross-validation procedure, where a feature was only added if it improved prediction performance by at least 1%.

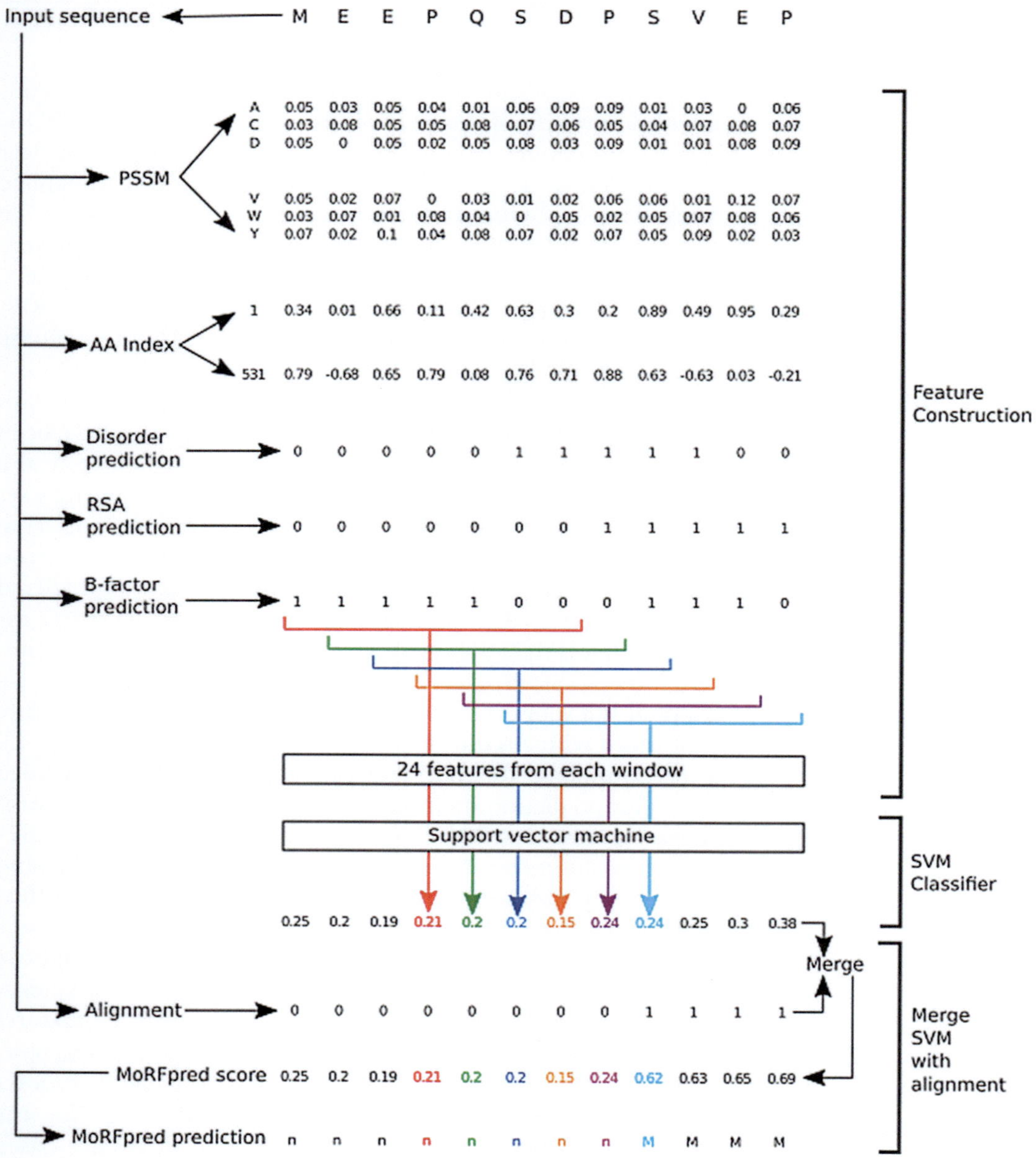

Fig. 1 Architecture of MoRFpred. The input sequence is used to generate sequence properties, from which input features are derived by windowed averaging. A support vector machine predicts MoRFs based on these input features. The SVM prediction is merged with similarity-based predictions to produce the final MoRFpred score, where scores above 0.5 are predicted MoRFs (M) and those less than 0.5 are predicted non-MoRFs (n)

Similarity-based predictions were done using a PSI-BLAST search against MoRF containing proteins in the training set. PSI-BLAST matches were selected based on an *e*-value threshold. An *e*-value = 0.5 was selected based on optimization of performance of the merged predictor. MoRF annotations from the training set are transferred to the query protein based on the PSI-BLAST alignment. These transferred annotations are merged with SVM-based prediction by adding one to the SVM prediction result and dividing by two, which ensures that the merged prediction for transferred annotations will be over the 0.5 threshold.

2.3 Predictive Quality

Prediction performance was assessed in the MoRFpred publication [17], using the true-positive and false-positive rates, overall accuracy (ACC), area under the ROC curve (AUC), and success rate. Success rate is a per-sequence measure of performance, where a sequence is considered successfully predicted if the MoRF residues have a higher average prediction score than the non-MoRF residues.

MoRFpred performance was assessed in comparison to ANCHOR and previously developed MoRF predictors. MoRFpred had ACC = 94.7% and the highest performance by success rate and AUC evaluations, with values of 71.8% and 67.3%, respectively. The original MoRF predictor had a very low false-positive rate, which artificially inflated its ACC value due to the large proportion of non-MoRF residues in the dataset. Adjusting the MoRFpred threshold to an equally low false-positive rate results in nearly double the true-positive rate of the original MoRF predictor.

2.4 Web Server

The MoRFpred web server is freely available at http://biomine.cs.vcu.edu/servers/MoRFpred/. The server can be accessed with an Internet connection and any modern web browser. All computations that are needed to complete predictions are performed on the server side.

On our web server, sequences submitted for prediction will be returned within 20 min of submission (*see* **Note 2**). The runtime of MoRFpred is dominated by the PSI-BLAST prediction, whose runtime varies with protein length and database similarity.

The main server page is where proteins are submitted for prediction. The web server only requires FASTA sequences of the proteins of interest to preform MoRFpred predictions. Up to five FASTA-formatted protein sequences may be entered into the large text entry field per submission. An e-mail address is required for each submission. All required programs for generating prediction features, including PSI-BLAST, and disorder, RSA, and B-factor predictions, are run automatically by scripts on the server. Upon completion of predictions for each submission, the server will send an e-mail notification with links to the prediction results.

2.5 Running MoRFpred

From the main server page, three steps are required to submit sequences to obtain the MoRFpred's predictions (Fig. 2, steps are highlighted with red numbers corresponding to the step):

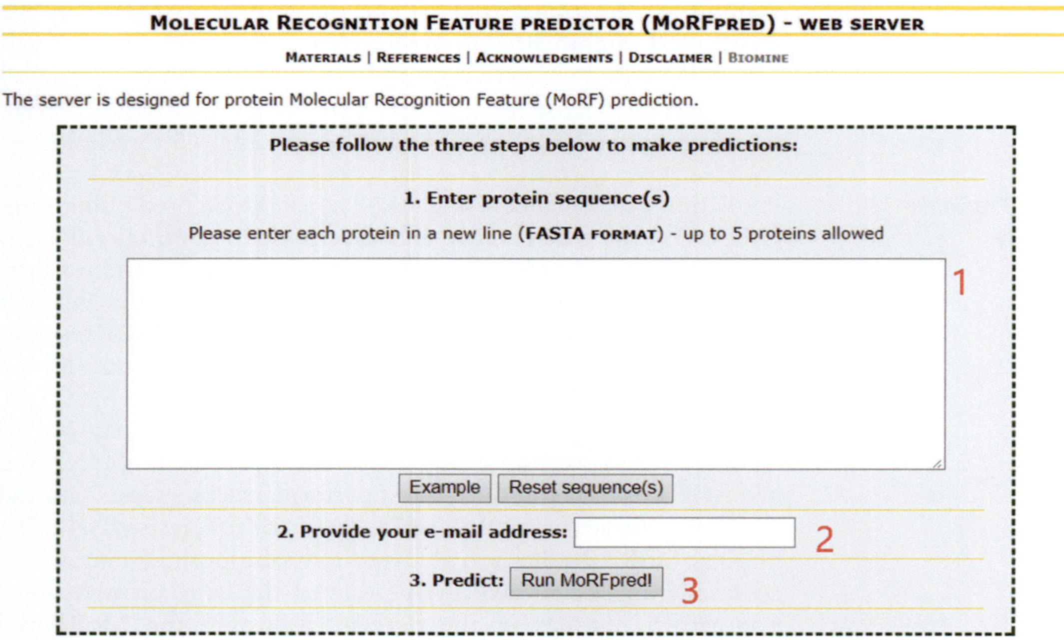

Fig. 2 Primary MoRFpred page, for submission of sequences for prediction. Red numbers indicate the sequence of steps required to submit a prediction

1. Copy your FASTA-formatted sequence (*see* **Note 3**) from its source file or web page and paste it into the text box (*see* **Notes 4** and **5**).
2. Enter an e-mail address. This is the address to which links to the prediction results will be sent.
3. Click "Run MoRFpred!". This submits the sequences to our server for MoRFpred predictions.

Once sequences are submitted for prediction, the browser is redirected to a status page that gives the current position of the submission in the server queue. This page will be automatically redirected to the results page when predictions are completed. The queue on the server is first come first serve, and if there is a large number of submissions, predictions may be delayed. Even if the web page is closed at this point, links to predictions will still be received through e-mail (*see* **Note 6**).

2.6 MoRFpred Results

The results page includes a link to the raw results (Fig. 3, red 1) as well as a color-coded text display of MoRFpred results (Fig. 3, red 2). The raw results (results.csv) file gives results for each submitted sequence, each in three lines, which are comma delimited:

1. The input sequence: the FASTA header followed by each residue of the input sequence.

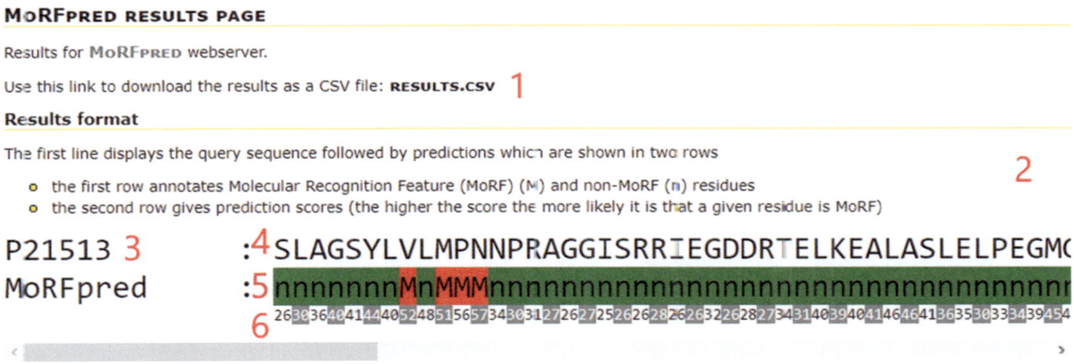

Fig. 3 MoRFpred prediction results page. Red numbers correspond to the primary features of the results page

2. Binary MoRFpred predictions: the string "MoRFpred" followed by one character for each residue in the input sequence—"O" for non-MoRF residues and "D" for MoRF residues.
3. The raw prediction output: the string "prob" followed by one floating point number between 0 and 1 for each residue of the input sequence. Predicted MoRF residues correspond to values greater than 0.5.

The color-coded text display of MoRFpred predictions includes the FASTA header of each input sequence (Fig. 3, red 3), and several aligned rows. The rows are aligned by residue from N-terminus to C-terminus. The rows are, from top to bottom, the input sequence (Fig. 3, red 4), binary MoRFpred results (Fig. 3, red 5) indicating non-MoRF residues (green "n"), and MoRF residues (red "M") and the raw prediction value (Fig. 3, red 6) multiplied by 10 and rounded, with alternating residues in black and white.

The notification e-mail contains links both to the results page (Fig. 4, red 1) and to the raw results file (Fig. 4, red 2). This e-mail can be saved to access results at a later time.

3 Case Studies

As subjects of the case studies we selected two proteins of different origin, human p53 (a 393-residue-long protein) and RNase E from *E. coli* (a 1061-residue-long protein). These two proteins have very different biological functions, are characterized by different levels of intrinsic disorder, and possess different numbers of MoRFs.

3.1 Case Study: p53

Because of its crucial biological roles in regulation of apoptosis, genomic stability, and inhibition of angiogenesis, as well as many

Predictions for MoRFpred job id: 20171009045915 are ready.

Upon the usage the users are requested to use the following citations:

Miri Disfani F, Hsu WL, Mizianty MJ, Oldfield C, Xue B, Dunker AK, Uversky VN, Kurgan LA, 2012. MoRFpred, a computational tool for sequence-based prediction and characterization of short disorder-to-order transitioning binding regions in proteins. Bioinformatics, 28(12): i75-i83.

You can find the results for this job at: http://biomine.cs.vcu.edu/webresults/MoRFpred/20171009045915/results.html 1

The CSV file can be found here: http://biomine.cs.vcu.edu/webresults/MoRFpred/20171009045915/results.csv 2

The webserver can be found here: http://biomine.cs.vcu.edu/servers/MoRFpred/

Thank you for using our webserver,
Biomine group

Fig. 4 Notification e-mail. Red numbers correspond to links to prediction results

mechanisms of anticancer activity, cellular tumor antigen p53 is one of the most studied proteins. The p53 signaling pathway is activated in response to a variety of stress signals. Activated p53 is accumulated in the nucleus, where its binding to specific DNA results in the induction or inhibition of a realm of different genes [37, 38], many of which are involved in apoptosis, growth arrest, or senescence [39–42]. In the unstressed mammalian cells, continuous ubiquitination of the non-phosphorylated p53 by double-minute-2 ubiquitin ligase (MDM2) [43] and subsequent proteasomal degradation ensure short lifetime and low levels of p53. There is also a negative feedback between the p53 and Akt pathways [44], where Akt is activated in cells exposed to various stimuli ranging from hormones to growth factors, and to extracellular matrix components [45], and controls the MDM2-mediated targeting of p53 for degradation [46]. Loss of p53 function due to mutations in this protein or some other alterations in the pathways leading to its activation and regulation is a common feature in the majority of human cancers [47]. Such mutations account for ~90% of cancer-related mutations in the *TP53* gene and are found in 50% of human cancers [48]. For example, up to 50% of advanced-stage prostate cancers contain mutations in p53 [49], and progression of prostate cancer to metastatic disease is characterized by the loss of p53 [50]. Furthermore, p53 levels may have prognostic value in urological oncology [51].

There are three major functional domains in human p53, the intrinsically disordered N-terminal regulatory domain (residues 1–92), the ordered central DNA-binding domain (DBD, residues 94–292) [52–54], and the intrinsically disordered C-terminal oligomerization and regulatory domain (residues 293–393) [55]. The regulatory domains can be further subdivided into functional subdomains/regions, such as transactivation domain 1 (TAD1)

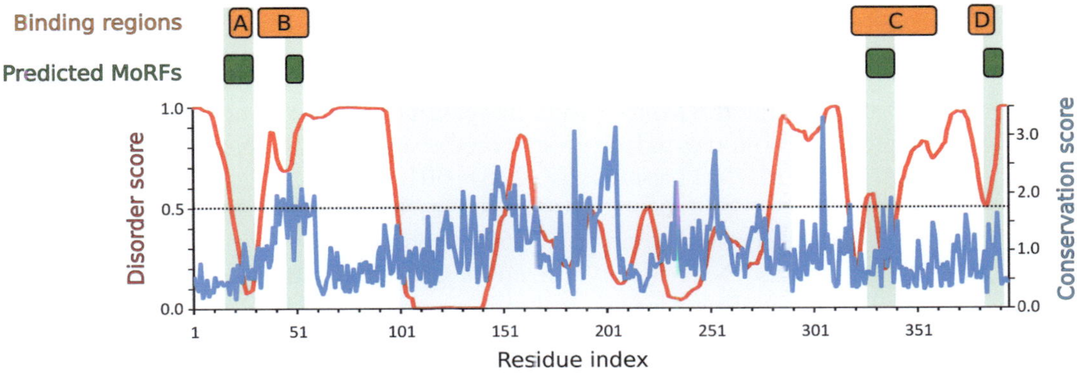

Fig. 5 Case study: p53. The correspondence between intrinsic disorder predictions (red line), sequence conservation (blue line), binding regions (orange boxes), and predicted MoRF regions (green boxes) is shown. Binding regions are discussed in the text. Sequence conservation is calculated from a set of p53 orthologs (OrthoDB) as the relative profile entropy over maximum entropy-weighted sequence (large values indicate greater conservation)

(residues 1–40), TAD2 (residues 40–60), and a proline-rich region, PR (residues 64–92), in the N-terminal regulatory domain, and tetramerization or oligomerization domain (OD; residues 325–356) and a regulatory C-terminal domain (CTD; residues 356–393) in the C-terminal regulatory domain [55, 56]. The N-terminal and C-terminal regulatory domains show exceptional binding promiscuity. Some of the illustrative examples of proteins interacting with the N-terminal transactivation region of p53 include CBP/p300, CSN5/Jab1, MDM2, RPA, TFIIH, and TFIID [43], whereas the CTD of p53 is engaged in interaction with 14-3-3, GSK3β, hGcn5, PARP-1, $S100B_{\beta\beta}$, TAF, TAF1, and TRRAP, to name a few [43]. Importantly, despite their crucial role in biological activities of p53, the regulatory regions of this protein are characterized by relatively poor evolutionary conservation, whereas the central DBD domain is highly conserved among different species. Irrespective of the general lack of conservation, there are four MoRFs in human p53 that overlap with or are included into the known binding sites of this protein (*see* Fig. 5).

The first MoRF (*see* Fig. 5, box A) coincides with the MDM2-binding site of p53. MDM2 is the E3 ubiquitin-protein ligase that is known as an important oncogene due to its overexpression in many human cancers, such as breast, colon, and prostate cancers, as well as hematologic malignancies and sarcomas [57]. MDM2 is most famous for its vital role in the p53 regulation via binding to a short stretch (residues 13–29) of the p53 TAD1 that prevents p53-driven activation or inhibition of various genes, via the MDM2-mediated p53 ubiquitination that targets this protein for the proteasomal degradation, and via active p53 transport out of the nucleus due to the presence of a nuclear export signal in MDM2

[58, 59]. Therefore, alteration of the p53-MDM2 interaction pathway is considered as a promising target for cancer therapy [57]. X-ray crystallographic studies of the p53-MDM2 complex revealed that the MDM2-binding region of p53 forms an α-helical structure bound to a deep groove on the surface of the N-terminal domain of MDM2 (residues 17–125) [60].

The second MoRF (*see* Fig. 5, box B) is included into the p53N fragment (residues 33–60) responsible for the p53 interaction with the N-terminal domain of the single-stranded DNA (ssDNA)-binding protein, replication protein A (RPA) [61]. This RPA70N domain is characterized by an oligonucleotide/oligosaccharide-binding fold typical for the ssDNA-binding domains, whereas the p53N fragment, which is disordered in isolation, forms two amphipathic helices, H1 and H2, following RPA70N binding [61]. Also, unlike other MoRFs in this protein, this MoRF displays a large amount of sequence conservation (*see* Fig. 5, conservation score).

The third MoRF (*see* Fig. 5, box C) is a part of the p53 tetramerization domain (325–356), structure of which represents a short β-strand (residues 326–333) followed by an α-helix (residues 335–355). These two structural elements are connected by a sharp turn facilitated by a conserved glycine residue (Gly334). Two monomers of the p53 tetramerization domain associate to form an antiparallel double-stranded sheet, and the antiparallel association of their helices forms a two-helical bundle. Four chains form a tetramer that can be described as a dimer of primary dimers [62].

The fourth MoRF (*see* Fig. 5, box D) is a part of the highly promiscuous C-terminal binding region of p53 (residues 374–388) that can bind to cyclin A [63], sirtuin [64], CBP [65], or S100ββ [66]. It was pointed out that upon interaction with different partners, this binding region of p53 displays all three major secondary structure types in the four complexes [67], where its core fragment becomes an α-helix when bound to S100ββ [66], a β-strand when bound to sirtuin [64], and a coil with two distinct backbone trajectories when bound to CBP [65] and cyclin A2 [63].

MoRFpred correctly identifies the four MoRF regions in p53 (*see* Fig. 5, green boxes), in spite of the significantly different conservation profiles of the four MoRF regions (*see* Fig. 5, blue lines). We note the relatively low conservation of the first, third, and fourth MoRF region and much higher conservation values for the second region. Interestingly, the red lines that identify the putative propensities for disorder, which were generated with VSL2B [68], correctly identify both termini of p53 as intrinsically disordered. However, they also register dips where the MoRFs are located. These dips are a by-product of the fact that MoRF regions become structured upon interacting with the protein partner, reducing the inherent propensity of these amino acids to be intrinsically disordered.

3.2 Case Study: RNase E

Endoribonucleases are hydrolytic enzymes that catalyze the endonucleolytic cleavage of RNA, have various specificities, are universally present in all organisms, and typically operate under tight cellular regulation. Endoribonucleases are involved in the maturation, modification, and degradation of different RNAs [69]. There are at least five endoribonucleases in *E. coli* (RNases I*, III, E, G, P). Among various activities attributed to RNase E are processing of transfer RNA, 9S ribosomal RNA, catalytic RNA of RNase P, transfer/messenger RNA (t/mRNA) that rescues stalled ribosomes [70–72], and general mRNA decay [73].

Being one of the larger *E. coli* proteins, RNase E consists of 1061 amino acid residues [74, 75]. There are two functionally different domains in this protein, the catalytic N-terminal domain (NTD; residues 1–498) and the regulatory C-terminal domain (CTD; residues 499–1061) [76–78]. Although the NTD is relatively conserved and has numerous homologues [79], there is little sequence conservation in the CTD [80], which is also characterized by low sequence complexity. The purified CTD was shown to be mostly disordered by a set of biophysical techniques, such as limited proteolysis, SDS–PAGE, SAXS, and far-UV CD [81]. Despite being highly disordered, the CTD was shown to interact with other degradosome components and with structured RNA [81]. In agreement with these experimental data, computational analysis clearly indicated that the NTD of RNase E was expected to be mostly structured, whereas the CTD had characteristics of a highly disordered protein [81].

The CTD is highly disordered, which is in agreement with the high values of the putative propensities for disorder generated for this protein with VSL2B [68] (*see* Fig. 6, red line). CTD is also characterized by the presence of four regions of increased structural propensity (labeled as segments A, B, C, and D, respectively), which correspond to MoRFs. The four MoRFs were correctly identified by the MoRFpred method (green boxes). Importantly, all these segments are related to various biological activities of RNase E, such as membrane targeting and CTD self-association (segment A corresponding to residues 565–585) or interactions with the components of the RNA degradosome, helicase (segment B, which is a portion of the arginine-rich domain (residues 628–843)) [78, 82], enolase (segment C (residues 833–850)) [81], and polynucleotide phosphorylase PNPase (segment D, RNase E residues 1021–1061) [81]. Like in the case of p53, some of the MoRF regions (*see* Fig. 6, segments C and D) are concomitant with a substantial decrease in the putative propensity for disorder (red line), but the remaining two regions do not register these dips. However, MoRFpred is still capable of identifying these MoRF regions, in spite of their high propensity for disorder and lack of conservation (blue line).

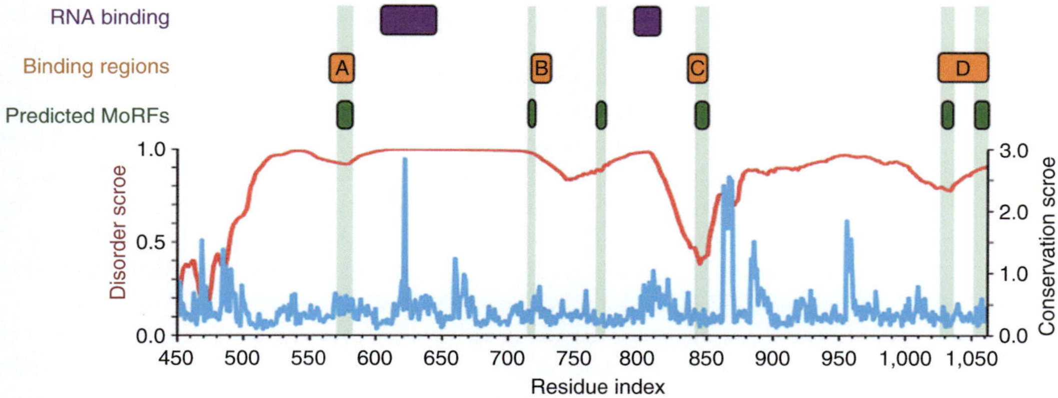

Fig. 6 Case study: RNase E. The correspondence between intrinsic disorder predictions (red line), sequence conservation (blue line), binding regions (orange boxes), and predicted MoRF regions (green boxes) is shown. Binding regions are discussed in the text. Sequence conservation is calculated from a set of RNase E orthologs (OrthoDB) as the relative profile entropy over maximum entropy-weighted sequence (large values indicate greater conservation)

4 Notes

1. The datasets can be downloaded from http://biomine.cs.vcu.edu/servers/MoRFpred/.
2. To partially compensate for the long runtime of the algorithm, up to five sequences can be submitted simultaneously to the web server. As soon as the results for one batch of up to five sequences are returned, another set of sequences can be submitted.
3. In FASTA format, each sequence is prefixed by a line beginning with ">" followed by some identifying text. The sequence should begin on the following line.
4. Up to five sequences can be submitted at a time. Ensure that each sequence has its own "FASTA header," which is a separate line beginning with ">."
5. The maximum length of each submitted sequence is 1000 residues.
6. It is advised to store or bookmark the link at this point. Predictions are stored on the server for at least 3 months, and keeping the link will allow return to the results pages. It also protects against lost predictions, in the case that an incorrect notification e-mail address was entered.

References

1. Wang C, Uversky VN, Kurgan L (2016) Disordered nucleiome: abundance of intrinsic disorder in the DNA- and RNA-binding proteins in 1121 species from Eukaryota, Bacteria and Archaea. Proteomics 16(10):1486–1498
2. Peng Z, Yan J, Fan X, Mizianty MJ, Xue B, Wang K, Hu G, Uversky VN, Kurgan L (2015) Exceptionally abundant exceptions: comprehensive characterization of intrinsic disorder in all domains of life. Cell Mol Life Sci 72 (1):137–151
3. Habchi J, Tompa P, Longhi S, Uversky VN (2014) Introducing protein intrinsic disorder. Chem Rev 114(13):6561–6588
4. Dunker AK, Babu MM, Barbar E, Blackledge M, Bondos SE, Dosztányi Z, Dyson HJ, Forman-Kay J, Fuxreiter M, Gsponer J, Han K-H, Jones DT, Longhi S, Metallo SJ, Nishikawa K, Nussinov R, Obradovic Z, Pappu RV, Rost B, Selenko P, Subramaniam V, Sussman JL, Tompa P, Uversky VN (2013) What's in a name? Why these proteins are intrinsically disordered. Intrinsically Disord Proteins 1(1):e24157
5. Brown CJ, Takayama S, Campen AM, Vise P, Marshall TW, Oldfield CJ (2002) Evolutionary rate heterogeneity in proteins with long disordered regions. J Mol Evol 55:104
6. Meszaros B, Tompa P, Simon I, Dosztanyi Z (2007) Molecular principles of the interactions of disordered proteins. J Mol Biol 372 (2):549–561
7. Trudeau T, Nassar R, Cumberworth A, Wong ET, Woollard G, Gsponer J (2013) Structure and intrinsic disorder in protein autoinhibition. Structure 21(3):332–341
8. Varadi M, Guharoy M, Zsolyomi F, Tompa P (2015) DisCons: a novel tool to quantify and classify evolutionary conservation of intrinsic protein disorder. BMC Bioinformatics 16 (1):153
9. Ait-Bara S, Carpousis AJ, Quentin Y (2015) RNase E in the gamma-Proteobacteria: conservation of intrinsically disordered noncatalytic region and molecular evolution of microdomains. Mol Genet Genomics 290(3):847–862
10. Davey NE, Cyert MS, Moses AM (2015) Short linear motifs – ex nihilo evolution of protein regulation. Cell Commun Signal 13(1):43
11. Mohan A, Oldfield CJ, Radivojac P, Vacic V, Cortese MS, Dunker AK, Uversky VN (2006) Analysis of molecular recognition features (MoRFs). J Mol Biol 362(5):1043–1059
12. Vacic V, Oldfield CJ, Mohan A, Radivojac P, Cortese MS, Uversky VN, Dunker AK (2007) Characterization of molecular recognition features, MoRFs, and their binding partners. J Proteome Res 6(6):2351–2366
13. Oldfield CJ, Cheng Y, Cortese MS, Romero P, Uversky VN, Dunker AK (2005) Coupled folding and binding with alpha-helix-forming molecular recognition elements. Biochemistry 44(37):12454–12470
14. Yan J, Dunker AK, Uversky VN, Kurgan L (2016) Molecular recognition features (MoRFs) in three domains of life. Mol BioSyst 12(3):697–710
15. Cheng Y, Oldfield CJ, Meng J, Romero P, Uversky VN, Dunker AK (2007) Mining α-helix-forming molecular recognition features with cross species sequence alignments. Biochemistry 46(47):13468–13477
16. Malhis N, Gsponer J (2015) Computational identification of MoRFs in protein sequences. Bioinformatics 31(11):1738–1744
17. Disfani FM, Hsu WL, Mizianty MJ, Oldfield CJ, Xue B, Dunker AK, Uversky VN, Kurgan L (2012) MoRFpred, a computational tool for sequence-based prediction and characterization of short disorder-to-order transitioning binding regions in proteins. Bioinformatics 28 (12):i75–i83
18. Malhis N, Jacobson M, Gsponer J (2016) MoRFchibi SYSTEM: software tools for the identification of MoRFs in protein sequences. Nucleic Acids Res 44:W488
19. Jones DT, Cozzetto D (2015) DISOPRED3: precise disordered region predictions with annotated protein-binding activity. Bioinformatics 31(6):857–863
20. Fang C, Noguchi T, Tominaga D, Yamana H (2013) MFSPSSMpred: identifying short disorder-to-order binding regions in disordered proteins based on contextual local evolutionary conservation. BMC Bioinformatics 14:300
21. Xue B, Dunker AK, Uversky VN (2010) Retro-MoRFs: identifying protein binding sites by normal and reverse alignment and intrinsic disorder prediction. Int J Mol Sci 11 (10):3725–3747
22. Puntervoll P, Linding R, Gemünd C, Chabanis-Davidson S, Mattingsdal M, Cameron S, Martin DMA, Ausiello G, Brannetti B, Costantini A, Ferrè F, Maselli V, Via A, Cesareni G, Diella F, Superti-Furga G, Wyrwicz L, Ramu C, McGuigan C, Gudavalli R, Letunic I, Bork P, Rychlewski L, Küster B, Helmer-Citterich M, Hunter WN, Aasland R, Gibson TJ (2003) ELM server: a

new resource for investigating short functional sites in modular eukaryotic proteins. Nucleic Acids Res 31(13):3625–3630

23. Meszaros B, Dosztanyi Z, Simon I (2012) Disordered binding regions and linear motifs--bridging the gap between two models of molecular recognition. PLoS One 7(10):e46829
24. Peng Z, Wang C, Uversky VN, Kurgan L (2017) Prediction of disordered RNA, DNA, and protein binding regions using DisoRDPbind. Methods Mol Biol 1484:187–203
25. Meszaros B, Simon I, Dosztanyi Z (2009) Prediction of protein binding regions in disordered proteins. PLoS Comput Biol 5(5):e1000376
26. Dosztanyi Z, Meszaros B, Simon I (2009) ANCHOR: web server for predicting protein binding regions in disordered proteins. Bioinformatics 25(20):2745–2746
27. Khan W, Duffy F, Pollastri G, Shields DC, Mooney C (2013) Predicting binding within disordered protein regions to structurally characterised peptide-binding domains. PLoS One 8(9):e72838
28. Berman HM, Westbrook J, Feng Z, Gilliland G, Bhat TN, Weissig H, Shindyalov IN, Bourne PE (2000) The protein data bank. Nucleic Acids Res 28(1):235–242
29. Dosztanyi Z, Csizmok V, Tompa P, Simon I (2005) IUPred: web server for the prediction of intrinsically unstructured regions of proteins based on estimated energy content. Bioinformatics 21(16):3433–3434
30. Ward JJ, McGuffin LJ, Bryson K, Buxton BF, Jones DT (2004) The DISOPRED server for the prediction of protein disorder. Bioinformatics 20(13):2138–2139
31. McGuffin LJ (2008) Intrinsic disorder prediction from the analysis of multiple protein fold recognition models. Bioinformatics 24(16):1798–1804
32. Mizianty MJ, Stach W, Chen K, Kedarisetti KD, Disfani FM, Kurgan L (2010) Improved sequence-based prediction of disordered regions with multilayer fusion of multiple information sources. Bioinformatics 26(18):i489–i496
33. Faraggi E, Xue B, Zhou Y (2009) Improving the prediction accuracy of residue solvent accessibility and real-value backbone torsion angles of proteins by guided-learning through a two-layer neural network. Proteins 74(4):847–856
34. Schlessinger A, Yachdav G, Rost B (2006) PROFbval: predict flexible and rigid residues in proteins. Bioinformatics 22(7):891–893
35. Kawashima S, Pokarowski P, Pokarowska M, Kolinski A, Katayama T, Kanehisa M (2008) AAindex: amino acid index database, progress report 2008. Nucleic Acids Res 36(Database issue):D202–D205
36. Tate RF (1954) Correlation between a discrete and a continuous variable. Point-Biserial correlation. Ann Math Statist 25(3):603–607
37. Zhao R, Gish K, Murphy M, Yin Y, Notterman D, Hoffman WH, Tom E, Mack DH, Levine AJ (2000) Analysis of p53-regulated gene expression patterns using oligonucleotide arrays. Genes Dev 14(8):981–993
38. Balint EE, Vousden KH (2001) Activation and activities of the p53 tumour suppressor protein. Br J Cancer 85(12):1813–1823
39. el-Deiry WS (1998) Regulation of p53 downstream genes. Semin Cancer Biol 8(5):345–357
40. Yu J, Zhang L, Hwang PM, Rago C, Kinzler KW, Vogelstein B (1999) Identification and classification of p53-regulated genes. Proc Natl Acad Sci U S A 96(25):14517–14522
41. Sax JK, El-Deiry WS (2003) p53-induced gene expression analysis. Methods Mol Biol 234:65–71
42. Fridman JS, Lowe SW (2003) Control of apoptosis by p53. Oncogene 22(56):9030–9040
43. Anderson CW, Appella E (2004) Signaling to the p53 tumor suppressor through pathways activated by genotoxic and nongenotoxic stress. In: Bradshaw RA, Dennis EA (eds) Handbook of cell signaling. Academic Press, New York, pp 237–247
44. Gottlieb TM, Leal JF, Seger R, Taya Y, Oren M (2002) Cross-talk between Akt, p53 and Mdm2: possible implications for the regulation of apoptosis. Oncogene 21(8):1299–1303
45. Nicholson KM, Anderson NG (2002) The protein kinase B/Akt signalling pathway in human malignancy. Cell Signal 14(5):381–395
46. Abraham AG, O'Neill E (2014) PI3K/Akt-mediated regulation of p53 in cancer. Biochem Soc Trans 42(4):798–803
47. Muller PA, Vousden KH (2013) p53 mutations in cancer. Nat Cell Biol 15(1):2–8
48. Soussi T, Beroud C (2001) Assessing TP53 status in human tumours to evaluate clinical outcome. Nat Rev Cancer 1(3):233–240
49. Bookstein R (1994) Tumor suppressor genes in prostatic oncogenesis. J Cell Biochem Suppl 19:217–223
50. Pencik J, Wiebringhaus R, Susani M, Culig Z, Kenner L (2015) IL-6/STAT3/ARF: the guardians of senescence, cancer progression

and metastasis in prostate cancer. Swiss Med Wkly 145:w14215

51. Wolff JM, Stephenson RN, Jakse G, Habib FK (1994) Retinoblastoma and p53 genes as prognostic indicators in urological oncology. Urol Int 53(1):1–5
52. Joerger AC, Ang HC, Veprintsev DB, Blair CM, Fersht AR (2005) Structures of p53 cancer mutants and mechanism of rescue by second-site suppressor mutations. J Biol Chem 280(16):16030–16037
53. Canadillas JM, Tidow H, Freund SM, Rutherford TJ, Ang HC, Fersht AR (2006) Solution structure of p53 core domain: structural basis for its instability. Proc Natl Acad Sci U S A 103 (7):2109–2114
54. Wang Y, Rosengarth A, Luecke H (2007) Structure of the human p53 core domain in the absence of DNA. Acta Crystallogr D Biol Crystallogr 63(Pt 3):276–281
55. Joerger AC, Fersht AR (2008) Structural biology of the tumor suppressor p53. Annu Rev Biochem 77:557–582
56. Uversky VN, Oldfield CJ, Midic U, Xie H, Xue B, Vucetic S, Iakoucheva LM, Obradovic Z, Dunker AK (2009) Unfoldomics of human diseases: linking protein intrinsic disorder with diseases. BMC Genomics 10(Suppl 1):S7
57. Bianco R, Ciardiello F, Tortora G (2005) Chemosensitization by antisense oligonucleotides targeting MDM2. Curr Cancer Drug Targets 5(1):51–56
58. Moll UM, Petrenko O (2003) The MDM2-p53 interaction. Mol Cancer Res 1 (14):1001–1008
59. Nag S, Qin J, Srivenugopal KS, Wang M, Zhang R (2013) The MDM2-p53 pathway revisited. J Biomed Res 27(4):254–271
60. Kussie PH, Gorina S, Marechal V, Elenbaas B, Moreau J, Levine AJ, Pavletich NP (1996) Structure of the MDM2 oncoprotein bound to the p53 tumor suppressor transactivation domain. Science 274(5289):948–953
61. Bochkareva E, Kaustov L, Ayed A, Yi GS, Lu Y, Pineda-Lucena A, Liao JC, Okorokov AL, Milner J, Arrowsmith CH, Bochkarev A (2005) Single-stranded DNA mimicry in the p53 transactivation domain interaction with replication protein A. Proc Natl Acad Sci U S A 102(43):15412–15417
62. Mora P, Carbajo RJ, Pineda-Lucena A, Sanchez del Pino MM, Perez-Paya E (2008) Solvent-exposed residues located in the beta-sheet modulate the stability of the tetramerization domain of p53--a structural and combinatorial approach. Proteins 71(4):1670–1685
63. Lowe ED, Tews I, Cheng KY, Brown NR, Gul S, Noble ME, Gamblin SJ, Johnson LN (2002) Specificity determinants of recruitment peptides bound to phospho-CDK2/cyclin A. Biochemistry 41(52):15625–15634
64. Avalos JL, Celic I, Muhammad S, Cosgrove MS, Boeke JD, Wolberger C (2002) Structure of a Sir2 enzyme bound to an acetylated p53 peptide. Mol Cell 10(3):523–535
65. Mujtaba S, He Y, Zeng L, Yan S, Plotnikova O, Sachchidanand SR, Zeleznik-Le NJ, Ronai Z, Zhou MM (2004) Structural mechanism of the bromodomain of the coactivator CBP in p53 transcriptional activation. Mol Cell 13 (2):251–263
66. Rustandi RR, Baldisseri DM, Weber DJ (2000) Structure of the negative regulatory domain of p53 bound to S100B(betabeta). Nat Struct Biol 7(7):570–574
67. Oldfield CJ, Meng J, Yang JY, Yang MQ, Uversky VN, Dunker AK (2008) Flexible nets: disorder and induced fit in the associations of p53 and 14-3-3 with their partners. BMC Genomics 9(Suppl 1):S1
68. Peng K, Radivojac P, Vucetic S, Dunker AK, Obradovic Z (2006) Length-dependent prediction of protein intrinsic disorder. BMC Bioinformatics 7:208
69. Ehretsmann CP, Carpousis AJ, Krisch HM (1992) Specificity of Escherichia coli endoribonuclease RNase E: in vivo and in vitro analysis of mutants in a bacteriophage T4 mRNA processing site. Genes Dev 6(1):149–159
70. Huang H, Liao J, Cohen SN (1998) Poly(A)- and poly(U)-specific RNA 3′ tail shortening by E. coli ribonuclease E. Nature 391 (6662):99–102
71. Kushner SR (2002) mRNA decay in Escherichia coli comes of age. J Bacteriol 184 (17):4658–4665 discussion 4657
72. Ow MC, Kushner SR (2002) Initiation of tRNA maturation by RNase E is essential for cell viability in E. coli. Genes Dev 16 (9):1102–1115
73. Steege DA (2000) Emerging features of mRNA decay in bacteria. RNA 6 (8):1079–1090
74. Casaregola S, Jacq A, Laoudj D, McGurk G, Margarson S, Tempete M, Norris V, Holland IB (1992) Cloning and analysis of the entire Escherichia coli ams gene. ams is identical to hmp1 and encodes a 114 kDa protein that migrates as a 180 kDa protein. J Mol Biol 228 (1):30–40
75. Claverie-Martin F, Diaz-Torres MR, Yancey SD, Kushner SR (1991) Analysis of the altered mRNA stability (ams) gene from Escherichia

coli. Nucleotide sequence, transcriptional analysis, and homology of its product to MRP3, a mitochondrial ribosomal protein from Neurospora crassa. J Biol Chem 266(5):2843–2851

76. Lopez PJ, Marchand I, Joyce SA, Dreyfus M (1999) The C-terminal half of RNase E, which organizes the Escherichia coli degradosome, participates in mRNA degradation but not rRNA processing in vivo. Mol Microbiol 33 (1):188–199
77. Cohen SN, McDowall KJ (1997) RNase E: still a wonderfully mysterious enzyme. Mol Microbiol 23(6):1099–1106
78. McDowall KJ, Cohen SN (1996) The N-terminal domain of the rne gene product has RNase E activity and is non-overlapping with the arginine-rich RNA-binding site. J Mol Biol 255(3):349–355
79. Wachi M, Umitsuki G, Shimizu M, Takada A, Nagai K (1999) Escherichia coli cafA gene encodes a novel RNase, designated as RNase G, involved in processing of the 5′ end of 16S rRNA. Biochem Biophys Res Commun 259(2):483–488
80. Kaberdin VR, Miczak A, Jakobsen JS, Lin-Chao S, McDowall KJ, von Gabain A (1998) The endoribonucleolytic N-terminal half of Escherichia coli RNase E is evolutionarily conserved in Synechocystis sp. and other bacteria but not the C-terminal half, which is sufficient for degradosome assembly. Proc Natl Acad Sci U S A 95(20):11637–11642
81. Callaghan AJ, Aurikko JP, Ilag LL, Gunter Grossmann J, Chandran V, Kuhnel K, Poljak L, Carpousis AJ, Robinson CV, Symmons MF, Luisi BF (2004) Studies of the RNA degradosome-organizing domain of the Escherichia coli ribonuclease RNase E. J Mol Biol 340(5):965–979
82. Taraseviciene L, Bjork GR, Uhlin BE (1995) Evidence for an RNA binding region in the Escherichia coli processing endoribonuclease RNase E. J Biol Chem 270(44):26391–26398

Chapter 20

Exploring Protein Conformational Diversity

Alexander Miguel Monzon, Maria Silvina Fornasari, Diego Javier Zea, and Gustavo Parisi

Abstract

The native state of proteins is composed of conformers in dynamical equilibrium. In this chapter, different issues related to conformational diversity are explored using a curated and experimentally based database called CoDNaS (Conformational Diversity in the Native State). This database is a collection of redundant structures for the same sequence. CoDNaS estimates the degree of conformational diversity using different global and local structural similarity measures. It allows the user to explore how structural differences among conformers change as a function of several structural features providing further biological information. This chapter explores the measurement of conformational diversity and its relationship with sequence divergence. Also, it discusses how proteins with high conformational diversity could affect homology modeling techniques.

Key words Conformational diversity, CoDNaS database, Conformers, Native state, Protein dynamics, Protein evolution

1 Introduction

Since the early crystallization studies on hemoglobin, it is known that two or more conformational states are required to sustain biological function. The native state is then better represented by an ensemble of alternative protein conformations in equilibrium. A wide range of protein movements between conformers have been explored. These range from large relative domain movements [1], secondary and tertiary element rearrangements [2], and loop displacements [3] to small residue rearrangements [4]. Structural differences between these conformers define the conformational diversity of the protein. In general, most proteins have a few well-defined conformational states. Human hemoglobin has two well-established T and R conformations [5], and the dimeric catabolite activator protein [6] has three, just to mention two examples. However, intrinsically disordered proteins (IDPs) have a native state with multiple conformers characterized by their high

Tobias Sikosek (ed.), *Computational Methods in Protein Evolution*, Methods in Molecular Biology, vol. 1851, https://doi.org/10.1007/978-1-4939-8736-8_20,

flexibility and mobility, defining very complex ensembles [7]. Whatever the mechanisms underlying conformational changes are, it is clear that in many cases a protein requires switching among different native structures to be functional. The conformational ensemble concept becomes then a key tool to explain an endless list of essential protein properties such as function [8–10], enzyme and antibody promiscuity [11], signal transduction [12], protein-protein recognition [13], origin of diseases [14], emergence of new protein functions [15], evolutionary rate [16], and order-disorder transitions [17], just to mention some of the most important.

This chapter describes how to explore protein conformational diversity using an experimentally based database. It gives some practical advice for the analysis of its data and it highlights the relevance of including conformational diversity in the study of protein evolution and homology modeling.

2 Methods

2.1 Discovering Protein Conformational Diversity Through Structure Redundancy

The study of protein conformational diversity can be addressed using redundant structures of the same protein obtained in different experimental conditions (e.g., with or without substrate or post-translational modifications, or at different pH values). This ensemble of structures provides snapshots of protein dynamism in their native state and could be considered as putative native conformations [18, 19]. Then, the sequence redundancy in the Protein Data Bank (PDB) [20] is an essential input to experimentally based studies of protein conformational diversity that provides insight into protein function. The continuous growth of the PDB during these last years has granted access to many protein native ensembles [21]. However, more efforts are needed to amend the incompleteness of the PDB in terms of its dynamical information content [22]. Different databases and methods have been developed over the last 10 years to take advantage of this redundancy (Table 1).

Table 1
Databases of protein conformational diversity

Database name	URL	Protein chains
PDBFlex [64]	http://pdbflex.org/	28,939
CoDNaS [24]	http://ufq.unq.edu.ar/codnas/	21,152
DynDom [67]	http://fizz.cmp.uea.ac.uk/dyndom/	1578
PSCDB [68]	http://idp1.force.cs.is.nagoya-u.ac.jp/pscdb/	839
MolMov [1]	http://www.molmovdb.org	230

2.2 Exploring Protein Conformational Diversity in the Native State Using the CoDNaS Database

CoDNaS is a database of protein conformations derived from experimental structures [23, 24]. CoDNaS contains a redundant collection of 3D structures for each protein obtained under different experimental conditions. It has an extensive annotation of the different conditions under which each conformer was obtained. CoDNaS also offers different structural similarity measures among conformers such as global RMSD and TM score, as well as local measures (e.g., RMSD per position expressed as Z-scores). In this way, CoDNaS facilitates the analysis of key information on small structural differences. CoDNaS allows users to easily relate the degree of conformational diversity with physical, chemical, and biological properties. The last version of the CoDNaS database (version 2.5) includes 73% of all available protein structures in the PDB (21,152 different protein chains, 320,144 structures, and 15.09 average conformers per protein), and possesses different tools to run sequence searches, display structural flexibility profiles, visualize structural alignments and bounded ligands, and allow users to browse the database by different structural classes.

2.2.1 Database Implementation, Biological Annotation, and External Links

Different conformers for each protein were identified and extracted from the PDB using the following protocol:

- BLASTClust [25] was run against all protein chains deposited in the PDB to obtain all available clusters at 95% of local sequence identity with a minimum coverage of 0.90 between all the sequences in the cluster. A limit at 95% was set to include putative sequence variations for a given protein. However, to avoid the inclusion of homologous structures in a given CoDNaS entry, UniProt accession numbers were used to check that all conformers belong to the same protein.
- The only considered clusters were those with at least two structures and with a resolution of less than 4.00 Å for each of the crystallographic structures.
- To estimate the structural dissimilarity between conformers in each cluster, C-alpha root mean square deviation (RMSD) using MAMMOTH (*see* **Note 3**) [26] was calculated for all the possible pairs of conformers for each protein. The maximum C-alpha RMSD value for each protein entry was registered as a measure of the conformational diversity extension.
- Additionally, all conformers for a given protein were clustered using a hierarchical procedure according to the RMSD values between them. This enables users to identify different conformational substates present in the native state of the protein.

– Furthermore, CoDNaS is cross-linked with other databases: UniProt [27]; Class, Architecture, Topology and Homology (CATH) [28]; Enzyme Commission [29]; MobiDB [30]; and Gene Ontology [31].

2.2.2 Working Case: Conformational Diversity of the Ephrin Type-A Receptor 4

Human ephrin type-A receptor 4 (EphA4) is a tyrosine kinase receptor. Eph receptors and their ephrin ligands are both anchored onto the plasma membrane and are subdivided into two subclasses (A and B) based on their sequence conservation and binding preferences [32]. In general, type-A receptors bind to ephrin but in particular EphA4 is the only receptor capable of binding to all nine ephrins and other small molecules with overlapped interfaces. Binding pattern in EphA4 can be explained exploring its ensemble of conformers. EphA4 has two groups of conformers: closed and open forms which have been biologically characterized and identified by molecular dynamic simulations and NMR studies [33]. Hence, open and closed conformations of the EphA4 can be easily explored in CoDNaS using the information provided by the hierarchical clustering based on the RMSD values between all pairs of conformers (Fig. 1). It is interesting to note the differences between the 29 conformers available in CoDNaS. Ten of them were obtained by nuclear magnetic resonance (NMR) and 19 by X-ray diffraction. It is possible to find this protein in CoDNaS searching by its UniProt accession number "P54764" and to access the entry page (protein pool identifier in CoDNaS is "2WO1_A"). The entry page includes a set of boxes with different information about the protein, such as protein overview, structural information, conformers, clusters of conformational states, and information about the pair of maximum conformational diversity. EphA4 has a maximum conformational diversity of RMSD = 3.23 Å between the structures 2WO3 chain A and 2LW8 chain A, model 7.

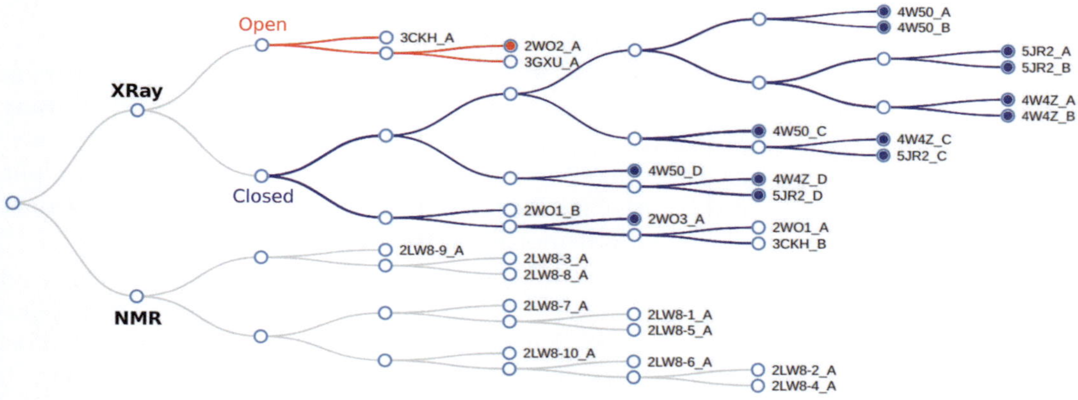

Fig. 1 Dendrogram of the EphA4 conformations. We can observe different conformational substates due to the experimental method used and transitions between open (red) and closed (blue) conformations. Filled nodes indicate that the conformer has ligand

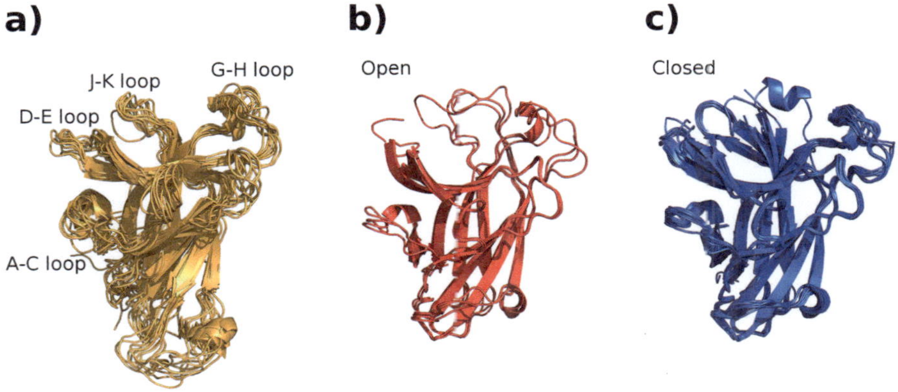

Fig. 2 Comparison between conformers of the EphA4 based on clustering information. (**a**) Superimposition of ten conformers from the NMR ensemble (PDB code = 2LW8). (**b**) Superimposition of 16 closed conformations (blue) of the EphA4. (**c**) Superimposition of three open conformations (red) of the EphA4

Figure 1 shows two main groups at the top, one containing all NMR conformers and the other containing all X-ray conformers (*see* also **Note 1**). Among the group of X-ray conformers, we can observe two branches which separate open and closed conformations of the EphA4. Filled nodes indicate conformers in complex with the ligand. Superimposition of these three different groups (NMR, X-ray closed, and X-ray open) reveals a high conformational variability in the regions of the B–C, D–E, G–H, and J–K loops (*see* Fig. 2) [34]. In particular, the flexibility of the D–E and J–K loops, which move upon binding to ephrin ligands, may be directly associated with EphA4 function and binding pattern.

2.3 Practical Issues Concerning Conformational Diversity

2.3.1 How Large Are Conformational Changes in Known Structural Space?

The extension of the conformational diversity was studied in a curated dataset (*see* **Note 2**) of ~5000 proteins with more than 5 conformers (*see* **Note 6**) per protein [35]. This study found three protein classes based on their dynamical behavior: rigid, malleable, and partially disordered proteins. Approximately 60% of the analyzed proteins are part of the first group, the rigid proteins. Conformational diversity of each protein was measured as the maximum RMSD (*see* **Note 5**) after an all-versus-all conformer pairwise comparisons. The RMSD distribution of rigid proteins has a peak in 0.8 Å, a value close to the crystallographic error which is near 0.5 Å (*see* Fig. 3). This result agrees with earlier studies that found a positive skewed distribution of RMSD [19, 36]. It also agrees with a previous work that found an average RMSD of 0.5 Å after comparisons between structures of the same protein in unbound states, a value slightly different from the observed between apo and substrate-bound forms [37]. Apparently, large-scale protein motions are not necessary to sustain biological function in the majority of the studied proteins. This observation is supported by the finding that even small changes between conformers could greatly affect catalytic parameters and biological behavior of enzymes [38, 39].

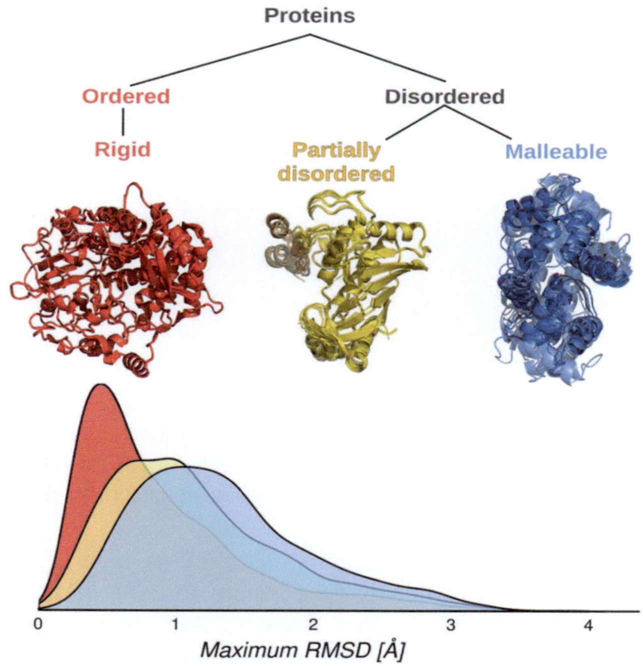

Fig. 3 The conformational diversity distribution can be represented by three main sets of proteins: rigid (all protein conformers without IDRs), partially disordered (with IDRs at least in one conformer and also in the pair of maximum conformational diversity), and malleable (with IDRs at least in one conformer of the protein but the pair of maximum conformational diversity remains ordered)

2.3.2 Which Kind of Proteins Have Larger Conformational Changes?

The tail of the distribution shown in Fig. 3 has mainly IDPs, malleable and partially disordered proteins in particular, and a minor proportion of globular or ordered proteins [35]. It is important to note that IDPs contain very flexible regions which several times appear as missing residues in the structures derived from crystallographic studies. Almost half of these IDPs show order-disorder transitions; that is, they have regions that are disordered in one group of conformers but ordered in alternative conformations. Surprisingly, regions gaining order upon ligand binding are almost as common as the ones gaining disordered regions upon binding. IDPs showing order-disorder transitions reach the highest RMSD values in their aligned ordered regions [17]. The high RMSD values between conformers are related to the increase of structural differences in the globular or ordered region of IDPs. These differences can be high due to very flexible loops or regions adopting variable conformations (e.g., malleable and partially disordered proteins).

In reference to globular or ordered proteins, large conformational movements have been previously described by M. Gerstein in the MolMov database [1]. Most of the changes comprise domains and/or fragments such as loops, normally as rigid bodies with hinge, shear, or more complex motions [40].

What is the relationship between sequence variation and conformational diversity?

Several studies highlight the existence of evolutionary signals coming from the conformational diversity of proteins [36]. For instance, it has been shown that proteins with large conformational diversity show lower evolutionary rates than proteins with more similar conformers [16]. It is reasonable to think that different conformers impose different structural constraints in protein evolution. Consequently, it is possible that the reduction in the rate of nonsynonymous mutations could be an effect of the increment in the structural constraints in the presence of different conformers. This is supported by the fact that 30% of the structurally constrained sites in a protein are conformer specific. In particular, conformers with the highest affinity to their ligands seem to be the ones that have more constraints on the divergence of their sequences [41].

Protein dynamism imposes different kinds of constraints to sequence variation. For example, residues playing a key role in hinge regions during collective movements, as well as other dynamically important positions, tend to be conserved throughout evolution [42, 43]. Also, there are coevolving residue pairs that facilitate the transition between conformers by cooperatively forming and breaking contacts [44]. Coevolving residue pairs could lead to covariation signals that are detectable in a multiple sequence alignment [45]. Since residues in contact have propensity to coevolve, detected covariation pairs can be used to predict protein tertiary structures [46]. In particular, coevolving residues that result from conformer specific contacts could be used to predict the structure of different conformers and transition states [47]. A couple of studies show that covariation methods are good at predicting contacts which are conserved in all the structures of a given protein [48, 49]. Therefore, covariation methods have a poor performance to predict conformer-specific contacts. However, this coevolutionary information can be used together with molecular dynamics to predict protein conformers when at least one structure and 2000 homologous sequences are available [50].

2.3.3 Importance of the Conformational Diversity in Homology Modeling

Template-based modeling (TBM) is based on the fact that homologous proteins with detectable sequence similarity possess similar 3D structures. Pioneering work by Chothia and Lesk found that structural divergence increases with evolutionary distance, measured as identity percentage, following a nonlinear relationship [51]. Very similar sequences show modest structural differences, which suddenly increase when percentage of sequence identity

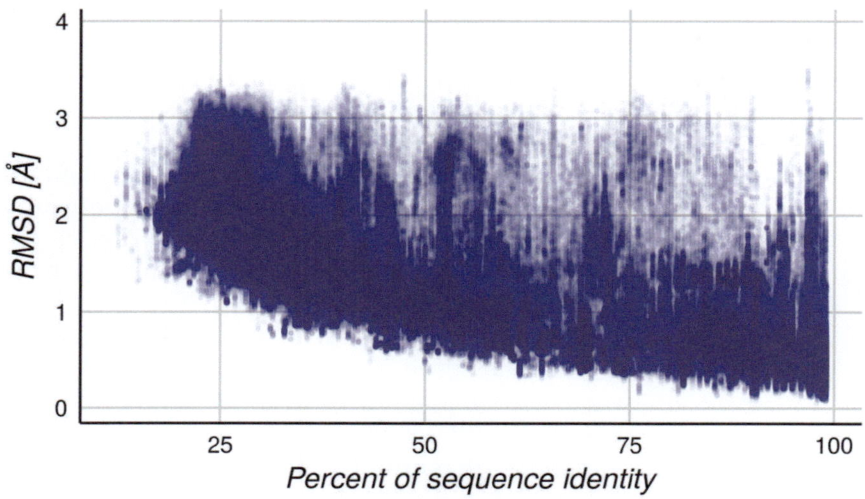

Fig. 4 RMSD versus percent of sequence identity. RMSD values were obtained from an all-versus-all comparison between two homologous proteins considering all their conformers. The figure contains about 3.5 million comparisons

drops below 30%. Their results and conclusions have been verified by numerous studies [52–56]. These studies found moderate-to-high correlation coefficients between different parameters related to structural and sequence similarity, i.e., RMSD versus identity percentage and evolutionary distance. They also found linear and nonlinear behavior, and an invariably low structural variation at 100% identity (~0.5 Å). However, when conformational diversity is taken into account the relationship between sequence and structural divergence is more complex [57]. Figure 4 shows how RMSDs between homologous proteins change as a function of identity percentage. This figure was derived from an all-versus-all pairwise alignment between all the conformers for 2024 proteins from 524 families. It is possible to observe that at around 100% identity (the conformational diversity of the protein) (*see* **Note 4**) several proteins show RMSDs as high as those reached by sequence divergence during evolution (say about 30–40% identity). This means that the structural divergence is a complex process since a given sequence (at 100% identity) could reach several angstroms of conformational (structural) diversity. Interestingly, if we split the population of proteins according to their corresponding degree of conformational diversity (rigids and highly dynamical proteins) we can observe in Fig. 5 that the rigid proteins could certainly be more suitable to TBM methodologies than highly dynamic ones. The rigid proteins show an average RMSD of 0.39 Å at 100% identity, meaning that more similar sequences have more similar structures. This last statement, basic to TBM reliability, apparently is not true for highly dynamical proteins (average RMSD at 100% 1.17 Å).

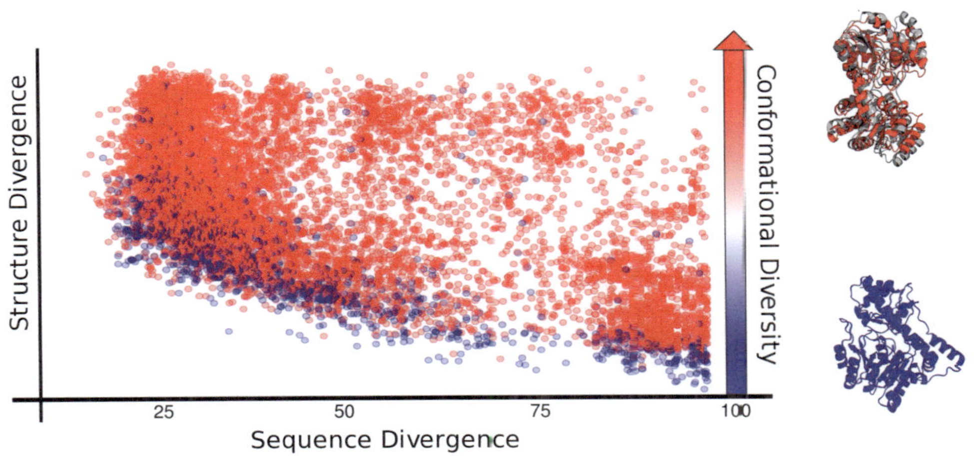

Fig. 5 Maximum RMSD versus sequence percent identity. Points refer to the maximum RMSD obtained from an all-versus-all comparison between conformers from two homologous proteins. Red dots are pairs of highly dynamic homologous proteins (conformational diversity >0.5 Å) and blue dots are pairs of rigid proteins (conformational diversity <=0.5 Å)

3 Notes

1. The differences among protein structures obtained by NMR and by X-ray diffraction are well established [58]. In conformational analysis, we should avoid mixing NMR and X-ray conformers in order to prevent biases in the RMSD values of our analysis. It is well known that NMR ensembles have larger RMSD values between models than conformers obtained by X-ray. However, having conformers obtained by these two methods provides us with a complementary source of information on protein conformational diversity by making sure that flexibility information is well reproduced in both cases [18].

2. X-ray resolution is another important aspect to take into account in the ensembles of structures. It is always recommended to use structures with a resolution of 2.5 Å or better for good backbone and side-chain estimations. Despite that, we can use structures with a resolution of 2.6–4 Å to estimate backbone RMSD. More information about structure resolution can be found at http://proteopedia.org/wiki/index.php/Resolution.

3. The problem of quantifying the differences between two structures of the same protein is nontrivial. There are several methods to calculate the global RMSD between a pair of structures and the magnitude of the RMSD value depends on the structural alignment algorithm used. A good review of structural

comparison methods can be found in Ref. [59]. In CoDNaS, MAMMOTH [26] was used to align each pair of conformers and estimate the global RMSD. MAMMOTH is a sequence-independent structural alignment program which provides the user with statistical reliability data of the results. MAMMOTH uses the MaxSub algorithm [60] to identify the maximal subset from a set of paired-up atoms which are spatially close. MAMMOTH defines "close" as a distance of <4 Å after superposition. This is the reason why MAMMOTH RMSD values are mainly between 0 and 4 Å. In addition to this, it is protein length independent.

4. Even if conformers belong to the same protein, the percentage of identity after a structural or sequential alignment could be less than 100%. Sometimes the structure of a protein is resolved after introducing some point mutations to the sequence. These mutations introduce mismatches in the alignment. Another source of mismatches is the gaps introduced by missing residues in structural alignment of crystallographic structures. The best way to check whether the structure corresponds to a specific protein or not is mapping the PDB code to the UniProt accession numbers. SIFTS offers residue-level mapping from PDB to UniProt [61]. It is possible to use the MIToS package to parse SIFTS XML files and use their residue-level mapping to guide rigid structural alignments [62].

5. It is well known that crystallographic contacts affect crystallographic structures. However, several works found that the RMSD pattern observed between many different crystallographic structures for the same protein follows a trend derived from the inherent flexibility of the protein rather than from the crystallization conditions [18, 19, 63, 64]. Mainly, they found that particular structural variations between different structures from the same protein are independent of crystallization conditions. Thus, different or identical ligands, even close homologous proteins, could or could not show structural variations derived from the experimental conditions used for structure determination. Also, Sikic and co-workers have found that loop flexibility is independent of crystal packing contacts [58]. Consequently, in general, the pattern of flexibility in a protein is robust to the structural bias introduced by crystal packing or crystallographic contacts. In previous works the effect of crystal contacts on backbone flexibility has been explored using normal mode analysis [65]. It was found that correlations between backbone flexibility profiles, predicted using simple structure-based methods and experimental profiles, have shown to be almost identical when sites involved in crystal contacts were included or removed. Besides these findings, we evaluated the influence of crystallographic contacts in the estimation of the

conformational diversity of a protein. For this purpose, a subset of monomeric proteins from CoDNaS were taken, with only one protein chain in the asymmetric unit (392 pairs of conformers) in order to remove heterocomplexes. Using UCSF Chimera [66] the number of crystallographic contacts at 4.5 Å of distance in both conformers of the pair showing the maximum RMSD for that protein was estimated. The number of contacts and their correlation with the maximum RMSD were studied. If the maximum RMSD were affected by crystallographic contacts, a high correlation (negative or positive) between mean number of contact atoms and RMSD value would have been expected. However, we obtained a negligible Spearman's correlation coefficient of 0.048.

6. To make a robust estimation of protein conformational diversity using the RMSD between conformers, it is important to take into account the number of conformers per protein. Proteins with more than five conformers are suggested in order to study backbone flexibility. Moreover, approximately 20 would be the suggested number for a reliable study on side-chain heterogeneity [18].

Acknowledgments

Authors would like to thank Paula Benencio for helping us with manuscript proofreading.

References

1. Gerstein M, Lesk AM, Chothia C (1994) Structural mechanisms for domain movements in proteins. Biochemistry 33:6739–6749
2. Gerstein M, Krebs W (1998) A database of macromolecular motions. Nucleic Acids Res 26:4280–4290
3. Gu Y, Li D-W, Brüschweiler R (2015) Decoding the mobility and time scales of protein loops. J Chem Theory Comput 11:1308–1314
4. Gora A, Brezovsky J, Damborsky J (2013) Gates of enzymes. Chem Rev 113:5871–5923
5. Perutz MF, Bolton W, Diamond R et al (1964) Structure of haemoglobin. An X-ray examination of reduced horse haemoglobin. Nature 203:687–690
6. Popovych N, Sun S, Ebright RH et al (2006) Dynamically driven protein allostery. Nat Struct Mol Biol 13:831–838
7. Dunker AK, Keith Dunker A, Silman I et al (2008) Function and structure of inherently disordered proteins. Curr Opin Struct Biol 18:756–764
8. Boehr DD, McElheny D, Dyson HJ et al (2006) The dynamic energy landscape of dihydrofolate reductase catalysis. Science 313:1638–1642
9. Tsai CJ, Del Sol A, Nussinov R (2009) Protein allostery, signal transmission and dynamics: a classification scheme of allosteric mechanisms. Mol BioSyst 5:207–216
10. Hilser VJ (2010) Biochemistry. An ensemble view of allostery. Science 327:653–654
11. James LC, Roversi P, Tawfik DS (2003) Antibody multispecificity mediated by conformational diversity. Science 299:1362–1367
12. Smock RG, Gierasch LM (2009) Sending signals dynamically. Science 324:198–203
13. Yogurtcu ON, Bora Erdemli S, Nussinov R et al (2008) Restricted mobility of conserved residues in protein-protein interfaces in molecular simulations. Biophys J 94:3475–3485

14. Lynch TJ, Bell DW, Sordella R et al (2004) Activating mutations in the epidermal growth factor receptor underlying responsiveness of non-small-cell lung cancer to gefitinib. N Engl J Med 350:2129–2139
15. Tokuriki N, Stricher F, Serrano L et al (2008) How protein stability and new functions trade off. PLoS Comput Biol 4:e1000002
16. Zea DJ, Miguel Monzon A, Fornasari MS et al (2013) Protein conformational diversity correlates with evolutionary rate. Mol Biol Evol 30:1500–1503
17. Zea DJ, Monzon AM, Gonzalez C et al (2016) Disorder transitions and conformational diversity cooperatively modulate biological function in proteins. Protein Sci 25:1138–1146
18. Best RB, Lindorff-Larsen K, DePristo MA et al (2006) Relation between native ensembles and experimental structures of proteins. Proc Natl Acad Sci U S A 103:10901–10906
19. Burra PV, Zhang Y, Godzik A et al (2009) Global distribution of conformational states derived from redundant models in the PDB points to non-uniqueness of the protein structure. Proc Natl Acad Sci U S A 106:10505–10510
20. Berman HM, Westbrook J, Feng Z et al (2000) The Protein Data Bank. Nucleic Acids Res 28:235–242
21. Wei G, Xi W, Nussinov R et al (2016) Protein ensembles: how does nature harness thermodynamic fluctuations for life? The diverse functional roles of conformational ensembles in the cell. Chem Rev 116:6516. https://doi.org/10.1021/acs.chemrev.5b00562
22. Marino-Buslje C, Monzon AM, Zea DJ et al (2017) On the dynamical incompleteness of the Protein Data Bank. Brief Bioinform. https://doi.org/10.1093/bib/bbx084
23. Monzon AM, Juritz E, Fornasari MS et al (2013) CoDNaS: a database of conformational diversity in the native state of proteins. Bioinformatics 29:2512–2514
24. Monzon AM, Rohr CO, Fornasari MS et al (2016) CoDNaS 2.0: a comprehensive database of protein conformational diversity in the native state. Database 2016:baw038
25. Altschul SF, Gish W, Miller W et al (1990) Basic local alignment search tool. J Mol Biol 215:403–410
26. Ortiz AR, Strauss CEM, Olmea O (2002) MAMMOTH (matching molecular models obtained from theory): an automated method for model comparison. Protein Sci 11:2606–2621
27. The UniProt Consortium (2017) UniProt: the universal protein knowledgebase. Nucleic Acids Res 45:D158–D169
28. Sillitoe I, Lewis TE, Cuff A et al (2015) CATH: comprehensive structural and functional annotations for genome sequences. Nucleic Acids Res 43:D376–D381
29. Bairoch A (2000) The ENZYME database in 2000. Nucleic Acids Res 28:304–305
30. Potenza E, Di Domenico T, Walsh I et al (2015) MobiDB 2.0: an improved database of intrinsically disordered and mobile proteins. Nucleic Acids Res 43:D315–D320
31. Ashburner M, Ball CA, Blake JA et al (2000) Gene ontology: tool for the unification of biology. The Gene Ontology Consortium. Nat Genet 25:25–29
32. Qin H, Shi J, Noberini R et al (2008) Crystal structure and NMR binding reveal that two small molecule antagonists target the high affinity ephrin-binding channel of the EphA4 receptor. J Biol Chem 283:29473–29484
33. Qin H, Lim L, Song J (2012) Protein dynamics at Eph receptor-ligand interfaces as revealed by crystallography, NMR and MD simulations. BMC Biophys 5:2
34. Bowden TA, Aricescu AR, Nettleship JE et al (2009) Structural plasticity of eph receptor A4 facilitates cross-class ephrin signaling. Structure 17:1386–1397
35. Monzon AM, Zea DJ, Fornasari MS et al (2017) Conformational diversity analysis reveals three functional mechanisms in proteins. PLoS Comput Biol 13:1–29
36. Parisi G, Zea DJ, Monzon AM et al (2015) Conformational diversity and the emergence of sequence signatures during evolution. Curr Opin Struct Biol 32:58–65
37. Gutteridge A, Thornton J (2005) Conformational changes observed in enzyme crystal structures upon substrate binding. J Mol Biol 346:21–28
38. Mesecar AD, Stoddard BL, Koshland DE Jr (1997) Orbital steering in the catalytic power of enzymes: small structural changes with large catalytic consequences. Science 277:202
39. Koshland DE (1998) Conformational changes: how small is big enough? Nat Med 4:1112–1114
40. Rashin AA, Rashin AHL, Jernigan RL (2010) Diversity of function-related conformational changes in proteins: coordinate uncertainty, fragment rigidity, and stability. Biochemistry 49:5683–5704
41. Juritz E, Palopoli N, Fornasari S et al (2013) Protein conformational diversity modulates sequence divergence. Mol Biol Evol 30:79–87
42. Liu Y, Bahar I (2012) Sequence evolution correlates with structural dynamics. Mol Biol Evol 29:2253–2263

43. Saldaño TE, Monzon AM, Parisi G et al (2016) Evolutionary conserved positions define protein conformational diversity. PLoS Comput Biol 12:e1004775
44. Jeon J, Nam H-J, Choi YS et al (2011) Molecular evolution of protein conformational changes revealed by a network of evolutionarily coupled residues. Mol Biol Evol 28:2675–2685
45. Codoñer FM, Fares MA (2008) Why should we care about molecular coevolution? Evol Bioinformatics Online 4:29–38
46. de Oliveira SHP, Shi J, Deane CM (2017) Comparing co-evolution methods and their application to template-free protein structure prediction. Bioinformatics 33:373–381
47. Morcos F, Jana B, Hwa T et al (2013) Coevolutionary signals across protein lineages help capture multiple protein conformations. Proc Natl Acad Sci U S A 110:20533–20538
48. Rodriguez-Rivas J, Marsili S, Juan D et al (2016) Conservation of coevolving protein interfaces bridges prokaryote–eukaryote homologies in the twilight zone. Proc Natl Acad Sci U S A 113:15018–15023
49. Zea DJ, Monzon AM, Parisi G, et al (2018) How is structural divergence related to evolutionary information?, Molecular Phylogenetics and Evolution, Available online 25 June 2018, ISSN 1055-7903, https://doi.org/10.1016/j.ympev.2018.06.033
50. Sfriso P, Duran-Frigola M, Mosca R et al (2016) Residues coevolution guides the systematic identification of alternative functional conformations in proteins. Structure 24:116–126
51. Chothia C, Lesk AM (1986) The relation between the divergence of sequence and structure in proteins. EMBO J 5:823–826
52. Koehl P, Levitt M (2002) Sequence variations within protein families are linearly related to structural variations. J Mol Biol 2836:551–562
53. Hubbard TJ, Blundell TL (1987) Comparison of solvent-inaccessible cores of homologous proteins: definitions useful for protein modelling. Protein Eng 1:159–171
54. Russell RB, Barton GJ (1994) Structural features can be unconserved in proteins with similar folds. An analysis of side-chain to side-chain contacts secondary structure and accessibility. J Mol Biol 244:332. https://doi.org/10.1006/jmbi.1994.1733
55. Wen B, Lampe JN, Roberts AG et al (2005) Evolutionary plasticity of protein families: coupling between sequence and structure variation. Proteins 61:535–544
56. Illergård K, Ardell DH, Elofsson A (2009) Structure is three to ten times more conserved than sequence--a study of structural response in protein cores. Proteins 77:499–508
57. Monzon AM, Zea DJ, Marino-Buslje C et al (2017) Homology modeling in a dynamical world. Protein Sci 26:2195
58. Sikic K, Tomic S, Carugo O (2010) Systematic comparison of crystal and NMR protein structures deposited in the protein data bank. Open Biochem J 4:83–95
59. Kufareva I, Abagyan R (2012) Methods of protein structure comparison. In: Orry AJW, Abagyan R (eds) Homology modeling: methods and protocols. Humana Press, Totowa, NJ, pp 231–257
60. Siew N, Elofsson A, Rychlewski L et al (2000) MaxSub: an automated measure for the assessment of protein structure prediction quality. Bioinformatics 16:776–785
61. Velankar S, Dana JM, Jacobsen J et al (2013) SIFTS: structure integration with function, taxonomy and sequences resource. Nucleic Acids Res 41:D483–D489
62. Zea DJ, Anfossi D, Nielsen M et al (2016) MIToS.jl: Mutual information tools for protein sequence analysis in the Julia language. Bioinformatics 33(4):564–565
63. Zoete V, Michielin O, Karplus M (2002) Relation between sequence and structure of HIV-1 protease inhibitor complexes: a model system for the analysis of protein flexibility. J Mol Biol 315:21–52
64. Hrabe T, Li Z, Sedova M et al (2016) PDBFlex: exploring flexibility in protein structures. Nucleic Acids Res 44:D423–D428
65. Maguid S, Fernández-Alberti S, Parisi G et al (2006) Evolutionary conservation of protein backbone flexibility. J Mol Evol 63:448–457
66. Pettersen EF, Goddard TD, Huang CC et al (2004) UCSF chimera--a visualization system for exploratory research and analysis. J Comput Chem 25:1605–1612
67. Lee RA, Razaz M, Hayward S (2003) The DynDom database of protein domain motions. Bioinformatics 19:1290–1291
68. Amemiya T, Koike R, Kidera A et al (2012) PSCDB: a database for protein structural change upon ligand binding. Nucleic Acids Res 40:D554–D558

Chapter 21

High-Throughput Antibody Structure Modeling and Design Using ABodyBuilder

Jinwoo Leem and Charlotte M. Deane

Abstract

Antibodies are proteins of the adaptive immune system; they can be designed to bind almost any molecule, and are increasingly being used as biotherapeutics. Experimental antibody design is an expensive and time-consuming process, and computational antibody design methods can now be used to help develop new therapeutics and diagnostics. Within the design pipeline, accurate antibody structure modeling is essential, as it provides the basis for antibody-antigen docking, binding affinity prediction, and estimating thermal stability. Ideally, models should be rapidly generated, allowing the exploration of the breadth of antibody space. This allows methods to replicate the natural processes of antibody diversification (e.g., V(D)J recombination and somatic hypermutation), and cope with large volumes of data that are typical of next-generation sequencing datasets. Here we describe ABodyBuilder and PEARS, algorithms that build and mutate antibody model structures. These methods take ~30 s to generate a model antibody structure.

Key words Antibody structure prediction, Side-chain prediction, Accuracy estimation, Developability

1 Introduction

In vertebrate organisms, antibodies are produced by B cells as part of the adaptive immune response. Through immunoglobulin gene recombination and somatic hypermutation, it is theoretically possible to generate ~10^{11} antibodies [1], each of which is specific for a foreign molecule (also known as an "antigen"). Antibodies bind their targets with high binding affinity, typically in the nanomolar range [2]. The level of antibody diversity, along with their unique binding properties, has led to an interest in designing antibodies for a wide range of applications, especially as novel biotherapeutics [e.g., 3–6].

Antibodies are comprised of four polypeptide chains: two "heavy" chains and two "light" chains (Fig. 1). Each chain is made up of multiple immunoglobulin domains, and is split into two regions: the variable (V) and constant (C). The V regions of the heavy (V_H) and light (V_L) chains combine to form the variable

Tobias Sikosek (ed.), *Computational Methods in Protein Evolution*, Methods in Molecular Biology, vol. 1851, https://doi.org/10.1007/978-1-4939-8736-8_21,

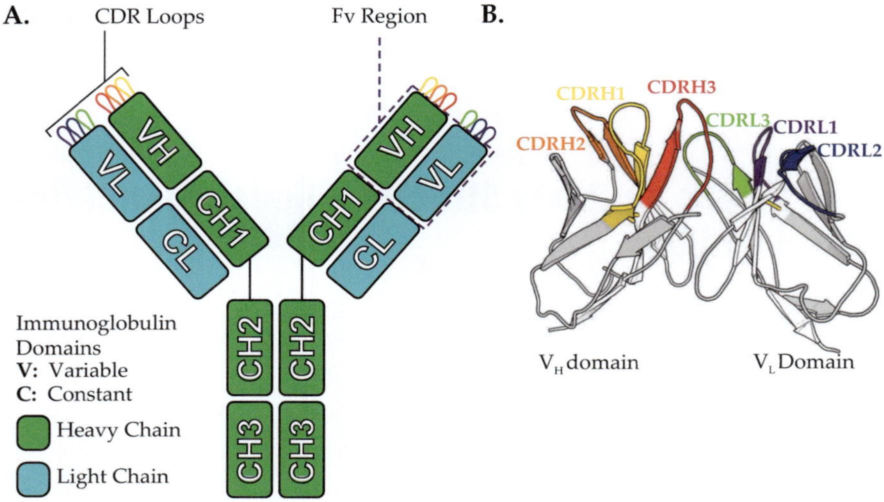

Fig. 1 Structure of an antibody molecule. (**a**) Antibodies are formed from two pairs of two protein chains: the heavy chains (green) and the light chains (cyan). Each chain has a series of immunoglobulin domains, known as the variable (V) or constant (C) regions. The two variable domains combine to form the variable fragment (Fv), and at the tip of the Fv are the CDR loops, which form the majority of the antigen-binding site. (**b**) The variable fragment has six CDR loops: CDRH1, CDRH2, and CDRH3 from the V_H domain, and CDRL1, CDRL2, and CDRL3 on the V_L

fragment (Fv), which is responsible for antigen recognition. Each V region has three loops, which have the highest degree of sequence and structural diversity between different antibodies [7]. These are known as the complementarity-determining region (CDR) loops. Three CDRs from the V_H (CDRH1, CDRH2, CDRH3) and three CDRs from the V_L (CDRL1, CDRL2, CDRL3) form the majority of the antigen-binding site. Despite variation in sequence, five of the six CDR loops (CDRH1, CDRH2, CDRL1, CDRL2, and CDRL3) are thought to adopt a limited number of conformations, known as the canonical classes [e.g., 8–10]. The remainder of the V domains is collectively known as the antibody's framework. Residues in the framework region and packing of the V_H and V_L domains (i.e., V_H–V_L orientation) can also affect antigen binding [11, 12].

Antibody design campaigns aim to engineer a new antibody structure that can bind a target of interest, or modify an antibody for enhanced function, such as affinity or thermostability. Several methods are used for experimental antibody design, such as phage display [13] and immunizing "humanized" mice [14]. However, these techniques require extensive resources [15]. To facilitate the progress of experimental work, computational techniques can help improve an antibody's affinity [16], increase its safety and stability [17–19], target new antigens [20], or explore new binding modes [21].

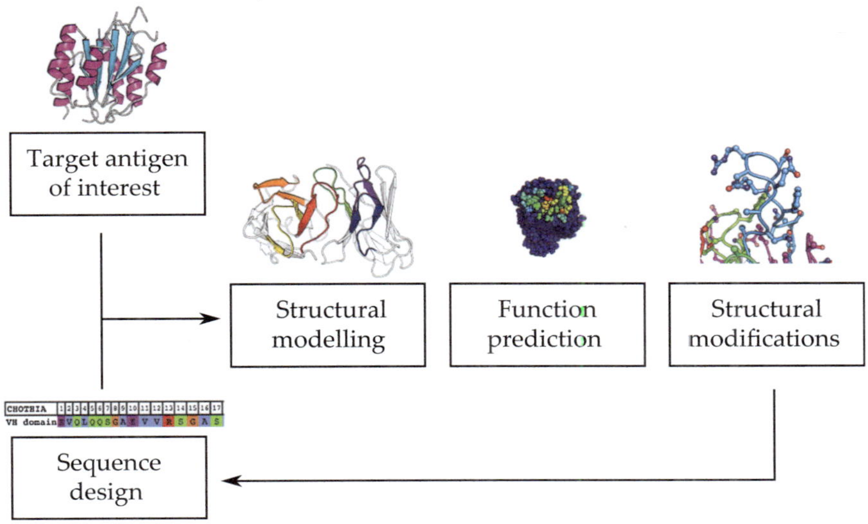

Fig. 2 Starting from an initial target sequence, it is imperative to build a model structure of the antibody [24–26]. Next, the model antibody structure is tested for a particular function; for example, they are docked to the target antigen for predicting binding affinity [27, 28]. From this newly formed complex, the antibody structure is allowed to mutate, leading to a new antibody sequence. This cycle is repeated, leading to multiple possible designs

Traditionally, antibodies have been computationally "redesigned" by mutation in silico. However, the ideal end goal would be to develop a complete antibody "design" methodology, where antibodies can be built de novo for a novel target [22, 23]. For both redesign and complete design, a pipeline must give an amino acid sequence that generates an antibody structure with desired properties, e.g., an antibody with sub-nanomolar binding affinity (Fig. 2) [29, 30]. Thus, computational design pipelines rely on several tools, including sequence annotation and analysis [31], structural modeling [24–26], function prediction [27, 28, 32], and mutation [33, 34].

Antibody structural modeling lies at the heart of computational antibody design [22, 23]. Although the antigen structure may be available (experimental or model), the sequence and the structure of the cognate antibody during the design phase are often unknown. Thus, it is imperative to build structural models to help understand how an antibody can interact with the antigen, e.g., via antibody-antigen docking [4, 27, 28]. Models allow users to investigate the impact of design choices (e.g., single amino acid mutations or grafting CDR loops) on specific design objectives (e.g., antibody stability and safety) [35].

Once the antibody structure is modeled, mutations can be introduced in silico, leading to directed antibody evolution. In many cases, the modeled backbone structure is retained; only the amino acid side chains are swapped and repacked [33, 34].

In this chapter, we outline the procedure for using ABodyBuilder for antibody structural modeling, and PEARS for single-amino-acid mutations. Both tools are suited for computational antibody redesign, though it is possible to use them for complete design problems.

1.1 Antibody Structural Modeling

Antibody structure prediction can cover a broad range of problems, such as CDR loop prediction [36–39] and predicting the orientation between the variable domains [40, 41]. This chapter specifically focuses on predicting the structure of the Fv [24, 25, 28], as this is the domain that is primarily responsible for antigen binding.

Antibody structure prediction is usually undertaken in a template-based manner as the frameworks of antibody structures are highly conserved. Most protocols follow a similar procedure, with minor variations; as an example, Fig. 3 shows an overview of the ABodyBuilder algorithm.

For a target antibody sequence, modeling programs first identify one or more template structure(s) to model the framework region. Templates can be selected from the Protein Data Bank (PDB) [44], or from a curated database, such as the Structural Antibody Database (SAbDab) [2]. The coordinates of the template structure(s) are copied and used as a scaffold for subsequent steps. Next, the orientation between the V_H and V_L domains is predicted. This can be done by using the V_H–V_L orientation of the template structure [24, 25, 45], machine learning techniques [40], or computational docking algorithms [26].

In the third stage, the CDR loops are modeled. This is often done by knowledge-based methods, such as FREAD [36, 46, 47]. Using a database of previously observed structural fragments, FREAD predicts the CDR loop structure based on sequence similarity to the target CDR sequence and anchor geometry [47]. If a suitable fragment is not available, CDR loops can be predicted by ab initio methods, such as MODELLER [48] and Rosetta [28, 37]. Programs such as Sphinx use both fragment-based and ab initio techniques for predicting the CDR loops, which is particularly useful for the CDRH3 loop [38]. For the CDRH1, CDRH2, CDRL1, CDRL2, and CDRL3 loops, it is possible to predict the canonical form of the loop based on sequence [10, 49].

Finally, the torsion angles of the side chains, known as the χ angles, are predicted using only the backbone information alone. Some modeling methods, such as ABodyBuilder, rely on dedicated side-chain prediction tools [24, 25, 42]. Other pipelines use a built-in side-chain prediction algorithm [43, 50], or a solvation model [51]. Following side-chain prediction, the model structure in some protocols undergoes energy minimization [42, 43].

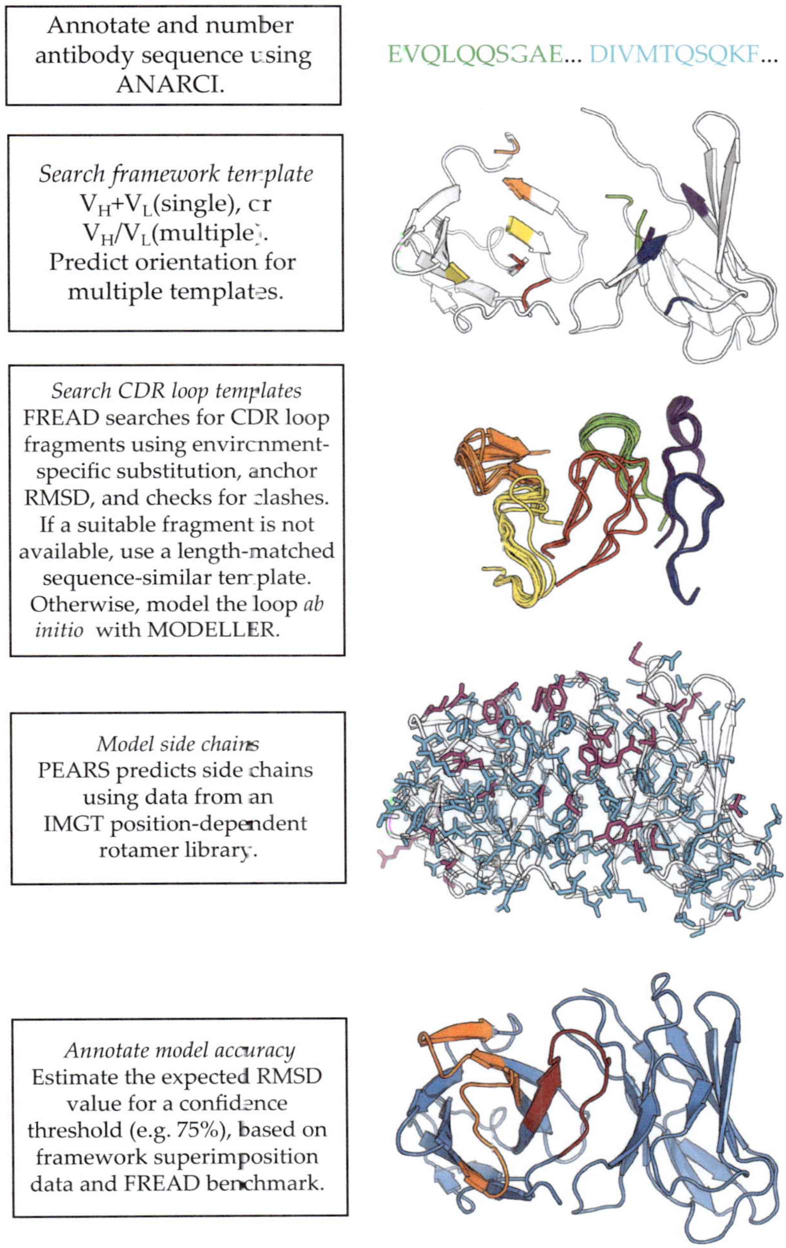

Fig. 3 Overview of the ABodyBuilder modeling methodology [24]. Most modeling methods follow a similar workflow with minor variations [25, 26, 42, 43]

Once the model is generated, ABodyBuilder annotates the expected model accuracy, and is currently the only freely available methodology that offers this functionality [24]. ABodyBuilder estimates the probability that a region, e.g., the framework or the CDRH1 loop, is modeled within a root mean square deviation

(RMSD) threshold, given the sequence identity or loop length. For example, ABodyBuilder reports the probability that the model's CDRL3 loop would be predicted with an RMSD within 1.0 Å. For design purposes, this allows users to understand the limitations of the model structure, and whether the model should be considered for further applications in silico. ABodyBuilder also flags positions that can cause potential developability issues [52], helping users eliminate some of the sources of error in antibody production.

1.2 Directed Evolution by Side-Chain Prediction

Side-chain prediction methods can be used to predict all the side chains on a model structure, or they can be used to introduce mutations [e.g., 33, 34]. It is assumed that in most cases, changing a single-amino-acid residue has little impact on the overall structure of a protein [33]. Thus, in silico mutation can be considered a specialized case of side-chain prediction.

In the traditional side-chain prediction problem, every residue's χ angle(s) must be predicted. In order to simplify the conformational search space, the χ angles are described in discrete forms, known as rotamers. Side-chain prediction methods generate predictions by sampling rotamers from rotamer libraries, which describe the probability of a rotamer for a given structural property. The most common structural property is the ϕ/ψ angles of the backbone [53, 54]. Other properties such as secondary structure [55] or an amino acid's position in a protein fragment [56] have also been used. For PEARS, our antibody-specific side-chain predictor, rotamer probabilities are dependent on their IMGT position [57]. Numbering schemes such as the IMGT scheme provide a method for comparing the amino acid sequences of two or more antibodies. In theory, a given position should represent a specific part of the immunoglobulin domain, and capture features such as the distribution of amino acids. While there are various schemes available [8, 58], the IMGT scheme is often preferred as it has a clear correlation to structure [57, 59].

2 Materials

2.1 Web Requirements

A WebGL and JavaScript-enabled web browser, such as Google Chrome, is recommended for ABodyBuilder (http://opig.stats.ox.ac.uk/webapps/abodybuilder) and PEARS (http://opig.stats.ox.ac.uk/webapps/pears). Currently, PEARS is part of the ABodyBuilder pipeline, though users can separately submit structures to PEARS for predicting the side chains, or mutating the antibody in silico.

2.2 Additional Software Requirements

To view the model structures locally, users are recommended to use PyMOL (https://sourceforge.net/projects/pymol/), which is available for Linux, Macintosh, and Windows. Users can download annotations, e.g., sequence liabilities, as CSV files, which can be opened in most spreadsheet applications (e.g., LibreOffice).

3 Methods

3.1 ABodyBuilder

3.1.1 Sequence Annotation

In the sequence submission form, submit the amino acid sequence of the target antibody. In order to model a paired antibody (including single-chain Fvs), submit sequences for both the heavy and light chains, while for single-domain antibodies (for example, V_HH antibodies), submit the sequence for one chain (*see* **Note 1**). In the text below, we describe the procedure for paired antibodies.

1. The submitted target sequence is numbered by ANARCI [31], which uses a database of hidden Markov models (HMMs; *see* **Note 1**) to number antibody sequences. During this process, the antibody's framework region and CDR loops are identified using the definitions from [9].

3.1.2 Framework Template Selection and Orientation Prediction

Once the sequence has been annotated by ANARCI, ABodyBuilder searches for a template framework structure from SAbDab [2].

1. ABodyBuilder identifies the template with the highest sequence identity to the target sequence across the framework region. If there is an antibody structure that is at least 80% sequence-identical for both chains, ABodyBuilder uses this structure as a single "global" template. Otherwise, it uses a "hybrid" template where two templates, one for the V_H and V_L, are used. *See* **Note 2** for example of template selections.
2. If ABodyBuilder finds a global template, its orientation is used. For hybrid templates, the orientation of the antibody with the highest global sequence identity is used. *See* **Note 2** for example of orientation selections.

3.1.3 Prediction of the CDR Loops

The CDR loops are predicted by a combination of FREAD [46, 47] and MODELLER [48]. The loops are predicted in the order of CDRL2, CDRH2, CDRL1, CDRH1, CDRL3, and then CDRH3. The ordering is based on our ability to predict each CDR loop individually, and the frequency of Cβ-Cβ contacts between CDR loops. The CDRL2 and CDRH2 loops are predicted first because they are usually modeled with the highest accuracy and there are no contacts between them. This is followed by CDRL1 and CDRH1 as they are the next best predicted loops, and then the CDRL3 and CDRH3.

1. FREAD is a database method; a CDR loop-specific database is used to predict each loop, i.e., a CDRL3-specific database is used to predict the CDRL3 loop. FREAD selects loops using an environment-specific substitution, anchor RMSD, and

checks for clashes with the scaffold (i.e., the framework region and existing CDR loops). If there are no suitable fragments in the CDR-specific database, FREAD uses an antibody-specific database, which includes fragments from all six CDR loops.

2. If FREAD does not find a suitable prediction, ABodyBuilder searches for a length-matched loop with the highest BLOSUM62 score to the target CDR loop sequence.
3. If a length-matched sequence-similar loop is not available, MODELLER is used to model the loop ab initio.

3.1.4 Side-Chain Prediction

Once the CDR loops are predicted on the template framework structure, the side chains of the model are predicted using PEARS. PEARS uses an IMGT position-dependent distribution of amino acid rotamers in antibody structures.

1. PEARS first builds the disulfide bridges in the antibody structure, typically between IMGT positions H23-H104 and L23-L104.
2. Next, PEARS identifies side chain types that are known to have a unimodal $\chi 1$ angle distribution (e.g., L116 tyrosine). The side chains at these positions are predicted first using rotamers with the same $\chi 1$ angle bin. If there are no suitable predictions, these positions are predicted in the next step.
3. The remaining side chains are predicted by dead-end elimination [60] and then graph decomposition, similar to other side-chain prediction methods [33, 61]. If no suitable predictions can be made, only a Cβ atom is placed.

3.1.5 Annotation of Model Structure and Download Links

Once ABodyBuilder completes the modeling process (~30 s), the user is immediately redirected to the results page, summarizing the templates that were used for the framework and the CDRs (Fig. 4). In addition, sequence alignments of the model and target sequences are provided. Users can choose to submit the structure for paratope prediction (Antibody i-Patch) [27] or epitope prediction (EpiPred) [32], or view the model structure for model accuracy and sequence liabilities (Fig. 4).

3.2 Pears

3.2.1 Structure Input Form

The first step requires the user to upload the structure of the antibody, with or without the antigen (Fig. 5), and specify the antibody chains, for example, "HL." To mutate residues in the antibody structure, the desired amino acid sequence of the antibody is then submitted (*see* **Note 3**). PEARS generates the mutated structure and the user is directed to a results page with the antibody, renumbered in the IMGT scheme, that will be available, along with a text file listing all the predicted χ angles.

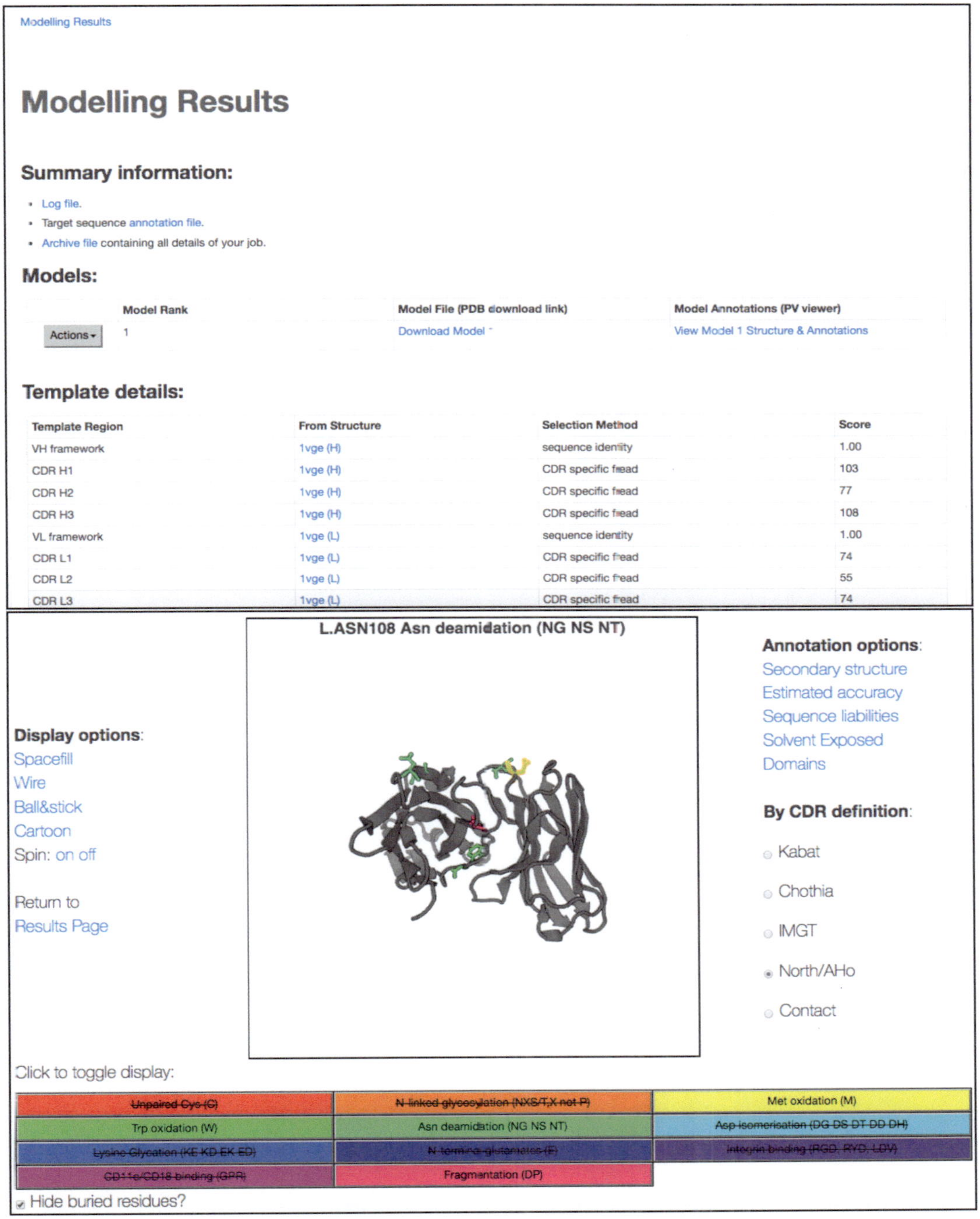

Fig. 4 Screenshots of the ABodyBuilder results and viewer pages. Once a model is built, users are directed to the results page (top) that lists the template structures that were used to model different regions of the antibody. The viewer page (bottom) shows the model using BioPV [62]

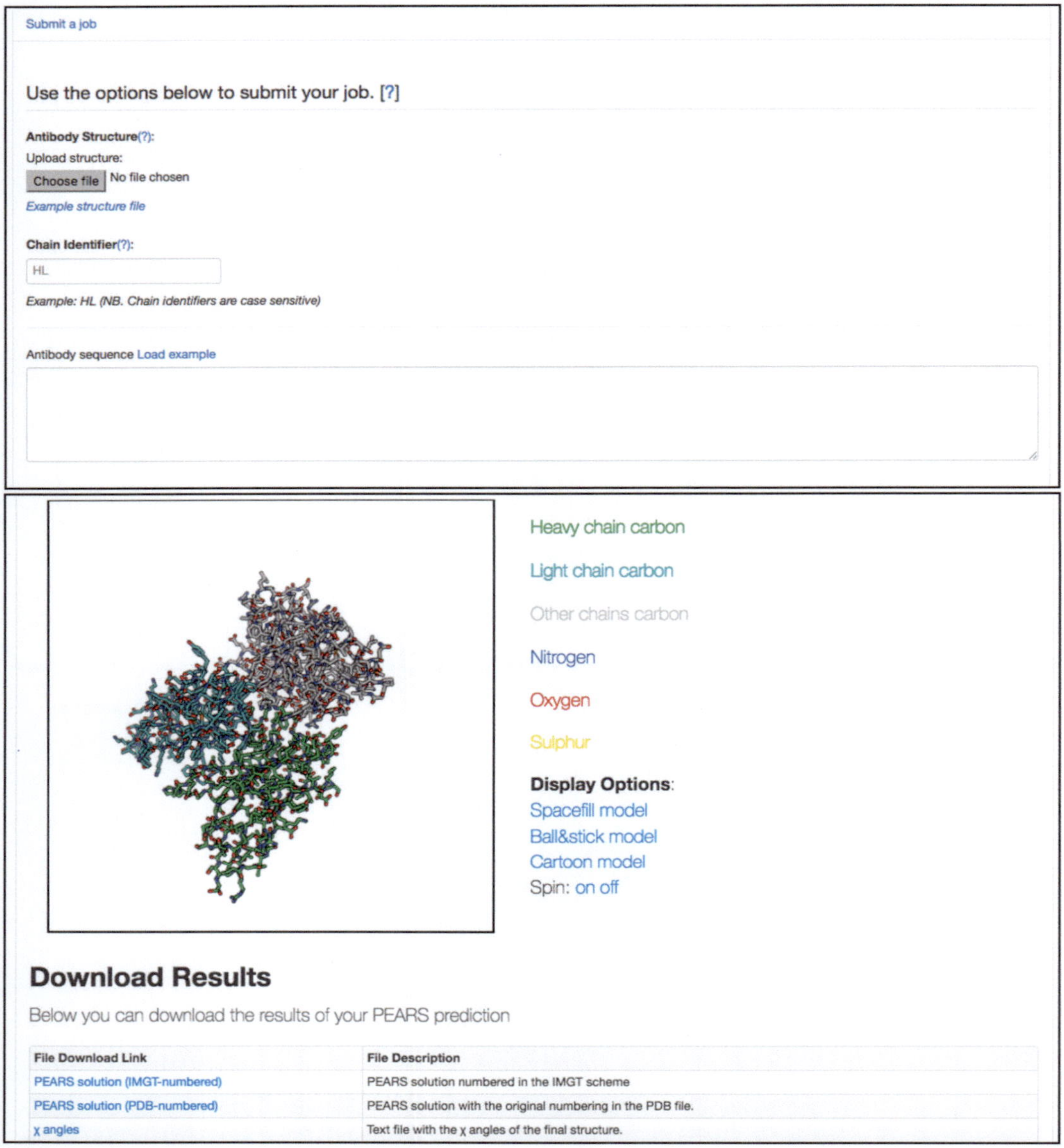

Fig. 5 Screenshot of the PEARS input and results pages. In the input page (top), users can submit a modified sequence of the antibody (*see* **Note 3**). The output page (bottom) shows the final prediction, and users can download a tab-separated file that lists the χ angles in the final model

3.2.2 Mutation of the Input Structure

When mutating the antibody structure, PEARS aligns the submitted sequence to the amino acid sequence in the structure (*see* **Note 3**). In the single-mutation case, PEARS simply uses the lowest energy rotamer to fit into the target position. Otherwise, it runs dead-end elimination and graph decomposition.

3.2.3 Resolving Clashes in the Structure

When PEARS predicts the side-chain structure of a target position, it uses a KD-tree algorithm to check for clashes. Two atoms are considered to clash if they are closer than 63% of the sum of their van der Waal's radii, which is similar to previously established cutoffs [34]. If clashes are detected, PEARS first adds Gaussian noise to the χ angles; if this does not resolve the clashes, no predictions are made, and the position is left with only a Cβ atom.

4 Notes

1. Sequences are initially numbered by ANARCI. If ANARCI cannot number the sequence (e.g., not a variable domain sequence), or it cannot detect the anchor residues of the CDR loops, ABodyBuilder immediately stops. We found that most antibodies that were not modeled by ABodyBuilder had failed at the ANARCI stage of the pipeline. Thus, users are advised to check that their sequence can be numbered using the ANARCI web application (http://opig.stats.ox.ac.uk/webapps/anarci) before running ABodyBuilder.
2. Below are example selections of global and hybrid templates.

Target antibody	V_H template (sequence identity)	V_L template (sequence identity)	Orientation template
1h0d:BA	1ejo:H (89%)	1ejo:L (86%)	1ejo:HL
12e8:HL	3nig:E (90%)	1i3g:L (95%)	1i3g:HL

In the case of 1h0d:BA, a single "global" template is used as both chains of 1ejo:HL have ≥80% sequence identity to the target sequence. For 12e8:HL, a "hybrid" template is used as the heavy chain of 1i3g:HL has 79.8% sequence identity to the target sequence. However, across both chains, 1i3g:HL has the highest global sequence identity, and is thus used for predicting the V_H–V_L orientation.

3. To mutate an antibody, we recommend submitting a sequence with identical length to the input structure, though PEARS can align sequences with mismatching lengths. Furthermore, the chain identifiers must refer to the antibody chains in the structure; otherwise the sequence alignment will fail. To fix existing side chains, use lowercase letters, and uppercase letters elsewhere. If multiple side-chain mutations are required, we recommend rerunning ABodyBuilder, though PEARS can handle multiple mutations at once.

References

1. Georgiou G, Ippolito GC, Beausang J, Busse CE, Wardemann H, Quake SR (2014) The promise and challenge of high-throughput sequencing of the antibody repertoire. Nat Biotechnol 32:158–168
2. Dunbar J, Krawczyk K, Leem J, Baker T, Fuchs A, Georges G, Shi J, Deane CM (2014) SAbDab: the structural antibody database. Nucleic Acids Res 42:D1140–D1146
3. Chames P, Van Regenmortel M, Weiss E, Baty D (2009) Therapeutic antibodies: successes, limitations and hopes for the future. Br J Pharmacol 157:220–233
4. Kuroda D, Shirai H, Jacobson MP, Nakamura H (2012) Computer-aided antibody design. Protein Eng Des Sel 25:507–521
5. Reichert JM (2017) Antibodies to watch in 2017. MAbs 9:167–181
6. Weiner GJ (2015) Building better monoclonal antibody-based therapeutics. Nat Rev Cancer 15:361–370
7. Schroeder HW, Cavacini L (2010) Structure and function of immunoglobulins. J Allergy Clin Immunol 125:41–52
8. Chothia C, Lesk A (1987) Canonical structures for the hypervariable regions of immunoglobulins. J Mol Biol 196:901–917
9. North B, Lehmann A, Dunbrack RL (2011) A new clustering of antibody CDR loop conformations. J Mol Biol 406:228–256
10. Nowak J, Baker T, Georges G, Kelm S, Klostermann S, Shi J, Sridharan S, Deane CM (2016) Length-independent structural similarities enrich the antibody CDR canonical class model. MAbs 8:751–760
11. Dunbar J, Fuchs A, Shi J, Deane CM (2013) ABangle: Characterising the VH-VL orientation in antibodies. Protein Eng Des Sel 26:611–620
12. Foote J, Winter G (1992) Antibody framework residues affecting the conformation of the hypervariable loops. J Mol Biol 224:487–499
13. McCafferty J, Griffiths AD, Winter G, Chiswell DJ (1990) Phage antibodies: filamentous phage displaying antibody variable domains. Nature 348:552–554
14. Lee E-C, Liang Q, Ali H, Bayliss L, Beasley A, Bloomfield-Gerdes T, Bonoli L, Brown R, Campbell J, Carpenter A, Chalk S, Davis A, England N, Fane-Dremucheva A, Franz B, Germaschewski V, Holmes H, Holmes S, Kirby I, Kosmac M, Legent A, Lui H, Manin A, O'Leary S, Paterson J, Sciarrillo R, Speak A, Spensberger D, Tuffery L, Waddell N, Wang W, Wells S, Wong V, Wood A, Owen MJ, Friedrich GA, Bradley A (2014) Complete humanization of the mouse immunoglobulin loci enables efficient therapeutic antibody discovery. Nat Biotech 32:356–363
15. Liu X, Taylor RD, Griffin L, Coker S-F, Adams R, Ceska T, Shi J, Lawson ADG, Baker T (2017) Computational design of an epitope-specific Keap1 binding antibody using hotspot residues grafting and CDR loop swapping. Sci Rep 7:41306
16. Lippow SM, Wittrup KD, Tidor B (2007) Computational design of antibody-affinity improvement beyond in vivo maturation. Nat Biotechnol 25:1171–1176
17. Choi Y, Hua C, Sentman CL, Ackerman ME, Bailey-Kellogg C (2015) Antibody humanization by structure-based computational protein design. MAbs 7:1045–1057
18. Miklos AE, Kluwe C, Der BS, Pai S, Sircar A, Hughes RA, Berrondo M, Xu J, Codrea V, Buckley PE, Calm AM, Welsh HS, Warner CR, Zacharko MA, Carney JP, Gray JJ, Georgiou G, Kuhlman B, Ellington AD (2012) Structure-based design of supercharged, highly thermoresistant antibodies. Chem Biol 19:449–455
19. Olimpieri PP, Marcatili P, Tramontano A (2015) Tabhu: tools for antibody humanization. Bioinformatics 31:434–435
20. Lewis SM, Wu X, Pustilnik A, Sereno A, Huang F, Rick HL, Guntas G, Leaver-Fay A, Smith EM, Ho C, Hansen-Estruch C, Chamberlain AK, Truhlar SM, Conner EM, Atwell S, Kuhlman B, Demarest SJ (2014) Generation of bispecific IgG antibodies by structure-based design of an orthogonal Fab interface. Nat Biotechnol 32:191–198
21. Dunbar J, Knapp B, Fuchs A, Shi J, Deane CM (2014) Examining variable domain orientations in antigen receptors gives insight into TCR-like antibody design. PLoS Comput Biol 10:1–10
22. Lapidoth GD, Baran D, Pszolla GM, Norn C, Alon A, Tyka MD, Fleishman SJ (2015) AbDesign: an algorithm for combinatorial backbone design guided by natural conformations and sequences. Proteins 83:1385–1406
23. Li T, Pantazes RJ, Maranas CD (2014) OptMAVEn – a new framework for the de novo design of antibody variable region models targeting specific antigen epitopes. PLoS One 9:1–17
24. Leem J, Dunbar J, Georges G, Shi J, Deane CM (2016) ABodyBuilder: automated

antibody structure prediction with data-driven accuracy estimation. MAbs 8:1259–1268

25. Marcatili P, Olimpieri PP, Chailyan A, Tramontano A (2014) Antibody structural modeling with prediction of immunoglobulin structure (PIGS). Nat Protoc 9:2771–2783
26. Sivasubramanian A, Sircar A, Chaudhury S, Gray JJ (2009) Toward high-resolution homology modeling of antibody Fv regions and application to antibody-antigen docking. Proteins 74:497–514
27. Krawczyk K, Baker T, Shi J, Deane CM (2013) Antibody i-Patch prediction of the antibody binding site improves rigid local antibody-antigen docking. Protein Eng Des Sel 26:621–629
28. Weitzner BD, Jeliazkov JR, Lyskov S, Marze N, Kuroda D, Frick R, Adolf-Bryfogle J, Biswas N, Dunbrack RL Jr, Gray JJ (2017) Modeling and docking of antibody structures with Rosetta. Nat Protoc 12:401–416
29. Huang P-S, Boyken SE, Baker D (2016) The coming of age of de novo protein design. Nature 537:320–327
30. Khoury GA, Smadbeck J, Kieslich CA, Floudas CA (2014) Protein folding and de novo protein design for biotechnological applications. Trends Biotechnol 32:99–109
31. Dunbar J, Deane CM (2016) ANARCI: antigen receptor numbering and receptor classification. Bioinformatics 32:298–300
32. Krawczyk K, Liu X, Baker T, Shi J, Deane CM (2014) Improving B-cell epitope prediction and its application to global antibody-antigen docking. Bioinformatics 30:2288–2294
33. Krivov GG, Shapovalov MV, Dunbrack RL (2009) Improved prediction of protein side-chain conformations with SCWRL4. Proteins 77:778–795
34. Nagata K, Randall A, Baldi P (2012) SIDEpro: a novel machine learning approach for the fast and accurate prediction of side-chain conformations. Proteins 80:142–153
35. Almagro JC, Teplyakov A, Luo J, Sweet RW, Kodangattil S, Hernandez-Guzman F, Gilliland GL (2014) Second antibody modeling assessment (AMA-II). Proteins 82:1553–1562
36. Choi Y, Deane CM (2011) Predicting antibody complementarity determining region structures without classification. Mol BioSyst 7:3327–3334
37. Finn JA, Koehler Leman J, Willis JR, Cisneros A, Crowe JE, Meiler J (2016) Improving loop modeling of the antibody complementarity-determining region 3 using knowledge-based restraints. PLoS One 11: e0154811
38. Marks C, Nowak J, Klostermann S, Georges G, Dunbar J, Shi J, Kelm S, Deane CM (2017) Sphinx: merging knowledge-based and ab initio approaches to improve protein loop prediction. Bioinformatics 33:1346–1353
39. Messih MA, Lepore R, Marcatili P, Tramontano A (2014) Improving the accuracy of the structure prediction of the third hypervariable loop of the heavy chains of antibodies. Bioinformatics 30:2733–2740
40. Bujotzek A, Dunbar J, Lipsmeier F, Schäfer W, Antes I, Deane CM, Georges G (2015a) Prediction of VH-VL domain orientation for antibody variable domain modeling. Proteins 83:681–695
41. Marze NA, Lyskov S, Gray JJ (2016) Improved prediction of antibody VL-VH orientation. Protein Eng Des Sel 29:409–418
42. Yamashita K, Ikeda K, Amada K, Liang S, Tsuchiya Y, Nakamura H, Shirai H, Standley DM (2014) Kotai antibody builder: automated high-resolution structural modeling of antibodies. Bioinformatics 30:3279–3280
43. Bujotzek A, Fuchs A, Qu C, Benz J, Klostermann S, Antes I, Georges G (2015b) MoFvAb: modeling the Fv region of antibodies. MAbs 7:838–852
44. Berman HM, Westbrook J, Feng Z, Gilliland G, Bhat TN, Weissig H, Shindyalov IN, Bourne PE (2000) The Protein Data Bank. Nucleic Acids Res 28:235–242
45. Maier JKX, Labute P (2014) Assessment of fully automated antibody homology modeling protocols in molecular operating environment. Proteins 82:1599–1610
46. Choi Y, Deane CM (2010) FREAD revisited: accurate loop structure prediction using a database search algorithm. Proteins 78:1431–1440
47. Deane CM, Blundell TL (2001) CODA: a combined algorithm for predicting the structurally variable regions of protein models. Protein Sci 10:599–612
48. Šali A, Blundell TL (1993) Comparative protein modelling by satisfaction of spatial restraints. J Mol Biol 234:779–815
49. Adolf-Bryfogle J, Xu Q, North B, Lehmann A, Dunbrack RL Jr (2015) PyIgClassify: a database of antibody CDR structural classifications. Nucleic Acids Res 43:D432–D438
50. Berrondo M, Kaufmann S, Berrondo M (2014) Automated aufbau of antibody structures from given sequences using Macromoltek's SmrtMolAntibody. Proteins 82:1636–1645
51. Zhu K, Day T, Warshaviak D, Murrett C, Friesner R, Pearlman D (2014) Antibody structure determination using a combination of homology modeling, energy-based

refinement, and loop prediction. Proteins 82:1646–1655

52. Jarasch A, Koll H, Regula JT, Bader M, Papadimitriou A, Kettenberger H (2015) Developability assessment during the selection of novel therapeutic antibodies. J Pharm Sci 104:1885–1898

53. Shapovalov MV, Dunbrack RL (2011) A smoothed backbone-dependent rotamer library for proteins derived from adaptive kernel density estimates and regressions. Structure 19:844–858

54. Towse C-L, Rysavy S, Vulovic I, Daggett V (2016) New dynamic rotamer libraries: data-driven analysis of side-chain conformational propensities. Structure 24:187–199

55. Lovell SC, Word JM, Richardson JS, Richardson DC (2000) The penultimate rotamer library. Proteins 40:389–408

56. Chinea G, Padron G, Hooft RWW, Sander C, Vriend G (1995) The use of position-specific rotamers in model building by homology. Proteins 23:415–421

57. Lefranc M-P, Pommié C, Ruiz M, Giudicelli V, Foulquier E, Truong L, Thouvenin-Contet V, Lefranc G (2003) IMGT unique numbering for immunoglobulin and T cell receptor variable domains and Ig superfamily V-like domains. Dev Comp Immunol 27:55–77

58. Kabat EA, Wu TT, Bilofsky H, Reid-Miller M, Perry HM (1983) Sequences of proteins of immunological interest, 3rd edn. National Institutes of Health, Bethesda

59. Lefranc M-P (2014) Immunoglobulin and T cell receptor genes: IMGT and the birth and rise of Immunoinformatics. Front Immunol 5:22

60. Desmet J, Maeyer MD, Hazes B, Lasters I (1992) The dead-end elimination theorem and its use in protein side-chain positioning. Nature 356:539–542

61. Miao Z, Cao Y, Jiang T (2011) RASP: rapid modeling of protein side chain conformations. Bioinformatics 27:3117–3122

62. Biasini M (2015) pv: v1.8.1

Chapter 22

In Silico-Directed Evolution Using CADEE

Beat Anton Amrein, Ashish Runthala, and Shina Caroline Lynn Kamerlin

Abstract

Recent years have seen an explosion of interest in both sequence- and structure-based approaches toward in silico-directed evolution. We recently developed a novel computational toolkit, CADEE, which facilitates the computer-aided directed evolution of enzymes. Our initial work (Amrein et al., IUCrJ 4:50–64, 2017) presented a pedagogical example of the application of CADEE to triosephosphate isomerase, to illustrate the CADEE workflow. In this contribution, we describe this workflow in detail, including code input/output snippets, in order to allow users to set up and execute CADEE simulations on any system of interest.

Key words Enzyme design, Directed evolution, Computational enzymology, Computational enzyme design, Empirical valence bond

1 Introduction

Directed evolution has revolutionized biotechnology, allowing enzyme activity to be modified in a targeted fashion with the requirement of minimal prior knowledge about the mechanistic features of the enzyme [1–5]. Nevertheless, despite constant methodological advances [6–8] there remain challenges with this approach, in no small part due to the vastness of the sequence space that needs sampling and the very small likelihood of identifying beneficial mutations [5, 9, 10]. Here, computational approaches can play important roles in focusing the multidimensional search space, in guiding experimental design, and, ultimately, in performing the directed evolution itself in silico. There have been a number of approaches that aim to address this problem using sequence- or structure-based approaches, and for reviews we refer the authors to, e.g., Refs. [11–13].

We recently developed a novel semiautomated approach for the computer-aided directed evolution of enzymes (CADEE) [14], based on Warshel's empirical valence bond (EVB) approach [15]. The EVB approach has been established as a powerful approach to investigate enzymatic systems, as illustrated in Fig. 1.

Tobias Sikosek (ed.), *Computational Methods in Protein Evolution*, Methods in Molecular Biology, vol. 1851,
https://doi.org/10.1007/978-1-4939-8736-8_22,

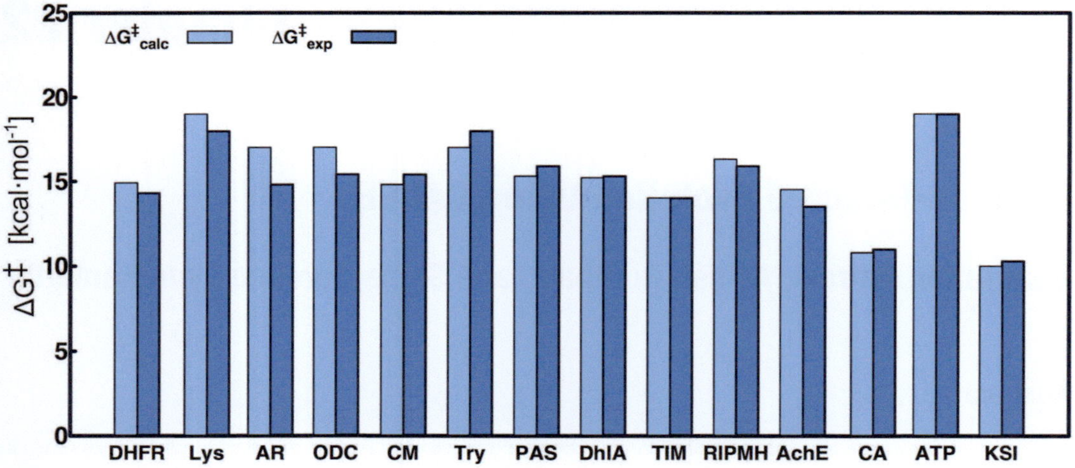

Fig. 1 Examples of various enzymes that have been studied with the EVB approach. The experimental activation free energies ($\Delta G^{\ddagger}_{exp}$) are shown in dark blue, and the calculated activation free energies ($\Delta G^{\ddagger}_{calc}$) are shown in sky blue. DHFR, Lys, AR, CM, Try, PAS, DhlA, TIM, RIPMH, AchE, ODC, CA, ATP, and KSI denote dihydrofolate reductase, lysozyme, aldose reductase, chorismate mutase, trypsin, a bacterial arylsulfatase, haloalkane dehalogenase, triosephosphate isomerase, a bacterial phosphonate monoester hydrolase, acetylcholine esterase, orotidine monophosphate decarboxylase, carbonic anhydrase, F1-ATPase, and ketosteroid isomerase, respectively. CC-BY adopted from Ref. [14], based on data originally presented in Refs. [16–19]

We present here a methodology guide to CADEE [14], to guide the user in setting up, deploying, and analyzing CADEE-based simulations. The main advantage of CADEE is that it removes the tedious manual computational setup and analysis steps that would otherwise be required when running large numbers of independent EVB simulations, thus greatly simplifying the simulation process. This framework is based on powerfully interwoven tools and a command-line interface. Specifically, CADEE assists the user with preparation of the simulations (cadee prep), performing the actual equilibration and EVB simulations (cadee dyn), and the subsequent analysis (cadee ana). In addition, due to CADEE's dependence on the EVB approach, the main workhorse of CADEE is the Q simulation package [20, 21], and CADEE uses a similar workflow to Q (for a description of this workflow, see the Q manual [21]). In terms of other external dependencies, CADEE utilizes the mpi4py [22] python package to distribute the computational workload during molecular dynamics simulations, and relies on Open Babel [23] to convert molecular structures into different formats. To automatize part of the analysis workflow, an in-house script collection (qscripts) is utilized. Finally, to allow protein structures to be automatically modified with residue substitutions, SCWRL [24] is required.

The EVB approach has been extensively used to study biochemical reactivity [15–17, 25–29] (Fig. 1), and it forms the main basis for CADEE, as it allows for extensive conformational sampling from multiple different starting conformations with low computational cost, thus accelerating the convergence of the calculated activation free energies. In addition, while the EVB approach requires calibration to a reference system [15, 17], once the EVB parabolas have been calibrated for the system of interest, the exact same parameters are then used *unchanged* to rapidly model the same reaction in different environments, for example in different enzymes or enzyme variants. This means that for computational studies with EVB, only a single frame of reference is required, and thus it removes the need for additional tedious reparameterization steps from the workflow. This makes EVB a perfect tool for in silico-directed evolution and the later screening stages of computational enzyme design [14, 26, 30–32]. The theoretical background for CADEE has been described in detail elsewhere, and we refer the reader to the original CADEE paper [14] for more details about the approach. In this contribution, we will specifically guide the user through the CADEE workflow with input files and pedagogical examples.

2 Background and Default Settings

In this section, we introduce important technical considerations and practical aspects that help users to understand how to perform CADEE simulations.

2.1 Spherical Boundary Conditions

In order to economize on computational time, CADEE simulations are performed using spherical rather than periodic boundary conditions, with the protein immersed into a spherical droplet of water described using the surface-constrained all-atom solvent model (SCAAS) [33], with long-range electrostatics treated using the local reaction field (LRF) approach [34]. In this model, all residues within the inner 85% of the droplet are fully mobile, residues within the outer 15% of the droplet are restrained to their crystallographic positions using a 10 kcal mol^{-1} $Å^{-2}$ positional restraint, and the motion of all residues outside the droplet is fully restricted with a 200 kcal mol^{-1} $Å^{-2}$ harmonic restraint to the initial crystallographic coordinates. Additionally, ionization of residues is restricted to the innermost 85% of the droplet (i.e., the mobile region) in order to avoid system instabilities. The use of such a setup both accelerates the conversion of the calculations and gives the user flexibility in terms of the sphere size used (allowing for example for extension of the sphere size where needed to study distant mutations or convergence of energies or dynamics). However, it also means that the user needs to benchmark different sphere sizes until the results become independent of the sphere size. The minimal usable sphere size varies for different systems but is typically in the range from 20 to 30 Å radius, depending on the system specifics.

2.2 Speed and Computing Resources

As CADEE is a distributed computing framework that runs a large number of individual tasks at once, it is important to keep computational overhead low. In CADEE, this overhead minimization is achieved with the following tricks:

1. In the case of multistep reactions, only the rate-limiting step is simulated to an initial approximation, and other steps are only simulated once a more limited selection of hits have been identified, to focus resources.
2. We have included standard settings intended for the best use of resources:
   (a) Hysteresis tends to be reduced in the runs due to equilibration at the approximate transition state along the reaction coordinate.
   (b) Short thermalization and 8 ns of tandem equilibration and EVB phases.
   (c) Four replicas each with 8 EVB snapshots = 32 data points for statistics.
   (d) The EVB calculations are initiated from structural snapshots collected every nanosecond of the initial equilibration; this allows post-calculation assignment of when the system has sufficiently equilibrated so that stable energetics are reached.
   (e) By default, each enzyme variant is simulated for a total of 50 ns.
3. In order to decrease the simulation time, we advise the user to start the EVB simulations at the transition state, propagating the trajectories to the reactant and product complexes. This both accelerates convergence (as the user is starting from a state with partial bonds to reacting atoms) and, provided that sufficient computational resources are available, reduces the real time of the simulations as trajectories can be propagated in both directions at once.
4. CADEE efficiency is high, thanks to a pleasingly parallel implementation, allowing hundreds to thousands of simulations to be performed in parallel.
5. While these high-efficiency defaults are best suited for production simulations, they are inconvenient for test simulations and the initial system setup. To overcome this, a special script can be employed, as described in Subheading 4.3.2.

2.3 Simulation Packages (Simpacks)

A simpack as used by CADEE is a tarball, which contains the input files for a CADEE simulation. Once a simulation has started, all the results are appended to the simpack. It is therefore not only an input, but also an output file. Once a simulation is started ("cadee dyn"), the contents of the simpack are copied to a temporary folder. The simulation is then spooled to the right position (skipping steps that have been computed previously) and then molecular dynamics simulations are performed. Once a simulation step has been completed, it will be compressed, collected, and appended to the simpack in intervals to reduce strain on the file system. A default CADEE simpack contains a total of 8 ns of equilibration time. Every 1000 ps thereof, a snapshot is used to perform a medium-length EVB simulation (each 520 ps total length).

By default, we suggest the user primarily relies on the medium-length EVB simulations for estimating the likely activation free energies for the constructs being tested (the longer the simulation time, the more likely the simulations have converged). As each simpack contains eight medium-length EVB runs, it is possible to allow retro-actively for additional N ns initial equilibration, by removing the data points of the first N medium-length EVB runs. The reasoning for this advanced *internal* setup (in comparison to traditional Q inputs and workflows) is that CADEE hides this complexity from the user and hence does not cripple productivity, but rather empowers the user during the analysis. For example, in our model system below, we have not accounted for (i.e., removed) the first two data points from the simpack, in order to give the system an additional 2 ns of initial equilibration time, resulting in a total equilibration time of 3 ns.

Technical details: Simpacks use the following nomenclature protocol: [variant-name]_[replica].tar, for example for the wild-type protein: "wt_0.tar, wt_1.tar, wt_2.tar, wt_3.tar" and for a histidine 104 to alanine variant: "H104A_0.tar, H104A_1.tar, H104A_2.tar, H104A_3.tar", etc. For our working example, CADEE creates four independent replicas (*seeds*) for each enzyme variant, leading to a total of 4 × 8 = 32 medium EVB runs. In addition, we have decided to manually remove the first 2 EVB simulations from each simpack, to allow for longer initial equilibration without increasing real simulation time, effectively yielding 4 × 6 = 24 medium EVB energy profiles (see also the previous paragraph). As a baseline, all simpacks must contain a topology file (mutant.top), a simulation-ready PDB file (mutant.pdb), and the FEP file (mutant.fep) which contains the EVB parameters for the different reacting states for the reaction being studied. If molecular dynamics simulations should be performed, the simpacks must contain numbered input files (*.inp), as per the following scheme: 01_* to 09_*: initialization and thermalization, 1000_eq.inp to 4660_fep.inp: equilibration and FEP files. Files containing the string ("_eq") are 50 ps equilibration runs (the reason for the short 50 ps equilibration runs being to allow data backup at least every hour of real time on all tested clusters). For more details about the content of the input files, refer to the Q manual [21].

3 CADEE Installation

This protocol has been written specifically in conjunction with CADEE version 0.9. However, we note that CADEE is constantly in development, and therefore we advise the reader to refer to the latest CADEE release and associated instructions, available along with the downloaded CADEE release https://github.com/kamerlinlab/cadee. For simplicity, here, we will guide the user through CADEE installation and setup using the model example of triosephosphate isomerase, as with our initial CADEE paper [14].

3.1 The Wild-Type Enzyme Reaction

CADEE relies on the user to have already characterized and validated the reaction mechanism of the enzyme of interest, and to have calibrated the EVB coupling and gas-phase shift parameters against relevant experimental or computational data (usually corresponding to the energetics of the reaction catalyzed by the wild-type enzyme, or the corresponding uncatalyzed reaction in aqueous solution), as described in Ref. [14]. It is therefore crucial that the system is carefully prepared, as the quality of data obtained from all subsequent steps builds on the correct modeling of the baseline reaction (i.e., for the EVB calibration).

3.2 Installation and System Requirements

CADEE has been written and tested on Linux machines. The parallel computing has been tested on a variety of Intel as well as AMD clusters, as these systems were accessible through the Swedish National Infrastructure for Computing (SNIC) at various sites in Linköping (NSC/Triolith https://www.nsc.liu.se/), Uppsala (UPPMAX/Tintin and Rackham https://www.uppmax.uu.se/), and Umeå (HPC2N/Akka, Abisko, and Kebnekaise https://www.hpc2n.umu.se/). We note that while the software was written to run on all SNIC-provided resources, we have not, as yet, tested it on other SNIC clusters. In addition, we have not used other resource managers than SLURM, as this is the primary resource manager on SNIC systems. For simplicity, we assume that the user will be using a Debian- or Ubuntu-based system, as those systems are under widespread use.

The CADEE installer may be downloaded from our official GitHub repository, which is located at http://www.github.com/kamerlinlab/cadee. For new users, we recommend following the CADEE installation instructions described in the following sections in order to get started.

3.3 How to Read this Chapter

Throughout this section, we assume that a modern implementation of the Bourne again shell (bash) is installed and used by the user. Code that needs to be typed into a terminal emulator is explicitly identified as a Code Input Snippet:

```
# Code Input Snippet (1)
/bin/bash --version
```

Note that lines ending with \\ imply that the command continues on the next line (therefore "Enter" should not be used or the command might not work as intended). Similarly, the corresponding output is explicitly identified as a Code Output Snippet:

```
# Code Output Snippet (1)
GNU bash, version 4.3.48(1)-release (x86_64-pc-linux-gnu)
Copyright (C) 2013 Free Software Foundation, Inc.
[...]
```

Here, lines containing [...] represent messages that have been removed to simplify and shorten output. And finally, all lines starting with "#" are comments.

3.4 Downloading and Installing Third-Party Programs

3.4.1 Installing a Compiler, MPI, git, and Python 2.7

As Q is to be compiled by the user, a FORTRAN compiler is necessary and we suggest the use of the free and open-source software (FOSS) compiler gfortran [35]. CADEE also requires a working MPI implementation, for example OpenMPI [36] and/or MPICH [37]. We further recommend installation of the versioning program Git (https://git-scm.com), as this allows for the installation instructions to be followed exactly. Since CADEE is a Python 2.7 [38] based framework that utilizes setup tools in the installation, Python and setup tools need to be installed. Finally, CADEE also requires Open Babel [23].

Assuming a Debian-based Linux distribution, the following commands will resolve the required and recommended software dependencies:

```
# Code Input Snippet (2)
sudo apt-get install gfortran openmpi-bin git openbabel
sudo apt-get install mpich gcc python2.7 python-pip
```

```
# Code Output Snippet (2)
# The software should be installed without errors.
```

3.4.2 Licensing and Downloading Q

Q [39] needs to be licensed, downloaded, and installed, as CADEE relies on the functional capabilities of this molecular simulation package (see http://xray.bmc.uu.se/~aqwww/q/ for further details). First-time users are advised to thoroughly familiarize themselves with Q, and to establish and try out simple reaction mechanisms with EVB before starting CADEE simulations, to facilitate the usage of CADEE with their own systems of interest.

3.4.3 Licensing and Downloading SCWRL4

SCWRL4 [24] needs to be licensed, downloaded, and installed, as CADEE utilizes the functional capabilities of this package to rapidly predict a likely side-chain orientation (rotamer). Users are advised to visit http://dunbrack.fccc.edu/scwrl4/ for instructions on the licensing, download, and installation of SCWRL4.

3.5 Download and Installation of CADEE

3.5.1 Downloading CADEE

CADEE can be either downloaded from https://github.com/kamerlinlab/cadee or obtained by git-cloning the repository to the user's computer:

```
# Code Input Snippet (3)
cd $HOME
mkdir -p Downloads
cd $HOME/Downloads
git clone https://github.com/kamerlinlab/cadee cadee
cd $HOME/Downloads/cadee
export CADEE_DIR="$PWD"
```

```
# Code Output Snippet (3)
# Example output:
Cloning into 'cadee'...
remote: Counting objects: 298, done.
remote: Compressing objects: 100% (34/34), done.
remote: Total 298 (delta 25), reused 40 (delta 18), pack-reused 246
Receiving objects: 100% (298/298), 479.34 KiB | 491.00 KiB/s, done.
Resolving deltas: 100% (118/118), done.
Checking connectivity... done.
```

3.5.2 Q and SCWRL Installation

First, SCWRL4 should be installed to a folder in $PATH, as also mentioned in the download instructions. Next, a copy of the Q executables has to be placed in the folder prepared for them ($CADEE_DIR/cadee/executables/q): Once Q has been compiled, qfep5, qdyn5, qprep5, and qcalc5 should be copied to $CADEE_DIR/cadee/executables/q/. Alternatively, the setup.py script will search in the $PATH for the Q executables.

3.5.3 CADEE Installation

Once all required dependencies are installed (*see* Subheading 3.4), one may proceed to install CADEE:

```
# Code Input Snippet (4)
python setup.py install --user
```

The setup first locates the executables of Q, Open Babel, and SCWRL4, and it will stop the installation if an error is detected, or if any of the following executables is not found: qdyn5, qfep5, qprep5, scwrl4, and babel. Additionally, Python libraries might need to be installed during the installation to make sure that CADEE will function properly (mpi4y, numpy).

```
# Code Output Snippet (4)
Welcome to CADEE Pre-Setup Check.
[...]
Installing cadee script to $HOME/.local/bin
[...]
Finished processing dependencies for cadee==0.9
```

4 Testing the CADEE Installation

4.1 First Start

Once CADEE has been installed successfully, the CADEE wrapper script ("cadee") will be available on the command line. This script has been written for the ease of users familiar with Q, as the syntax is maintained between the two programs.

```
# Code Input Snippet (5)
cadee --help
```

If the output is similar to the following lines, CADEE has been installed successfully:

```
# Code Output Snippet (5)

   ___    _    ___  ___  ___
  / __|  /_\  |   \| __|| __|  Computer-Aided Directed
 | (__  / _ \ | |) | _| | _|   Enzyme Evolution
  \___|/_/ \_\|___/|___||___|
         v. 0.9

(C) Copyright 2017 Beat Anton Amrein & Shina Caroline Lynn
Kamerlin
Usage:
        cadee [ prep(p) | dyn(d) | ana(a) | tool(t) ]
  Multi Core Tasks:
        mpirun -n X cadee dyn
        mpiexec -n X cadee dyn
        X == Number of cores to use; 2+.
```

In case the output does not resemble the above output (e.g., "cadee: command not found"), the installation has failed (or is incomplete), and we suggest users refer to the troubleshooting described in Subheading 8.2.

4.2 Preparing a CADEE Simulation

As described in the introduction, to use CADEE, the valence bond states describing different reacting species for the reaction of interest need to be pre-parameterized and calibrated for running the EVB simulations that underlie CADEE (for details about the EVB approach, see, e.g., Refs. [15, 17]). To simplify CADEE usage and understanding, and to allow for easy CADEE testing, we have included a set of sample EVB input files for the user (see $CADEE_DIR/example). For more information about the theoretical background, we refer the user to earlier publications [14, 15, 17].

In order to run properly, CADEE requires the following files:

1. A structure file in PDB format ($CADEE_DIR/example/**wt.pdb**), comprising the wild-type enzyme with correct ionization states for ionizable residues, solvated in a water droplet (generated by Qprep5): The initial coordinates are typically obtained from the Protein Data Bank [40, 41] and then adjusted to be compatible with Q.
2. A "FEP file" ($CADEE_DIR/example/**wt.fep**), i.e., a file containing the force field parameters for the different EVB states, for the purposes of the simulation setup: Note that Qdyn does not distinguish between general free energy perturbation and specific EVB calculations when reading input.
3. The qprep5 input file ($CADEE_DIR/example/**wt.qpinp**), which was used to generate the initial simulation-ready PDB file: CADEE will use this file to check the system configuration and to prepare the relevant topologies.
4. The full path to the folder containing all topology and parameter files needed to perform the simulation ($CADEE_DIR/example/**libraries/**).

4.2.1 Preparing a CADEE Simulation

To begin with our working example, a simulation may be **prep**ared with CADEE's "prep" keyword:

```
# Code Input Snippet (6)
mkdir testing_example
cd testing_example
cp -r $CADEE_DIR/example/*.
cadee prep wt.pdb wt.fep wt.qpinp ./libraries/ --template $CADEE_DIR/\\
simpack_templates/simpack_template_0.05ns_15ps_2.5ps_32.5ps.tar.bz2
```

In this example, the sample input files included with the CADEE distribution ($CADEE_DIR/testing_example/wt) contain the benchmark simulations of the reaction with the wild-type enzyme. If the "cadee prep" command is initiated without the "--libmut" or "--alascan" arguments, CADEE will create a subfolder "wt" and prepare a "wild-type" simulation, i.e., no mutations are introduced and the EVB FEP file is not modified. This is useful, because the wild-type reaction is the first to be tested by CADEE, and if the wild-type reaction does not run properly, then there is no point in trying out other enzyme variants.

Note also that without the --template argument, CADEE would prepare inputs worth 12 ns of simulation time. This default setup will be used later in these protocols. For the sake of testing CADEE in a short time period, we use instead a 0.1 ns template, for demonstration purposes (the energetics obtained are thus not meaningful in and of themselves, and the simulations are rather merely illustrative of how to prepare and execute CADEE).

```
# Code Output Snippet (6)
[...]
INFO:root:No parameters provided. Will prepare simpacks from input; "wt".
INFO:prep.create_inputs:Creating input files for wt.
INFO:root:Packing wt:
INFO:root:Pack # 0, Seed: 582993
INFO:root:Pack # 1, Seed: 399763
INFO:root:Pack # 2, Seed: 669722
INFO:root:Pack # 3, Seed: 851033
Success! You find your simpacks in $CADEE_DIR/testing_example/wt.
```

In case the subfolder "wt" exists, CADEE will warn the user about this. In many cases, the wt.qpinp files then need to be adapted to a CADEE-specific format, using absolute filenames and coordinates: The "cadee prep" command will try to automatically perform these changes and create a new file, inserting ".new" before the file extension; for example, in the example used here, this would be "wt.**new**.qpinp" (if this file already exists, "cadee prep" stops and asks the user to remove "wt.new.qpinp"). Only then are the wild-type simpacks created and finally packed. The very last line indicates that the simpacks are ready to use. *Caution*: The simpacks have been prepared, but not yet computed. Instructions for the computation are provided in Subheading 4.3.1. Note that instead of deleting the old "wt.new.qpinp" the input line may be adjusted, and instead of wt.qpinp, wt.**new**.qpinp may be used:

```
# Code Input Snippet (7)
cadee prep wt.pdb wt.fep wt.new.qpinp./libraries/
```

The above input, will then, in turn, generate the following output:

```
# Code Output Snippet (7)
[...]
CRITICAL:root:Cannot continue: Folder $CADEE_DIR/testing_example/wt exists.
Please (re)move it.
```

A directory listing of the folder indicated on the last line reveals:

```
ls $CADEE_DIR/testing_example/wt
wt_0.tar wt_1.tar  wt_2.tar  wt_3.tar
```

CADEE sequentially names the simpacks [mutant-name]_[X].tar, with X in [0,1,2,3,. . .], where X denotes the replica number for each individual trajectory (for example wt_0.tar, wt_1.tar, wt_2.tar wt_3.tar, for the wild-type enzyme). *Caution:* If the simpack is untarred, it is important to make sure that this is done inside an empty (sub-)folder, because simpacks do not contain any directory structure, only files. It should however not be necessary for users to unpack simpacks, as this is performed internally by CADEE.

4.2.2 Preparing for Molecular Dynamics Simulations

Put simply, a simpack contains input files required to run qdyn5 and qfep5, both of which are utilities that are needed in order to perform and analyze EVB simulations. When CADEE is computing simpacks, no new files are generated in the simpack folder, but the simpacks simply increase in size from a couple of megabytes to gigabytes. It is therefore crucial that the folder containing the simpacks holds enough free storage to accommodate this. A simpack contains all restart information needed, and if a run is interrupted and later restarted, the simpack alone is enough to restart the CADEE simulation. Simpacks should in principle not be corrupted, except if CADEE has stopped ungracefully, for example if a simulation runs out of storage. Clearly, however, in the event that simpacks are corrupted, then the corrupted simpacks need to be repaired before proceeding (*see* the troubleshooting description in Subheading 8.3).

4.3 Performing Ensemble Simulations Using CADEE

4.3.1 Efficient Way: Saving CPU Time—"CADEE Dyn"

CADEE includes scripts to automatize parallel computation, and in a standard CADEE simulation several simpacks are computed at once. When in doubt, the user should refer to their supercomputing support team, as scripts may need to be adapted to the computing resources available. For simplicity, we assume that an interactive session is available and/or the user is able to run CADEE locally. For the working example presented here, a four-core allocation is required.

The following command will copy the wild-type simulation that was prepared above in Subheading 4.2.1 into a new folder and start the simulation. *Note*: Depending on the architecture and speed of the computing resource, this computation may need up to 2 h to complete on a four-core machine.

```
# Code Input Snippet (8)
mkdir -p $HOME/global/cadee_tutorial
cp -r $CADEE_DIR/testing_example/* $HOME/global/cadee_tutorial
# you may need to adjust mpirun
mpirun.mpich -np 5 cadee dyn $HOME/global/cadee_tutorial/wt | tee cadee.log
```

This command will launch "cadee dyn" with four working tasks (plus one for input/output). Note that the log file will be only written to standard out (the console) by default, and when using the "| tee cadee.log" part of above command, the log is additionally written to cadee.log. We note that the resulting simpack includes all input files and output files; that is, "cadee dyn" will not generate special output files, but instead the simpack files will become larger (for more about simpacks, *see* Subheading 2.3). Depending on the mpi implementation used, the "mpirun.mpich" command needs to be adjusted (possible commands include "srun," "mpiexec," or "mpirun"). The command above should create output similar to:

```
# Code Output Snippet (8)
[...]
Rank 0: Started @ 1506234059.03 MPI Info:  enabled: True rank: 0 size: 5
0 - 170924 08:20:59,032 - dyn - INFO - Settings: Path:
$HOME/global/cadee_tutorial/wt, Alpha: None, Hij: None, Force mapping: False.
0 - 170924 08:20:59,032 - dyn - INFO - Add input file wt_2.tar.
0 - 170924 08:20:59,032 - dyn - INFO - Add input file wt_3.tar.
0 - 170924 08:20:59,032 - dyn - INFO - Add input file wt_1.tar.
0 - 170924 08:20:59,032 - dyn - INFO - Add input file wt_0.tar.
0 - 170924 08:20:59,033 - dyn - INFO - Prioritized.
0 - 170924 08:20:59,075 - cadee.dyn.tools - INFO - Committed cadee.db.
0 - 170924 08:20:59,076 - dyn - INFO - Number of simpacks left on queue 3.
0 - 170924 08:20:59,076 - dyn - INFO - Number of simpacks left on queue 2.
0 - 170924 08:20:59,076 - dyn - INFO - Number of simpacks left on queue 1.
0 - 170924 08:20:59,076 - dyn - INFO - Number of simpacks left on queue 0.
1 - 170924 08:20:59,076 - dyn - INFO - Working on
$HOME/global/cadee_tutorial/wt/wt_3.tar.
2 - 170924 08:20:59,076 - dyn - INFO - Working on
$HOME/global/cadee_tutorial/wt/wt_2.tar.
3 - 170924 08:20:59,076 - dyn - INFO - Working on
$HOME/global/cadee_tutorial/wt/wt_1.tar.
4 - 170924 08:20:59,076 - dyn - INFO - Working on
$HOME/global/cadee_tutorial/wt/wt_0.tar.
2 - 170924 08:20:59,197 - dyn.traj - INFO - Next qdyn simulation step initialized.
1 - 170924 08:20:59,198 - dyn.traj - INFO - Next qdyn simulation step initialized.
3 - 170924 08:20:59,198 - dyn.traj - INFO - Next qdyn simulation step initialized.
```

```
2 - 170924 08:20:59,227 - dyn.traj - INFO - 01_dyn_seed.inp
3 - 170924 08:20:59,236 - dyn.traj - INFO - 01_dyn_seed.inp
1 - 170924 08:20:59,246 - dyn.traj - INFO - 01_dyn_seed.inp
4 - 170924 08:20:59,256 - dyn.traj - INFO - Next qdyn simulation step initialized.
4 - 170924 08:20:59,313 - dyn.traj - INFO - 01_dyn_seed.inp
0 - 170924 08:20:59,878 - dyn - INFO - Sleeping @ 0.8 s.
0 - 170924 08:21:22,216 - dyn - INFO - Slept for 22.3 seconds.
2 - 170924 08:21:22,162 - dyn.traj - WARNING - Found HOT ATOM'
2 - 170924 08:21:22,199 - dyn.traj - INFO - 02_dyn_no_shake.inp
3 - 170924 08:21:22,247 - dyn.traj - WARNING - Found HOT ATOM'
3 - 170924 08:21:22,285 - dyn.traj - INFO - 02_dyn_no_shake.inp
0 - 170924 08:21:22,817 - dyn - INFO - Sleeping @ 23.8 s.
0 - 170924 08:21:23,719 - dyn - INFO - Slept for 0.9 seconds.
4 - 170924 08:21:23,707 - dyn.traj - WARNING - Found HOT ATOM'
4 - 170924 08:21:23,747 - dyn.traj - INFO - 02_dyn_no_shake.inp
0 - 170924 08:21:24,320 - dyn - INFO - Sleeping @ 25.3 s.
0 - 170924 08:21:24,923 - dyn - INFO - Slept for 0.6 seconds.
1 - 170924 08:21:24,857 - dyn.traj - WARNING - Found HOT ATOM'
1 - 170924 08:21:24,895 - dyn.traj - INFO - 02_dyn_no_shake.inp
0 - 170924 08:21:25,424 - dyn - INFO - Sleeping @ 26.4 s.
0 - 170924 08:22:20,905 - dyn - INFO - Slept for 55.4 seconds.
2 - 170924 08:22:20,817 - dyn.traj - INFO - 03_dyn_warm_1.inp
[...]
1 - 170924 08:32:58,208 - dyn - INFO - Backup timing 0.049s, MB:   9.65 Speed:
196.77 MB/s
1 - 170924 08:32:58,212 - dyn.traj - INFO - 1190_eq.inp
[...]
1 - 170924 09:01:01,811 - dyn.traj - INFO - 1450_fep.inp
[...]
0 - 170924 09:01:46,782 - dyn - INFO - Sleeping @ 2447.7 s.
0 - 170924 09:02:05,615 - dyn - INFO - Slept for 18.8 seconds.
1 - 170924 09:02:05,766 - dyn - INFO - Backup timing 0.145s, MB:  13.45 Speed:
92.55 MB/s
0 - 170924 09:02:05,815 - dyn - INFO - Sending shutdown message to 1
0 - 170924 09:02:05,815 - dyn - INFO - Number of simpacks left on queue 0.
0 - 170924 09:02:05,915 - dyn - INFO - Worker 1 was removed from worker-list:
There are 0 (out of 4) left[...]
0 - 170924 09:02:05,916 - dyn - INFO - Preparing to end this Simulation! Syncing...
0 - 170924 09:02:05,916 - cadee.dyn.tools - INFO - Committed cadee.db.
0 - 170924 09:02:05,916 - dyn - INFO - Database connection closed.
0 - 170924 09:02:05,916 - dyn - INFO - Removing Temporary Files...
0 - 170924 09:02:05,916 - dyn - INFO - DONE. Exiting
```

As implied in the log excerpt above, the parallel dynamics simulation was started with five MPI ranks; the first process is used as "master" for distributing work, I/O control, and logging the job. All ranks > = 1 are slave processes, performing the actual number crunching. To lower the load on the file system, parallel input/output is limited to 8 simpacks reading/writing simultaneously, except when more than 128 cores are used (then #cores/16 are allowed). The temporary storage is assumed to be set in the environment variable $CADEE_TMP. If this variable is not set, /scratch/, then /tmp and then /dev/shm are used. *Caution:* It is a user responsibility to make sure that enough space is available in the folder for temporary files.

4.3.2 Saving Wall-Clock Time

In certain scenarios, it is important to get results fast, and to use the available resources for speed, not for efficiency, such as when a wild-type reaction needs to be prototyped for a certain enzyme. In such a case, CADEE ships scripts which need to be adjusted to the user's machine. These scripts are located in $CADEE_DIR/cadee/tools/pcadee.sh and $CADEE_DIR/cadee/tools/srunq.sh, respectively. Once adapted to the computer system of interest, they can be launched by:

```
# Code Input Snippet (9)
mkdir -p $HOME/global/cadee_tutorial_wallclock
cp -r $CADEE_DIR/testing_example/* $HOME/global/cadee_tutorial_wallclock
$CADEE_DIR/cadee/tools/pcadee.sh $HOME/global/cadee_tutorial_wallclock/wt
```

```
# Code Output Snippet (9)

Simpack Folder $HOME/global/cadee_tutorial_wallclock/wt
Will use 4 per simpack.
Will run at most 1 simpacks at one time.
This will use 4 cores from 4.

  Will Distribute Jobs and Start Work in 1 Second
   ===============================================
$HOME/global/cadee_tutorial_wallclock/wt/wt_1.tar
wt_1.tar @0 > Start: Son Sep 24 09:45:04 CEST 2017
wt_1.tar @0 > You supplied a SIMPACK:
$HOME/global/cadee_tutorial_wallclock/wt/wt_1.tar
wt_1.tar @0 > Unpacking Simpack
($HOME/global/cadee_tutorial_wallclock/wt/wt_1.tar) to tmpdir (/tmp/8788).
wt_1.tar @0 >
wt_1.tar @0 >
wt_1.tar @0 > ###########
wt_1.tar @0 > # CONFIG: #
wt_1.tar @0 > ###########
wt_1.tar @0 >  bkp int: 540
wt_1.tar @0 >  simpack: $HOME/global/cadee_tutorial_wallclock/wt/wt_1.tar
wt_1.tar @0 >  cores:   4
wt_1.tar @0 >  exe:     mpiexec -n 4 $HOME/bin/qdyn5p
wt_1.tar @0 >   md5sum: 6219dabb4f56f72bb914c1cb159f79a2
$HOME/bin/qdyn5p
```

```
wt_1.tar @0 >  workdir: /tmp/8788
wt_1.tar @0 >
wt_1.tar @0 >
wt_1.tar @0 >
wt_1.tar @0 > Working Directory;  localhost:/tmp/8788
wt_1.tar @0 > Preparing 01_dyn_seed ...
wt_1.tar @0 > Running MD Simulation on 01_dyn_seed.inp ...
wt_1.tar @12 > Finished: 01_dyn_seed.log
wt_1.tar @12 > Zipping.
wt_1.tar @12 > Backup Skipped.
wt_1.tar @12 > Working Directory;  localhost:/tmp/8788
wt_1.tar @12 > Preparing 02_dyn_no_shake ...
wt_1.tar @12 > Running MD Simulation on 02_dyn_no_shake.inp ...
wt_1.tar @35 > Finished: 02_dyn_no_shake.log
wt_1.tar @35 > Zipping.
[...]
wt_1.tar @517 > Running MD Simulation on 1260_fep.inp ...
wt_1.tar @542 > Finished: 1260_fep.log
wt_1.tar @542 > Zipping.
wt_1.tar @542 > Backup Complete, Duration: 0, [ Son Sep 24 09:54:06 CEST 2017 ]
[...]
wt_1.tar @967 > Running MD Simulation on 1450_fep.inp ...
wt_1.tar @986 > Finished: 1450_fep.log
wt_1.tar @986 > Zipping.
wt_1.tar @986 > Backup Skipped.
wt_1.tar @986 > Backup Complete, Duration: 0, [ Son Sep 24 10:01:31 CEST 2017 ]
[...]
wt_1.tar @986 > All OK.
wt_1.tar @986 > End: Son Sep 24 10:01:31 CEST 2017
wt_1.tar @986 > Duration: 986 s
Cleanup Done.
$HOME/global/cadee_tutorial_wallclock/wt/wt_1.tar
[...]
$HOME/global/cadee_tutorial_wallclock/wt/wt_2.tar
$HOME/global/cadee_tutorial_wallclock/wt/wt_2.tar
[...]
No Simpacks left. Terminating after 4424.
```

5 Pedagogical Examples

In this section, we demonstrate the commands required to reproduce the pedagogical examples from our original CADEE [14] publication. While both the preparation and analysis of the example can be performed on a laptop computer, we strongly advise the user to use a supercomputing cluster to perform the dynamic simulations: As each simpack needs several days to finish and this example implies computation of approximately 1100 simpacks, laptop and office computers will be inadequate to perform all required computations in a reasonable timeframe. Please note that CADEE counts residues sequentially, starting at 1, renumbering based on missing residues. The PDB structure used in this example (1ney) is missing the first residue. To convert a CADEE residue number in the sequence number +1 has to be added. For better readability, we use CADEE residue numbering throughout the whole text.

5.1 Example: Alanine Scan

To prepare simpacks for an alanine scan is straightforward, once the system has been correctly prepared and benchmarked. The argument needed to run an alanine scan using CADEE is "--alascan." Optional parameters are "--radius" (mutate all residues within a certain radius around the center of the simulation sphere) and "--nummuts" (prepare alanine scan inputs for the N innermost residues, increasing the radius around the simulation center).

```
# Code Input Snippet (10)
cp -r $CADEE_DIR/example $CADEE_DIR/pedagogical_example
cd $CADEE_DIR/pedagogical_example
cadee prep wt.pdb wt.fep wt.qpinp libraries --alascan --nummuts 48
```

For example, in the original CADEE paper [14], we used --nummuts 48 to make sure that a full compute node of the Abisko cluster at HPC2N in Umeå [42] was used.

```
# Code Output Snippet (10)
Determining Radius needed to accommodate 48 mutants. Please
wait...Done! Radius is 14.24
[...]
INFO:root:Preparing alascan.
INFO:prep.alascan:Won't mutate residue, contains FEPatoms: 164
INFO:prep.alascan:Look up center_xyz in qprep5inp -> 23.311
42.835 14.513
INFO:prep.alascan:Won't mutate residue, contains FEPatoms: 495
INFO:prep.alascan:Mutate ('ARG', '97') to ALA
[...]
INFO:prep.alascan:Mutate ('VAL', '6') to ALA
INFO:prep.create_inputs:Creating input files for LEU235ALA .
[...]
INFO:prep.create_inputs:Creating input files for LEU92ALA .
INFO:root:Packing LEU235ALA:
INFO:root:Pack # 0, Seed: 936223
INFO:root:Pack # 1, Seed: 601835
INFO:root:Pack # 2, Seed: 449120
INFO:root:Pack # 3, Seed: 757307
INFO:root:Packing SER95ALA:
INFO:root:Pack # 0, Seed: 857607
```

```
INFO:root:Pack # 1, Seed: 337917
INFO:root:Pack # 2, Seed: 561337
INFO:root:Pack # 3, Seed: 188871
[...]
Success! You find your simpacks in $CADEE_DIR/pedagogical_ex-
ample/ala_scan .
```

This code snippet will generate the simpacks for 48 protein variants (including WT) and it will create four seeds per input (a total of 192 simpacks).

Next, to actually perform the computational alanine scan, the simpacks should be copied to a location with plenty of storage, a minimum of 3 gigabytes/simpack is recommended, and the simulation may then be started:

```
# Code Input Snippet (11)
cp -r $CADEE_DIR/pedagogical_example $HOME/global
mpirun.mpich -np 193 cadee dyn $HOME/global/pedagogical_example/ala_scan
# Code Output Snippet (11)
Rank 0: Started @ 1506241452.51 MPI Info:  enabled: True rank: 0 size: 5
0 - 170924 10:24:12,510 - dyn - INFO - Settings: Path:
$HOME/global/pedagogical_example/ala_scan, Alpha: None, Hij: None, Force
mapping: False.
0 - 170924 10:24:12,531 - dyn - INFO - Add input file LEU12ALA_2.tar.
[...]
0 - 170924 10:24:12,531 - dyn - INFO - Add input file CYS40ALA_1.tar.
0 - 170924 10:24:12,532 - dyn - INFO - Prioritized.
0 - 170924 10:24:12,565 - cadee.dyn.tools - INFO - Committed cadee.db.
0 - 170924 10:24:12,565 - dyn - INFO - Number of simpacks left on queue 191.
0 - 170924 10:24:12,565 - dyn - INFO - Number of simpacks left on queue 190.
0 - 170924 10:24:12,565 - dyn - INFO - Number of simpacks left on queue 189.
[...]
[# All 192 (=48x4) simpacks need to be processed.]
[# We recommended to use 192 cores and approx. 10 days wallclock time]
[# Alternatively, the job can be split into smaller parts, for example 4x48]
[...]
```

Simulating one simpack on modern hardware usually takes between 1 week and 10 days. Older CPUs, and/or increasing the simulation sphere increases the simulation time up to 2 weeks or longer. After the simulation is finished, an analysis tool is available to specifically analyze the alanine scan, and to prepare the next simulation. The activation free energies can be evaluated using the standard EVB mapping procedure, with user-defined off-diagonal elements (H_{ij}) and gas-phase shifts (α). In the present case, we are using triosephosphate isomerase as our model system, following our initial CADEE paper [14]. The off-diagonal and gas-phase shift parameters have been previously published in the Supporting

Information of Ref. [14], and were calibrated to 60.0 (H_{ij}) and 229.0 (α) kcal mol^{-1}, respectively. CADEE will automatically prepare the free energy mapping of all EVB simulations performed, provided that the aforementioned EVB parameters have been defined on the command line.

```
# Code Input Snippet (12)
mpirun.mpich -n 2 cadee dyn \\
HOME/global/pedagogical_example/ala_scan \\
-hij 60.0 -alpha 229.0 -force
```

This will add the mapping output to cadee.db. In case a new EVB free energy mapping should be enforced, the --force flag can be used. This might be advisable, if the free energy mapping is interrupted ungracefully, or when the initial EVB parameters provided as input need to be corrected. CADEE will then simply first remove old EVB mapping results and subsequently restart the mapping all over again (apart from the mapping files, no other files will be deleted by --force_map).

```
# Code Output Snippet (12)
Rank 0: Started @ 1506257656.62 MPI Info:  enabled: True rank: 0 size: 2
0 - 170924 14:54:16,618 - dyn - INFO - Settings: Path:
$HOME/global/pedagogical_example/ala_scan, Alpha: 229.0, Hij: 60.0, Force
mapping: True.
0 - 170924 14:54:16,618 - dyn - INFO - Add input file ARG97ALA_0.tar.
[...]
0 - 170924 14:54:16,935 - cadee.dyn.tools - INFO - Committed cadee.db.
0 - 170924 14:54:16,936 - dyn - INFO - Number of simpacks left on queue 191.
1 - 170924 14:54:16,937 - dyn - INFO - Working on $HOME/global/pedagogica-
l_example/ala_scan/ARG97ALA_1.tar.
[...]
1 - 170924 14:54:57,647 - dyn.ana - INFO - deleting old .qana.mapped files...
0 - 170924 14:54:58,204 - dyn - INFO - Sleeping @ 41.6 s.
0 - 170924 14:55:02,512 - dyn - INFO - Slept for 4.3 seconds.
1 - 170924 14:55:02,504 - dyn.ana - INFO - ARG97ALA 1 medium 1190_eq dGa:
14.67 dG0: 9.25
0 - 170924 14:55:03,121 - dyn - INFO - Sleeping @ 46.5 s.
0 - 170924 14:55:07,629 - dyn - INFO - Slept for 4.5 seconds.
1 - 170924 14:55:07,619 - dyn.ana - INFO - ARG97ALA 1 medium 1650_eq dGa:
14.97 dG0: 9.34
0 - 170924 14:55:08,130 - dyn - INFO - Sleeping @ 51.5 s.
0 - 170924 14:55:13,139 - dyn - INFO - Slept for 5.0 seconds.
[...]
0 - 170924 19:55:46,294 - dyn - INFO - Sleeping @ 89.7 s.
0 - 170924 19:55:59,815 - dyn - INFO - Slept for 13.5 seconds.
1 - 170924 19:55:59,775 - dyn - INFO - Backup timing 13.982s, MB:   0.42 Speed:
0.03 MB/s
```

```
0 - 170924 19:57:23,050 - dyn - INFO - Sending shutdown message to 1
0 - 170924 19:57:23,050 - dyn - INFO - Number of simpacks left on queue 0.
0 - 170924 19:57:23,151 - dyn - INFO - Worker 1 was removed from worker-list:
There are 0 (out of 1) left
[...]
0 - 170924 19:57:23,151 - dyn - INFO - Preparing to end this Simulation!
Syncing...
0 - 170924 19:57:23,289 - cadee.dyn.tools - INFO - Committed cadee.db.
0 - 170924 19:57:23,289 - dyn - INFO - Database connection closed.
0 - 170924 19:57:23,289 - dyn - INFO - Removing Temporary Files...
0 - 170924 19:57:23,291 - dyn - INFO - DONE. Exiting
```

5.2 Example: Automated Analysis of a CADEE Alanine Scan

The analyse.py script is used to perform automated analysis of CADEE alanine scans. This script can be called by:

```
# Code Input Snippet (13)
cadee ana alanize cadee.db
firefox index.html
```

```
# Code Output Snippet (13)
# Firefox will be started and a web ui can be used.
```

This will create a file *index.html*, which can be downloaded and opened in the browser of the user's choice. The interface (ui), which is shown also in Fig. 2, is intuitive to use; the user may select the residues to be mutated next directly through this interface, and the html interface will present the user with information on how to proceed.

The web user interface allows for the clicking of a residue and then choosing the action that should be performed on it. The input to run cadee.py will then be displayed in the "CADEE command" section.

5.3 Example: Manual Analysis of CADEE Simulations

In some cases, the user might desire the raw data from a CADEE simulation to perform analysis with their own post-processing scripts, and avoid information overload from the cadee.db files (which are in sqlite3 format). For those cases, we provide a script to convert the activation energies or the free energies to the comma-separated value (csv) file format. The corresponding data can then be opened with any relevant spreadsheet software (*see* also Fig. 3).

```
# Code Input Snippet (14)
cadee ana csv cadee.db activation_barriers.csv #dG*
cadee ana csv_exo cadee.db free_energy.csv #ddG
/bin/ls
```

Fig. 2 A screenshot of the web user interface for the analysis of alanine scans. CC-BY adapted from Ref. [14]

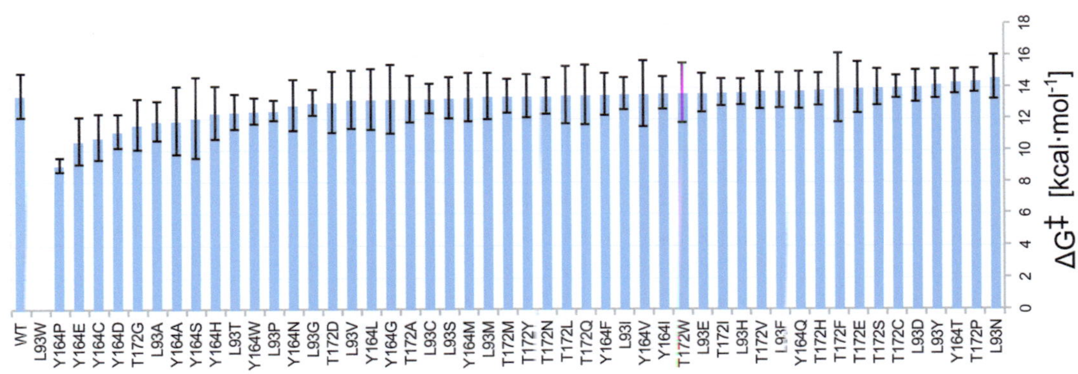

Fig. 3 The initial alanine scan performed on triosephosphate isomerase (TIM). The alanine scan was prepared with the –nummut argument as follows: After the initial system setup, 48 residues distributed radially around the center of the simulation sphere that were neither alanine nor glycine were selected, and hydrogen atoms and heavy atoms other than backbone atoms and C_β were removed. Each variant was subsequently re-solvated and the CADEE simulation was started. From the displayed data, three positions were selected to start the next simulations: L93, Y164, and T172, respectively, in positions 92, 163, and 171 in CADEE (see main text). CC-BY adopted from Ref. [14]

```
# Code Output Snippet (14)
Exporting Barrier
Success... Wrote activation_barriers.csv...
Exporting deltaG
Success... Wrote free_energies.csv...
activation_barriers.csv free_energy.csv
```

5.4 Example: Concatenating cadee.db Files

In some cases, the user might desire to merge cadee.db files (which are in sqlite3 format). For those cases, we provide a script to concatenate two or more cadee.db files.

```
# Code Input Snippet (15)
cadee ana cat cadee.db cadee.db1
/bin/ls *.db
```

```
# Code Output Snippet (15)
[...]
0 - 170924 20:27:49,213 - cadee.dyn.tools - INFO - Committed
cadee.db.
cadee1.db
cadee2.db concat_cadee.db
```

5.5 Example: Point Saturation Mutagenesis

Once a promising hotspot site has been identified, one way to continue the CADEE analysis is to perform computational combinatorial saturation mutagenesis on it (Fig. 4). This can be simplified by using a reduced set of amino acids for the calculations (see Reetz [43]), and CADEE supports different amino acid libraries for this purpose. A list of the amino acid libraries implemented into CADEE is shown in Table 1.

In the current working example, we have saturated three positions to all 20 natural amino acids, which were initiated as follows:

```
# Code Input Snippet (16)
cd $CADEE_DIR/pedagogical_example
cadee prep wt.pdb wt.fep wt.qpinp libraries --libmut 92:\\
SATURATE --libmut 163:SATURATE --libmut 171:SATURATE
```

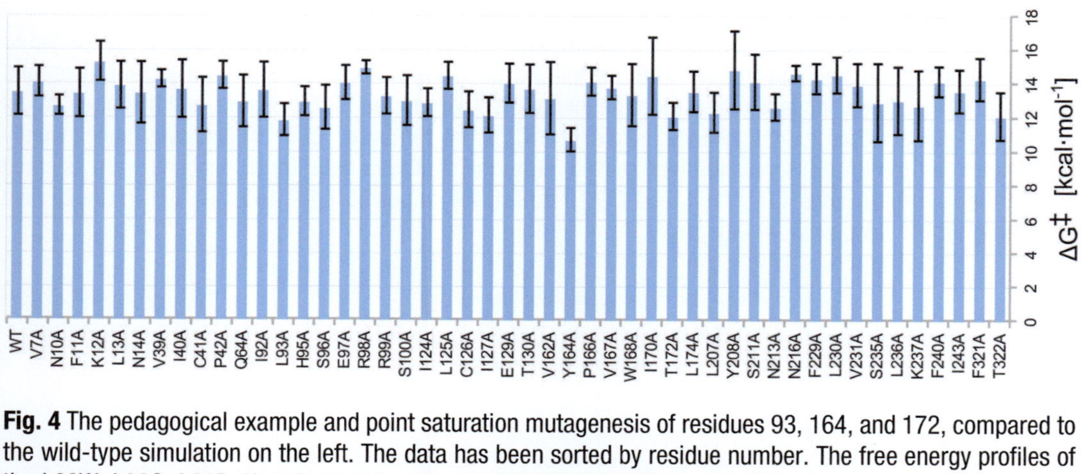

Fig. 4 The pedagogical example and point saturation mutagenesis of residues 93, 164, and 172, compared to the wild-type simulation on the left. The data has been sorted by residue number. The free energy profiles of the L93W, L93Q, L93R, Y164R, T172R, L93K, Y164K, and T172K variants did not converge. CC-BY adopted from Ref. [14]

Table 1
We present here CADEE's built-in amino acid libraries, with both the associated shortcut and the one-letter amino acid codes for each library residue, respectively[a]

Shortcut	Library used
All, saturation	ARNDCEQGHILKMFPSTWYV
NDT	FLIVYHNDCRSG
Special	CGP
Hydrophobic	AVILMFYW
Minus, negative, charged−	DE
Plus, positive, charged+	RHK
Charged	DERHK
Neutral	STNQ
Custom sequence of one-letter amino acid codes	
"AVILME"	AVILME
"AAVILME"	Error: (cannot use A 2×)

[a]When "cadee prep" is launched with the --libmut argument, the designated amino acid position will be mutated to a library of amino acids. For example "cadee prep . . . --libmut 92:ALL" will saturate position 92, mutating it to all 20 natural amino acids. Or "cadee prep . . . –libmut 92:SPECIAL 163:MINUS" will prepare a combinatorial saturation mutagenesis run (to 93(wt/C/P/G) with 163(wt/D/E), respectively, ([3 + 1] × [2 + 1]) = 12 mutants)

In the original CADEE paper [14], we used --nummuts 48 to make sure that a full compute node of the Abisko cluster at HPC2N in Umeå [42] could be used. The --libmut keyword calls SCWRL4 to prepare inputs for the arbitrary point mutations. As arguments, CADEE expects a residue number followed by a colon and then a keyword or a sequence of one-letter amino acid codes. The above input would hence mutate the residues at positions 92, 163, and 171 to all 20 natural amino acids. This command results in $3 \times 20 \times 4 = 240$ simpacks. Note that throughout the main textm we are using CADDEE integral residue numbers; these correspond to residues 93, 164, and 173 is the PDB structure.

```
# Code Output Snippet (16)
INFO:root:Preparing libmut - LIBrary MUTatagenesis.
92:SATURATE (+ native/wt)
[...]
INFO:prep.pyscwrl:Clash-Score was: 5.262, will now re-run and
allow Scwrl4 to modify residues [90, 92]
INFO:prep.pyscwrl:Clash-Score new: 0.0 ==>  Keep.
[...]
```

The aforementioned Code Output Snippet describes what CADEE is doing: First, it creates a mutant and computes a "Clash-Score" (which is exponential to the overlap of the van der Waals radii). CADEE then allows Scwrl4 to realign those residues in the clash zone (i.e., residues 90 and 92) and rerun Scwrl4. Upon encountering no steric clashes, CADEE keeps the new configuration (for details about the Scwrl4 calculations, see the Scwrl4 publication [24]).

```
163:SATURATE (+ native/wt)
[...]
INFO:root:Working on $CADEE_DIR/pedagogical_example/libmut/Y163W.
INFO:prep.pyscwrl:Clash-Score was: 0.218, will now re-run and allow Scwrl4 to modify
residues [163, 183, 667]
INFO:prep.pyscwrl:Clash-Score new: 0.460648886108 ==> Rollback!
[...]
```

In this case, as the clash was minor, allowing SCWRL4 to realign residues 163, 183, and 667 did not help the Clash-Score, so CADEE reverts to the original alignment ("Rollback!").

```
INFO:root:Working on $CADEE_DIR/pedagogical_example/libmut/L092D.
INFO:prep.pyscwrl:No clashes detected.
[...]
```

In the above case, changing leucine to an aspartic acid did not cause clashes. CADEE hence continues with this suggestion.

```
171:SATURATE (+ native/wt)
[...]
INFO:prep.create_inputs:Creating input files for Y163R.
[...]
INFO:prep.create_inputs:Creating input files for T171H.
INFO:root:Packing L092V:
INFO:root:Pack # 0, Seed: 446064
INFO:root:Pack # 1, Seed: 900050
INFO:root:Pack # 2, Seed: 641446
INFO:root:Pack # 3, Seed: 151899
INFO:root:Packing Y163M:
INFO:root:Pack # 0, Seed: 76740
INFO:root:Pack # 1, Seed: 404917
INFO:root:Pack # 2, Seed: 499416
INFO:root:Pack # 3, Seed: 751226
[...]
Success! You find your simpacks in $CADEE_DIR/pedagogical_ex-
ample/libmut.
```

Finally, the generated simpacks need to be computed with "cadee dyn" (not shown).

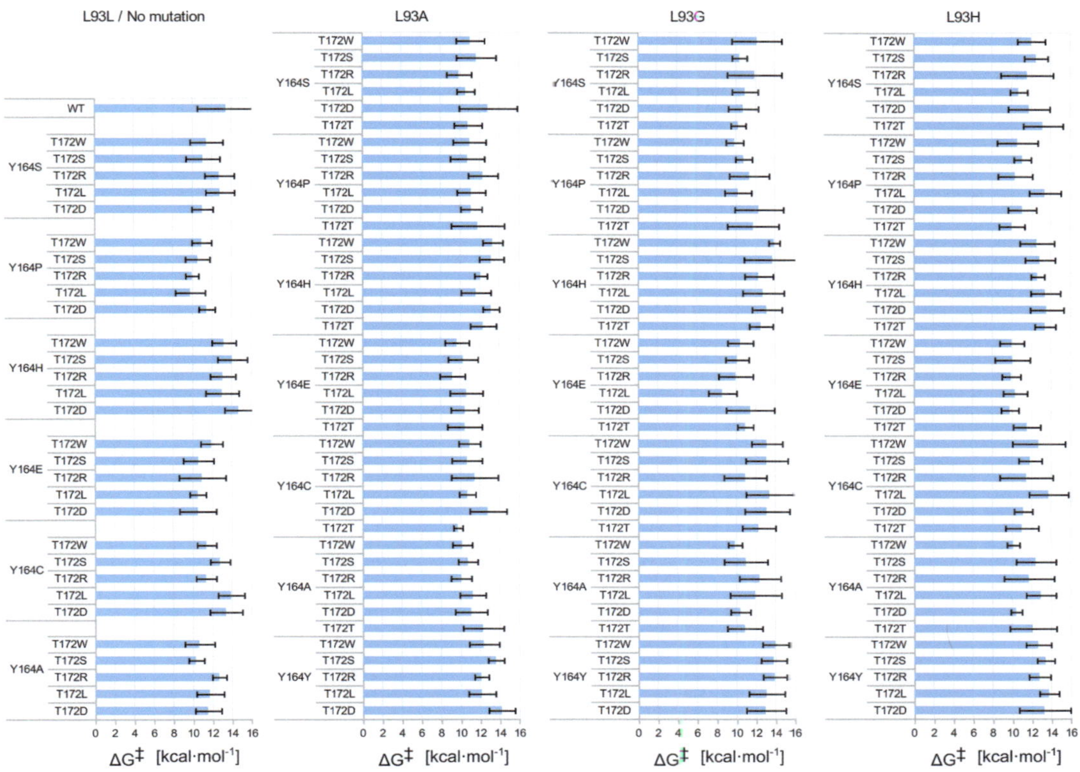

Fig. 5 Data obtained through partial combinatorial saturation mutagenesis at positions 93(A/G/H), 164(S/P/H/E/C/A), and 172(W/S/R/L/D). $\Delta G^{\ddagger}$ denotes the calculated activation free energies for each variant, and the error bars denote the standard deviation over $4 \times 6 = 24$ EVB trajectories per variant. As displayed, some variants have very large uncertainty in the calculated values. These instabilities can be caused by different factors or combinations of factors (for example structural instabilities caused by the insertion of the new residue or insufficient equilibration time). To improve the equilibration, the trend within data collection could be studied and longer simulations conducted, or the current ones extended, or additional simulations performed. CC-BY adopted from Ref. [14]

5.6 Example: Combinatorial Saturation Mutagenesis

After a reduced set of interesting amino acids have been selected by the user, combinatorial saturation mutagenesis can be performed (Fig. 5) to screen if the subsequent mutations are additive or, when introduced at the same time, cause a higher effect than the individual mutations (hysteresis). Note that CADEE was written with the aim of testing the saturation mutagenesis of several different residues together with a single command: "cadee prep . . . --libmut." We have therefore decided to use the results obtained from individual point saturation at each of the three positions, choosing a subset of amino acids to be tested at each hot spot:

```
# Code Input Snippet (17)
mv libmut point_saturation
cadee prep wt.pdb wt.fep wt.qpinp libraries --libmut 92:'AGH'\\
163:'SPHECA' 171:'WSRLD'
```

This would prepare 4 × 7 × 6 = 168 enzyme variants (×4 = 672 simpacks), as we demonstrated in the case of the published triosephosphate isomerase data [14].

```
# Code Output Snippet (17)
INFO:root:Preparing libmut - LIBrary MUTatagenesis.
92:AGH (+ native/wt)
163:SPHECA (+ native/wt)
171:WSRLD (+ native/wt)
INFO:root:[(92, ['A', 'G', 'H']), (163, ['S', 'P', 'H', 'E',
'C', 'A']), (171, ['W', 'S', 'R', 'L', 'D'])]
[...]
INFO:root:Working on $CADEE_DIR/pedagogical_example/libmut/
Y163A.
INFO:prep.pyscwrl:No clashes detected.
[...]
INFO:root:Working on $CADEE_DIR/pedagogical_example/libmut/
Y163S-T171W.
INFO:prep.pyscwrl:Clash-Score was: 130.2, will now re-run and
allow Scwrl4 to modify residues [163, 171, 519, 547, 549,
551, 730, 767, 771]
INFO:prep.pyscwrl:Clash-Score new: 0.0 ==> Keep.
[...]
INFO:root:Working on $CADEE_DIR/pedagogical_example/libmut/
L092A-Y163S-T171W.
INFO:prep.pyscwrl:Clash-Score was: 130.2, will now re-run and
allow Scwrl4 to modify residues [92, 163, 171, 519, 547,
549, 551, 730, 767, 771]
INFO:prep.pyscwrl:Clash-Score new: 0.0 ==> Keep.
[...]
INFO:root:Working on $CADEE_DIR/pedagogical_example/libmut/
L092H-Y163C-T171R.
INFO:prep.pyscwrl:Clash-Score was: 50.87, will now re-run and
allow Scwrl4 to modify residues [90, 92, 163, 171, 551,
562, 730, 767, 771]
INFO:prep.pyscwrl:Clash-Score new: 0.0 ==> Keep.
[...]
INFO:root:Working on $CADEE_DIR/pedagogical_example/libmut/
L092H-Y163A-T171D.
INFO:prep.pyscwrl:Clash-Score was: 5.893, will now re-run and
allow Scwrl4 to modify residues [90, 92, 163, 171, 730, 767]
INFO:prep.pyscwrl:Clash-Score new: 0.0 ==> Keep.
[...]
INFO:prep.create_inputs:Creating input files for L092G.
[...]
INFO:prep.create_inputs:Creating input files for Y163C-T171L.
[...]
```

```
INFO:prep.create_inputs:Creating input files for L092H-Y163S-
T171D.
[...]
INFO:root:Packing L092A-Y163S:
INFO:root:Pack # 0, Seed: 328529
INFO:root:Pack # 1, Seed: 546508
INFO:root:Pack # 2, Seed: 813942
INFO:root:Pack # 3, Seed: 634507
INFO:root:Packing L092A-Y163H-T171R:
INFO:root:Pack # 0, Seed: 270226
INFO:root:Pack # 1, Seed: 520637
INFO:root:Pack # 2, Seed: 90092
INFO:root:Pack # 3, Seed: 579659
INFO:root:Packing Y163A:
INFO:root:Pack # 0, Seed: 858212
INFO:root:Pack # 1, Seed: 538674
INFO:root:Pack # 2, Seed: 412398
INFO:root:Pack # 3, Seed: 690991
[...]
Success! You find your simpacks in $CADEE_DIR/pedagogical_ex-
ample/libmut.
```

Finally, the generated simpacks need to be computed with "cadee dyn" (not shown).

5.7 CADEE Customization

CADEE provides a straightforward and fast way to generate and test hundreds to thousands of mutants of a well-parameterized EVB reaction. To generate simpacks, CADEE relies on "simpack-templates": Currently, one simpack-template is included as a default, and a second one has been used in Subheading 4.2. We strongly recommend that users examine the existing templates and adjust them as per user requirements: Additional templates and documentation (readme.md) are available in $CADEE_DIR/simpack_templates/.

6 Limitations of CADEE

In the case of multistep reaction profiles where only the initial rate-limiting step was subjected to CADEE evolution, we recommend taking the best CADEE hits and running EVB on all other reaction steps to ensure that the proposed residue substitutions do not cause a change in rate-limiting step. We also recommend that additional (and longer) simulations should be run for the best hits identified to both improve the quality of the predictions obtained and reduce the risk of false positives due to too short sampling time.

6.1 Specific Limitations of CADEE v 0.9:

- Thermalization (initial system heating) and equilibration are performed at the putative transition state, i.e., at $\lambda = 0.5$ along the reaction coordinate.
- The automatic EVB mapping does not currently support more complex functional forms for the off-diagonal term.
- Rank0 is reserved for input/output and not computing calculations.
- The --trajcsv argument is not officially supported.
- Temperature averages are not extracted from Qdyn6 log files.
- Arbitrary mutations are only obtained using SCWRL4; other tools are not currently available, but can be implemented by adding additional modules.

7 How to Deploy CADEE Effectively

Protein structures are complex, and efficiently predicting functionally beneficial mutations therefore requires a well-defined strategy. The following protocol could be used to evaluate the residue alterations that result from the CADEE analysis. Here, if multiple domains are encoded in the sequence, the scores could be individually estimated for each of the domains.

1. Selecting and curating the native enzyme structure:
   (a) This step is the most important step to define the ultimate accuracy of the simulations, as the structural topology is used by CADEE for the EVB analysis. Hence, the selected protein structure should be the experimentally solved structure or the closest predicted model with the well-defined near-native topology for structurally continuous or discontinuous domains(s), to accurately define the active-site contours.
   (b) The active-site charge and proton configuration for the reactive residues should be exactly the same as the enzyme's native state actually interacting with the reactant, as it would accurately guide the EVB computation by considering the polarization effects of such charges, and result in biologically meaningful $\Delta\Delta G$ scores.
2. Pre-analyzing the considered structure before the CADEE analysis:
   (a) Selecting the top-ranked set of homologs on the basis of screening the protein sequence and structure databases with the information retrieved from additional datasets, viz. structural classification of proteins (SCOP [44]), evolutionary classification of protein domains (ECOD [45]),

protein family (Pfam [46]), conserved domain database (CDD [47]), SMART [48], and TIGRFAM [49], by considering the culling options for reducing the spurious redundant hits: This entire information affirms the consideration of a good candidate that has diversified in its sequence evolution, although it retains the structural topology of the considered sequence, especially when a well-annotated database does not exist.

(b) Evaluating the co-localized residues with sequence and structural penalty-based profiles, and iterating this step for the final top-ranked hits, as the erroneous alignment shift errors usually incurred due to the simultaneous consideration of distant and closest hits is easily resolved. An easy way to detect accuracy is the sequence overlap of the functionally conserved and active-site residues.

(c) Evaluating the residue propensity at each locus in correlation with the mean scores scattered across the chain in the constructed profile to further correlate with the CADEE results.

(d) Evaluating the biophysical restraints imposed on the secondary/tertiary structure by altering a given residue.

(e) Estimating whether the considered mutation overlaps with the structural state observed in the closest hits through tools like PSIPRED [50, 51].

(f) Robustly evaluating the stability of the sequence, if some mutation(s) are considered. This is because the native energetic Z-score of a structure should be well predicted prior to the test, and if a mutation forcefully decreases that level, the extent of decrement could also be devastating for the model, or if it increases it won't simply be meaningless. Currently, CADEE evaluates the structural topology for the localized atomic clashes through SCWRL4 to produce the best set of $\Delta\Delta G$ scores. If any of these steps is not implemented properly and the protocol accumulates a few errors that could subsequently extrapolate to alteration of the protein structure, especially in/near the demarcated sphere zone, CADEE will ultimately yield meaningless results and therefore care is crucial.

(g) Evaluating the CADEE results through coevolving residue propensities and substitution matrices would allow the CADEE-generated top-ranked scores to be justified with the underlying biochemical mechanism defining the improved activity or stability.

(h) The resulted mutations should be biochemically considered and evaluated for elucidating their plausible role in either stabilizing the enzyme structure or improving the enzymatic activity.

8 Troubleshooting

8.1 CADEE Installation

A successful CADEE installation ends with this line:

```
Finished processing dependencies for cadee
```

If this does not closely resemble the CADEE installation output, something with the setup did not work. The user is advised to double-check that all the dependencies are installed, and try to (re-) install CADEE again.

8.2 CADEE First Start

If the first start of the CADEE script fails, two things could have gone wrong:

1. Problem: Setup failed. The following script can be used to check if CADEE is installed properly.

```
# Code Input Snippet (18)
ok=1
msg="\n\n\n\n\n"
python -c 'import cadee'
if [[ $? -ne 0 ]]
then
  msg="$msg\nUnable to load cadee module!"
  ok=0
fi
cadee > /dev/null
if [[ $? -eq 127 ]]
then
  msg="$msg\nUnable to locate 'cadee' in \$PATH."
  ok=0
fi
if [[ $ok -eq 1 ]]
then
  msg="$msg\nCADEE is installed."
else
  msg="$msg\nCADEE is *NOT* installed!
DIAGNOSIS:"
fi
echo -e "\n\n$msg"
# Code Output Snippet (18)
CADEE is installed.
```

2. Problem: Debian or Ubuntu are being used, and the Python script path is not in the user's $PATH. If the above script produces the output "Unable to locate 'cadee' in $PATH.", but not "Unable to load cadee module!" then the issue might be the $PATH variable.

```
# Code Input Snippet (19)
# ONLY FOR DEBIAN/UBUNTU
echo $PATH | grep "$HOME/.local/bin" || echo 'Please fix $PATH.'
# Code Output Snippet (19)
Please fix $PATH.
```

Solution: If the user is using Debian or Ubuntu, the Python script path might be missing in the PATH. This can be fixed by adding the following line to the end of the $HOME/.profile file:

PATH="$HOME/.local/bin:$PATH" (using the user's favorite text editor)

Alternatively, the following script can be executed:

```
# Code Input Snippet (20)
# Ubuntu / Debian ONLY
if [ -d "$HOME/.local/bin" ] ; then
  ok=0
  echo $PATH | grep -q "$HOME/.local/bin:" && ok=1
  echo $PATH | grep -q ":$HOME/.local/bin" && ok=1
  if [ $ok -eq 0 ]
  then
    echo 'export PATH="$HOME/.local/bin:$PATH"' >> $HOME/.profile
    source $HOME/.profile
    echo 'Added $HOME/.local/bin to .profile.'
  else
    echo 'Stop: $HOME/.local/bin is already in your $PATH.'
  fi
else
  echo 'Stop: $HOME/.local/bin is not a directory. '
fi
# Code Output Snippet (20)
'Added $HOME/.local/bin to .profile.
```

8.3 Simpacks

For some computer architectures, the compute time needed to perform one simulation exceeds the wall-clock limit. CADEE is hence able to restart and continue simulations, and the user can simply resubmit the original submission file, to continue the simulation with the same command, as there is no special restart flag. To detect unfinished simpacks, the most straightforward way is to compare the simpack sizes (/bin/ls –lS). Sometimes, however, a node may have crashed, or a hard disk quota may have been hit, and hence a simpack may be faulty and not finish even with enough wall-clock time available. In those cases, it is advisable to untar the simpack and repack it. A script to do this is:

```
# Code Input Snippet (21)
cadee tool repair_simpack /$HOME/global/cadee_tutorial/wt/\\
wt_0.tar
# Code Output Snippet (21)
[...]
1. Searching duplicate logfiles:
2. Searching duplicate energy files:
3. Searching missing restartfiles:
4. Searching damaged logfiles:
5. Search for logfiles lacking 'terminated normally':
6. Search for gzipped logfiles lacking 'terminated normally':
/$HOME/global/cadee_tutorial/wt/wt_0.tar:
No Problems with this simpack. Awesome!
```

If the script stops because of a bad tar archive (and it is certain that it is a simpack issue) the --force flag may be used to force repacking of the archive. CAUTION: If the parameter is not actually a simpack, the script will behave unpredictably, and may lead to data loss, especially if applied with the --force flag. The flag is especially helpful in crashes caused by disk space shortages. When used, the faulty simpack tarballs are unpacked, faulty files are deleted, and a new, uncorrupt archive is written back to the disk. *Caution:* The original simpack will be overwritten during this process.

8.3.1 Simpack Customization

A simpack contains all files necessary to perform an EVB simulation with Q. A minimal simpack hence contains files for the (1) initialization, (2) thermalization/heat-up, (3) equilibration, (4) free energy perturbation/empirical valence bond computation, and (5) empirical valence bond free energy mapping. More detailed information about simpacks and how they can be customized can be found in $CADEE_DIR/simpack_templates/readme.md.

9 Overview and Conclusions

We recently developed a comprehensive and widely automated toolkit for the computer-aided directed evolution of enzymes (CADEE), freely available for download from Github at the following link: https://github.com/kamerlinlab/cadee. The theoretical background to CADEE has been described in detail in Ref. [14]. The current contribution provides detailed protocols for different types of simulations supported by CADEE, as well as relevant snippets of code input/output, and when used in conjunction with the original CADEE publication [14] provides a comprehensive overview of the current scope, limitations, and future prospects of CADEE.

Acknowledgments

The European Research Council provided financial support under the European Community's Seventh Framework Programme (FP7/2007-2013)/ERC Grant Agreement 306474. SCLK would also like to thank the Knut and Alice Wallenberg Foundation and the Royal Swedish Academy of Sciences for a Wallenberg Academy Fellowship, and the Swedish Research Council for providing support through project grant 2015-04928. All calculations were performed on the Abisko cluster at the HPC2N center in Umeå and on the Triolith cluster at the NSC in Linköping, thanks to a generous supercomputing allocation provided by the Swedish National Infrastructure for Computing (SNIC grant 2015/16-12). In addition, we would like to thank Arina Gromova for extensive testing of CADEE, Fabian Steffen-Munsberg for initial testing, and Miha Purg for helpful discussions about qscripts/qtools.

References

1. Bornscheuer UT (1998) Directed evolution of enzymes. Angew Chem Int Ed 37:3105–3108
2. Bull AT, Ward AC, Goodfellow M (2000) Search and discovery strategies for biotechnology: the paradigm shift. Microbiol Mol Biol Rev 64:573–606
3. Tao H, Cornish VW (2002) Milestones in directed enzyme evolution. Curr Opin Chem Biol 6:858–864
4. Currin A, Swainston N, Day PJ, Kell DB (2015) Synthetic biology for the directed evolution of biocatalysts: navigating sequence space intelligently. Chem Soc Rev 44:1172–1239
5. Packer MS, Liu DR (2015) Methods for the directed evolution of proteins. Nat Rev Genet 16:79–394
6. Arnold FH, Volkov AA (1999) Directed evolution of biocatalysts. Curr Opin Chem Biol 3:54–59
7. Jäckel C, Kast P, Hilvert D (2008) Protein design by directed evolution. Annu Rev Biophys 37:153–173
8. Currin A, Swainston N, Day PJ, Kell DB (2015) Synthetic biology for the directed evolution of protein biocatalysts: navigating sequence space intelligently. Chem Soc Rev 44:1172–1239
9. Romero PA, Arnold FH (2009) Exploring protein fitness landscapes by directed evolution. Nat Rev Mol Cell Biol 10:866–876
10. Gumulya Y, Sanchis J, Reetz MT (2012) Many pathways in laboratory evolution can lead to improved enzymes: how to escape from local minima. ChemBioChem 13:1060–1066
11. Barrozo A, Borstnar R, Marloie G, Kamerlin SCL (2012) Computational protein engineering: bridging the gap between rational design and laboratory evolution. Int J Mol Sci 13:12428–12460
12. Kiss G, Çelebi-Ölçum N, Moretti R, Baker D, Houk KN (2012) Computational enzyme design. Angew Chem Int Ed 52:5700–5725
13. Romero-Rivera A, Garcia-Borràs M, Osuna S (2017) Computational tools for the evaluation of laboratory-engineered biocatalysts. Chem Commun 53:284–297
14. Amrein BA, Steffen-Munsberg F, Szeler I, Purg M, Kulkarni Y, Kamerlin SCL (2017) CADEE: computer-aided directed evolution of enzymes. IUCrJ 4:50–64
15. Warshel A, Weiss RM (1980) An empirical valence bond approach for comparing reactions in solutions and in enzymes. J Am Chem Soc 102:6218–6226
16. Warshel A, Sharma PK, Kato M, Xiang Y, Liu H, Olsson MHM (2006) Electrostatic basis for enzyme catalysis. Chem Rev 106:320–3235
17. Kamerlin SCL, Warshel A (2010) The EVB as a quantitative tool for formulating simulations and analyzing biological and chemical reactions. Faraday Discuss 145:71–106
18. Luo J, van Loo B, Kamerlin SCL (2012) Examining the promiscuous phosphatase activity of Pseudomonas aeruginosa arylsulfatase: a

comparison to analogous phosphatases. Proteins Struct Funct Bioinf 80:1211–1226

19. Barrozo A, Duarte F, Bauer P, Carvalho ATP, Kamerlin SCL (2015) Cooperative electrostatic interactions drive functional evolution in the alkaline phosphatase superfamily. J Am Chem Soc 137:9061–9076
20. Q Official Website. http://xray.bmc.uu.se/~aqwww/q
21. Manual for the molecular Dynamics package Q. http://xray.bmc.uu.se/~aqwww/q/documents/qman5.pdf
22. MPI4Py. https://pypi.python.org/pypi/mpi4py
23. O'Boyle NM, Banck M, James CA, Morley C, Vandermeersch T, Hutchison GR (2011) Open babel: an open chemical toolbox. J Cheminform 3:33–33
24. Krivov GG, Shapovalov MV, Dunbrack RL (2009) Improved prediction of protein side-chain conformations with SCWRL4. Proteins Struct Funct Bioinf 77:778–795
25. Frushicheva MP, Cao J, Chu ZT, Warshel A (2010) Exploring challenges in rational enzyme design by simulating the catalysis in artificial Kemp eliminase. Proc Natl Acad Sci 107:16869–16874
26. Frushicheva MP, Cao J, Warshel A (2011) Challenges and advances in validating enzyme design proposals: the case of Kemp eliminase catalysis. Biochemistry 50:3849–3858
27. Kamerlin SCL, Warshel A (2011) The empirical valence bond model: theory and applications. WIREs Comput Mol Sci 1:30–45
28. Amrein BA, Bauer P, Duarte F, Janfalk Carlsson Å, Naworyta A, Mowbray SL, Widersten M, Kamerlin SCL (2015) Expanding the catalytic triad in epoxide hydrolases and related enzymes. ACS Catal 5:5702–5713
29. Ben-David M, Sussman JL, Maxwell CI, Szeler K, Kamerlin SCL, Tawfik DS (2015) Catalytic stimulation by restrained active-site floppiness—the case of high density lipoprotein-bound serum paraoxonase-1. J Mol Biol 427:1359–1374
30. Roca M, Vardi-Kilshtain A, Warshel A (2009) Toward accurate screening in computer-aided enzyme design. Biochemistry 48:3046–3056
31. Frushicheva MP, Mills MJL, Schopf P, Singh MK, Prasad RB, Warshel A (2014) Computer aided enzyme design and catalytic concepts. Curr Opin Chem Biol 21:56–62
32. Carvalho ATP, Barrozo A, Doron D, Kilshtain AV, Major DT, Kamerlin SCL (2014) Challenges in computational studies of enzyme structure, function and dynamics. J Mol Graph Model 54:62–79
33. King G, Warshel A (1989) A surface constrained all-atom solvent model for effective simulations of polar solutions. J Chem Phys 91:3647–3661
34. Lee FS, Warshel A (1992) A local reaction field method for fast evaluation of long-range electrostatic interactions in molecular simulations. J Chem Phys 97:3100–3107
35. Stallman RM (2009) GCC developer community, using the Gnu compiler collection: A Gnu manual for Gcc version 4.3.3. CreateSpace. p 636
36. Gabriel E, Fagg GE, Bosilca G, Angskun T, Dongarra JJ, Squyres JM, Sahay V, Kambadur P, Barrett B, Lumsdaine A, Castain RH, Daniel DJ, Graham RL, Woodall TS (2004) Open MPI: Goals, concept, and design of a next generation MPI implementation. In: Kranzlmüller D, Kacsuk P, Dongarra J (eds) Recent Advances in Parallel Virtual Machine and Message Passing Interface: 11th European PVM/MPI Users' Group Meeting Budapest, Hungary, September 19–22, 2004. Proceedings. Springer Berlin Heidelberg, Berlin, Heidelberg, pp 97–104
37. Gropp W (2002) MPICH2: A New Start for MPI Implementations. In: Proceedings of the 9th European PVM/MPI Users' Group Meeting on recent advances in parallel virtual machine and message passing interface, Springer-Verlag, p 7
38. Python Software Foundation. Python Language Reference, version 2.7. http://www.python.org/
39. Marelius J, Kolmodin K, Feierberg I, Åqvist J (1998) Q: A molecular dynamics program for free energy calculations and empirical valence bond simulations in biomolecular systems. J Mol Graph Model 16:213–225
40. Berman HM, Westbrook J, Feng Z, Gilliland G, Bhat TN, Weissig H, Shindyalov IN, Bourne PE (2000) The Protein Data Bank. Nucleic Acids Res 28:235–242
41. Berman HM, Henrick K, Nakamura H (2003) Announcing the worldwide Protein Data Bank. Nat Struct Mol Biol 10:980–980
42. HPC2N. http://www.hpc2n.umu.se/
43. Reetz MT, Wu S (2008) Greatly reduced amino acid alphabets in directed evolution: making the right choice for saturation mutagenesis at homologous enzyme positions. Chem Commun 21:5499–5501
44. Murzin AG, Brenner SE, Hubbart T, Chothia C (1995) SCOP: a structural classification of proteins database for the investigation of sequences and structures. J Mol Biol 247:536–540

45. Cheng H, Schaeffer RD, Liao Y, Kinch LN, Pei J, Shi S, Kim BH, Grishin NV (2014) ECOD: an evolutionary classification of protein domains. PLoS Comput Biol 10: e1003926
46. Finn RD, Bateman A, Clements J, Coggill P, Eberhardt RY, Eddy SR, Heger A, Hetherington K, Holm L, Mistry J, Sonnhammer ELL, Tate J, Punta M (2014) Pfam: the protein families database. Nucleic Acids Res 42:D222–D230
47. Marchler-Bauer A, Lu S, Anderson JB, Chitsaz F, Derbyshire MK, DeWeese-Scott C, Fong JH, Geer LY, Geer RC, Gonzales NR, Gwadz M, Hurwitz DI, Jackson JD, Ke Z, Lanczycki J, Lu F, Marchler GH, Mullokandov M, Omelchenko MV, Robertson CL, Song JS, Thanki N, Yamashita RA, Zhang D, Zhang N, Zheng C, Bryant SH (2011) CDD: a conserved domain database for the functional annotation of proteins. Nucleic Acids Res 39(Database):D225–D229
48. Ponting CP, Schultz J, Milpetz F, Bork P (1999) SMART: identification and annotation of domains from signalling and extracellular protein sequences. Nucleic Acids Res 27:229–232
49. Haft DH, Selengut JD, White O (2003) The TIGRFAMs database of protein families. Nucleic Acids Res 31:371–373
50. Jones DT (1999) Protein secondary structure prediction based on position-specific scoring matrices. J Mol Biol 292:195–202
51. Buchan DWA, Minneci F, Nugent TCO, Bryson K, Jones DT (2013) Scalable web services for the PSIPRED Protein Analysis Workbench. Nucleic Acids Res 41(W1): W340–W348

Correction to: Enhancing Statistical Multiple Sequence Alignment and Tree Inference Using Structural Information

Joseph L. Herman

Correction to:
Chapter 10 in: Tobias Sikosek (ed.),
***Computational Methods in Protein Evolution*,**
Methods in Molecular Biology, vol. 1851,
https://doi.org/10.1007/978-1-4939-8736-8_10

The published version of this book included errors in code listings in Chapter 10. These code listings have been corrected and text has been updated.

The updated online version of this chapter can be found at
https://doi.org/10.1007/978-1-4939-8736-8_10

Tobias Sikosek (ed.), *Computational Methods in Protein Evolution*, Methods in Molecular Biology, vol. 1851,
https://doi.org/10.1007/978-1-4939-8736-8_23, © Springer Science+Business Media, LLC, part of Springer Nature 2019

Index

Tobias Sikosek (ed.), *Computational Methods in Protein Evolution*, Methods in Molecular Biology, vol. 1851, https://doi.org/10.1007/978-1-4939-8736-8,

MIX
Papier aus verantwortungsvollen Quellen
Paper from responsible sources
FSC® C105338
www.fsc.org

If you have any concerns about our products,
you can contact us on
ProductSafety@springernature.com

In case Publisher is established outside the EU,
the EU authorized representative is:
Springer Nature Customer Service Center GmbH
Europaplatz 3, 69115 Heidelberg, Germany

Printed by Libri Plureos GmbH
in Hamburg, Germany